Applications of Gels for Enhanced Oil Recovery

Guest Editors

Ming Qu

Tuo Liang

Mingguo Peng

MDPI Basel • Beijing • Wuhan • Barcelona • Belgrade • Novi Sad • Cluj • Manchester

Guest Editors

Ming Qu
SANYA Offshore Oil & Gas Research Institute
Northeast Petroleum University
Sanya
China

Tuo Liang
College of Petroleum Engineering
Xi'an Shiyou University
Xi'an
China

Mingguo Peng
School of Petroleum and Natural Gas Engineering
School of Energy
Changzhou University
Changzhou
China

Editorial Office
MDPI AG
Grosspeteranlage 5
4052 Basel, Switzerland

This is a reprint of the Special Issue, published open access by the journal *Gels* (ISSN 2310-2861), freely accessible at: https://www.mdpi.com/journal/gels/special_issues/28LBTDH25S.

For citation purposes, cite each article independently as indicated on the article page online and as indicated below:

Lastname, A.A.; Lastname, B.B. Article Title. *Journal Name* **Year**, *Volume Number*, Page Range.

ISBN 978-3-7258-6796-7 (Hbk)
ISBN 978-3-7258-6797-4 (PDF)
https://doi.org/10.3390/books978-3-7258-6797-4

Contents

Preface

We are honored, as Guest Editors, to present this Reprint, "Applications of Gels for Enhanced Oil Recovery." The ongoing development of mature oil fields and the discovery of unconventional resources have led to increasingly demanding reservoir environments, posing greater challenges for gel technology. This volume aims to showcase the latest groundbreaking research in the field, encompassing innovative gel formulations designed for extreme conditions, a deeper understanding of gel percolation mechanisms, and the application of emerging interdisciplinary approaches such as nanotechnology. We extend our sincere gratitude to all the authors who contributed their excellent work and to the reviewers for their professional efforts. We hope that this Reprint will serve as a valuable reference for researchers and engineers and further stimulate advancements in gel technology for enhanced oil recovery.

Ming Qu, Tuo Liang, and Mingguo Peng
Guest Editors

gels

MDPI

Article

Thermochemical Degradation of a Polyacrylamide Gel as a Dual-Function Strategy for Enhanced Oil Recovery and Reservoir Remediation

Jiaying Wang , Renbao Zhao *, Yuan Yuan, Yunpeng Zhang, Guangsen Zhu, Jingtong Tian, Haiyang Zhang, Haitao Ren, Guanghui Zhou and Bin Liao

State Key Laboratory of Petroleum Resources and Engineering, China University of Petroleum (Beijing), Beijing 102249, China; wangjiayingswpu@163.com (J.W.); yuany9939@163.com (Y.Y.); 2024215221@student.cup.edu.cn (Y.Z.); 2023210410@student.cup.edu.cn (G.Z.); 2023210409@student.cup.edu.cn (J.T.); 2022215216@student.cup.edu.cn (H.Z.); 15935903697@163.com (H.R.); zhouguanghui0603@163.com (G.Z.); 18079745126@163.com (B.L.)

* Correspondence: zhaorb@cup.edu.cn

Abstract: The accumulation of residual hydrolyzed polyacrylamide (HPAM) gel or molecular-based solutions in reservoirs after polymer flooding poses dual challenges: irreversible formation damage and long-term environmental risk issues. However, existing research mainly focuses on treating polymers in surface-produced water, neglecting both in situ decomposition of residual polymer gel or molecular-based solutions in reservoirs and the degradation of HPAM gels under high temperatures from in situ combustion (ISC). This work investigates the thermochemical behavior of HPAM gel during ISC and its dual-function role in enhanced oil recovery (EOR) and reservoir remediation. It was demonstrated that the residual gel and/or molecular-based solutions undergo efficient degradation, serving as an in situ fuel that significantly reduces the activation energy for crude oil oxidation by up to 58.4% in the low-temperature stage and 75.2% in the high-temperature stage. Factors influencing the gel's degradation and the combustion process, including its molecular weight, ionic type, and crude oil viscosity, were systematically evaluated. Optimal conditions achieved over 90% gel degradation. Combustion tube experiments validated the dual benefits of this approach: an incremental oil recovery of 68.6% and an average HPAM gel removal efficiency of 64.8%. This work presents a novel strategy for utilizing retained gels in situ to simultaneously enhance oil recovery and mitigate gel-induced formation damage, offering significant insights for the management of mature gel-treated reservoirs.

Keywords: polymer gel-treated reservoirs; in-situ degradation; thermochemical conversion; activation energy reduction; reservoir remediation; oil recovery

1. Introduction

Polymer flooding [1–3] is considered a mature enhanced oil recovery technique designed to improve sweep efficiency over waterflooding, where the adverse viscosity ratio (μ_o/μ_w) between oil and water leads to unstable displacement [4]. The polymer flooding method can further enhance the oil recovery factor and has shown remarkable performance in many industrial-scale applications as the oil field is developed into a high water cut period, which is mainly attributed to its better mobility control during the displacement process [5]. The viscous polymer solution could significantly reduce the viscous fingering phenomenon, allowing the displacing fluid to advance more evenly in the reservoir pores.

Therefore, more sweep efficiency is achieved, accompanied by more crude oil recovered. The technical and economic success of polymer flooding has led to its widespread global implementation, with over 70 field projects currently operational worldwide, particularly in China, Canada, and Oman, demonstrating incremental recoveries of 5–20% STOIIP [6–10]. HPAM has been the most commonly applied polymer in the field [11]. Recent research has shown that modifying polyacrylamide with sodium p-styrenesulfonate (SSS) can effectively enhance its high-temperature resistance—exhibiting excellent stability even at 180 °C—by increasing the gyration radius, reducing the diffusion coefficient, stabilizing hydrogen bonds, and optimizing the spatial network structure (e.g., widened branches and regulated pore parameters) [12]. This modification provides a new approach for improving polyacrylamide's adaptability to deep high-temperature reservoirs, addressing the degradation issue of conventional polyacrylamide under extreme thermal conditions.

However, with the continuous injection of polymer solution or gels, especially in the later stage of polymer EOR, serious problems have gradually emerged [13]. Long-term injection of polymer gels decreases the effective permeability of the reservoir and plugs the formation pores, resulting in irreversible reservoir damage [14,15]. Specifically, in the industrial polymer injection blocks of the Daqing Oilfield, the polymer gel retention rate was found to be in the range of 70% to 80% at the end of polymer injection. Subsequently, after 5 to 6 years of water flooding, the polymer gel residue rate was observed to have decreased by 12.8% to 17.6%. However, it should be noted that 60% of the polymer is still retained underground [16]. Moreover, in Block II4-5 of the Shuanghe North Fault Block in the Henan Oilfield, a total of 3628.48 tons of polymer has been injected into the oil reservoirs through prepared solutions of different concentrations with surface and/or produced formation water. When the subsequent water flooding was carried out for 10 months, the amount of polymer remaining in the formation was measured to be 3103.80 tons, and correspondingly, with a retention rate of 85.54% [17]. Critically, retained polymers pose substantial environmental risks. Following the injection of polymers into reservoirs for enhanced oil recovery, a portion of the polymers becomes retained through mechanisms, including adsorption, thereby blocking pore throats. Another fraction may migrate to adjacent aquifers due to interlayer crossflow, while the remainder is produced alongside oil–water effluents at production wells (Figure 1). After polyacrylamide (PAM) enters the subsurface environment, its ecological risks mainly stem from the migration and diffusion of residual acrylamide monomer (AM) in the product. Acrylamide is a carcinogenic, mutagenic, and reprotoxic monomer [18]. It can cause biological damage by affecting the central nervous system. Subsurface AM can enter groundwater along with partially hydrolyzed polyacrylamide from wellbore leakage or interlayer migration. AM in groundwater has extremely low reactivity, which is verified when no significant biodegradation occurs, and sandy sediments have difficulty adsorbing it, allowing it to exist stably for a long time [19]. Therefore, it would endanger aquatic organisms dependent on groundwater, which may have a serious impact on their growth, reproduction, and survival [20,21], thereby affecting the surface ecosystem through the food chain.

The natural degradation of synthetic polymers exhibits exceptionally prolonged durations spanning centuries, primarily due to the absence of microbial degradation mechanisms, inherent molecular structural resistance [22], and constraints imposed by oxygen-deprived environmental conditions. Particularly within oil reservoir formations, these polymers persist without complete conversion into environmentally neutral substances. Consequently, during this extended period, the detrimental impact of residual HPAM gel on the environment persists. Polymer flooding achieves only limited enhancement in oil recovery due to inherent constraints in sweep efficiency modification. This restricted effectiveness leaves substantial residual hydrocarbon resources retained within the polymer-flooded oil reservoirs, maintaining significant potential for subsequent extraction through other recovery strategies. However, the residual polymers seriously decrease the injection capacity and trigger great challenges for further exploitation during the subsequent reservoir development period of crude oil, and they pose a detriment to the substance environment. Nevertheless, currently, the negative impacts of residual polymers have not been paid sufficient attention. Therefore, the removal of underground polymers remains a topic that has been scarcely explored, with few related studies so far.

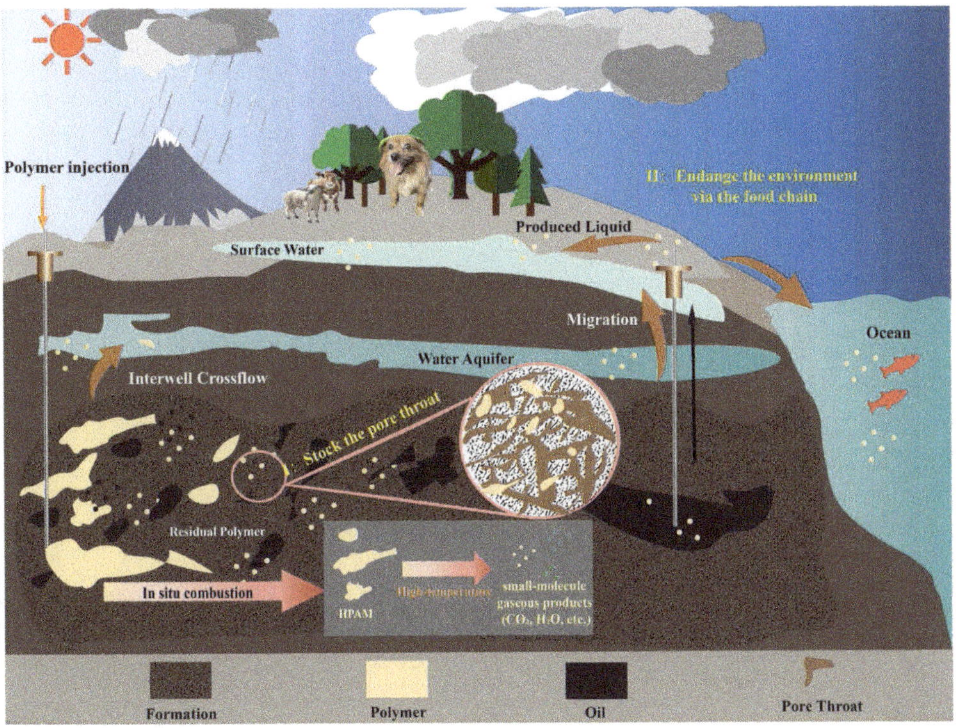

Figure 1. Schematic diagram of potential ecological risks of underground residual HPAM and proposed solution pathways.

ISC [23–25] is one of the thermal EOR methods that could be applicable in a variety of reservoirs. Air or other oxygen-enriched gas is injected into the formation and ignited during the process. Heavy components undergo complex thermal cracking and oxidation reactions, including low-temperature oxidation (LTO), fuel deposition (FD), and high-temperature oxidation (HTO), generating a huge amount of heat as well. Propelled by the heat and produced gases, the crude oil is efficiently displaced and recovered, thereby enhancing the oil recovery. In this work, ISC is proposed as a method that can both enhance oil recovery and remove the underground residual polymers. The high-temperature environment causes the polymers to be degraded or even burned, thereby restoring part of the reservoir permeability and improving the seepage performance of the reservoir. At the same time, a large amount of gas generation could increase the internal pressure of the reservoir and drive the crude oil to flow towards the production well, thereby achieving a higher recovery factor. Therefore, it is of great significance to conduct experimental research on the oxidation characteristics of polymers during the ISC process. It helps to deeply explore the potential and feasibility of ISC in removing the remaining underground polymers.

Currently, there has been a certain amount of research on the pyrolysis behaviors of polyacrylamide [26,27]. However, most of these studies have been focused on the pyrolysis process in bulk or simple systems, lacking in-depth analysis of the oxidative pyrolysis of HPAM gel in porous media. Self-designed reaction equipment was utilized in this work. A comprehensive investigation was carried out into the oxidation and pyrolysis characteristics exhibited by HPAM gel within porous media. The activation energy of various samples during the combustion process was quantitatively calculated. Furthermore, the feasibility of the degradation and removal of underground residual polymer gel during the ISC process was explored. In addition, an in-depth examination was also conducted to clarify the impact of the molecular weight, the ionic type of the gel, and the viscosity of crude oil throughout the ISC process. Moreover, the conductivity method was proposed to determine the content of HPAM gel in oil sand, so as to ascertain the degradation efficiency of HPAM gel under various conditions.

While extensive research has been reported on polymer gel and/or molecular-based solution applications in EOR, critical gaps remain in using in situ degradation to achieve both reservoir remediation and incremental recovery. Previous studies mostly focus on chemical degradation mechanisms under ideal conditions or isolated evaluations of EOR performance. None, however, explores the synergetic potential of turning trapped polymer into agents that boost recovery. This work demonstrates that ISC can effectively remediate residual polymer gel and/or solution contamination in post-polymer-flooded reservoirs. The experimental results validate the ISC method as a promising solution for addressing both polymer gel-induced formation damage and groundwater contamination in mature polymer-treated reservoirs. Suggesting the ISC technique could not only enhance oil recovery but also provide an environmentally sustainable approach to reservoir remediation.

2. Results and Discussion

2.1. The Oxidation Behavior and Residues of HPAM

During the kinetic cell experiment, real-time monitoring and recording were meticulously implemented to track the variations in temperature and the changes in the concentration of the effluent gases. The variation in carbon dioxide (COx) concentrations and temperature is widely used as a key indicator for monitoring the reaction progress of ISC and polymer pyrolysis [26,28]. They were finally analyzed and graphically represented in Figure 2, highlighting distinct combustion stages and their associated reaction characteristics.

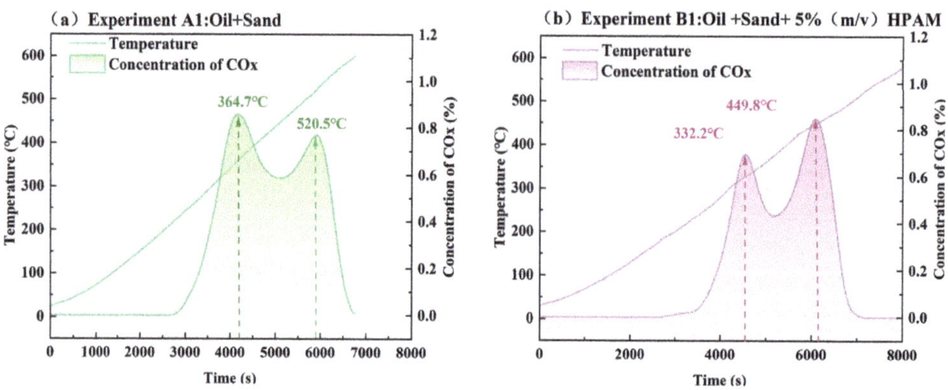

Figure 2. Temperature and concentration of COx variations in Experiment A1 (**a**) and B1 (**b**).

Figure 2 illustrates the temperature and concentration of COx variations in Experiments A1 (a) and B1 (b) under identical conditions. Both profiles reveal dual COx concentration peaks associated with LTO and high-temperature HTO processes. The first peak (364.7 °C in A1 vs. 332.2 °C in B1) corresponds to low-temperature oxidation, where heavy hydrocarbons undergo sequential reactions including vaporization, oxygen addition, and peroxide formation, generating coke deposits alongside CO_X emissions. The subsequent peak (520.5 °C in A1 vs. 449.8 °C in B1) arises from high-temperature oxidation, characterized by thermal cracking of residual hydrocarbons into CO_X and H_2O through exothermic pathways.

The combustion process in both experiments exhibited exothermic behavior, characterized by distinct temperature and COx concentration profiles. Notably, Experiment B1 demonstrated significantly enhanced exothermicity, as evidenced by a hump in temperature elevations at both COx peaks compared to Experiment A1. This suggests that HPAM gel amplifies the heat release during the combustion process. Furthermore, the presence of HPAM resulted in a notable reduction in peak temperatures: the first COx peak occurred at 332.2 °C in B1, 32.5 °C lower than the 364.7 °C observed in A1, while the second peak shifted from 520.5 °C in A1 to 449.8 °C in B1, a reduction of 70.7 °C. It indicates that the HPAM gel alters the combustion kinetics, likely by lowering the activation energy required for oxidation reactions, which will be described in Section 2.2.1.

These findings suggest that the residual HPAM gel within the porous media reduced both the low-temperature oxidation and high-temperature oxidation thresholds, with a more pronounced decrease in the HTO stage. This is likely due to nitrogen-containing radicals (e.g., $-NH_2$) from HPAM pyrolysis lowering the oxidation activation energy and accelerating reactions, while enhanced radical activity at high temperatures significantly shortens the induction period. Figure 3 illustrates a possible evolution mechanism of HPAM gel during this process. In the LTO range, the C–N and C–C bonds in the side chain break first due to their relatively low bond energies. The primary products at this stage are ammonia, amide groups, and a small amount of carbon dioxide. As the temperature rises to the HTO range, the C–C bonds in the backbone and C=O bonds, as well as N–H bonds, undergo cleavage, generating large quantities of carbon oxides and short-chain alkanes such as methane. Additionally, the exothermic decomposition of HPAM itself releases heat, further promoting oxidation and synergistically reducing both LTO and HTO temperatures, which is also consistent with a previous experimental study on the pyrolysis of polyacrylamide [29,30].

Figure 4 shows the visual characteristics of the samples in Experiments A1 and B1 before and after the combustion. The oil sand in A1 exhibited a dark brown color before combustion. Samples were finally retrieved from the reactor upon ending the experiment.

It was distinctly discernible that the sample from Experiment A1 had turned white after combustion, signifying the complete combustion of the crude oil with no residual oil. In contrast, the sample from Experiment B1 manifested an off-white appearance after combustion. This phenomenon could potentially be ascribed to the fact that during the experimental process, a small amount of coke, which was generated as a byproduct of HPAM decomposition, remained entrapped within the oil sand matrix.

The degradation efficiency of HPAM gel in Experiment B1 was calculated to be 86.7%. This result indicates that the residual HPAM gel in the oil sand can be effectively degraded during the ISC process. This finding confirms that ISC can substantially remove residual polymers in reservoirs that have undergone polymer flooding, thereby mitigating issues related to polymer clogging and contamination, offering a dual benefit of enhanced oil recovery and environmental cleanup.

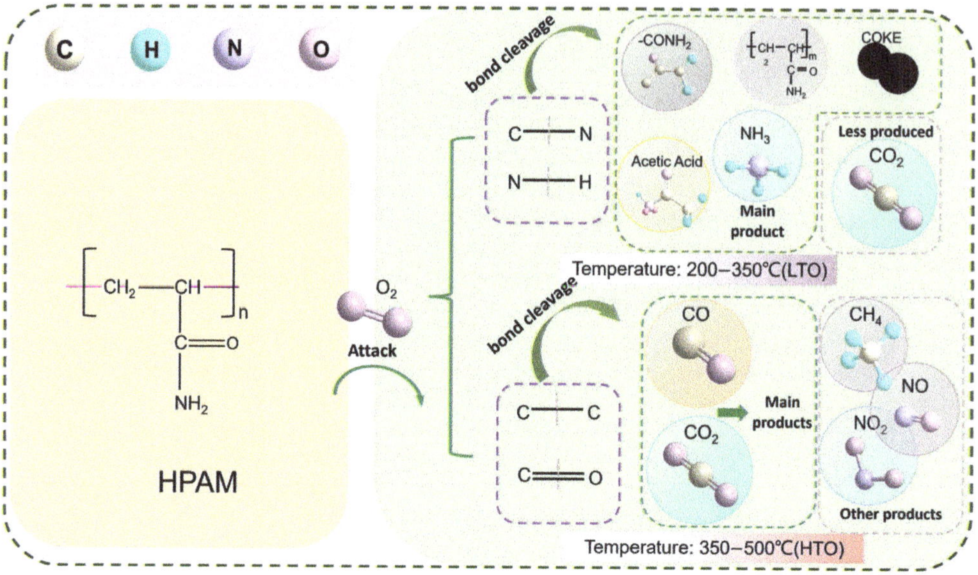

Figure 3. HPAM molecular evolution in different temperature intervals.

Figure 4. Appearance of samples in A1 and B1 before and after combustion.

2.2. Oxidation and Degradation Characteristics of HPAM Gel Under Different Porous Media

Previous studies have confirmed that residual polymer gels retained in porous media are not only combustible but also exhibit a degree of enhancement for in situ combustion processes in oil reservoirs. Subsequently, a detailed investigation of the impact of residual polymers in porous media on the activation energy of crude oil during the oxidation reaction process was carried out. The potential role of HPAM gel as a sacrificial reaction promoter in the in situ combustion process was further quantitatively evaluated. Meanwhile, oxidation behaviors of HPAM gel with distinct molecular properties and crude oils of varying viscosities were compared under controlled in situ combustion conditions.

2.2.1. Influence of the HPAM Gel on the Active Energy of the Oil

Figure 5 illustrates the temperature profiles and COx concentration trends observed during Experiments A1–A3 and B1–B3, providing insights into the oxidation dynamics and the influence of HPAM gel on reaction kinetics.

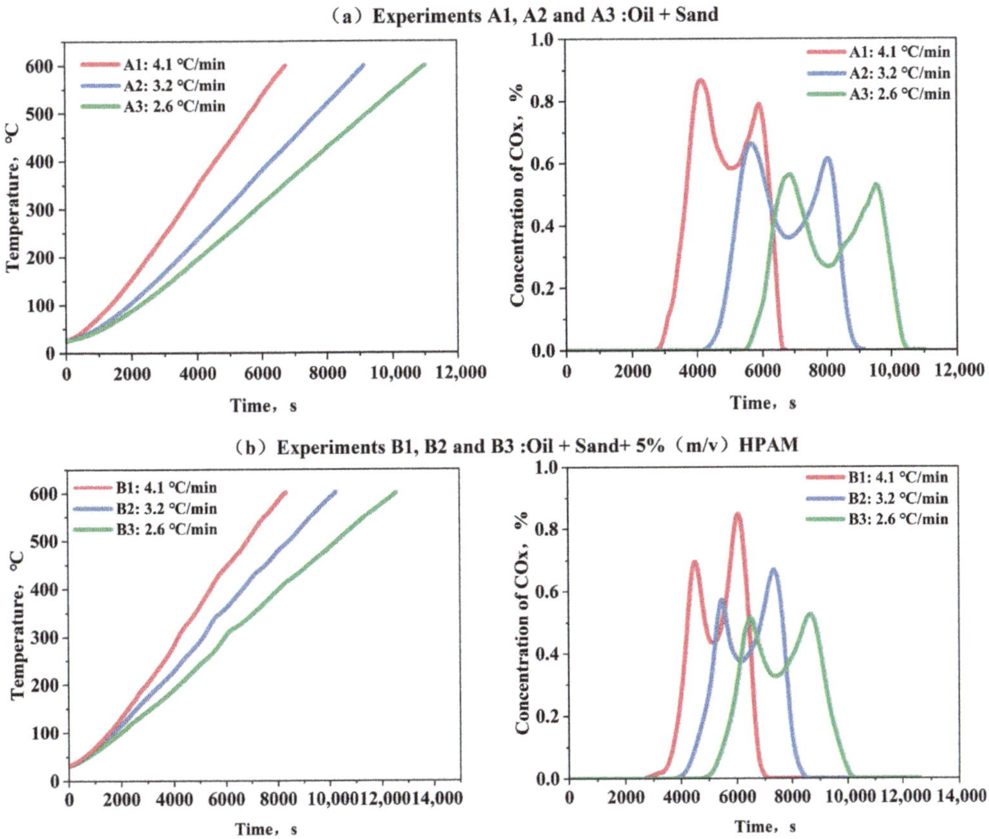

Figure 5. Temperature and concentration variations in Experiments A1 to A3 (**a**) and B1 to B3 (**b**).

A comparison in Figure 5 reveals that in the latter, each set of temperature variation curves exhibits two distinct peaks corresponding to the two peaks in concentration variation. This indicates a more pronounced exothermic reaction in the oil sand containing HPAM gel within the porous medium. Additionally, the carbon oxide concentration variations across all six experiments exhibit two peaks, which can be associated with the LTO and HTO processes of the oil. Specifically, the first peak in Group A is higher than the second, whereas the second peak in Group B is elevated. This suggests that HPAM gel primarily promotes the HTO process, leading to increased carbon oxide production during HTO compared to LTO. Using the Friedman method, the activation energies of Groups A and B were calculated and presented in Tables 1 and 2, respectively, with their activation energy fingerprints shown in Figure 6.

Analysis of the activation energy data for Group A reveals that a conversion rate ranging from 0.005 to 0.27 corresponds to a temperature interval of 268.7–359.5 °C, aligning with the LTO stage where oxygen addition and cracking reactions are dominant. Here, activation energy ranges from 33.213 to 96.475 kJ/mol. At conversion rates between 0.27 and

0.58 (359.5–445.1 °C), crude oil cracking reactions become dominant, with activation energies ranging from 99.612 to 291.423 kJ/mol. When the conversion rate exceeds 0.89, the crude oil enters the high-temperature oxidation stage (445.1–523.5 °C), with activation energies ranging approximately from 291.423 to 745.637 kJ/mol.

Table 1. Calculation results of active energy of oil (Experiment A) in porous media.

Conversion, Fraction	Temperature, °C	Maximum, KJ/mol	Average, KJ/mol
0.005~0.27	268.7~359.5	96.475	54.200
0.27~0.58	359.5~445.1	291.423	210.727
0.58~0.89	445.1~523.5	745.637	239.285
Average	/	/	172.815

Table 2. Calculation results of active energy of oil with HPAM gel (Experiment B) in porous media.

Conversion, Fraction	Temperature, °C	Maximum, KJ/mol	Average, KJ/mol
0.005~0.15	242.6~317.3	40.125	24.724
0.15~0.66	337.3~424.6	43.654	34.731
0.66~0.85	424.6~451.7	185.124	63.889
Average	/	/	41.114

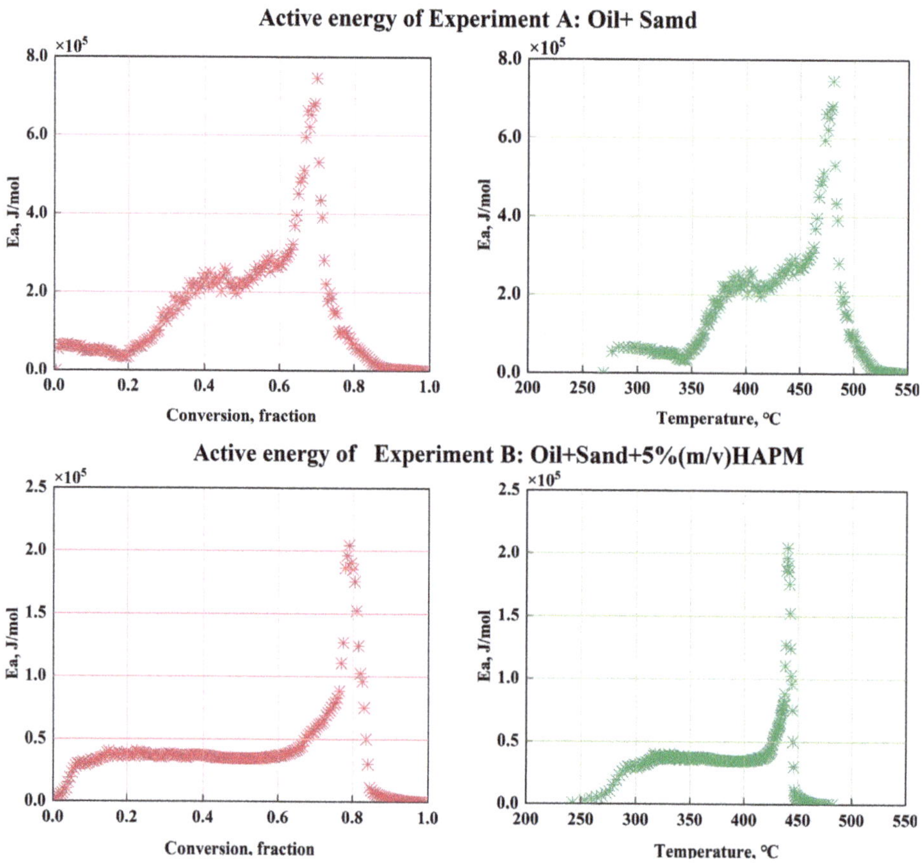

Figure 6. Activation energy fingerprints in groups A and B.

Compared with Group A, the activation energy results in Group B reveal significant reductions in both LTO and HTO processes. The maximum activation energy of low-temperature oxidation is diminished from 96.475 kJ/mol to 40.125 kJ/mol, with a reduction rate of 58.4%. The maximum activation energy in the high-temperature oxidation stage is lowered from 745.637 kJ/mol to 185.124 kJ/mol, with a reduction rate of 75.2%. This promoting effect can be attributed to the heat released during HPAM gel decomposition, which facilitates the oxidation process and reduces the activation energy required for crude oil oxidation. These findings provide quantitative evidence of the gel's promoting role in the ISC process.

2.2.2. Oxidation Characteristics of HPAM Gel with Different Properties

To investigate the influence of HPAM gel properties on the ISC process of crude oil, six sets of experiments were conducted. Experiments C1, C2, and C3 were designed to study the effect of molecular weight on oxidation behavior in simulated oil sand reservoirs, with each experiment utilizing HPAM of different molecular weights. Experiments D1, D2, and D3 focused on the impact of HPAM ionic types, specifically cationic polyacrylamide, anionic polyacrylamide, and nonionic polyacrylamide, respectively.

1. Oxidation characteristics of HPAM gel with different molecular weights

Figure 7 presents the experiment results of the mixed samples of oil sand with HPAM of three different molecular weights (Experiments C1, C2, and C3). Each set of experimental results is presented in the form of curves showing the variations in temperature and COx concentration over time.

It is observed that as the molecular weight of HPAM increases, the temperature required to generate COx also rises. Specifically, in Experiment C1 with the HPAM of molecular weight ranging from 8 to 10 MDa, COx production began at approximately 171.4 °C. For Experiment C2, with an HPAM of 10–12 MDa molecular weight, this initiation temperature increased to 186.9 °C. In Experiment C3, with HPAM with a molecular weight of 12–14 MDa, the temperature further rose to 211.1 °C. A similar trend was observed for the first and second COx concentration peaks. The first peak temperature increased from 297.0 °C in C1 to 303.9 °C in C2 and 310.9 °C in C3. The second peak temperature showed an even more pronounced increase, rising from 424.0 °C in C1 to 436.3 °C in C3. These trends can be attributed to the structural properties of higher-molecular-weight HPAM, which feature longer molecular chains, tighter entanglement, and stronger intermolecular forces. As a result, more energy is required to break and degrade these chains, leading to higher initiation and peak temperatures.

These results indicate that the lower-molecular-weight HPAM gel has a more pronounced promoting effect on the ISC process of crude oil. Its simpler molecular structure facilitates easier chain breakage and degradation, making it more effective at promoting oxidation and reducing reaction temperatures. Figure 8 exhibits the appearance of the samples from Experiments C1, C2, and C3 before and after combustion. All three samples exhibited similar appearances, displaying white or off-white coloration, which indicates complete or near-complete degradation of the hydrocarbons and HPAM gel. The degradation efficiencies of the samples were measured based on the conductivity method described earlier, and the test results are shown in Table 3.

Table 3. The calculation results of degradation efficiencies for HPAM gel in the experiments C1, C2, and C3.

No.	Experiments	Degradation Efficiency, %
1	C1 (HPAM of 8–10 MDa)	91.5 ± 0.4
2	C2 (HPAM of 10–12 MDa)	89.1 ± 0.2
3	C3 (HPAM of 12–14 MDa)	86.7 ± 0.2

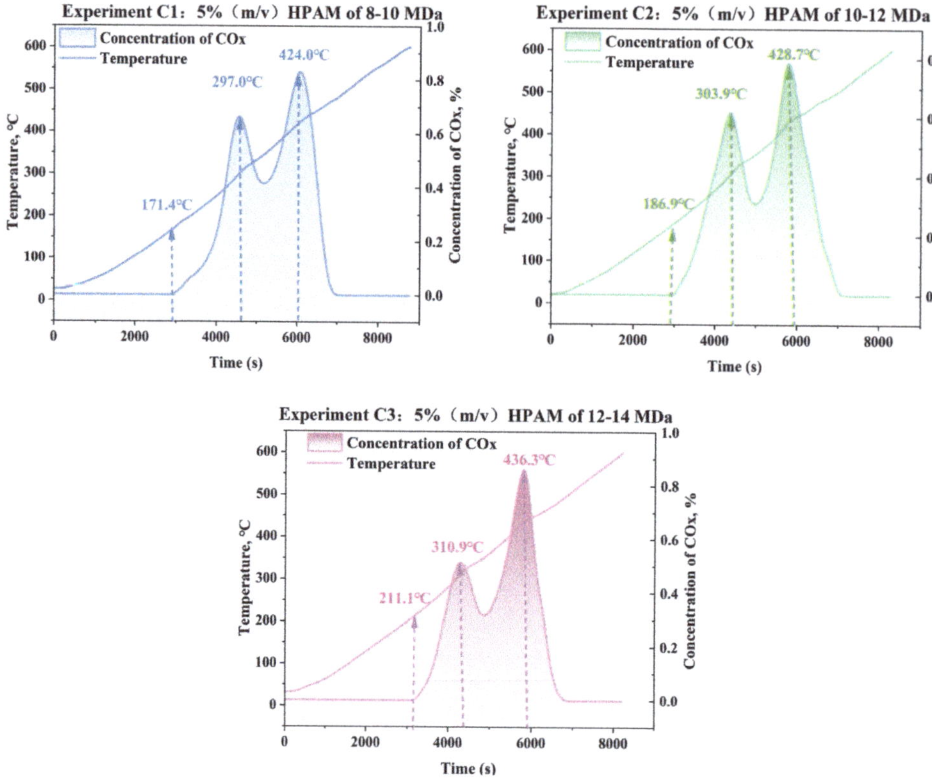

Figure 7. Temperature and concentration profiles of Experiments C1–C3 with HPAM of varying molecular weights.

Figure 8. Appearance of samples from Experiments C1, C2, and C3 before and after combustion.

The results demonstrate that samples with different molecular weights of HPAM gel undergo significant degradation during the ISC process in porous media. There is a clear inverse relationship between the degradation efficiency and molecular weight. That means that when the molecular weight increases, the degradation rate decreases. For example, when the molecular weight of HPAM increases from between 8 and 10 million to between 12 and 14 million, the degradation rate has a noticeable drop, falling from 91.5% to 86.7%. This shows that the higher the molecular weight, the more difficult it is for the HPAM gel to be degraded under the same treatment conditions. The possible reason is that high-molecular-weight HPAM has longer molecular chains, more complex structures, and more stable chemical bonds, requiring higher energy or more severe conditions to trigger its degradation reaction, thus leading to the decline in the degradation rate. Consequently, residual gel with a lower molecular weight of HPAM in the reservoir is more susceptible to cracking during the ISC process, making it a more effective candidate for enhancing oxidation efficiency.

The result is consistent with the research of Shatat [31]. In their work, it was confirmed that high-molecular-weight HPAM has a more stable structure and relatively lower degradation efficiency compared with low-molecular-weight HPAM.

2. Oxidation characteristics of HPAM gel with different ionic types

Gel with different ionic types of HPAM is expected to have different degrees of influence on the ISC process of crude oil. Figure 9 presents the temperature and concentration curves of HPAM with cationic (Experiment D1), anionic (Experiment D2), and nonionic (Experiment D3) properties in porous media, illustrating how these structural variations affect the oxidation behavior.

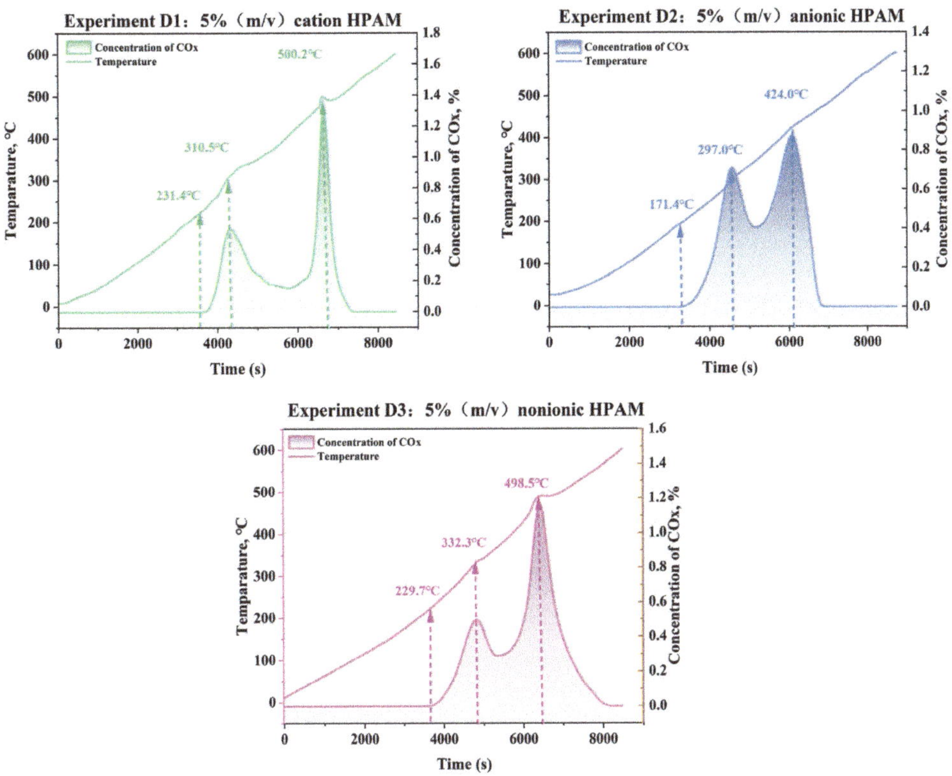

Figure 9. Temperature and concentration profiles of Experiments D1–D3 with varying types of HPAM.

It shows that anionic HPAM gel exhibits the lowest degradation initiation temperature (171.4 °C) during the ISC process, whereas cationic and nonionic HPAM gel require higher initiation temperatures (231.4 °C and 229.7 °C, respectively), indicating greater energy requirements for reaction activation. While anionic HPAM (D2) exhibits the earliest degradation in porous media, its COx concentrations show comparable levels between LTO (0.7%) and HTO (0.5–0.9%). In contrast, cationic (D1) and nonionic HPAM (D3) display pronounced promotion selectivity: their HTO peaks (1.2–1.4%) surpass LTO peaks (0.5%) by 140–180%, indicating superior high-temperature promotion efficiency despite requiring higher initiation temperatures.

This can be attributed to the cationic groups with positive charges on the molecular chain of cationic HPAM, which strengthen intermolecular forces, making molecular chain movement relatively difficult and thus enhancing thermal stability to some extent, leading to a relatively high degradation temperature. Conversely, the polar groups in the anionic HPAM molecular chain increase polarity and water absorption, predisposing it to hydrolysis and molecular chain breakage at high temperatures. For nonionic HPAM gel, the lack of ionic groups results in relatively weak intermolecular forces and a relatively active amide group. Consequently, the thermal stability of anionic and nonionic HPAM gel is relatively low, and so are their degradation temperatures.

Similarly, a comparison of the appearances and degradation efficiency measurements was carried out on the samples of Experiments D1, D2, and D3 after oxidation, as shown in Figure 10 and Table 4.

Figure 10. Appearances of samples in D1, D2, and D3 before and after combustion.

Table 4. The calculation results of degradation efficiencies for HPAM gel in the experiments D1, D2, and D3.

No.	Experiments	Degradation Efficiency, %
1	D1 (Cation HPAM)	88.4 ± 0.2
2	D2 (Anionic HPAM)	89.1 ± 0.2
3	D3 (Nonionic HPAM)	87.9 ± 0.1

Post-oxidation analysis revealed comparable visual characteristics across all three sample groups (D1, D2, and D3), with degradation efficiencies of 88.4%, 89.1%, and 87.9%, respectively. These results demonstrate that the residual HPAM gel could achieve degradation efficiencies higher than 85% in oil reservoirs during the ISC process, with the ionic type (anionic/cationic/nonionic) showing <1.5% variation in degradation outcomes under identical thermal conditions. The consistent degradation performance across HPAM variants suggests that ISC effectively mitigates polymer retention issues regardless of molecular charge characteristics, providing a robust remediation strategy for polymer gel-treated reservoirs.

Results show that differences in molecular weight resulted in a variation in degradation efficiency of up to 5%, whereas differences in ionic type led to a remarkably small variation of less than 1.5%. This comparison underscores that molecular weight serves as the primary controlling factor for HPAM degradation, to a greater extent than ionic type. The difference can be attributed to the distinct underlying factors each property influences. Molecular weight primarily governs the overall structural entanglement and thermal stability of the polymer chains, thereby dominating the fundamental difficulty of degradation and the ultimate efficiency. In contrast, the ionic type introduces different functional groups into the side chains. The varying thermal stability and reactivity of these specific functional groups are responsible for the differences observed in the initial pyrolysis temperatures. These findings are consistent with those reported in the literature [26].

However, once the temperature exceeds a critical threshold (e.g., upon entering the high-temperature oxidation stage above approximately 500 °C), all HPAM types undergo backbone scission and deep oxidation. The ISC process provides a high-temperature, oxygen-rich environment with abundant radical activity. Under these severe conditions, initial decomposition differences are overridden and all polymer structures are extensively oxidized, leading to consistently high degradation efficiencies exceeding 85% across all HPAM types investigated.

2.2.3. Influence of Oil Viscosity on the Degradation and Oxidation of HPAM Gel

Figure 11 shows the temporal evolution of temperature and COx concentration profiles of E1, E2, and E3 containing crude oils of varying viscosity (128–15,310 mPa·s).

Analysis of reaction initiation temperatures reveals a positive correlation between crude oil viscosity and LTO thresholds: lower-viscosity oils (128 mPa·s) presented more activity with a lower initiation oxidation temperature of 175.9 °C, while higher viscosity oils require an oxidation temperature of 243.1 °C, which is much higher.

Additionally, when compared with the higher-viscosity crude oil, another notable difference emerges in the porous media environment of the lower-viscosity crude oil. Lower-viscosity mixtures exhibit distinct reaction behavior, showing that the COx concentrations during the low-temperature oxidation process are 2.1–3.5 times higher than those during the high-temperature oxidation process. This difference can be attributed to the fundamental properties of different oils. Light oil normally possesses a lower flash point (around 175 °C) and also a much lower ignition point (around 300–320 °C) than heavy oil in bulk conditions, which are presented in the ranges of 220–240 °C and 320–340 °C, respectively. The coke combustion temperature or HTO process, however, shows less of a difference, varying in the range of 460–480 °C for different oils.

Subsequently, post-oxidation analysis of Experiments E1–E3 shows HPAM gel degradation efficiencies quantified through conductivity measurements, which are shown in Table 5.

Table 5. Post-oxidation analysis of Experiments E1–E3 quantified HPAM degradation efficiencies.

Experiment NO.	Oil Viscosity, mPa·s	Degradation of HPAM Gel, %
E1	15,310	93.1 ± 0.2
E2	2500	89.9 ± 0.3
E3	128	85.2 ± 0.1

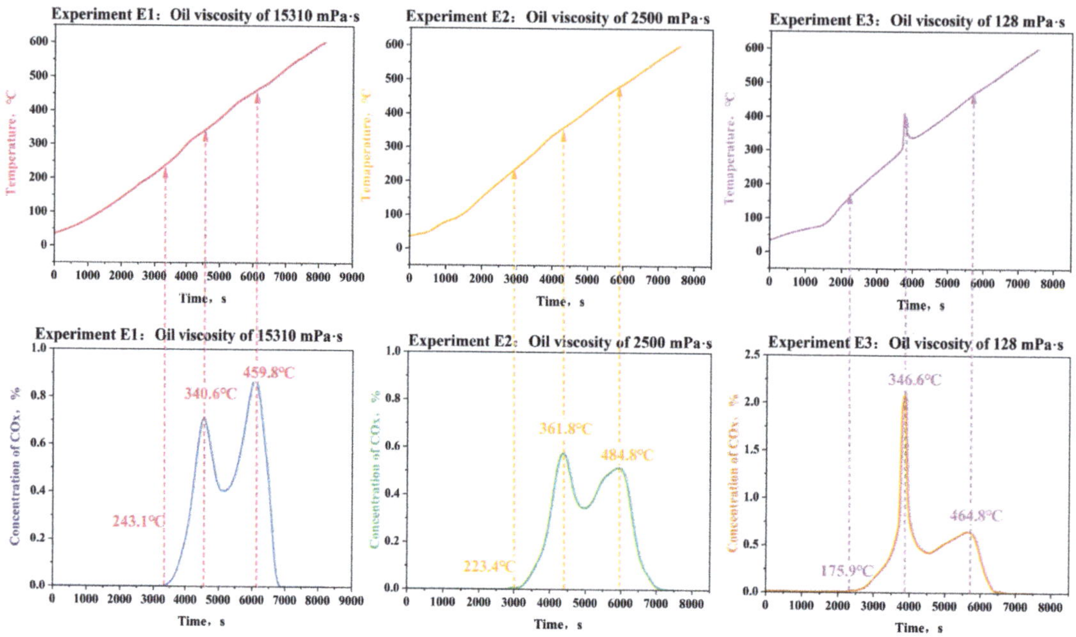

Figure 11. Temperature and concentration profiles of Experiments E1–E3 with oil of varying viscosities.

As the results indicate, there is a positive correlation between the crude oil viscosity and the degradation efficiency of HPAM gel in the system. Oil sand mixtures with higher viscosity demonstrate superior degradation performance, with E1 (15,310 mPa·s) achieving 93.1%, compared to 85.2% in E3 (128 mPa·s). The result could be attributed to the higher content of heavy components (e.g., asphaltenes and resins) in high-viscosity crude oil. During the oxidation, these heavy components preferentially form coke deposits, which extend the duration of high-temperature oxidation, thereby providing sustained thermal energy for HPAM molecular chain scission. Simultaneously, active free radicals (e.g., ·OH and O^-) generated from heavy component oxidation could initiate oxidative degradation of HPAM backbones through thermo-chemical synergy. Furthermore, the higher calorific value of viscous crude oil may enhance energy supply efficiency during the oxidation reaction, further facilitating polymer decomposition. Notably, these findings highlight the potential advantage of high-viscosity crude oil in effectively removing HPAM contaminants during the ISC processes, offering critical insights for optimizing polymer pollution remediation strategies in heavy oil reservoirs.

This result indicates that during the in situ combustion process of crude oil with different viscosities, HPAM gel can be degraded to a large extent, ranging from 85.2% to 93.1%. The higher the viscosity of the crude oil, the more conducive it is to the degradation of HPAM during the ISC process.

2.3. *Analysis of Enhanced Oil Recovery*

Combustion tube experiments were conducted to simulate in situ combustion for enhanced oil recovery in post-polymer-flooded reservoirs. The experimental setup featured a reservoir permeability of 65×10^{-3} μm^2 and a porosity of 32.2%, with a residual polymer content of 100 μg/g after polymer flooding. The ignition temperature was set at 600 °C. The stable propagation of the combustion front was further evidenced by consistently maintained COx generation levels in effluent gas concentrations, as shown in Figure 12, with an average value of 8.7%. Figure 12 also presents the experiment's differential pressure and oil recovery factor curves. Figure 13 shows the temperature profiles at different locations along the combustion tube, showing peak temperatures of 502.7 °C (T3), 482.2 °C (T4), 441.6 °C (T5), and 81.7 °C (T6), indicating stable combustion front propagation toward the production well. At 1500 s, as the combustion front neared temperature measurement well T2, a notable increase in the pressure differential commenced. This phenomenon served as a clear indication of the formation of an oil bank, highlighting a critical stage in the combustion process. The maximum differential pressure of 1947 kPa occurred at 2741 s, coinciding with rapid oil production at the production well. The ultimate recovery factor was calculated to be 68.6%. Figure 14 shows the morphological characteristics of oil sand post-ISC. To determine residual HPAM in post-ISC oil sand, triplicate sampling and testing were conducted at six designated points indicated in the figure, ensuring measurement accuracy. The HPAM gel removal efficiency results are presented in Figure 12.

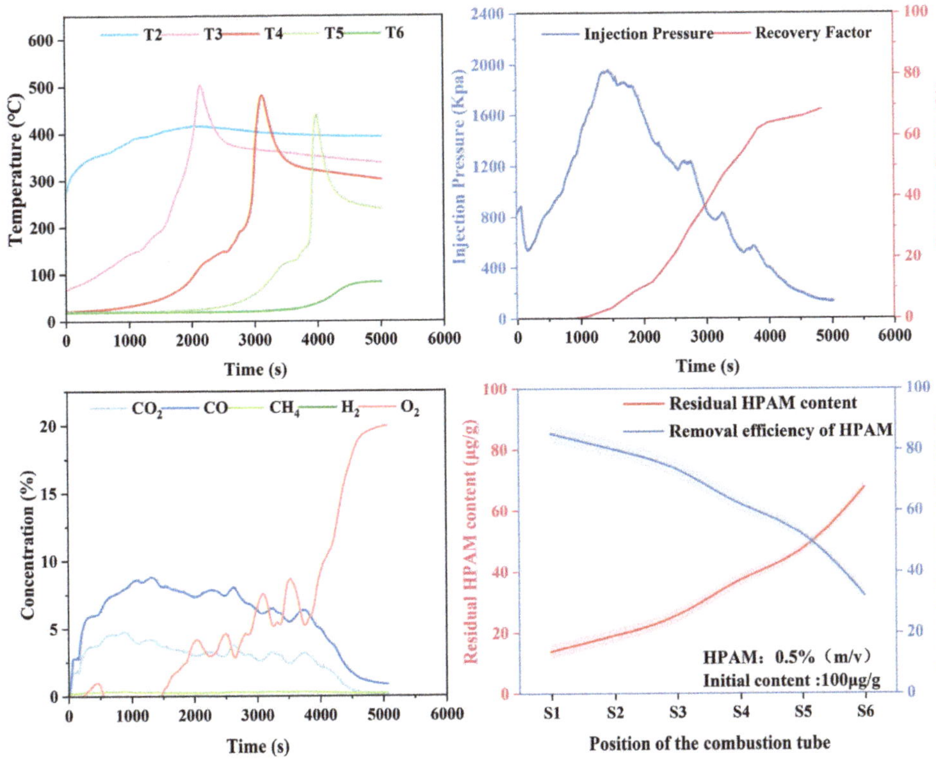

Figure 12. Results of combustion tube experiments: temperature, pressure, recovery factor, and HPAM gel degradation.

Notably, the highest HPAM gel degradation efficiency of 85.3% was recorded at well T1, in stark contrast to the lowest value of 32.1% observed at well T6. The data obtained reveal a clear inverse relationship between the HPAM degradation efficiency and the distance from the injection well. This correlation aligns with the gradual temperature decrease in the combustion front along the tube. It can be attributed to the enhanced thermal decomposition of HPAM gel at elevated temperatures, which in turn leads to a reduction in residual polymer concentrations. Even though T6 was characterized by low temperature, the continuous generation of high-temperature gases and the movement of crude oil during the ISC process enabled the partial extraction of HPAM through the production well. As a result, despite the incomplete advancement of the combustion front to the production wellbore, the removal efficiency of HPAM at T6 remained above 30%.

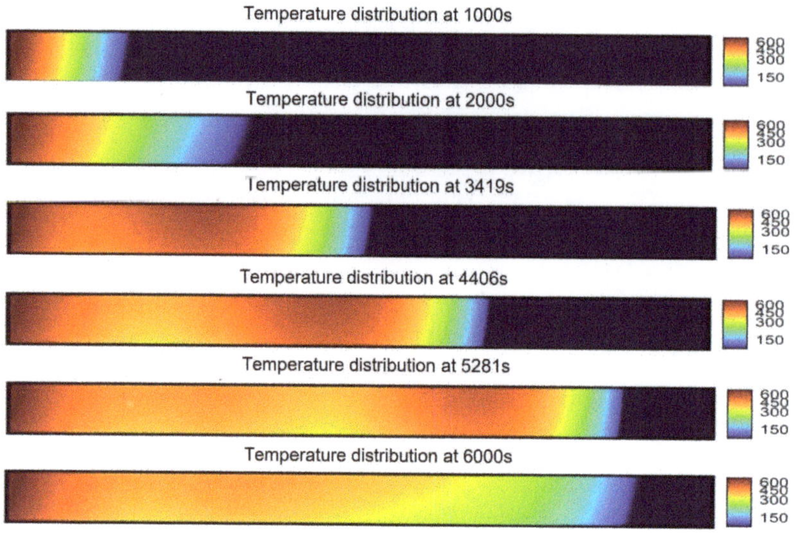

Figure 13. Temperature distribution along the tube at different times.

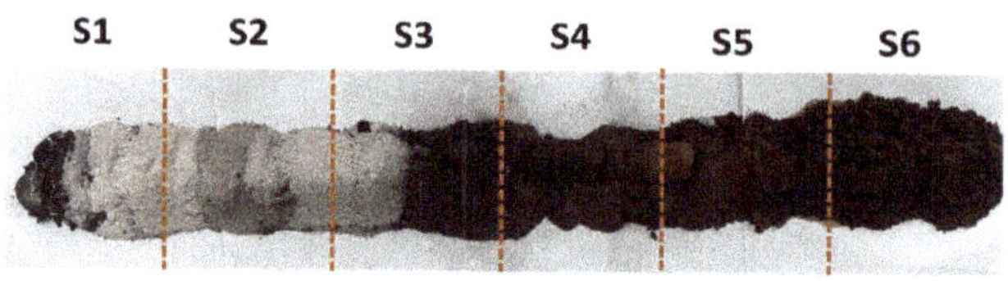

Figure 14. Morphological characteristics of oil sand and sampling points for residual HPAM gel analysis.

Combustion tube experiments showed that HPAM gel had an average degradation efficiency of 64.8% and enabled an oil recovery factor of 68.6%, far exceeding most lab-scale ISC results. These findings indicate that during the ISC process, HPAM gel serves dual functions. On the one hand, it acts as a fuel, promoting the enhancement of crude oil recovery during the process. On the other hand, under the high-temperature conditions of the reservoir and during the secondary displacement process, HPAM undergoes a high efficiency of degradation, effectively eliminating HPAM gel contamination in the subsurface reservoir.

2.4. Further Discussion

Figure 15 provides a schematic representation of the EOR mechanisms through ISC in polymer gel-treated reservoirs.

During the ISC process, distinct thermal zones typically develop within the formation, sequentially including the combusted zone, combustion front, coke zone, steam zone, oil bank, and undisturbed zone [32]. In the combustion zone, residual HPAM gel undergoes substantial thermal degradation. The combustion front is where pyrolysis reactions predominantly occur. At the critical temperature threshold of around 300 °C, the gel first undergoes oxidative carbonization to form coke deposits. This carbon-rich residue serves as essential fuel for sustaining the stable propagation of the combustion front. Subsequently, the increasing temperature promotes further cracking reactions, where HPAM gel decomposes into carbon oxides, water, and nitrogen oxides. These gaseous products combine with hydrocarbon oxidation byproducts to form the steam zone, effectively displacing crude oil toward production wells while remaining largely sequestered in the formation due to flow resistance. The mobilized oil accumulates in the oil bank region, exhibiting reduced viscosity and enhanced mobility. Partial dissolution of combustion gases into the oil phase facilitates production recovery. The undisturbed zone retains both original reservoir fluids and undegraded HPAM gel residues.

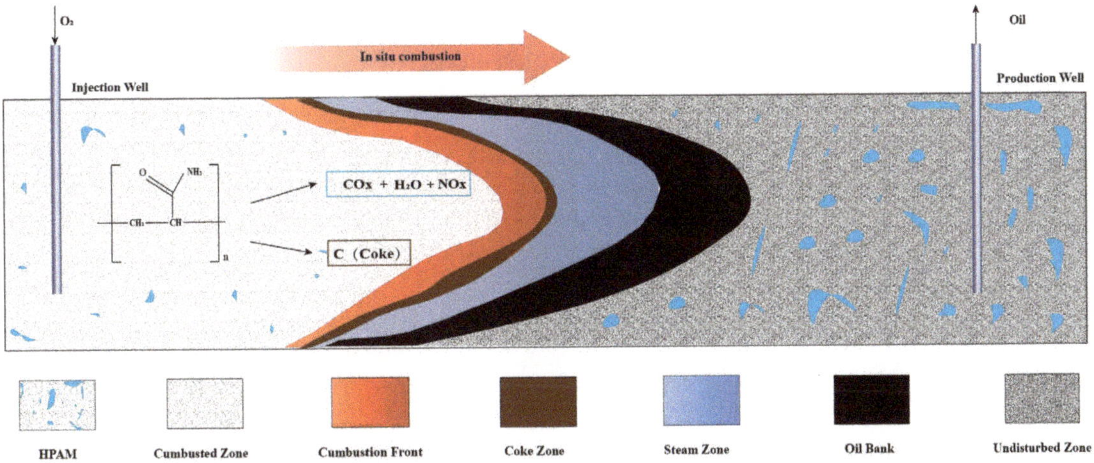

Figure 15. The EOR mechanisms through ISC in polymer gel-treated reservoirs.

This work demonstrates that residual HPAM gel performs dual functions in ISC processes: (1) As a supplemental fuel source through controlled degradation, it enhances the stability of the combustion front and improves sweep efficiency; and (2) in situ decomposition addresses subsurface polymer contamination while eliminating pore-blocking residues. The generated gases contribute to oil displacement while remaining geologically sequestered, presenting an environmentally sustainable EOR approach without additional ecological burdens.

Although the ISC following polymer gel application demonstrates predictable recovery improvement [33], this work analyzes the role of polyacrylamide gel during ISC from a reaction kinetics perspective. The application of ISC in polymer gel-treated reservoirs remains a complex engineering challenge, which requires further investigation into the relationship of the EOR effect with the reservoir characteristics, including porosity, permeability, oil saturation, and so on. Future research should prioritize developing predictive models that integrate chemical kinetics with reservoir engineering principles for comprehensive process optimization.

3. Conclusions

The thermochemical degradation behavior of residual HPAM during ISC in post-polymer flooding reservoirs was systematically investigated in this work, and its synergistic effects on EOR and reservoir remediation were studied. Key findings are as follows.

1. Residual HPAM undergoes effective thermochemical degradation during ISC, acting as an in situ fuel to accelerate crude oil oxidation. Quantitative analysis shows that

HPAM reduces the activation energy by up to 58.4% in the LTO stage and 75.2% in the HTO stage.
2. ISC achieves dual benefits: it removes residual HPAM while enhancing oil recovery. Experimental results demonstrate an average HPAM removal efficiency of 64.8% across the reservoir, with incremental oil recovery by 68.6%.
3. High-viscosity oil, low-molecular-weight HPAM, and anionic HPAM facilitate HPAM degradation during ISC, with the degradation efficiency ranging from 85.2% to 93.1% under different conditions.

These findings confirm that ISC is a viable follow-up EOR technique for post-polymer flooding reservoirs, offering a sustainable solution to mitigate permeability damage from residual HPAM while boosting oil recovery.

4. Materials and Methods

4.1. Materials

The materials used in the experiment consisted of HPAM gel and oil sand mixtures, which simulated the porous medium reservoir environment. The oil sand mixtures were prepared via uniform mixing of quartz sand and crude oil. Tables 6 and 7 show different types of HPAM and crude oil with different viscosities used in the experiments.

Table 6. Characteristics of polyacrylamide polymers with varying properties.

No.	Name	CAS	Manufacturer	Molecular Weight
1	Cationic Polyacrylamide	9003-05-8	Macklin, Shanghai, China	12 MDa
2	Anionic Polyacrylamide	9003-05-8	Macklin, Shanghai, China	12 MDa
3	Nonionic Polyacrylamide	9003-05-8	Macklin, Shanghai, China	12 MDa
4	Anionic Polyacrylamide	9003-05-8	Macklin, Shanghai, China	8–10 MDa
5	Anionic Polyacrylamide	9003-05-8	Macklin, Shanghai, China	10–12 MDa
6	Anionic Polyacrylamide	9003-05-8	Macklin, Shanghai, China	12–14 MDa

Table 7. Physicochemical properties of crude oil samples from different oilfields.

Sample No.	Producing Oilfield	Viscosity, mPa·s	Density, $g\cdot(cm^3)^{-1}$
O1	Nanyang Oilfield	15,100	0.96
O2	Xinjiang Oilfield	1500	0.95
O3	Huabei Oilfield	128	0.90

All HPAMs used in this work are derivatives of polyacrylamide (repeating unit: $-CH_2-CH(CONH_2)-$) with different charge properties, and their chemical structures are shown in Figure 16. The charge origins are clarified as follows.

The chemical structure of AP contains two main repeating units: neutral amide units ($-CH_2-CH(CONH_2)-$) and anionic carboxylate units ($-CH_2-CH(COO^-)-$). The negative charge originates from the hydrolysis of amide groups ($-CONH_2$) in PAM. Cationic units were based on the PAM backbone; AP was modified by introducing cationic monomer units, including Methacryloxyethyltrimethyl ammonium chloride ($C_9H_{18}ClNO_2$). The positive charge was derived from the quaternary ammonium groups ($-N^+(CH_3)_3$) in the modified monomer. The chemical structure of NP was dominated by neutral amide units ($-CH_2-CH(CONH_2)-$) with almost no ionizable groups.

To objectively evaluate the combustion characteristics of HPAM and crude oil during the ISC process without additional catalytic effects, the gels used in the combustion cell experiments and combustion tube in this work are physical HPAM gels with concentrations of 5% and 0.5% (m/v), respectively. The preparation method was as follows: Prepare

100 mL of deionized water, stir at low speed, and slowly add 5 g/0.5 g of HPAM powder simultaneously. Stir continuously for 12 h to achieve complete dissolution. After stopping stirring, stand them at room temperature for 24–48 h to allow the molecular chains to fully extend and entangle, forming a uniform gel.

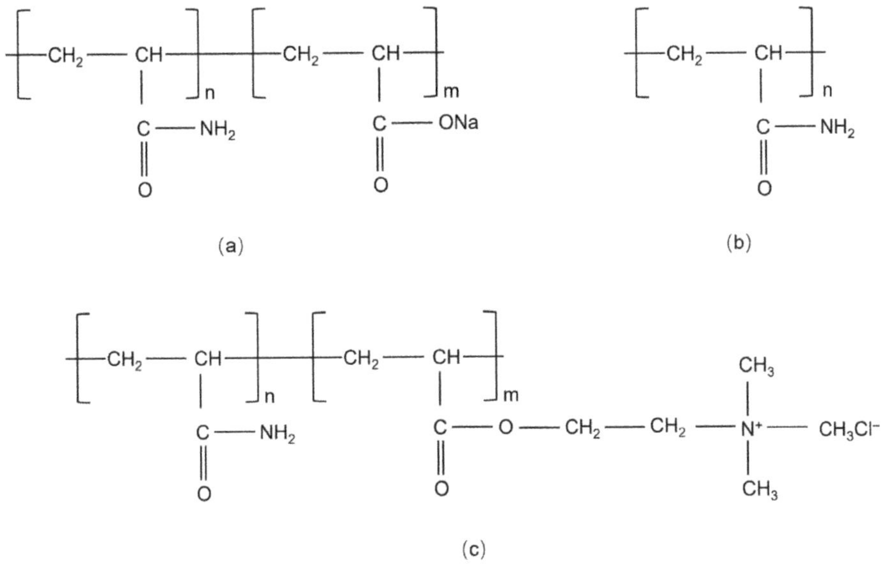

Figure 16. Molecular structure diagrams of anionic polyacrylamide (**a**), nonionic polyacrylamide (**b**), and cationic polyacrylamide (**c**).

4.2. Apparatus and Methodology

4.2.1. Reaction Kinetic Experiment

Kinetic cell experiments were employed to study reaction kinetics and analyze the oxidation behavior of crude oil and HPAM gel within porous media. The schematic of the kinetic cell experimental apparatus is presented in Figure 17. The cylindrical reactor, with a height of 10 cm, features an inner diameter of 2.0 cm and an outer diameter of 3.0 cm. The integrated experimental setup consists of a gas injection system, a temperature-controlled reaction chamber, a temperature detector, a real-time gas concentration monitoring system, a tail gas scrubber, and computerized data acquisition. A detailed photograph of the system is provided in Figure 18.

The experimental design comprised two baseline groups in active energy determination: Group A comprised a porous media system constructed with crude oil to simulate the original reservoir conditions. Group B contained an oil sand system with HPAM gel to simulate post-polymer-flood reservoir conditions. Each experiment of Group A and Group B was repeated with three distinct heating rates (4.1 °C/min, 3.2 °C/min, and 2.6 °C/min) to determine the system's activation energy.

Parametric studies included Groups C, D, and E, which investigated HPAM gel-containing oil sand mixtures with controlled variables. Specifically, three different molecular weights of HPAM were considered in Group C. Group D evaluated ionic types, including cationic polyacrylamide, anionic polyacrylamide, and nonionic polyacrylamide. Group E maintains a constant anionic HPAM (10–12 MDa) while testing crude oils with different viscosities. The experimental scheme is listed in Tables 8 and 9.

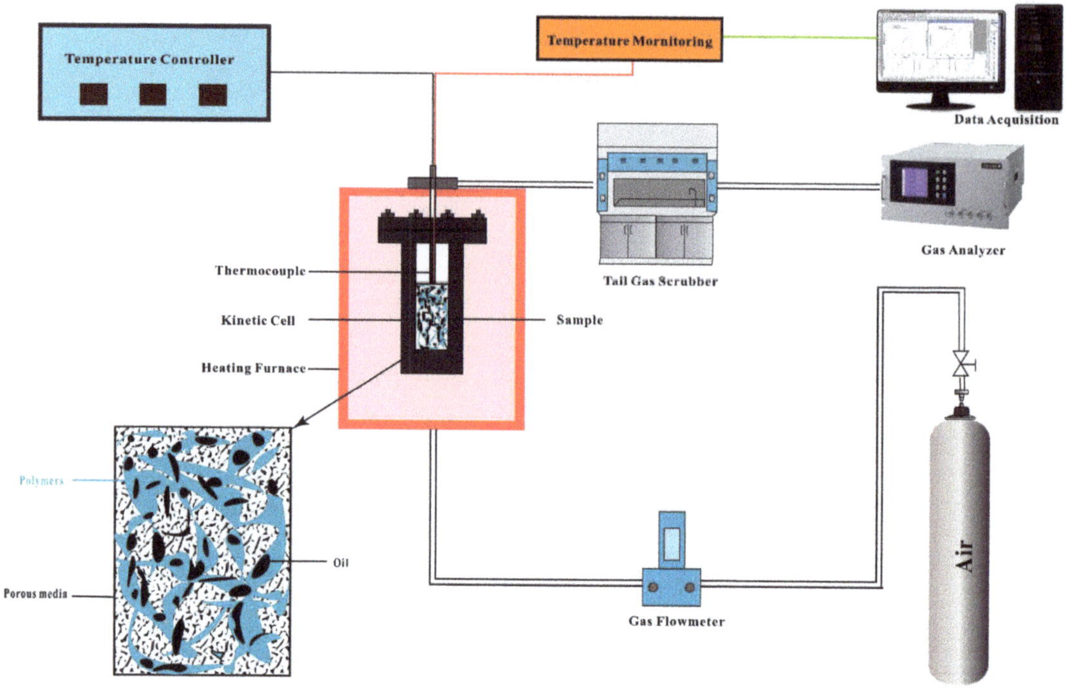

Figure 17. Schematic diagram of the kinetic cell experimental process.

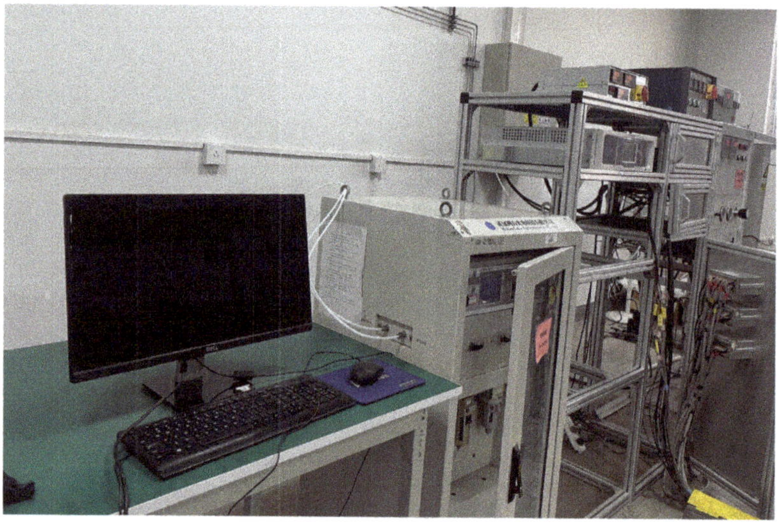

Figure 18. Photograph of the kinetic cell experiment setup.

The experimental procedure is described as follows: The kinetic cell reactor was cleaned and dried. Oil sand samples were prepared: Group A (crude oil: quartz sand = 1:20) and Group B (polymer-contained oil sand = 0.5:1:20), both stirred for 20 min. After placing a filter at the reactor bottom, 5.0 g of quartz sand was loaded. The 10.5 g sample was then added and gently compacted with a glass rod to limit height increase to $\leq$1.0 cm, followed by 10.0 g quartz sand secured with asbestos. Following gasket sealing and gas line connection, a stepwise pressurization test was set at 2.0 L/min while maintaining a pressure differential below 0.5 MPa. System integrity was confirmed when the stabilized flow rate remained under 0.1 L/min. Air injection commenced, and after >5 min of stable O_2 concentration, pressures were fine-tuned (differential <0.5 MPa). Heating from 25 °C to 600 °C at a specified rate enabled real-time monitoring of $CO_2/O_2/CO/CH_4/H_2$. Heating terminated automatically at 600 °C, with post-cooling disassembly and analysis of temperature/gas concentration data.

Table 8. Experimental design for Group A and Group B.

NO.	Sample	Ratio (Oil: Sand: [HPAM If Applicable])	Heating Rate, °C·min^{-1}
A1	Oil + Sand	1:20	4.1
A2	Oil + Sand	1:20	3.2
A3	Oil + Sand	1:20	2.6
B1	Oil + Sand + HPAM	1:20:0.25	4.1
B2	Oil + Sand + HPAM	1:20:0.25	3.2
B3	Oil + Sand + HPAM	1:20:0.25	2.6

Table 9. Experimental design for KC with different HPAM types and crude oils.

NO.	Molecular Weight of Polyacrylamide, MDA	NO.	Type of Polyacrylamide	NO.	Oil Code	Oil Viscosity, mPa·s
C1	8–10	D1	Cationic Polyacrylamide	E1	O1	15,310
C2	10–12	D2	Anionic Polyacrylamide	E2	O2	1500
C3	12–14	D3	Nonionic Polyacrylamide	E3	O3	128

4.2.2. Enhanced Oil Recovery Experiment

The combustion tube experimental setup is a commonly used method for studying in situ combustion EOR techniques. The combustion tube, as shown in Figure 19, was self-designed and comprised an air injection system, an ignition control system, the combustion tube itself, a gas analyzer, and a data monitoring system. The combustion tube was equipped with six temperature-monitoring wells (labeled as T1, T2, T3, T4, T5, and T6), where thermocouples were set inside to monitor temperature variations at different positions.

4.2.3. Activation Energy

Reactant molecules have to overcome a specific energy barrier, known as the activation energy, to engage in effective collisions and then convert into product molecules. In the ISC process, reactants like crude oil molecules and oxygen molecules must acquire adequate energy to leap over this energy barrier, thereby triggering the oxidation reaction. In this work, the activation energy is calculated based on the Friedman formula:

$$ln\frac{d\alpha}{dt} = -\frac{E_a}{RT} + lnAf(a) \quad (1)$$

where α is the generation concentration of carbon oxides in the effluent gas and is obtained through real-time monitoring by a gas analyzer. T refers to the temperature of the reaction cell, which is monitored in real-time by a thermocouple. t is the reaction time and $f(a)$ represents the mechanism function.

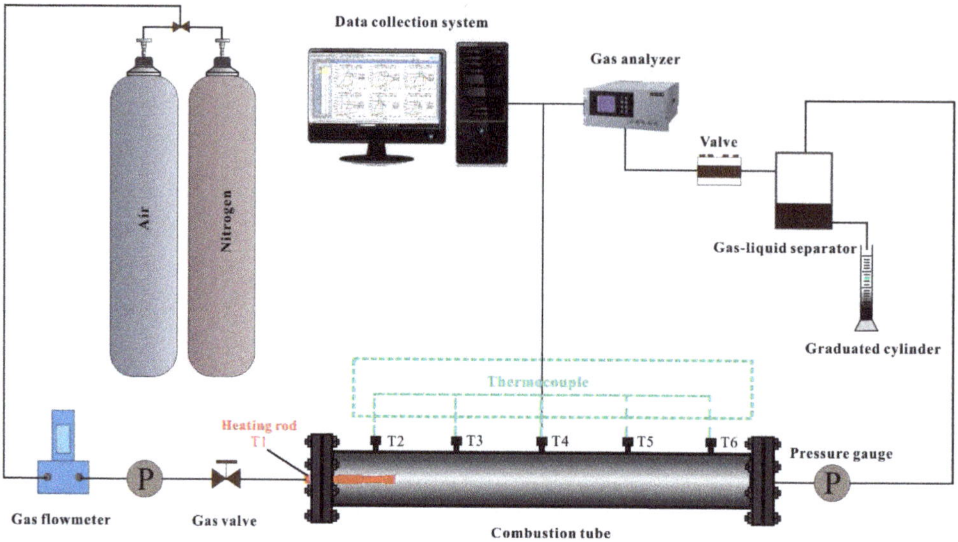

Figure 19. Schematic diagram of the combustion tube experimental process.

Due to the complexity of the reactions of the samples utilized in the experiment, no mature method was established to accurately determine the mechanism function. Therefore, the iso-conversion method was adopted in this work to ascertain the activation energy. Consequently, for a given sample, three sets of combustion cell experiments with different heating rates needed to be carried out. Thus, three curves of $ln\frac{d\alpha}{dt}$ varying with $-\frac{1}{T}$ can be plotted, which are then utilized for the calculation of the activation energy. In total, six sets of experiments were implemented in this part to complete the determination and comparison of the activation energies of the two groups of samples. Specifically, Experiments A1 to A3 (Group A) used oil sand samples, while Experiments B1 to B3 (Group B) used the same oil sand with added residual HPAM. Each experiment was carried out at heating rates of 4.1 °C/min, 3.2 °C/min, and 2.6 °C/min.

4.2.4. Determination of Residual HPAM

A linear relationship ($R^2 > 0.98$) was established between concentration (0.1–1.0 g/L) and conductivity of the HPAM gel. Hence, conductivity analysis was implemented for post-combustion residual HPAM gel quantification. The method demonstrated a detection limit of 0.01 g/L with sensitivity <5% RSD. Initially, the standard curves for different types of HPAM were established. HPAM solutions were prepared with concentrations of 0.1 g/L, 0.2 g/L, 0.3 g/L, 0.5 g/L, and 1.0 g/L. Standard HPAM solutions were prepared using deionized water and allowed to equilibrate for 30 min before measurement. Following complete dissolution (magnetic stirring at 500 rpm for 15 min), solution conductivity was measured using a calibrated conductivity meter with temperature compensation. The conductivity data were then calibrated to reference temperature (25 °C) by (2):

$$\rho_o = \frac{\rho_t}{1 + \alpha(t_1 - t_0)} \tag{2}$$

where t_1 is the testing temperature, °C t_0 is the reference temperature 25 °C, and α is the temperature coefficient of the solution, taken as 0.022 here. ρ_t and ρ_o are the corresponding conductivities at t_1 and t_0, respectively.

The experimental measurements were repeated three times to ensure the accuracy and reliability of the data. Subsequently, the average value of the obtained conductivity data was computed and recorded. Based on the validated dataset, calibration curves for cationic polyacrylamide, anionic polyacrylamide, and nonionic polyacrylamide solutions were generated, and they are graphically presented in Figure 20. The corresponding linear regression equations for these curves, which quantitatively describe the relationship between conductivity and concentration, are designated as Equations (3), (4), and (5), respectively.

$$y = 787.07x - 9.0043R^2 = 0.9917 \quad (3)$$

$$y = 1084.7x + 28.368R^2 = 0.9871 \quad (4)$$

$$y = 831.01x + 22.436R^2 = 0.9848 \quad (5)$$

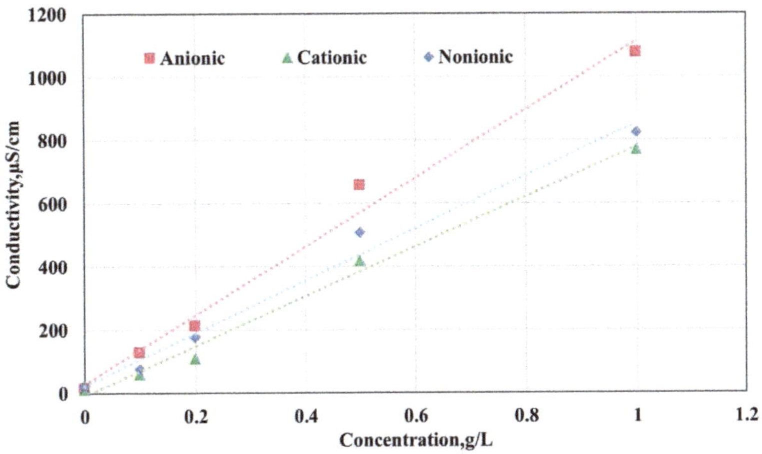

Figure 20. Standard curves of gels with different concentrations.

To precisely quantify the residual HPAM gel content in the post-combustion samples, each sample was dissolved in a specific volume of deionized water and continuously agitated for 30 min to ensure complete dissolution of residual HPAM gel. Subsequently, the resultant solution was filtered, and the conductivity of the filtrate was measured using a calibrated conductivity meter. Thereafter, the residual HPAM gel concentration and mass were determined by Equations (3)–(5). The degradation efficiency, denoted as ED, was calculated by the formula presented in Equation (6):

$$ED = \frac{c_1 \times v_1}{m} \times 100\% \quad (6)$$

where c_1 is the concentration of the residual HPAM after the combustion, v_1 is the volume of the solution in which the residual sample is dissolved, and m is the original mass of HPAM mixed with the oil sand.

Author Contributions: J.W. conducted experimental data analysis and drafted the manuscript; R.Z. secured funding for the research; Y.Y., Y.Z., G.Z. (Guangsen Zhu), J.T., H.Z., H.R., and G.Z. (Guanghui Zhou) contributed to data collection, the literature review, and critical revision of the manuscript; B.L. provided conceptual guidance and supervised the overall research process. All authors have read and agreed to the published version of the manuscript.

Funding: This research was funded by the Joint Fund Program of the National Natural Science Foundation of China, Grant No.: 52274052.

Institutional Review Board Statement: Not applicable.

Informed Consent Statement: Not applicable.

Data Availability Statement: The corresponding author will provide requested data upon inquiry. For additional details or specific data requests, please directly contact the corresponding author. This work was prepared without the use of AI tools, and the author bears complete responsibility for the publication's content.

Conflicts of Interest: The authors declare no conflicts of interest.

Abbreviations

The following abbreviations are used in this manuscript:

EOR	Enhanced oil recovery
ISC	In situ combustion
HPAM	Hydrolyzed polyacrylamide
PAM	Polyacrylamide
AM	Acrylamide monomer
LTO	Low-temperature oxidation
HTO	High-temperature oxidation

References

1. Juarez-Morejon, J.L.; Bertin, H.; Omari, A.; Hamon, G.; Cottin, C.; Morel, D.; Romero, C.; Bourdarot, G. A new approach to poly mer flooding: Impact of the early polymer injection and wettability on final oil recovery. In Proceedings of the SPE Europec featured at 80th EAGE Conference and Exhibition, Copenhagen, Denmark, 11–14 June 2018.
2. Presser, D.J.; Cafaro, V.G.; Cafaro, D.C. Optimal Production Strategies for the Development of Mature Oil Fields through Polymer Flooding. *Ind. Eng. Chem. Res.* **2021**, *60*, 10235–10253. [CrossRef]
3. Abedi, B.; Castano, E.P.M.; Heidaryan, E.; Shadloo, M.S. Pore-scale visualization on polymer flooding: Application of singular value decomposition-based image analysis method. *J. Porous Media* **2020**, *23*, 531–543. [CrossRef]
4. Dafaalla, M.; Azad, M.S.; Ayirala, S.; Alotaibi, M.; Fahmi, M.; Saleh, S.; Shehri, D.A.; Mahmoud, M. Potential of polymer's viscos ity and viscoelasticity for accessible oil recovery during low salinity polymer flooding in heterogeneous carbonates. *Fuel* **2025**, *379*, 133008. [CrossRef]
5. Choi, B.I.; Jeong, M.S.; Lee, K.S. Temperature-dependent viscosity model of HPAM polymer through high-temperature res ervoirs. *Polym. Degrad. Stab.* **2014**, *110*, 225–231. [CrossRef]
6. Beteta, A.; Sorbie, K.S.; Johnson, G. Immiscible Viscous Fingering at the Field Scale: Numerical Simulation of the Captain Polymer Flood. *SPE J.* **2023**, *28*, 3165–3179. [CrossRef]
7. Fedorov, K.; Pospelova, T.; Kobyashev, A.V.; Gilmanov, A.; Kovalchuk, T.N.; Shevelev, A. Determination of Adsorption-Retention Constants and Inaccessible Pore Volume for High-Molecular Polymers. In Proceedings of the SPE Russian Petroleum Technology Conference, Moscow, Russia, 12–14 October 2021.
8. Guo, H.; Wang, Z.; Dang, S.; Wen, R.; Lyu, X.; Liu, H.; Yang, M. What is Learned from Polymer Flooding Practices in Offshore Reservoirs? In Proceedings of the Offshore Technology Conference, Houston, TX, USA, 1–4 May 2023.
9. Seright, R.S.; Wang, D.; Lerner, N.; Nguyen, A.; Sabid, J.; Tochor, R. Can 25-cp polymer solution efficiently displace 1,600-cpoil during polymer flooding? *SPE J.* **2018**, *23*, 2260–2278. [CrossRef]
10. Edwards, R.; Aitkulov, A.; Redwine, C.; Cunha, K. Viscous oil polymer flood milne point field case history concept to full field implementation. In Proceedings of the SPE Improved Oil Recovery Conference, Online, 25–29 April 2022.
11. Thakuria, C.; Amri, M.; Saqri, K.; Jaspers, H.; Zuhaimi, K.H.K. Performance Review of Polymer Flooding in a Major Brown Oil Field of Sultanate of Oman. In Proceedings of the SPE Enhanced Oil Recovery Conference, Kuala Lumpur, Malaysia, 2–4 July 2013.
12. Yu, S.; Guo, T.; Cao, G. Characteristics of sodium p-styrenesulfonate modified polyacrylamide at high temperature under dual scale boundary. *Phys. Fluids* **2025**, *37*, 073115. [CrossRef]
13. Bao, M.; Chen, Q.; Li, Y.; Jiang, G. Biodegradation of partially hydrolyzed polyacrylamide by bacteria isolated from production water after polymer flooding in an oil field. *J. Hazard. Mater.* **2010**, *184*, 105–110. [CrossRef] [PubMed]

14. Hao, T.; Zhong, L.; Liu, J.; Sun, H.; Zhu, T.; Zhang, H.; Wu, S. Mechanistic Study on the Decrease in Injectivity during Salt-Resistant Polymer Flooding. *ACS Omega* **2022**, *7*, 11293–11304. [CrossRef] [PubMed]
15. Seright, R.S.; Wang, D. Polymer flooding: Current status and future directions. *Pet. Sci.* **2023**, *20*, 910–921. [CrossRef]
16. Gao, S. Influence of The Residual Polymer on The Enhanced Recovery After the Polymer Flooding. *Pet. Geol. Oilfield Dev. Daqing* **2014**, *33*, 112–116. [CrossRef]
17. Miao, J.; Zhao, F.; Li, S.; Yu, Y.; Chen, M. Presence and distribution of residual polymer in formation. *Spec. Oil Gas Reserv.* **2005**, *12*, 88–90. [CrossRef]
18. Stuart, M.; Lapworth, D.; Crane, E.; Hart, A. Review of risk from potential emerging contaminants in UK groundwater. *Sci. Total Environ.* **2012**, *416*, 1–21. [CrossRef] [PubMed]
19. Binet, S.; Bru, K.; Klinka, T.; Touzé, S.; Motelica-Heino, M. Water and acrylamide monomer transfer rates from a settling ba sin to groundwaters. *Environ. Sci. Pollut. Res.* **2014**, *22*, 6431–6439. [CrossRef] [PubMed]
20. Wang, X.Y. *Impact of Polymer Flooding on Ecological and Development Environment in Dagang Oil-Field and Its Mechanism*; Northeast Petroleum University: Daqing, China, 2022.
21. Sun, W.; Zhou, Q.; Wang, R.; Liu, M.; Yu, W. Preliminary study on joint degradation mechanism of three poly acrylamide degrading bacteria. *J. Yangtze Univ. (Nat. Sci. Ed.)* **2024**, *21*, 1–6. [CrossRef]
22. Andrady, A.L. Assessment of Environmental Biodegradation of Synthetic Polymers. *J. Macromol. Sci. Part C Polym. Rev. Part C Polym. Rev.* **1994**, *34*, 25–76. [CrossRef]
23. Xu, Z.; Ding, Y.; Tao, L.; Hu, Z.; Zhang, X.; Bai, J.; Shi, W.; Li, J.; Li, S. Low-carbon development strategy to achieve heat conversion in heavy oil reservoirs: In-situ combustion. *Geoenergy Sci. Eng.* **2024**, *240*, 213055. [CrossRef]
24. Zhao, R.; Yu, S.; Yang, J.; Heng, M.; Zhang, C.; Wu, Y.; Zhang, J.; Yue, X.A. Optimization of well spacing to achieve a stable combustion during the THAI process. *Energy* **2018**, *151*, 467–477. [CrossRef]
25. Deniz-Paker, M.; Cinar, M. An Experimental Investigation of In Situ Combustion in High Permeability Contrast Systems. *J. Energy Resour. Technol.* **2022**, *144*, 073007. [CrossRef]
26. Steudel, A.; Friedrich, F.; Lieske, W.; Baille, W.; Konig, D.; Schuhmann, R.; Emmerich, K. Simultaneous thermal analysis of cationic, nonionic and anionic polyacrylamide. *Heliyon* **2019**, *5*, e02973. [CrossRef]
27. Liu, J.; Zhang, B.; Yan, Z.; Xu, Z.; Zhang, Z.; Wu, X.; Wang, Q. Study on thermal decomposition and denitration performance of polyacrylamide. *Appl. Chem. Ind.* **2024**, *53*, 1104–1108. [CrossRef]
28. Zhao, R.; Wang, J.; Men, Z.; He, J.; Sun, Z.; Wang, T.; Li, X.; Yuan, Y.; Xu, H.; Zhang, H. Experimental investigation on cyclic steam stimulation assisted modified THAI to enhance oil recovery in steam-treated heavy oil. *Energy* **2024**, *307*, 11. [CrossRef]
29. Silva, M.E.S.R.; Dutra, E.R.; Mano, V.; Machado, J.C. Preparation and thermal study of polymers derived from acrylamide. *Polym. Degrad. Stab.* **2000**, *67*, 491–495. [CrossRef]
30. Yang, M.H. The Two-Stages Thermal Degradation of Polyacrylamide. *Polym. Test.* **1998**, *17*, 191–198. [CrossRef]
31. Shatat, R.S.; Niazi, S.K.; Ariffin, A. Synthesis and Characterization of Different Molecular Weights Polyacrylamide. *IOSR J. Appl. Chem.* **2017**, *10*, 67–73. [CrossRef]
32. Guan, W.; Ma, D.; Liang, J.; Li, C.; Xi, C.; Zhang, X. Experimental Research on Thermodynamic Characteristics of In-situ Combustion Zones in Heavy Oil Reservoir. *Acta Pet. Sin.* **2010**, *31*, 100–104. [CrossRef]
33. Zhao, F.; Zhu, G.; Li, G.; Jiang, Y.; Liu, L. Feasibility Study on Further Enhanced Oil Recovery by Isc of Remaining Oil after Polymer Flooding. *SSRN Electron. J.* **2022**, *12*, 1–15. [CrossRef]

gels

MDPI

Article

Mechanisms of Mobility Control and Enhanced Oil Recovery of Weak Gels in Heterogeneous Reservoirs

Zhengxiao Xu [1,*], Ming Sun [1], Lei Tao [1,*], Jiajia Bai [1], Wenyang Shi [1], Na Zhang [2] and Yuyao Peng [1]

1 School of Petroleum and Natural Gas Engineering, Changzhou University, Changzhou 213164, China; ming132687@163.com (M.S.); baijiajiaa@163.com (J.B.); w.shi@cczu.edu.cn (W.S.); pyy15839421072@outlook.com (Y.P.)

2 School of Petrochemical Engineering, Changzhou University, Changzhou 213164, China; zhangna@cczu.edu.cn

* Correspondence: xzx@cczu.edu.cn (Z.X.); taolei2637@163.com (L.T.)

Abstract: At present, most oilfields in China have entered the late, high-water-cut stage, commonly facing declining single-well productivity and increasingly pronounced reservoir heterogeneity. Prolonged waterflooding has further exacerbated permeability contrast, yielding complex, hard-to-produce residual-oil distributions. Accordingly, the development of efficient enhanced oil recovery (EOR) technologies has become a strategic priority and an urgent research focus in oil and gas field development. Weak gels, typical non-Newtonian fluids, exhibit both viscous and elastic responses, and their distinctive rheology shows broad application potential for crude oil extraction in porous media. Targeting medium–high-permeability reservoirs with high water cut, this study optimized and evaluated a weak gel system. Experimental results demonstrate that the optimized weak gel system achieves remarkable oil displacement performance. The one-dimensional dual-sandpack flooding tests yielded a total recovery of 72.26%, with the weak gel flooding stage contributing an incremental recovery of 14.52%. In the physical three-dimensional model experiments, the total recovery reached 46.12%, of which the weak gel flooding phase accounted for 16.36%. Through one-dimensional sandpack flow experiments and three-dimensional physical model simulations, the oil displacement mechanisms and synergistic effects of the optimized system in heterogeneous reservoirs were systematically elucidated from macro to micro scales. The optimized system demonstrates integrated synergistic performance during flooding, effectively combining mobility control, displacement, and oil-washing mechanisms. Macroscopically, it effectively strips residual oil in high-permeability zones via viscosity enhancement and viscoelastic effects, efficiently blocks high-permeability channels, diverts flow to medium-permeability regions, and enhances macroscopic sweep efficiency. Microscopically, it mobilizes residual oil via normal stress action and a filamentous transport mechanism, improving oil-washing efficiency and increasing ultimate oil recovery. This study demonstrates the technical feasibility and practical effectiveness of the optimized weak gel system for enhancing oil recovery in heterogeneous reservoirs, providing critical technical support for the efficient development of medium–high-permeability reservoirs with high water cut.

Keywords: heterogeneity; medium–high permeability; high water cut; viscoelasticity; synergistic effect

https://doi.org/10.3390/gels11110854

1. Introduction

As many Chinese oilfields have matured into the high-water-cut stage, operators face declining single-well productivity and accentuated reservoir heterogeneity. Prolonged waterflooding exacerbates permeability contrast, redistributes residual oil into poorly swept zones, and makes capillary-trapped oil increasingly difficult to mobilize [1,2]. Consequently, the development and deployment of enhanced oil recovery (EOR) technologies have become strategic priorities in oil and gas field development. The core mechanism involves overcoming microscopic flow barriers dominated by capillary and viscous forces, thereby efficiently mobilizing residual oil and improving ultimate recovery [3–6]. Conventional water flooding, while being an economically viable primary development method in early stages, leads to progressive reservoir degradation with continued injection [7,8]. The process intensifies reservoir heterogeneity, amplifies permeability contrasts, and may induce formation damage [9–11].

Polymer flooding is a relatively well-established and widely applied chemical EOR method [12]. It involves dissolving water-soluble polymers—such as hydrolyzed polyacrylamide (HPAM), hydrophobically associating polymers, and polyacrylamide-based copolymers—in the injected water [13–17]. The principal mechanisms are improvement in the oil–water mobility ratio and enhancement in sweep efficiency at both macroscopic and microscopic scales [18–22]. Weak gel flooding [5,23–26] relies fundamentally on the use of crosslinkers to bridge polymer molecular chains, forming a three-dimensional network with excellent viscoelasticity and structural strength. At the macroscopic scale, mobility control is achieved by increasing the viscosity of the aqueous phase; at the pore scale, elastic displacement is critical, whereby stored and released elastic energy effectively mobilizes residual oil, lowers residual oil saturation, and improves overall displacement efficiency, thereby providing an effective solution for oilfield development [27–30]. However, under complex reservoir conditions, weak gel systems that are not tailored to specific reservoir properties may yield displacement performance far below expectations or even negative outcomes. In heterogeneous reservoirs, poorly matched rheological parameters can undermine mobility control, leaving low-permeability zones unswept and causing channeling through high-permeability zones, thereby exacerbating interlayer conflicts [31–33]. In homogeneous reservoirs, excessively high viscoelasticity can sharply increase injection pressure, induce in situ shear degradation, and lead to severe energy loss [34].

Current research has clarified the mechanisms of macroscopic mobility control and microscopic displacement by weak gels, with notable advances in improving sweep efficiency at both scales. Ali H. S. et al. [35], through a review of project cases, reported a strong correlation between optimal polymer viscosity and both oil viscosity and permeability contrast, emphasizing that matching polymer viscosity to specific reservoir parameters is essential for successful polymer flooding. Alisheva Z. et al. [36] experimentally demonstrated the interaction between polymer viscoelasticity and wettability: under strongly water-wet conditions, elasticity enhances oil displacement efficiency, whereas in weakly water-wet systems, it inhibits oil droplet deformation; moreover, the direction of normal stress changes with increasing elasticity. Xuwei L. et al. [37] employed numerical simulation and the response surface method to optimize polymer selection, concluding that low-viscosity polymer solutions with a high residual resistance factor (RRF) are more suitable for heterogeneous heavy-oil reservoirs. Di Q et al. [38], based on nuclear magnetic resonance (NMR) imaging studies, found that weak gels exhibit an integrated migration pattern during subsequent water flooding processes and that different polymer-weak gel displacement modes have a decisive influence on the distribution of residual oil within core samples. Hatzignatiou G D et al. [39] conducted experimental screening and evaluation of various polymers and weak gels, investigating the effects of parameters such as filterability, injectivity, gelation time, gel strength, and gel shrinkage ratio on oil recovery. The study demonstrated that the Type A gel exhibited excellent stability, high strength, and the ability to withstand the pressures exerted during subsequent water flooding. Yajuan T et al. [40] developed a temperature-tolerant weak gel system suitable for high-salinity injection water environments to address challenges associated with highly mineralized injection fluids. Through dual-sandpack parallel oil displacement experiments, the study verified that this weak gel system exhibited excellent migration capability and plugging performance.

Although recent research has advanced weak gel flooding, the dynamic interaction between weak gels, complex pore structures, and fluid distributions in three-dimensional physical simulations, particularly those representative of heterogeneous reservoirs, remains insufficiently characterized. Against this backdrop, this study targets medium–high-permeability reservoirs with high water cut. A physical three-dimensional model was constructed to reproduce the target reservoir properties, and the viscosities of weak gel solutions at different concentrations were measured to optimize a formulation suited to the reservoir conditions. Through dual-sandpack flow experiments and physical three-dimensional model displacement tests, as well as by integrating macroscopic displacement performance, pressure data, and fractional-flow behavior, we comprehensively evaluated the displacement efficiency of the optimized weak gel system. Furthermore, depth-of-field microscopy and scanning electron microscopy (SEM) were used to analyze, at the pore scale, the weak gel displacement mechanism and the mobilization of residual oil. The optimized system exhibits a synergistic effect during flooding by integrating mobility control, displacement, and oil-washing mechanisms. At the macroscopic scale, it enhances fluid viscosity and exerts viscoelastic effects to efficiently displace residual oil in high-permeability zones while selectively plugging high-permeability channels to divert fluid into medium-permeability regions, thereby improving macroscopic sweep efficiency. At the microscopic scale, the system mobilizes residual oil through normal stress effects and a filamentous transport mechanism, enhancing oil-washing efficiency and ultimately increasing final recovery. This study provides a practical and theoretical basis for enhancing oil recovery using optimized weak gel systems in medium–high-permeability, high-water-cut reservoirs.

2. Results and Discussion

2.1. Optimization of Polymer Concentration in Weak Gel System

Given the measured crude oil viscosity of 37.2 mPa·s at 60 °C in the target reservoir, this study optimized the weak gel system by evaluating five concentrations: 500, 1000, 1500, 2000, and 2500 mg/L. Viscosities of the polymer solutions were measured using an Anton Paar MCR 302 rheometer, and the resulting viscosity curves for each concentration are shown in Figure 1. The optimal concentration was selected based on the ratio of polymer solution viscosity to crude oil viscosity under reservoir conditions. This approach ensures effective mobility control and is expected to contribute to enhanced oil recovery performance.

As shown in Figure 1, the polymer solution viscosity increases approximately linearly with concentration over the tested range (500–2500 mg/L). According to established guidelines for the polymer-to-crude viscosity ratio [41,42], the target crude is classified as medium–high viscosity, for which an optimal ratio of 2:1–4:1 is recommended. At a polymer concentration of 1000 mg/L, the measured viscosity is 137 mPa·s, about 3.68 times that of the crude oil, meeting the target range; therefore, 1000 mg/L was selected as the optimal concentration for the reservoir conditions. A trace amount of crosslinker was then added, and the mixture was stirred at 700 r/min for 1 h. After preparation, the system was aged for 24 h to obtain the optimized weak gel system.

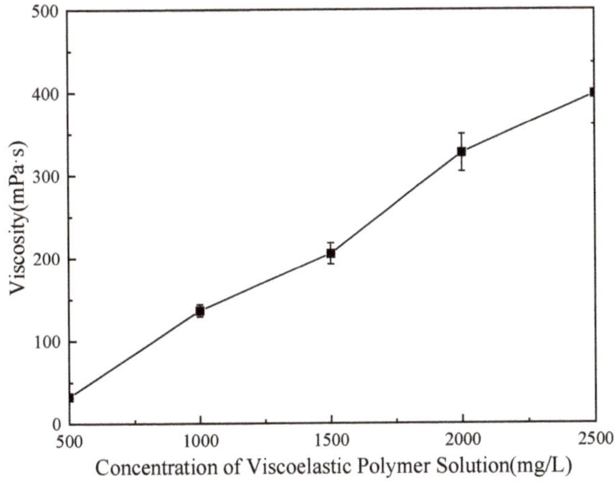

Figure 1. Viscosity curve of the weak gel at different concentrations.

2.2. Dual-Sandpack Weak Gel Flooding Experiment

The dual-sandpack models, pre-saturated with oil and water, were connected in parallel and subjected to the following sequential flooding process: First, water flooding was conducted at a flow rate of 1 mL/min until the water cut exceeded 90%. This was followed by the injection of 0.3 PV of a weak gel solution at a concentration of 1000 mg/L while maintaining the flow rate at 1 mL/min. Finally, water flooding was resumed at the same flow rate until the water cut reached approximately 95%, marking the end of the experiment. Throughout the experiment, the volumes of oil and water produced from both sandpacks were measured in 10 min intervals, with graduated cylinders being replaced promptly after each measurement. Based on these measurements, water cut, oil recovery, and fractional flow were calculated and are plotted in Figures 2 and 3. Meanwhile, pressure sensors were used to acquire the injection–production differential pressure data once per second. The collected pressure data were processed to generate the differential pressure curve presented in Figure 4.

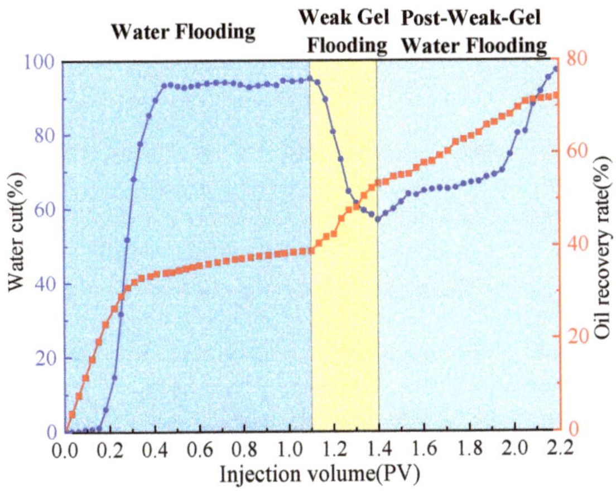

Figure 2. Water cut and recovery rate during weak gel flooding in dual-sandpack models.

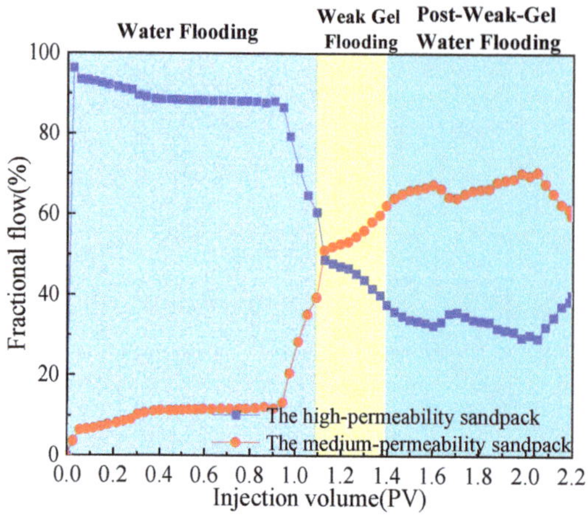

Figure 3. Fractional flow from both sandpacks during weak gel flooding in dual-sandpack models.

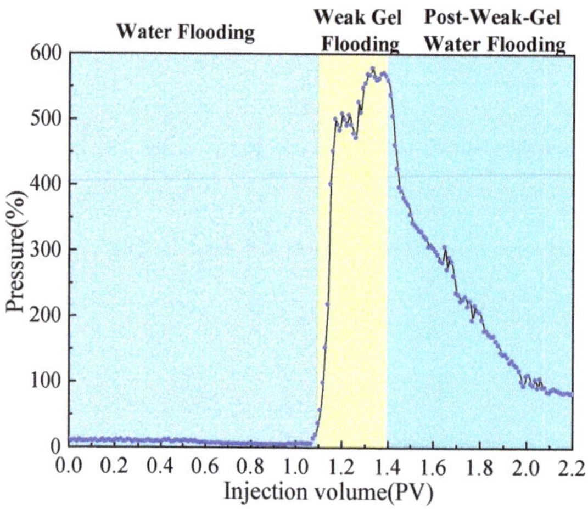

Figure 4. Injection–production differential pressure during weak gel flooding in the dual-sandpack models.

As shown comprehensively in Figures 2–4, the oil recovery rate during the water flooding stage was 38.91%, reflecting a significant imbalance in the displacement efficiency between the medium- and high-permeability zones under high-water-cut conditions. During this stage, the injection–production differential pressure remained essentially stable. The water cut increased rapidly in the initial phase and stabilized around 0.4 PV of fluid injected, while the rate of increase in oil recovery became slower. Fractional-flow trends indicate that the injected fluid initially entered the high-permeability sandpack, after which the fractional flow from this unit gradually decreased. Upon weak gel injection, the viscoelastic solution preferentially accessed high-permeability channels, impeding the preferential water flow paths. Owing to viscoelastic deformation, additional normal stresses were generated, producing a microscopic, selective, water-blocking, oil-promoting effect [6,23]. These stresses enabled displacement of previously unswept oil in the medium-permeability zones through pore throats, mobilizing residual oil in the medium-permeability sandpack, which experimentally manifested as a sustained increase in flow from this sandpack at later times. The underlying mechanism is effective plugging and mobility control by the weak gel via viscoelastic effects and physical retention.

In the subsequent water-flooding stage, an additional 18.83% incremental oil recovery was achieved, although the rate of increase was lower than during weak gel flooding. The injection–production differential pressure declined, while the water cut rose initially, then decreased, and ultimately increased again. The fractional flow from the medium-permeability sandpack first increased and then slightly decreased. These results indicate that weak gel flooding not only significantly increased production but also modified the distribution of the remaining oil, creating a more balanced displacement environment for the subsequent waterflood. This further confirms that the adsorption and retention of the weak gel in pore channels provided sustained benefits. In conclusion, within the one-dimensional dual-sandpack model, the weak gel significantly enhanced oil recovery by improving mobility control, increasing sweep efficiency in medium-permeability zones, and effectively mobilizing residual oil.

2.3. *Physical Three-Dimensional Model Experiment of Weak Gel Flooding*

2.3.1. Analysis of Experimental Results from Physical Three-Dimensional Model

A three-dimensional physical model was constructed to simulate the target reservoir's geological characteristics. The injection–production pattern employed one injection well and one production well, positioned at diagonal corners of the model, each located 5 cm from the adjacent boundaries. Permeability was distributed with the upper layer having a permeability of 500 mD and that of the lower layer was 1500 mD. The experimental process consisted of the following stages: (1) Water flooding: brine was injected at a constant rate of 10 mL/min until the water cut exceeded 90%. (2) Weak gel flooding: a weak gel solution with a concentration of 1000 mg/L was injected at 0.3 PV and a rate of 10 mL/min. (3) Post-weak-gel water flooding: brine injection was resumed at 10 mL/min until the water cut reached approximately 95%, marking the end of the experiment. During the experiment, the oil and water production volumes at the outlet of the physical three-dimensional model were recorded every 10 min, with graduated cylinders being replaced promptly after each measurement. After the experiment, the data were processed to calculate the water cut and oil recovery rate, and the results are presented in Figure 5. Differential pressure data were recorded by pressure sensors in 1 s intervals, and the injection–production pressure difference curve is plotted in Figure 6. Based on parameters such as injection volume, pressure, and injection time, a Hall plot was generated and is presented in Figure 7.

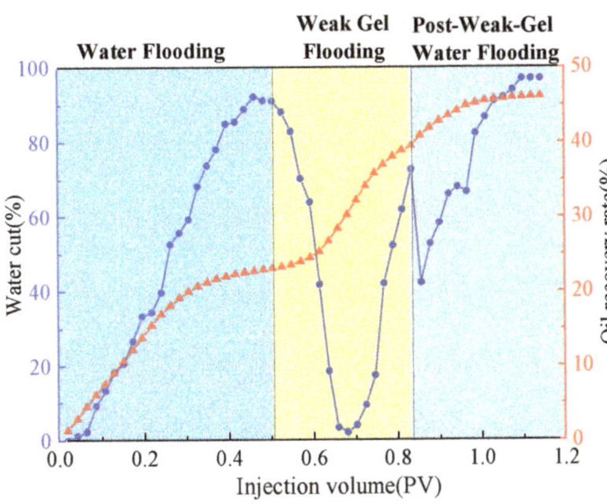

Figure 5. Variation in water cut and oil recovery rate during weak gel flooding in physical three-dimensional model.

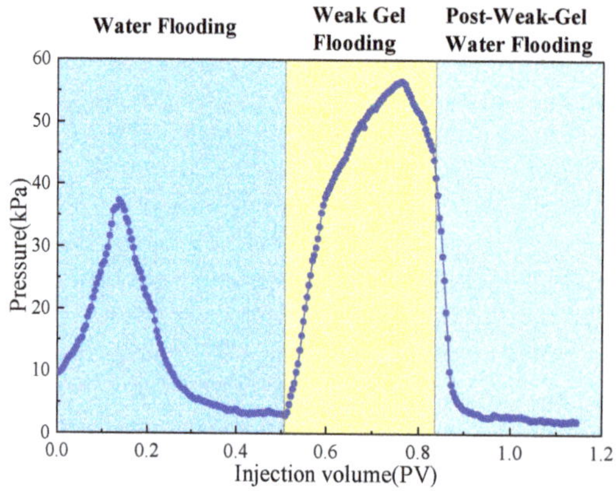

Figure 6. Variation in injection–production differential pressure during weak gel flooding in physical three-dimensional model.

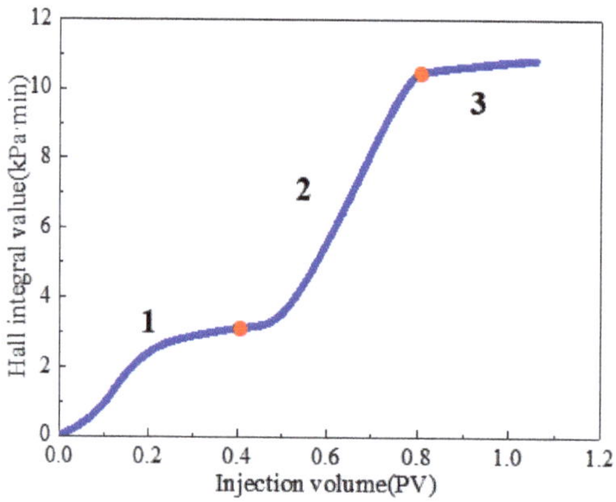

Figure 7. Hall plot for weak gel flooding in physical three-dimensional model.

As shown in Figures 5 and 6, the cumulative oil recovery reached 46.12%. Analysis of the Hall plot (Figure 7) indicates that during Segment 1 (water flooding) recovery was 23.08%; the Hall plot rose initially and then approached a steady slope, accompanied by a rapid increase in water cut and a brief pressure rise before stabilization. During Segment 2 (weak gel flooding), recovery reached 16.36%; the water cut displayed a V-shaped trajectory with a minimum of 2.08%. The injection–production pressure increased rapidly before this minimum and then declined, indicating that the weak gel initially entered and partially plugged high-permeability channels, diverting flow to medium-permeability zones until weak gel breakthrough and/or shear-thinning reduced flow resistance and redirected flow toward high-permeability paths. In Segment 3 (post-weak-gel water flooding), an additional 6.68% was recovered; the water cut rose rapidly but at a decreasing rate and pressure remained stable, suggesting that the weak gel flood had already mobilized most of the remaining oil, leaving limited potential for further production.

2.3.2. Analysis of Oil Displacement Efficiency in Physical Three-Dimensional Model

Following the weak gel flooding in the physical three-dimensional model, the model was opened for post-test visual inspection to assess displacement performance. As shown in Figure 8, a significant oil displacement effect was observed particularly near the injection and production wells, with dashed lines in the figure delineating the sweep behavior of the weak gel flooding. A distinct flow channel formed by the weak gel was clearly visible between the two wells, indicating that the optimized weak gel system exhibited high oil-washing efficiency and strong sweep capability.

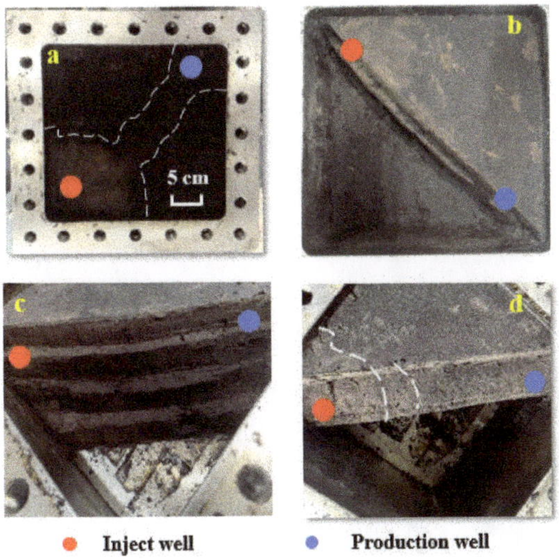

Figure 8. Actual oil displacement effect of weak gel flooding in physical three-dimensional model: (**a–d**) present the four stages of the post-flooding excavation process from the three-dimensional physical model after weak gel flooding.

Post-flood sampling was conducted to assess displacement effectiveness in the physical three-dimensional model. Five sand-layer samples were collected along the line between the injection and production wells (Figure 9), specifically taken from the medium-permeability zone near the injection well, the high-permeability zone near the injection well, the transitional medium–high permeability zone at the midpoint between the injection and production wells, the medium-permeability zone near the production well, and the central high-permeability zone near the production well.

Figure 9. Sampling distribution for post-flooding analysis in physical 3D model of weak gel flooding.

This sampling strategy enabled a comprehensive analysis of crude oil mobilization and viscoelastic weak gel migration across permeability layers and spatial regions during flooding, providing critical insights into displacement characteristics and fluid propagation behavior in heterogeneous reservoirs.

2.3.3. Microscopic Residual Oil Analysis and Post-Flooding Scanning Electron Microscopy (SEM) Characterization

The five sand samples shown in Figure 9 were examined with a depth-of-field microscope; the resulting micrographs are shown in Figure 10. The sample from the high-permeability zone near the production well was air-dried and subsequently analyzed by scanning electron microscopy (SEM) to characterize the weak gel's microscopic morphology. The SEM results are presented in Figure 11.

Figure 10. Depth-of-field microscopic analysis of physical 3D Model samples after weak gel flooding.

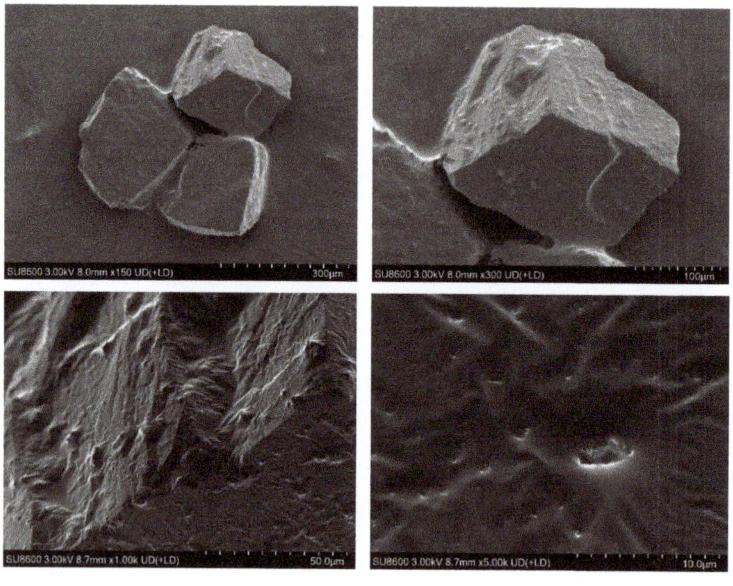

Figure 11. SEM images of the samples.

Based on the depth-of-field micrographs and SEM images, together with the observed displacement effects during sampling, the following conclusions were drawn: the medium-permeability layers exhibited higher residual-oil saturation with lower moisture, whereas the high-permeability zones showed markedly higher displacement efficiency with greater moisture and visible residual weak gel. These observations indicate that the weak gel preferentially entered and blocked high-permeability channels, mobilizing movable residual oil along these pathways. During the water-flooding stage, the injected fluid primarily swept the high-permeability zones, resulting in limited coverage of the upper medium-permeability regions.

In the SEM images magnified at 5000× (Figure 11), the sand particle surfaces appear smooth-textured overall, lacking the rough crystalline structures typical of mineral grains. This suggested that the original surfaces were covered by a residual layer. Further observation revealed that this coating exhibited distinct "flow-induced elongated textures" and wrinkled membrane-like morphologies, which were identified as residual film structures formed by the weak gel and crude oil remaining on the particle surfaces after the flooding process.

In the center of the image, a concave feature approximately 2–3 μm in size is observed, with its margins showing a distinct edge-brightening ("highlighted edges") effect, likely due to charge accumulation caused by electron-beam interaction with agglomerated weak gel. This observation further supports the presence of residual weak gel on the sand grain surface. Taken together with the surface morphology, namely, the wrinkled, membrane-like structures and highlighted edges, it was preliminarily concluded that an oil film or a mixed oil/viscoelastic weak gel residual layer existed on the sample surface. These morphological characteristics suggest that the weak gel underwent flow deposition during displacement, which, while enhancing oil recovery efficiency, may have also contributed to localized entrapment of residual oil.

Based on the experimental data, Hall plot analysis, depth-of-field microscopy, and SEM imaging, the mechanisms of residual oil mobilization by viscoelastic weak gel flooding are summarized as follows: Mobilization arises from mobility control coupled with physicochemical synergistic effects. The weak gel's viscoelasticity and the additional normal stresses generated by extensional deformation promote displacement of residual oil in high-permeability zones and, to a lesser extent, in medium-permeability zones. This establishes flow barriers that force fluid diversion and produce macroscopic fluid redistribution. At the pore scale, elastic effects and normal stresses disturb the equilibrium at the oil–water–rock three-phase interface, mobilizing additional residual oil. The weak gel exhibits shear thinning and extensional thickening behavior: the lower apparent viscosity in high-permeability channels enhances penetration and plugging efficiency, whereas higher apparent viscosity in medium-permeability zones increases displacement pressure [8,26]. This dual rheology enables the weak gel to enter large pores to impose flow resistance while displacing residual oil from small pores, thereby improving microscopic sweep efficiency [43,44]. The weak gel thus enables multiscale mobilization: macroscopically, it balances sweep coverage; microscopically, it expels residual oil from blind pores via elastic deformation, reduces oil–water interfacial tension, promotes crude-oil emulsification, and mobilizes dispersed residual oil through filamentous stretching. The wrinkled residual weak gel observed in high-permeability zones evidences flow deposition that forms microscopic barriers, ultimately enhancing oil recovery.

In summary, the oil recovery rates at various stages calculated from the dual-sandpack experiments and three-dimensional physical simulation tests are presented in Table 1.

Table 1. Comparison of oil recovery rates from weak gel flooding experiments in dual-sandpack and physical three-dimensional models.

Model Type	Water-Flooding Stage %	Weak-Gel-Flooding Stage %	Post-Weak-Gel Water-Flooding Stage %
Dual-Sandpack Experiment	38.91	14.52	18.83
Physical Three-Dimensional Model Experiment	23.08	16.36	6.68

3. Conclusions

The dual-sandpack results indicate that the weak gel (concentration: 1000 mg/L) flooding stage yielded an additional 14.52% oil recovery, and the subsequent waterflooding stage contributed a further 18.83%, demonstrating strong displacement performance and sustained EOR effects. By increasing aqueous phase viscosity and reducing the oil–water mobility ratio, the weak gel preferentially entered and blocked high-permeability channels, diverting flow into the medium-permeability sandpack and thereby improving sweep efficiency. At the pore scale, normal-stress effects effectively mobilized residual oil, with physical retention and chemical adsorption acting synergistically.

The physical three-dimensional model flooding experiment systematically evaluated the displacement performance and sweep behavior of a weak gel (concentration: 1000 mg/L) in a heterogeneous reservoir. The weak-gel-flooding stage increased oil recovery by 16.36%, with an additional 6.68% obtained during subsequent waterflooding. Macroscopically, a distinct flow channel developed between the injector and producer, corroborated by clear color contrasts in the sand layers, indicating differential oil mobilization. Microscopically, normal-stress and viscoelastic effects displaced residual oil from medium- and high-permeability zones by establishing a mobility-control barrier that enhanced fluid diversion. Through flow-deposition, the weak gel formed microscopic plugs, reduced oil–water interfacial tension, promoted crude oil emulsification, and mobilized and transported dispersed residual oil via filamentous displacement, collectively improving oil-washing efficiency.

This study elucidates, at both macroscopic and microscopic scales, the displacement mechanisms and synergistic effects of the optimized weak gel system in heterogeneous reservoirs. The system integrates three coupled functions, mobility control, oil displacement, and oil washing, demonstrating its feasibility and effectiveness for enhancing oil recovery in heterogeneous formations. The results provide reliable technical support for improving recovery in medium–high-permeability heterogeneous reservoirs with high water cut.

4. Materials and Methods

4.1. Materials and Equipment

The synthetic formation water was prepared to match the ionic composition of field brine from the target reservoir, with per-liter ion concentrations listed in Table 2. Three grades of quartz sand, 60–80 mesh, 80–100 mesh, and 100–120 mesh, were employed to simulate the reservoir geology. These sands were compacted and packed to construct both dual-sandpack models and a physical three-dimensional model with a positive rhythmic permeability distribution. The dual-sandpack models exhibited permeabilities of 500 mD and 1500 mD, while the physical three-dimensional model was designed with an upper-layer permeability of 500 mD and a lower-layer permeability of 1500 mD. Each sandpack model measured 2.5 cm in diameter and 60 cm in length, with a maximum pressure tolerance of 30 MPa. The physical three-dimensional model featured an internal chamber measuring 30 cm × 30 cm × 20 cm, capable of withstanding pressures up to 10 MPa and temperatures up to 80 °C. The model incorporated multiple integrated monitoring points for real-time data acquisition, meeting the experimental requirements.

Table 2. Ionic composition of formation water.

Component	Na_2SO_4	NaCl	$CaCl_2$	KCl	$MgCl_2 \cdot H_2O$
Quality (g)	9.976	10.582	0.419	0.183	1.506

Experimental instruments comprised an electric stirrer; an MCR 302 rheometer (Anton Paar, Graz, Austria); an OHR-M2G precision pressure sensor (HORIBA, Ltd., Kyoto, Japan); an ISCO dual-plunger metering pump (Model 100DX, IL, Tridan, Danville, IL, USA) for injecting the aqueous urea solution; a depth-of-field microscope DVM2500 (Leica Microsystems, Wetzlar, Germany); and a scanning electron microscope GeminiSEM 450 (Carl Zeiss Microscopy GmbH, Oberkochen, Germany).

4.2. *Optimization of Polymer Concentration*

Using the pre-optimized concentration appropriate for the reservoir conditions as a baseline, viscosities of polymer solutions at multiple concentrations were systematically measured, the optimal concentration was identified, and a weak gel system tailored to the target reservoir was formulated.

4.2.1. Preparation of Polymer Solutions

A 5000 mg L^{-1} weak gel stock solution was prepared by dissolving 2.5 g of the oil-displacing agent in 497.5 g of distilled water. Using a magnetic stirrer initially set to 400 r/min, the polymer was added gradually along the vortex wall; the stirring speed was then increased to 700 r/min and maintained for 1 h to ensure complete dissolution. The stock solution was allowed to stand for 24 h. It was subsequently diluted with synthetic formation water to target concentrations of 500, 1000, 1500, 2000, and 2500 mg/L. During dilution, the mixture was stirred continuously at 300 r min^{-1} for 15 min to obtain the target weak gel solutions.

4.2.2. Viscosity Measurement of Weak Gel Solutions at Various Concentrations

The measuring rotor of the Anton Paar MCR 302 rheometer was installed by aligning the vertical grooves on the rotor tip with those on the instrument interface. After mounting, torque, angle, and temperature calibrations were performed. Using a pipette, the solution was carefully loaded into the center of the fixture to avoid air bubbles. Silicone oil or a solvent trap was applied to prevent evaporation. The upper fixture was gradually lowered to the preset gap. Prior to each sample measurement, the fixture was thoroughly cleaned and dried with an appropriate solvent. The measurement was conducted at 60 °C with a shear rate of 7.34 s^{-1} over a duration of 3 min. Thirty data points were recorded for each weak gel solution. The viscosity values at different concentrations (500, 1000, 1500, 2000, and 2500 mg/L) were collected to establish the relationship between polymer concentration and viscosity. The discrete distribution of viscosity values was assessed by calculating the measurement error. A trace amount of crosslinker was then added, and the mixture was stirred and allowed to stand, resulting in the formation of the optimized weak gel system.

4.3. Weak-Gel-Flooding Sandpack Experiment

Dual-sandpack models with permeabilities of 500 mD and 1500 mD were prepared and connected in parallel (schematic shown in Figure 12). After packing with quartz sand, the models were vacuumed and sequentially saturated with brine and crude oil. The experimental procedure comprised the following stages: (1) Water flooding: brine was injected at a constant rate of 1.0 mL/min until the water cut exceeded 90%. (2) Weak gel flooding: a weak gel solution (1000 mg/L concentration) was injected at 0.3 PV and a rate of 1.0 mL/min. (3) Post-weak-gel water flooding: brine injection was resumed at 1.0 mL/min until the water cut reached approximately 95%, marking the end of the experiment. The produced fluid at the outlet was visually observed, and experimental data (including production rates and differential pressures) were recorded. Subsequently, the data were systematically analyzed to evaluate the oil displacement performance of the weak gel and investigate the mechanisms of residual oil mobilization.

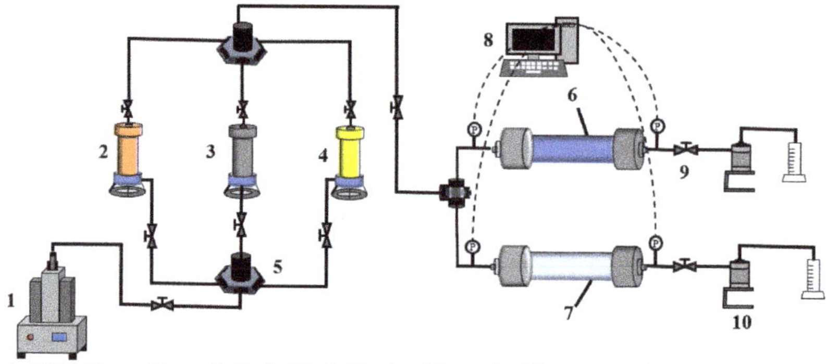

Figure 12. The dual-sandpack oil displacement experiment.

4.4. Physical Three-Dimensional Model Experiment

A physical three-dimensional model was constructed to simulate the target reservoir's geological characteristics. The injection–production pattern employed one injection well and one production well, positioned at diagonal corners of the model, each located 5 cm from the adjacent boundaries. Permeability was distributed in a positive rhythmic pattern (Figure 13), with the upper layer having a permeability of 500 mD and that of the lower layer being 1500 mD. Quartz sand was proportionally packed and compacted layer-by-layer to ensure uniform density and formation homogeneity. After packing, the model was sealed with a cover plate and securely fastened to guarantee overall containment. The model was subsequently saturated with brine and crude oil. A schematic of the experimental procedure is provided in Figure 14. The experimental process consisted of the following stages. (1) Water flooding: brine was injected at a constant rate of 10 mL/min until the water cut exceeded 90%. (2) Weak gel flooding: a weak gel solution with a concentration of 1000 mg/L was injected at 0.3 PV and a rate of 10 mL/min. (3) Post-Weak-gel water flooding: brine injection was resumed at 10 mL/min until the water cut reached approximately 95%, marking the end of the experiment.

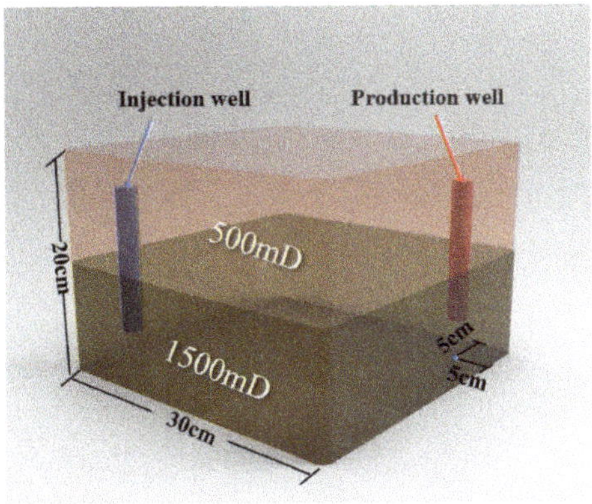

Figure 13. The well configuration and permeability distribution in the physical three-dimensional model.

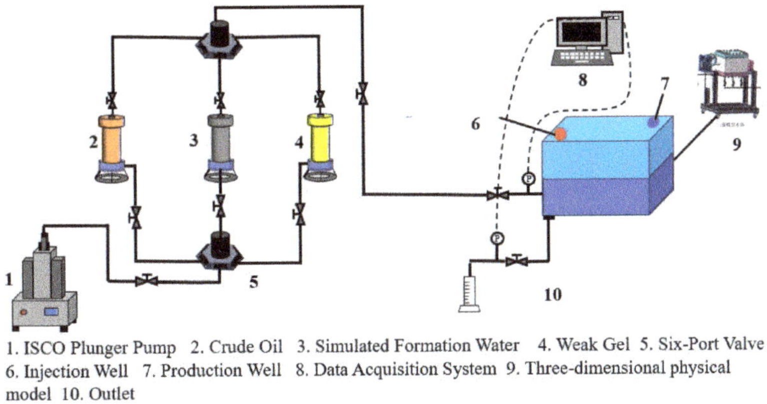

Figure 14. The oil displacement process in the physical three-dimensional model.

Author Contributions: Conceptualization, Z.X. and M.S.; methodology, L.T. and W.S.; formal analysis, L.T. and J.B.; data curation, M.S. and Y.P.; visualization, J.B. and W.S.; writing—original draft, Z.X. and M.S.; writing—review and editing, Z.X., M.S. and L.T.; project administration, Z.X. and N.Z.; funding acquisition, Z.X. and L.T. All authors have read and agreed to the published version of the manuscript.

Funding: This project was supported by the Open Research Fund of State Key Laboratory of Deep Oil and Gas (grant No. SKLDOG2024-KFYB-14), the Natural Science Research Project of Jiangsu Higher Education Institutions (grant No. 23KJB440001), the National Natural Science Foundation of China (grant No. 52404017), and the Natural Science Foundation of Jiangsu Province (grant No. BK20240968), Changzhou Longcheng Talent Program-Leading innovative talent introduction and cultivation project (grant No. CQ20240107). The valuable comments of the anonymous reviewers are sincerely appreciated.

Institutional Review Board Statement: Not applicable.

Informed Consent Statement: Not applicable.

Data Availability Statement: The data presented in this study are available on request from the corresponding author.

Conflicts of Interest: The authors declare no conflicts of interest.

References

1. Bai, B.; Zhou, J.; Yin, M. A comprehensive review of polyacrylamide polymer gels for conformance control. *Pet. Explor. Dev. Online* **2015**, *42*, 525–532. [CrossRef]
2. Amirian, E.; Dejam, M.; Chen, Z. Performance forecasting for polymer flooding in heavy oil reservoirs. *Fuel* **2018**, *216*, 83–100. [CrossRef]
3. Xu, Z.-X.; Li, S.-Y.; Li, B.-F.; Chen, D.-Q.; Liu, Z.-Y.; Li, Z.-M. A review of development methods and EOR technologies for carbonate reservoirs. *Pet. Sci.* **2020**, *17*, 990–1013. [CrossRef]
4. Xu, Z.; Ding, G.; Tao, L.; Shi, W.; Bai, J.; Dang, F. Foam stabilization mechanism of core-shell particles: Insights from the gas-liquid interface theory. *Capillarity* **2025**, *16*, 5–17. [CrossRef]
5. Yao, E.; Yu, G.; Li, B.; Zhao, L.; Li, Y.; Bai, H.; Zhou, F. High-Temperature-Resistant, Low-Concentration Water-Controlling Composite Cross-Linked Polyacrylamide Weak Gel System Prepared from Oilfield Sewage. *ACS Omega* **2022**, *7*, 12570–12579. [CrossRef]
6. Wei, B.; Zheng, N.; Xue, Y.; Hou, J.; Liu, Y.; Guo, Z.; Qin, X.; Du, Q. Study on multi-scale oil displacement mechanism polymer/nanoparticle composite flooding. *Front. Chem.* **2025**, *13*, 1605416. [CrossRef]
7. Liu, P.; Zhang, J.; Wang, J.; Zhong, H.; Wang, R.; Li, W. Study on Water-Driving Law and Remaining Oil Distribution Pattern in Ultra-low Permeability Reservoir. *E3S Web Conf.* **2023**, *406*, 02036. [CrossRef]
8. Ge, L.; Chen, X.; Wang, G.; Zhang, G.; Li, J.; Liu, Y.; Xiao, L.; Wen, Y.; Yuan, W.; Qu, M.; et al. Analysis of the Distribution Pattern of Remaining Oil and Development Potential after Weak Gel Flooding in the Offshore LD Oilfield. *Gels* **2024**, *10*, 236. [CrossRef]
9. Tangparitkul, S.; Sukee, A.; Jiang, J.; Tapanya, C.; Fongkham, N.; Yang, H. Role of interfacial tension on wettability-controlled fluid displacement in porous rock: A capillary-dominated flow and how to control it. *Capillarity* **2023**, *9*, 55–64. [CrossRef]
10. Mahmud, W.M. Impact of salinity and temperature variations on relative permeability and residual oil saturation in neutral-wet sandstone. *Capillarity* **2022**, *5*, 23–31. [CrossRef]
11. He, J.; Jin, X.; Liu, X.; Yuan, L.; Liu, R.; Chen, S.; Wu, H.; Yang, W.; Wang, J.; Zhang, H.; et al. Comprehensive Study on Viscosity-Increasing and Oil Displacement Characteristics of Functional Polymer. *Processes* **2025**, *13*, 1859. [CrossRef]
12. Kakati, A.; Kumar, G.; Sangwai, J.S. Low Salinity Polymer Flooding: Effect on Polymer Rheology, Injectivity, Retention, and Oil Recovery Efficiency. *Energy Fuels* **2020**, *34*, 5715–5732. [CrossRef]
13. Li, Z.; Sun, Y.; Chen, Y.; Liu, Z.; Tang, Y.; Yang, W. Capillary-driven processing in carbon fiber-reinforced polymer composites: From multiscale modeling to advanced manufacturing. *Capillarity* **2025**, *16*, 77–86. [CrossRef]
14. Minakov, V.; Pryazhnikov, M.I.; Neverov, A.L.; Sukhodaev, P.O.; Zhigarev, V.A. Wettability, interfacial tension, and capillary imbibition of nanomaterial-modified cross-linked gels for hydraulic fracturing. *Capillarity* **2024**, *12*, 27–40. [CrossRef]
15. Lu, X.; Cao, B.; Xie, K.; Cao, W.; Liu, Y.; Zhang, Y.; Wang, X.; Zhang, J. Enhanced oil recovery mechanisms of polymer flooding in a heterogeneous oil reservoir. *Pet. Explor. Dev. Online* **2021**, *48*, 169–178. [CrossRef]
16. Al-Busaidi, K.; Al-Maamari, R.S.; Karimi, M.; Mahrouqi, D.A.; Hinai, H.A. Impact of brine chemistry on viscoelastic properties and geochemical interactions of low salinity polymer at rock-brine interfaces. *J. Mol. Liq.* **2025**, *426*, 127372. [CrossRef]
17. Zhi, J.; Liu, Y.; Chen, J.; Bo, L.; Qu, G.; Jiang, N.; He, W. Preparation and Performance Evaluation of a Temperature and Salt Resistant Hydrophobic Associative Weak Polymer Gel System. *Molecules* **2023**, *28*, 3125. [CrossRef]
18. Skauge, A.; Zamani, N.; Gausdal Jacobsen, J.; Shaker Shiran, B.; Al-Shakry, B.; Skauge, T. Polymer Flow in Porous Media: Relevance to Enhanced Oil Recovery. *Colloids Interfaces* **2018**, *2*, 27. [CrossRef]
19. Liu, Y.; Zhong, H.; Peng, X.; Zhao, C. Factors Influencing Micro Sweep Efficiency When Flooding a Formation with Viscoelastic Polymer Solutions. *Chem. Technol. Fuels Oils* **2015**, *51*, 160–167. [CrossRef]
20. Sun, X.; Suo, L.; Huang, Y.; Wang, H.; Yu, H.; Xu, C.; Xu, J.; Qin, X.; Sun, W.; Cao, Y.; et al. Study on the Occurrence Characteristics of the Remaining Oil in Sandstone Reservoirs with Different Permeability after Polymer Flooding. *Polymers* **2024**, *16*, 1902. [CrossRef]
21. Yang, Y.; Tao, L.; Iglauer, S.; Hejazi, S.H.; Yao, J.; Zhang, W.; Zhang, K. Quantitative Statistical Evaluation of Micro Residual Oil after Polymer Flooding Based on X-ray Micro Computed-Tomography Scanning. *Energy Fuels* **2020**, *34*, 10762–10772. [CrossRef]
22. Dai, Y.; Lu, F.; Tang, Y.; Wang, Y.; He, X.; Wang, T.; Wu, J. The Simulation of Ester Lubricants and Their Application in Weak Gel Drilling Fluids. *Gels* **2024**, *10*, 178. [CrossRef]
23. Long, Y.; Zhang, C.; Yin, D.; Huang, T.; Zhang, H.; Yue, M.; Huang, X. Rheological Properties of Weak Gel System Cross-Linked from Chromium Acetate and Polyacrylamide and Its Application in Enhanced Oil Recovery After Polymer Flooding for Heterogeneous Reservoir. *Gels* **2024**, *10*, 784. [CrossRef]
24. Yang, S.; Yu, W.; Zhao, M.; Ding, F.; Zhang, Y. A Review of Weak Gel Fracturing Fluids for Deep Shale Gas Reservoirs. *Gels* **2024**, *10*, 345. [CrossRef] [PubMed]
25. Elsharafi, M.O.; Bai, B. Experimental work to determine the effect of load pressure on the gel pack permeability of strong and weak preformed particle gels. *Fuel* **2017**, *188*, 332–342. [CrossRef]

26. Wang, J.; Kang, B.; Zhang, L.; Darowska, B.J.; Xu, P. Macroscopic Profile Modification and Microscopic Displacement Mechanism of Weak Gel Flowing in Porous Media. *J. Chem.* **2016**, *2016*, 2379362. [CrossRef]
27. Zeynalli, M.; Mushtaq, M.; Al-Shalabi, E.W.; Alfazazi, U.; Hassan, A.M.; Al Ameri, W. A comprehensive review of viscoelastic polymer flooding in sandstone and carbonate rocks. *Sci. Rep.* **2023**, *13*, 17679. [CrossRef]
28. Clarke, A.; Howe, A.M.; Mitchell, J.; Staniland, J.; Hawkes, L.A. How Viscoelastic-Polymer Flooding Enhances Displacement Efficiency. *SPE J.* **2016**, *21*, 675–687. [CrossRef]
29. Liu, L.; Wang, L.; Song, H.; Bai, M. Investigation on influences of polymer solution properties on stress distribution and deformation of residual oil. *Eng. Appl. Comput. Fluid Mech.* **2020**, *14*, 401–410. [CrossRef]
30. Wang, J.; Liu, H. A novel model and sensitivity analysis for viscoelastic polymer flooding in offshore oilfield. *J. Ind. Eng. Chem.* **2014**, *20*, 656–667. [CrossRef]
31. Liu, L.; Zhao, M.; Pi, Y.; Fan, X.; Cheng, G.; Jiang, L. Experimental Study on Enhanced Oil Recovery of the Heterogeneous System after Polymer Flooding. *Processes* **2023**, *11*, 2865. [CrossRef]
32. Ameli, F.; Moghbeli, M.R.; Alashkar, A. On the effect of salinity and nano-particles on polymer flooding in a heterogeneous porous media: Experimental and modeling approaches. *J. Pet. Sci. Eng.* **2018**, *174*, 1152–1168. [CrossRef]
33. Azad, S.; Trivedi, J.J. Novel viscoelastic model for predicting the synthetic polymer's viscoelastic behavior in porous media using direct extensional rheological measurements. *Fuel* **2019**, *235*, 218–226. [CrossRef]
34. Delshad, M.; Kim, D.H.; Magbagbeola, O.A.; Huh, C.; Pope, G.A.; Tarahhom, F. Mechanistic Interpretation and Utilization of Viscoelastic Behavior of Polymer Solutions for Improved Polymer-Flood Efficiency. In Proceedings of the SPE Symposium on Improved Oil Recovery, Tulsa, OK, USA, 20–23 April 2008.
35. Shakir, H.A.; Abdulhameed, R.F.; Hilal, H.A.A.; Alfarge, D.; Aljarah, A.M. Factors impacting the performance of polymer-based EOR in oil reservoirs. *Chem. Pap.* **2023**, *77*, 4859–4875. [CrossRef]
36. Alisheva, Z.; Nadirov, K.; Dujaili, A.N.A.; Bimbetova, G.; Nadirova, Z.; Zhantasov, M.; Tileuberdi, N.; Dauletuly, A. Integrated Strategies for Controlling Water Cut in Mature Oil Fields in Kazakhstan. *Polymers* **2025**, *17*, 829. [CrossRef] [PubMed]
37. Liang, X.; Shi, L.; Cheng, L.; Wang, X.; Ye, Z. Optimization of polymer mobility control for enhanced heavy oil recovery: Based on response surface method. *J. Pet. Sci. Eng.* **2021**, *206*, 109065. [CrossRef]
38. Di, Q.; Zhang, J.; Hua, S.; Chen, H.; Gu, C. Visualization experiments on polymer-weak gel profile control and displacement by NMR technique. *Pet. Explor. Dev. Online* **2017**, *44*, 294–298. [CrossRef]
39. Hatzignatiou, D.G.; Giske, N.H.; Stavland, A. Polymers and polymer-based gelants for improved oil recovery and water control in naturally fractured chalk formations. *Chem. Eng. Sci.* **2018**, *187*, 302–317. [CrossRef]
40. Tang, Y.; Wang, H.; Gao, S.; Han, D.; Du, J.; Wang, R.; Liu, Y.; Qu, G. Study on Performance Evaluation of Profile Control System with High Temperature Resistant Weak Gel Prepared from the Produced Fluid. *Chem. Technol. Fuels Oils* **2023**, *59*, 192–201. [CrossRef]
41. Zotelle, A.C.; Souza, A.W.; Pires, P.J.; Soares, E.J.; Siqueira, R.N. Viscosity ratio effects on fluid displacement pattern and recovery efficiency on porous media. *J. Braz. Soc. Mech. Sci. Eng.* **2023**, *45*, 149. [CrossRef]
42. Cao, W.; Xie, K.; Lu, X.; Liu, Y.; Zhang, Y. Effect of profile-control oil-displacement agent on increasing oil recovery and its mechanism. *Fuel* **2019**, *237*, 1151–1160. [CrossRef]
43. Zhao, L.; Sun, X.; Zhang, H.; Xu, C.; Sui, X.; Qin, X.; Zeng, M. Study on the Impact of Microscopic Pore Structure Characteristics in Tight Sandstone on Microscopic Remaining Oil after Polymer Flooding. *Polymers* **2024**, *16*, 2757. [CrossRef]
44. Wang, J.; Wang, R.; Liu, P.; Xu, H. Gel-Forming and Plugging Performance Evaluation of Emulsion Polymer Crosslinking System in Fractured Carbonate Rock. *Processes* **2022**, *10*, 1550. [CrossRef]

gels

Article

Synergy of Low Injection Resistance and High Plugging in a High-Strength Nanobentonite/Amphoteric Polymer Gel

Huaizhu Liu [1,2], Guiqiang Fei [1,3,*], Kaiping Tian [1,3,*], Zhao Zhu [1,3], Donghang Ji [1,3] and Houyong Luo [1,3,*]

1 Department of Chemistry and Chemical Engineering, Shaanxi University of Science and Technology, Xi'an 710021, China; zcy_liuhzh@petrochina.com.cn (H.L.); zhuzhao@sust.edu.cn (Z.Z.); jidonghang@sust.edu.cn (D.J.)
2 Tangshan Jiyou Ruifeng Chemical Limited Company, Tangshan 063500, China
3 Xi'an Key Laboratory of Auxiliary Chemistry and Technology for Oil and Gas Fields, Shaanxi University of Science and Technology, Xi'an 710021, China
* Correspondence: feiguiqiang@sust.edu.cn (G.F.); tiankaiping@sust.edu.cn (K.T.); luohouyong@sust.edu.cn (H.L.)

Abstract: Long-term water flooding development has exacerbated reservoir heterogeneity, and traditional polymer gels are unable to simultaneously meet the requirements of high injectability and strong plugging strength. If the viscosity of the polymer is high, its injectability will be poor; on the contrary the viscosity is low, the plugging strength will be poor, which severely restricts the oil recovery effect. This study synthesized an NBAP through free radical polymerization followed by a substitution reaction, and a plugging system (NBAP-B1) was subsequently formed by combining the polymer with a Cr^{3+} crosslinking agent. Rheological experiments demonstrated that the system exhibited significant shear thinning behavior, as well as excellent temperature and salt resistance. Gelation experiments indicated that the NBAP-B1 system featured controllable gelation time (20~150 h) and high gelation strength (J grade), along with excellent resistance to both high temperature and high salinity. Microscopic analysis revealed that the gel formed by NBAP-B1 possessed a dense and uniform three-dimensional network structure. Injection and plugging experiments demonstrated that NBAP-B1 exhibited optimal injectability and outstanding plugging performance. Additionally, profile control and displacement tests revealed a 18.37% enhancement in oil recovery efficiency by water flooding after the application of NBAP-B1 for conformance control. Collectively, these results demonstrate that the NBAP exhibits significantly superior performance compared to single component systems. It combines excellent injectability with high strength plugging capability, offering an effective approach for enhancing oil recovery in low permeability reservoirs.

Keywords: low-permeability reservoir; injectability; plugging strength; plugging system; enhanced oil recovery

1. Introduction

Petroleum plays an indispensable and foundational role in areas such as energy supply, transportation, and chemical production, exerting a profound influence on society economic development [1–3]. However, amid continuously growing global energy demand, the petroleum supply is facing significant challenges, with low extraction efficiency emerging as a critical bottleneck constraining the development of the petroleum industry [4–6]. Research indicates that low-permeability reservoirs already account for more than 70% of newly developed oil fields globally, yet their average recovery rate remains below 30%, resulting in significant resource wastage [7,8]. As a primary stimulation technique for enhancing production in such reservoirs, fracturing technology aims to improve water-flooding efficiency by increasing reservoir permeability [9–13]. However, long-term fracturing operations often lead to the formation of preferential flow channels, which exacerbate reservoir heterogeneity and cause issues such as water channeling and ineffective recycling of injected water [14–16]. Taking the Gudao Block of Shengli Oilfield as an example, during the ultrahigh water cut development stage, the comprehensive water cut can exceed 90%, while the oil recovery rate remains below 10%. Notably, the proportion of high water-cut wells ranges from 40% to 60%. Therefore, addressing the issue of high water cut has become a critical scientific challenge that must be resolved to enhance oil recovery in low-permeability reservoirs [17–21].

As an effective method for addressing high water cut in oil reservoirs, profile control and water shutoff technology relies on the injection of chemical agents to selectively block high-permeability channels, alter fluid flow paths, and increase sweep efficiency, thereby displacing remaining oil from low-permeability zones [22,23]. Conventional profile control and water shutoff agents commonly in use include gels, gelants, microspheres, and resins, among other traditional materials [24–26]. However, as reservoir development conditions become increasingly complex, the application of these conventional agents in low-permeability reservoirs (permeability < 10 mD) faces significant technical challenges. Specifically, although high-viscosity plugging systems can form high-strength gel blockages, their injection becomes difficult in low-permeability reservoirs due to the fine pore-throat structures, significantly limiting the conformance volume [27–30]. On the other hand, low-viscosity systems offer favorable injectivity but often fail to achieve effective plugging due to insufficient gelation strength [30–33]. This inherent contradiction between injectivity and plugging strength significantly restricts the effectiveness of conventional plugging agents in low-permeability reservoirs [34–37].

In recent years, numerous studies have sought to enhance plugging agent performance through two main approaches, one of which involves the development of composite material systems. For instance, Song et al. developed P(AA-AM-SA)/TiO_2 nanocomposite microspheres, which demonstrated exceptional plugging performance (plugging rate > 95.3%) under high-temperature and high-salinity conditions, resulting in a 12.89% increase in oil recovery [38]. The incorporation of nanomaterials has significantly enhanced the mechanical strength of polymers; however, the multi-component nature of these systems often leads to high costs and operational complexity [39–42]. Another approach involves chemical modification. For example, Pu et al. [43] synthesized an amphoteric polyacrylamide (LHPAM), which exhibits notably low viscosity and favorable injectivity. After crosslinking with Cr^{3+}, the system achieved a plugging efficiency exceeding 90%. However, such gels often fail to maintain long-term stability under high-temperature and high-salinity conditions. To address this limitation, this study proposes a novel nanocomposite plugging agent designed to simultaneously enhance injectivity and plugging strength. The approach involves incorporating both anionic and cationic functional groups into the polymer backbone. This dual functionalization induces polymer chain coiling through ionic interactions, resulting in low viscosity and excellent injectivity; Meanwhile, bentonite nanoparticles are anchored to the polymer backbone via chemical bonding, significantly enhancing the mechanical strength of the three-dimensional network within the polymer structure. This unique molecular design leverages the synergistic effect of nanomaterial reinforcement and zwitterionic modification, successfully addressing the long-standing challenge of achieving both low viscosity and high mechanical strength in conventional plugging agents for low-permeability reservoirs. To date, there have been limited reports on the use of low-viscosity yet high-strength polymer gels for profile control and water shutoff in such reservoirs to enhance oil recovery. The proposed method demonstrates significant potential for application in the development of unconventional reservoirs.

Building on this foundation, the objective of this study is to develop a low-viscosity, high-strength nanopolymer gel system to effectively enhance oil displacement efficiency in low-permeability reservoirs. Using acrylamide (AM) as the monomer, the polymer was functionalized via free-radical polymerization and substitution reactions to introduce 3-chloro-2-hydroxypropyltrimethylammonium chloride (cationic group) and sodium 2-chloroethylsulfonate (anionic group). Bentonite nanoparticles were subsequently incorporated, resulting in a nanocomposite termed NBAP (Nanocomposite Bentonite/Amphoteric Polyacrylamide). and was further formulated with polyvinyl alcohol and Cr^{3+} crosslinker to form a nanopolymer plugging system designated as NBAP-B1. The structure of the material was characterized using Fourier transform infrared spectroscopy (FTIR) and scanning electron microscopy (SEM). Its rheological properties, gelation behavior, injectivity, and plugging performance were evaluated through rheological tests, gelation experiments, and physical simulation studies. These analyses elucidate the synergistic effect of low injection resistance and high plugging efficiency in NBAP-B1 and reveal its mechanism for profile control and oil displacement in low-permeability reservoirs. The findings of this study provide important theoretical insights and technical support for the efficient development of low-permeability reservoirs, offering significant potential for enhancing oil recovery.

2. Results and Discussion

2.1. Structural Characterization

2.1.1. FTIR

Figure 1 presented the Fourier transform infrared (FTIR) spectroscopy results of the different samples. A broad absorption band was observed in the range of 3400–3500 cm^{-1}, which is primarily attributed to the stretched vibration of the NH_3^+ group within the polymer molecules. It is noteworthy that the hydroxyl group (OH) in 3-chloro-2-hydroxypropyltrimethylammonium chloride present in both CNPA and NBAP may also contribute to the spectral features observed in this region. The characteristic peak at 2920 cm^{-1} corresponds to the symmetric stretched vibration of methylene groups (-CH_2-) on the polymer backbone. Notably, CNPA and NBAP exhibit a distinct absorption band near 2760 cm^{-1}, originated from the stretched vibration of the -CH_3 groups in 3-chloro-2-hydroxypropyltrimethylammonium chloride. A comparison between bentonite-contained samples (NBAP and NBNP) and those without bentonite (CNPA and NPA) revealed a distinct absorption band at 1037 cm^{-1} in the former, which corresponds to the asymmetric stretched vibration of the Si–O–Si bond in bentonite. The presence of these characteristic peaks confirms the successful synthesis of all target samples.

2.1.2. SEM

Figure 2 presents a comparative analysis of the morphological characteristics of the samples obtained by scanned electron microscopy (SEM). Microstructural analysis revealed that the control samples without bentonite (the amphoteric polymer CNPA and the anionic polymer NPA) exhibited a porous surface morphology. Notably, CNPA displayed more pronounced pore dimensions, which may be attributed to the steric hindrance effects of the zwitterionic functional groups leading to a looser arrangement of molecular chains. Following the incorporation of bentonite, the surface porosity of both NBAP and NBNP samples was significantly reduced and virtually eliminated, indicating that the layered structure of bentonite effectively filled the defects within the polymer matrix. It is noteworthy that a comparison of the surface morphology between zwitterionic polymers (CNPA and NBAP) and anionic polymers (NPA and NBNP) revealed that the former exhibited more pronounced surface roughness and complex topological structures. This can be attributed to enhanced intermolecular interactions—such as electrostatic forces and hydrogen bonding—resulting from the zwitterionic functional groups, which led to a higher degree of phase separation; In contrast, the latter displayed relatively smoother surface characteristics, suggesting that the anionic functional groups promote a more homogeneous molecular distribution primarily through electrostatic repulsion.

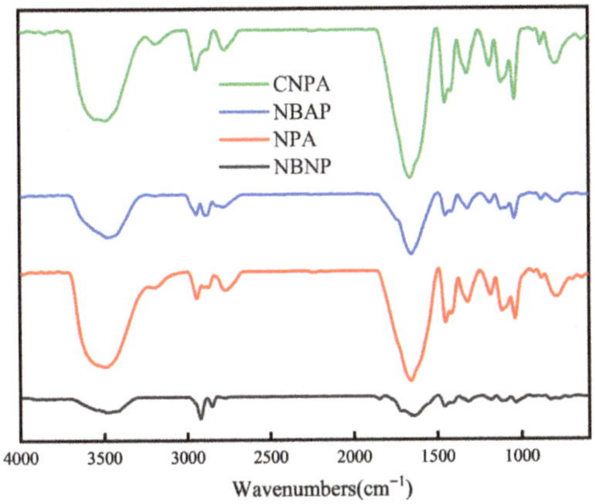

Figure 1. FTIR spectra of the different samples.

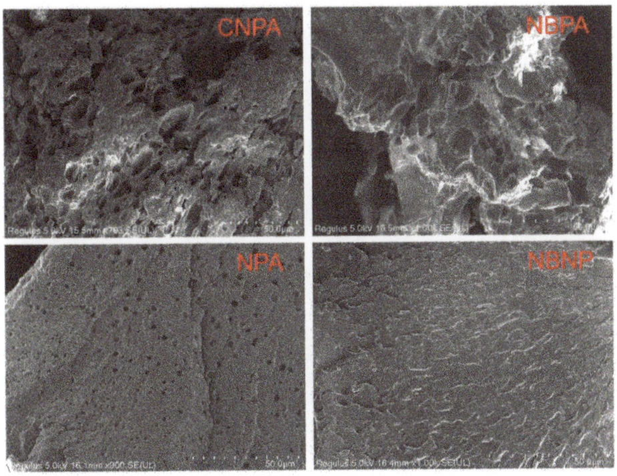

Figure 2. SEM characterization of the different samples.

2.2. Rheological Properties

The effects of various conditions on the viscosity of the plugging systems were evaluated to analyze their rheological behavior. The results are presented in Figure 3. Figure 3a illustrates the relationship between polymer dosage and viscosity. The viscosity of all plugging systems increased with higher polymer loading. This behavior can be attributed to enhanced intermolecular entanglement resulting from the increased polymer chain concentration, which consequently elevated flow resistance. Moreover, the viscosity of NBAP-B1 was significantly lower than that of CNPA-B1 and NBNP-B1. This reduction can be explained by charge neutralization between the bentonite nanosheets and zwitterionic functional groups in NBAP-B1, which restricted the extension of polymer chains. In contrast, CNPA-B1 (without bentonite) and NBNP-B1 (where electrostatic repulsion dominates between anionic groups and bentonite) exhibited greater molecular chain mobility, resulting in higher viscosity. Figure 3b depicts the relationship between shear rate and viscosity. The viscosity of the systems decreased with increasing shear rate, indicating shear-thinning behavior. Furthermore, NBAP-B1 exhibited superior shear stability compared to CNPA-B1 and NBNP-B1. This enhanced performance is attributed to the dynamically reversible crosslinked network formed between the zwitterionic functional groups and bentonite nanosheets in NBAP-B1, which enables rapid structural reorganization under shear stress. In contrast, CNPA-B1 (lacking bentonite reinforcement) and NBNP-B1 (relying solely on anionic interactions) demonstrated weaker network recovery capabilities. Figure 3c,d illustrated the effects of temperature and salinity on the viscosity of the plugging systems, respectively. The viscosity gradually decreased with increasing temperature or salinity. Among them, NBNP-B1 exhibited the most pronounced viscosity reduction, while NBAP-B1 and CNPA-B1 showed relatively minor decreases. This difference can be attributed to the fact that the zwitterionic structures in both NBAP-B1 and CNPA-B1 mitigate the effects of temperature and salinity variations through intramolecular and intermolecular charge balance. In contrast, NBNP-B1 relies solely on electrostatic repulsion from anionic groups, making it more susceptible to thermal dissociation and charge shielding under high-salinity conditions.

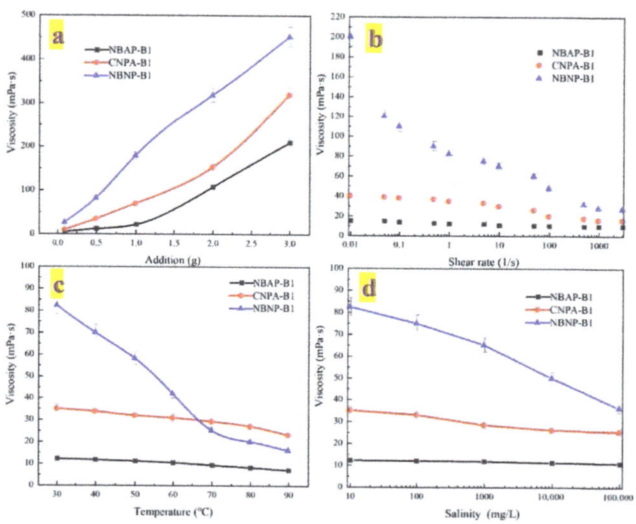

Figure 3. Rheological properties of different plugging systems: (**a**) polymer dosage, (**b**) shear rate, (**c**) temperature, and (**d**) salinity.

2.3. Gelation Performance

2.3.1. Gelation Time

The influence of various conditions on the gelation time of the plugging systems is presented in Figure 4. Figure 4a illustrates the relationship between polymer dosage and gelation time. As the polymer loading increased, the gelation time of the system gradually decreased. Among the different plugging systems, the gelation time followed the order: NBAP-B1 < CNPA-B1 < NBNP-B1. This behavior can be attributed to the synergistic effect between the zwitterionic groups (coexisting positive and negative charges) and bentonite in NBAP-B1, which accelerated the formation of a crosslinked network. In comparison, gelation in CNPA-B1 relied solely on the self-assembly of zwitterionic moieties, resulting in a moderately slower process. In contrast, NBNP-B1 exhibited the longest gelation time due to strong electrostatic repulsion among anionic groups, which imposed the greatest resistance to network formation. Figure 4b demonstrates the effect of Cr^{3+} crosslinker dosage on gelation time. As the concentration of Cr^{3+} crosslinker increased, the gelation time of NBAP-B1 and CNPA-B1 gradually decreased. In contrast, NBNP-B1 exhibited a non-monotonic trend, with gelation time initially decreasing and then increasing. This difference arises because the charge balance of zwitterionic groups in NBAP-B1 and CNPA-B1 enables uniform crosslinking with Cr^{3+} ions. In contrast, at high Cr^{3+} concentrations, the anionic groups in NBNP-B1 undergo excessive crosslinking, resulting in an overly dense network structure that restricts molecular chain mobility and consequently prolongs the gelation time. Figure 4c shows that as temperature increased, the gelation time of NBAP-B1 initially decreased and then increased, while that of CNPA-B1 rose gradually, and NBNP-B1 exhibited a consistent decrease. The non-monotonic change in gelation time for NBAP-B1 first decreasing and then increasing—can be explained by the synergistic effect between the zwitterionic groups and bentonite. Moderate heating promotes crosslinking, whereas excessively high temperatures disrupt the charge balance, ultimately delaying gel formation; The gradual increase in gelation time for CNPA-B1 is attributed to the insufficient thermal stability of the purely zwitterionic network. The consistent decrease in gelation time observed for NBNP-B1 is primarily driven by the thermally activated chelation between anionic groups and Cr^{3+} ions. Figure 4d shows that as salinity increased, the gelation time of NBAP-B1 decreased, while that of CNPA-B1 initially decreased and then increased. In contrast, NBNP-B1 exhibited a gradual increase in gelation time. The decrease in gelation time for NBAP-B1 is due to the compression of the electric double layer by salt ions, which enhances the crosslinking efficiency between zwitterionic groups and Cr^{3+}; The gelation time of CNPA-B1 decreased initially and then increased with rising salinity. This non-monotonic trend can be explained by the fact that low salt concentrations promote chain expansion of the zwitterionic polymer, whereas high salt levels weaken charge-driven interactions, ultimately impairing crosslinking; The gelation time of NBNP-B1 increased consistently due to the suppression of crosslinking caused by charge shielding of anionic groups under high salinity. Furthermore, a comprehensive analysis of Figure 4 reveals that the gelation time of the plugging systems ranges from 20 to 150 h, indicating that it can be controllably adjusted to meet specific reservoir plugging requirements.

2.3.2. Gelation Strength

The influence of various conditions on the gelation strength of the plugging systems was investigated, and the corresponding results are presented in Figure 5. Figure 5a illustrates the relationship between polymer dosage and gelation strength. As the polymer concentration increased, the gelation strength of the system gradually enhanced. This enhancement can be attributed to the increased number of polymer chains, which provide more crosslinking sites and promote intermolecular entanglement, thereby facilitating the formation of a denser three-dimensional network structure and ultimately improving the gelation strength. Figure 4b illustrates the effect of Cr^{3+} crosslinker concentration on gelation strength. As the crosslinker dosage increased, the gelation strength of NBNP-B1 and CNPA-B1 initially rose and then declined, whereas that of NBAP-B1 increased initially and subsequently plateaued. This divergence occurs because excessive Cr^{3+} in NBNP-B1 and CNPA-B1 causes over-crosslinking, resulting in network embrittlement and consequent strength reduction. In contrast, the bentonite reinforcement and charge-balancing effect of zwitterionic groups in NBAP-B1 enable the maintenance of a stable crosslinked structure even at high Cr^{3+} concentrations. As shown in Figure 4c,d, the gelation strength of all systems initially increased and then decreased with rising temperature or salinity. This non-monotonic behavior occurs because moderate increases in temperature enhance molecular chain mobility, while higher salinity compresses the electric double layer—both of which improve crosslinking efficiency initially. However, beyond a critical point, elevated temperatures disrupt the network integrity, and excessive salinity induces charge shielding that weakens intermolecular interactions, ultimately leading to a decline in gelation strength. Furthermore, a comparison of gelation strength among the different systems revealed the following order: NBNP-B1 < CNPA-B1 < NBAP-B1. NBAP-B1 achieved the highest gelation strength, reaching grade J (the strongest level), whereas CNPA-B1 and NBNP-B1 attained maximum strengths of grade H and grade F, respectively. This difference is primarily due to the synergistic effect between zwitterionic groups and bentonite nanosheets in NBAP-B1, which facilitates the formation of the most dense and stable three-dimensional network. CNPA-B1 relies solely on zwitterionic self-assembly, resulting in a moderately effective yet suboptimal network structure. In contrast, the electrostatic repulsion between anionic groups and bentonite in NBNP-B1 leads to the weakest structural integrity among the three systems.

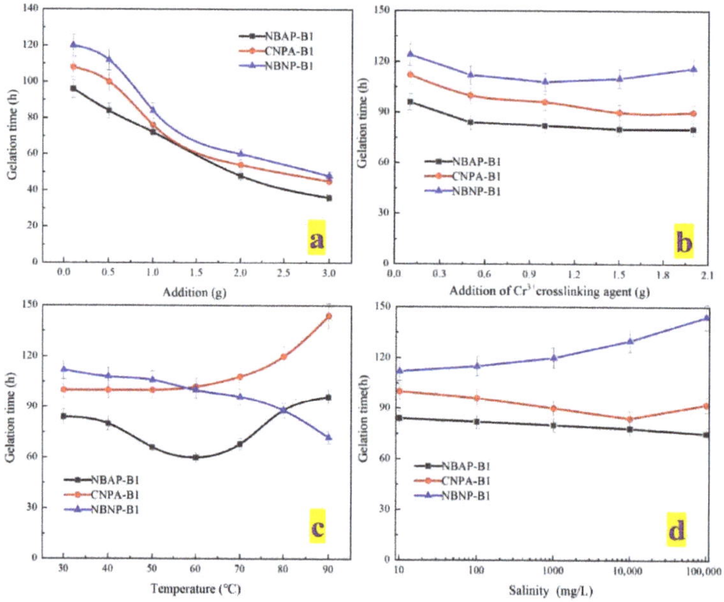

Figure 4. Gelation time of different plugging systems: (**a**) polymer dosage, (**b**) crosslinker concentration, (**c**) temperature, and (**d**) salinity.

2.3.3. Gel Microstructure

Figure 6 displays the microstructures of the gels formed by the different plugging systems. As shown in the image, the NBAP-B1 gel exhibits a uniform surface with a small number of fine pores, whereas the NBNP-B1 gel possesses a loose structure with larger voids. This enhanced structural integrity in NBAP-B1 arises from the strong electrostatic attraction between the cationic quaternary ammonium groups and the negatively charged surfaces of bentonite, combined with hydrogen bonding between the anionic sulfonate groups and the hydroxyl groups on bentonite. These synergistic interactions facilitate the formation of a dense and robust crosslinked network. In contrast, the loose structure of NBNP-B1 results primarily from electrostatic repulsion between the anionic groups and the negatively charged bentonite surfaces. It is particularly noteworthy that although CNPA-B1 contains no bentonite, its zwitterionic nature still enables the formation of a relatively uniform fine-network structure due to intramolecular self-assembly of oppositely charged groups. However, the absence of a nano-reinforcing phase leads to significantly inferior structural strength compared to NBAP-B1.

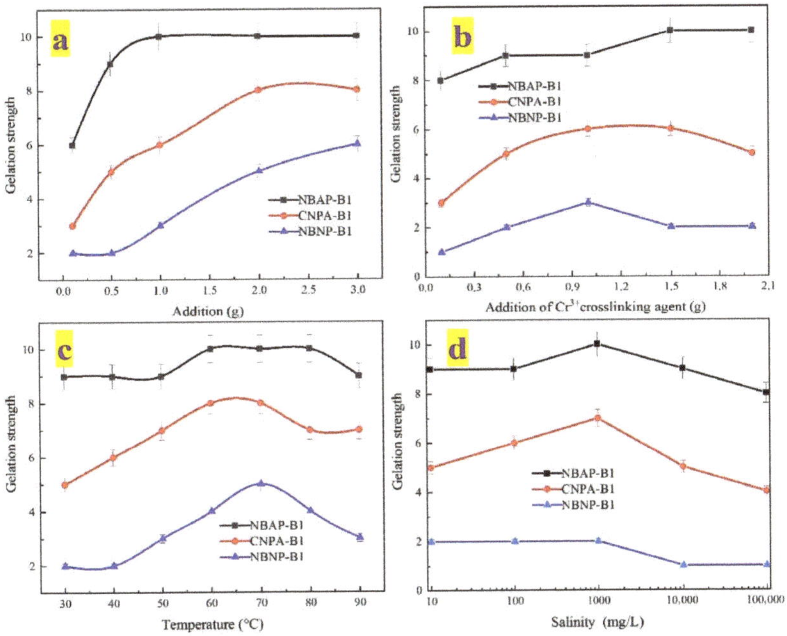

Figure 5. Gelation strength of different plugging systems: (**a**) polymer dosage, (**b**) crosslinker concentration, (**c**) temperature, and (**d**) salinity.

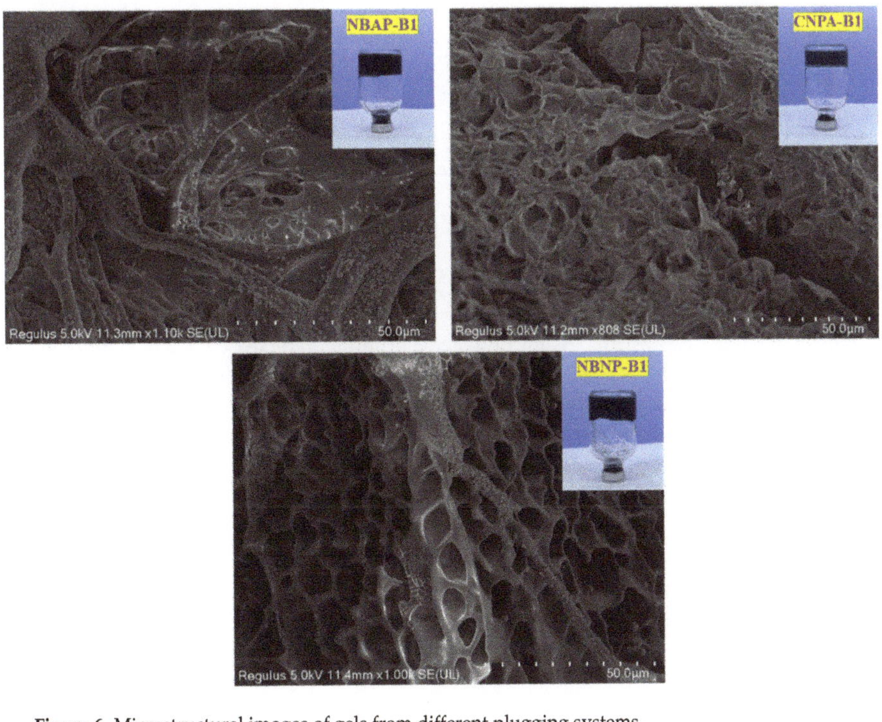

Figure 6. Microstructural images of gels from different plugging systems.

2.4. Injectability

A higher resistance factor indicates poorer injectability. The effects of various conditions on the injectability of the plugging systems are presented in Figure 7. As shown in Figure 7a, the resistance factor of the plugging systems gradually increased with higher polymer loading. This increase can be attributed to enhanced intermolecular chain entanglement and a strengthened network structure resulting from higher polymer concentration, which collectively elevate the flow resistance of the plugging system. Meanwhile, Figure 7b demonstrates the influence of permeability on injectability. The resistance factor gradually decreased as permeability increased. This reduction occurs because higher permeability results in larger or better-connected pore channels within the reservoir, allowing polymer molecules to pass through more readily with reduced entrapment and clogging, thereby lowering the resistance factor. Furthermore, the resistance factors of the plugging systems followed the order: NBAP-B1 < CNPA-B1 < NBNP-B1. In NBAP-B1, the zwitterionic polymer chains exhibit electrostatic repulsion toward both the negatively charged nano-bentonite and the rock surfaces. This causes the polymer chains to adopt a more coiled conformation, reducing effective adsorption and bridging at pore throats, thereby resulting in the lowest flow resistance.

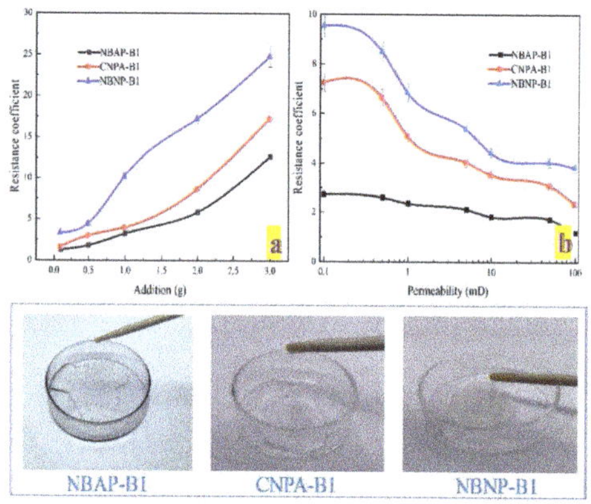

Figure 7. Resistance factor of different plugging systems: effects of polymer concentration (**a**) and permeability (**b**).

2.5. Plugging Performance

2.5.1. Breakthrough Pressure

Figure 8 illustrates the relationship between polymer dosage and breakthrough pressure in the plugging systems. As the polymer concentration increased, the breakthrough pressure of the systems also rose. This increase can be attributed to the enhanced intermolecular entanglement and adsorption resulting from higher polymer concentration, which promotes the formation of a denser and more robust plugging network. Consequently, a higher pressure is required to penetrate this reinforced structure. Furthermore, the breakthrough pressure of the plugging systems followed the order: NBNP-B1 < CNPA-B1 < NBAP-B1. The superior performance of NBAP-B1 can be attributed to the combined effects of strong adsorption and complexation capabilities from its zwitterionic functional groups, as well as the physical filling and bridging actions provided by bentonite. These synergistic interactions enhance the strength and density of the gel network, resulting in the highest breakthrough pressure. This is because the anionic functional groups in NBNP-B1 experience electrostatic repulsion from both the negatively charged nano-bentonite and the rock surfaces, hindering the formation of an effective seal. As a result, NBNP-B1 exhibits the lowest breakthrough pressure; In contrast, the zwitterionic polymer in NBAP-B1 facilitates strong adsorption and bridging of nanoparticles via its cationic groups interacting with negatively charged surfaces, leading to the formation of a dense plugging barrier. This mechanism accounts for its highest breakthrough pressure.

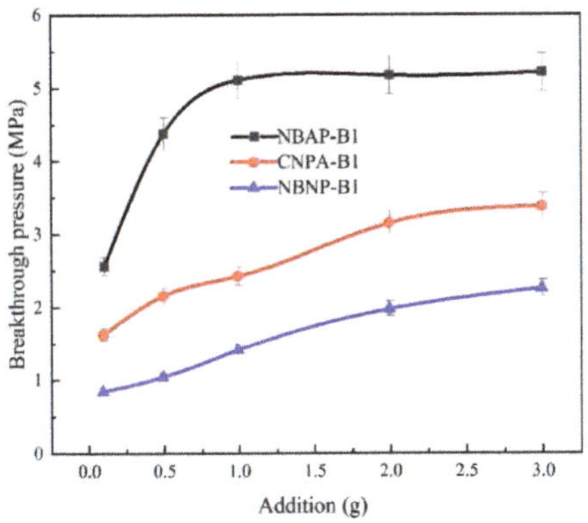

Figure 8. Breakthrough pressure of different plugging systems.

2.5.2. Plugging Efficiency

The influence of various conditions on the plugging efficiency of the plugging systems is presented in Figure 9. As shown in Figure 9a, the plugging efficiency gradually increased with higher polymer loading. This improvement can be attributed to enhanced intermolecular adsorption, bridging effects, and a denser network structure resulting from the increased polymer concentration, which collectively contribute to more effective pore throat blockage and consequently higher plugging efficiency. As observed in Figure 9b, the plugging efficiency of the systems initially increased and then decreased with rising core permeability. This trend occurs because in low-permeability cores, the narrow pore throats restrict the penetration of polymer molecules, limiting effective plugging. Conversely, in high-permeability cores, the larger pore throats make the plugging structure more susceptible to erosion and mechanical failure under flow conditions; Medium-permeability cores possess pore throat dimensions that best match the size of polymer chains/aggregates, thereby enabling optimal bridging and retention effects, which results in the highest plugging efficiency. Furthermore, the plugging efficiency of the different systems followed the order: NBNP-B1 < CNPA-B1 < NBAP-B1. The superior performance of NBAP-B1 is due to the strong adsorption and nanoparticle bridging facilitated by its cationic groups interacting with negatively charged surfaces, resulting in the densest and most stable plugging layer. In comparison, CNPA-B1 relies solely on the self-assembly and bridging of the zwitterionic polymer itself, leading to moderately effective but inferior plugging strength; In contrast, the anionic polymer in NBNP-B1 experiences electrostatic repulsion from the negatively charged components, hindering effective retention and plugging, thereby resulting in the lowest plugging efficiency.

2.6. Oil Displacement Performance

As shown in Figure 10, the profile control and oil displacement process of the plugging system NBAP-B1 can be divided into three distinct stages. The first stage involves initial water flooding, during which the maximum injection pressure reached 1.43 MPa, resulting in an oil recovery rate of 35.3%. In the second stage, the NBAP-B1 system was injected and allowed to gel, leading to a rise in the maximum injection pressure to 1.58 MPa and an increase in the oil recovery rate to 45.94%. During the third stage, subsequent water flooding was conducted, with the maximum injection pressure stabilizing at 1.57 MPa and the recovery rate further rising to 53.67%. During the initial water flooding process, due to the heterogeneity of the core, injected water primarily flowed through the high-permeability zones (10 mD), effectively displacing crude oil from these regions. After the injection of the plugging system, the agent predominantly entered and blocked the high-permeability zones, while a small amount of the low-viscosity polymer penetrated into the low-permeability zones (5 mD). This resulted in an increase in injection pressure and an enhancement in oil recovery by 10.64%. Following gelation, subsequent water flooding was carried out, during which the plugging system mitigated core heterogeneity by diverting injected water into low-permeability zones. This led to a notable increase in injection pressure, effectively displacing trapped crude oil from these previously under-swept areas and thereby improving overall sweep efficiency. As a result, the cumulative enhancement in oil recovery reached 18.37%.

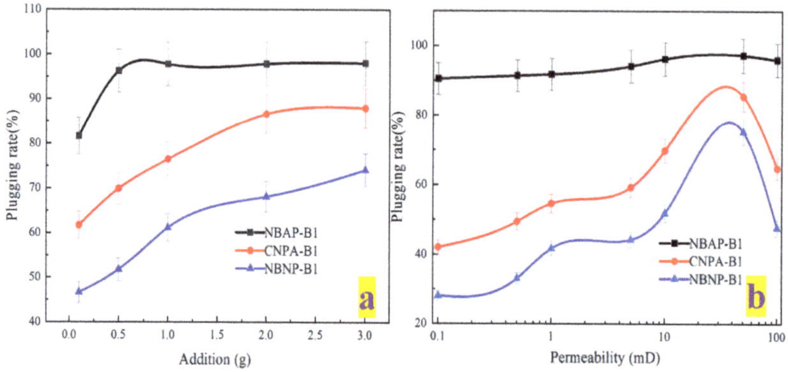

Figure 9. Plugging efficiency of different plugging systems: effects of polymer concentration (**a**) and permeability (**b**).

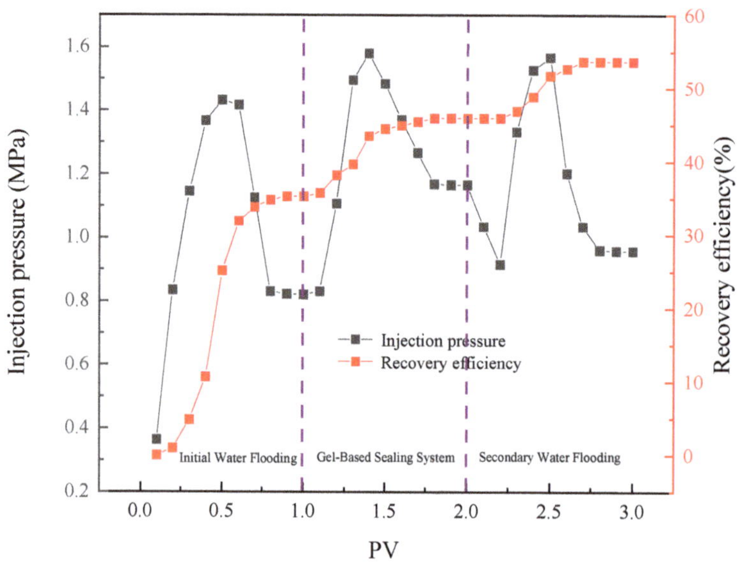

Figure 10. Oil Displacement efficiency of the plugging system NBAP-B1 at different stages.

3. Conclusions

This study successfully developed a nanobentonite-reinforced amphoteric polyacrylamide composite gel (NBAP-B1), achieving three key innovations in the field of deep profile control and displacement materials through molecular structure design and nanocomposite technology:

First, we innovatively adopted a sequential reaction process to simultaneously incorporate anionic/cationic functional groups into the polymer backbone, constructing a unique dual-electronic molecular structure. This zwitterionic characteristic provides the material with exceptional salt tolerance and, through synergy with nanobentonite, forms a dense three-dimensional network unattainable in conventional polymers.

Second, we broke through the conventional approach of using nanomaterials merely as fillers. By combining chemical grafting with physical compounding, nanobentonite platelets were transformed into integral components of the polymer network. This design not only significantly enhances the mechanical strength of the gel but also endows the system with the unique property of "low initial viscosity-high strength evolution," successfully resolving the technical challenge of balancing injectability and blocking performance during deep fluid diversion.

Most importantly, this research provides a precisely tunable solution for enhanced oil recovery in low-permeability reservoirs. The adjustable gelation time (20–150 h) and adaptive plugging characteristics demonstrated by the NBAP system enable customized design according to actual reservoir conditions. While achieving efficient fluid diversion, it enhances oil recovery by 18.37%, showing promising application prospects. These results indicate its promising potential for conformance control applications.

4. Materials and Methods

4.1. Materials

Acrylamide (AM, purity > 99%) was obtained from Tianjin Komiou Chemical Reagent Co., Ltd. (Tianjin, China); *N,N′*-Methylenebisacrylamide, ammonium persulfate, sodium hydroxide, and absolute ethanol (purity > 99.9%) were supplied by Tianjin Tianli Chemical Reagent Co., Ltd. (Tianjin, China).; Anhydrous chromium chloride, 3-chloro-2-hydroxypropyltrimethylammonium chloride, sodium 2-chloroethanesulfonate, thiourea, potassium bromide, and other related reagents (purity > 99.9%) were supplied by Shanghai Macklin Biochemical Technology Co., Ltd. (Shanghai, China); Polyvinyl alcohol (PVA) was obtained from Sigma-Aldrich (Shanghai, China)Trading Co., Ltd. (Shanghai, China) with a molecular weight ranging from 31,000 to 50,000 and a purity greater than 99.9%. All chemicals used were of analytical grade. Nanobentonite (purity > 96%, particle size: 200–500 nm) was obtained from Tianjin Yandong Haotian Mineral Products Co., Ltd. (Tianjin, China). The simulated oil used in this study was authentic crude oil sourced from the Chang 7 reservoir in Block Xi 233 of the Changqing Oilfield. It has a viscosity of 1.59 mPa·s and a density of 0.74 g/cm^3 at 65 °C. The crude oil is predominantly composed of C5-C10 hydrocarbons, which account for approximately 80% of its composition. Hydrocarbons in the C11–C20 range constitute approximately 20% of the composition, while components above C20 are present in minor amounts. Notably, no components $\geq$ C29 were detected, consistent with classification as a light crude oil.

4.2. Synthesis of Bentonite/Amphoteric Polyacrylamide Composite Polymer NBAP

20 g of AM were dissolved in 80 g of deionized water under stirring. Subsequently, 0.2 g of ammonium persulfate was added to the solution, and the reaction was carried out at 65 °C under continuous stirring for 45 min. A total of 1 g of 3-chloro-2-hydroxypropyltrimethylammonium chloride and 2 g of sodium 2-chloroethanesulfonate

were dissolved in 10 g of deionized water. The resulting solution was then added to the previously prepared mixture. The pH of the system was adjusted to 12 using a NaOH solution, and the temperature was raised to 70 °C. The reaction proceeded under stirring for 6 h to obtain the amphoteric polymer, denoted as CNPA. Subsequently, 0.2 g of nano-bentonite was added to the mixture, and the reaction was continued at 70 °C for an additional 6 h. The product was then purified via flocculation using a water-ethanol mixture, followed by repeated washing and filtration. After drying at 80 °C, the solid was ground into a fine powder to obtain the nano-bentonite/amphoteric polyacrylamide composite polymer, designated as NBAP. The synthetic route is illustrated in Figure 11. For comparison with NBAP, a nano-bentonite/anionic polyacrylamide composite (designated as NBNP) and a pure anionic polymer (denoted as NPA) were synthesized following the same procedure but without the addition of 3-chloro-2-hydroxypropyltrimethylammonium chloride.

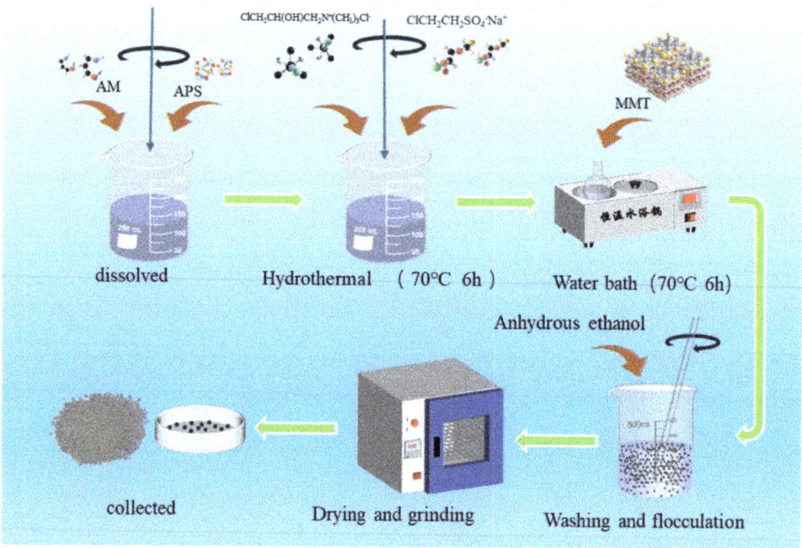

Figure 11. Schematic illustration of the polymer synthesis.

4.3. Structural Characterization of NBAP

The functional groups of the NBAP composite were characterized using an INVENIO Fourier transform infrared (FTIR) spectrometer (Bruker Corporation, Bremen, Germany) to verify the successful synthesis of the polymer. Potassium bromide (KBr) was dried using a heating lamp and subsequently ground in a mortar. The sample was then added at a mass ratio of 1:100 (sample to KBr) for further homogenization. The mixture was thoroughly ground with a pestle to achieve a uniform particle size of less than 2.5 μm and then pressed into pellets using a hydraulic press; The resulting pellet was finally transferred to the FTIR spectrometer for scanning. Data were acquired and analyzed using dedicated software to identify characteristic peaks corresponding to various functional groups. The scanning was performed over a wavenumber range of 4400 to 400 cm^{-1}.

The microstructure of the synthesized NBAP sample was characterized using a Regulus8100 field-emission scanning electron microscope (FE-SEM; Hitachi, Tokyo, Japan). Specifically, the sample was mounted on a conductive stage, freeze-fractured in liquid nitrogen, and sputter-coated with a thin layer of gold. The prepared specimen was then transferred to the FE-SEM chamber for observation to analyze the influence of nanoparticles on the polymeric morphology.

4.4. Formulation of a Nanopolymer-Based Plugging Agent System

NBAP-B1 and polyvinyl alcohol (PVA) were first dissolved separately in deionized water under stirring. The two solutions were then combined, followed by the addition of a pre-prepared Cr^{3+} crosslinking agent, ammonium persulfate solution, and the oxygen scavenger thiourea. After thorough mixing, the resulting solution was transferred into glass vials, which were sealed and placed in an oven. Gelation behavior was monitored at specific time intervals under controlled temperature conditions. The composition of each formulation is summarized in Table 1. The Cr^{3+} crosslinking agent was prepared in the laboratory according to the following specific formulation: Add 1 g $[Cr(H_2O)_6]Cl_3$ and 1.95 g CH_3COOH to 100 mL of water solution, added NaOH to adjust the pH to above 12, passed nitrogen for 30 min, reacted at 50 °C for 2 h, and storeed it in a sealed container at room temperature for 24 h. The amphoteric polymer plugging system CNPA-B1 and the bentonite/anionic polymer plugging system NBNP-B1 were prepared following the identical procedure.

Table 1. Dosage of Individual Components in the Plugging System.

Number	Water (g)	One of the Following: NBAP, CNPA, or NBNP. (g)	PVA (g)	Cr^{3+} Crosslinking Agent (mL)	10% Ammonium Persulfate Solution (g)	Thiourea (g)
1	100	0.1	0.5	0.5	0.2	0.02
2	100	0.5	0.5	0.5	0.2	0.02
3	100	1	0.5	0.5	0.2	0.02
4	100	2	0.5	0.5	0.2	0.02
5	100	2.5	0.5	0.5	0.2	0.02

4.5. Rheological Property Evaluation

This study investigated the adaptability of plugging systems under various conditions typical of low-permeability reservoirs. Following the experimental procedure outlined in Section 4.4, three plugging systems with different compositions—namely NBAP-B1, CNPA-B1, and NBNP-B1—were prepared. The rheological properties of each system were systematically characterized using an NDJ-5S rotational viscometer (Brookfield Engineering Laboratories, Inc., USA). The effects of polymer type (NBAP, CNPA, or NBNP), dosage (0.1 g, 0.5 g, 1 g, 2 g, or 3 g), temperature (30~90 °C), and shear rate (0.01~3000 s^{-1}) on the viscosity of the plugging systems were systematically investigated. Additionally, simulated formation water with varying salinities was prepared by dissolving sodium chloride in deionized water. The plugging systems were formulated using this brine instead of pure deionized water to evaluate the effect of salinity (100 mg/L, 1000 mg/L, 10,000 mg/L, and 100,000 mg/L) on their viscosity. It is noteworthy that when investigating the influence of a specific factor, all other parameters were held constant. Specifically, the polymer dosage, temperature, shear rate, and salinity were fixed at 0.5 g, 30 °C, 1 s^{-1}, and 0 mg/L, respectively.

4.6. Gelation Performance Evaluation

This study investigated the gelation time and gelation strength of various plugging systems to comprehensively evaluate their gelation performance. Following the experimental procedure described in Section 4.4, plugging systems with different compositions (NBAP-B1, CNPA-B1, and NBNP-B1) were prepared. Each system was transferred into glass vials, which were then placed at different temperatures. The gelation time and gelation strength were monitored and recorded accordingly. The gelation strength was characterized using the Marathon gelation strength Codes method. To facilitate comparison of strength variations among different systems, the gelation levels were quantified numerically, as detailed in Table 2. The effects of polymer type (NBAP, CNPA, or NBNP), polymer dosage (0.1, 0.5, 1, 2, or 3 g), Cr^{3+} crosslinker volume (0.1, 0.5, 1, 2, or 3 mL), and temperature (30–90 °C) on the gelation performance of the plugging systems were systematically investigated. Additionally, simulated formation water with varying salinities was prepared by dissolving sodium chloride in deionized water. The plugging systems were formulated using this brine instead of pure deionized water to evaluate the influence of salinity—specifically at concentrations of 100, 1000, 10,000, and 100,000 mg/L—on the gelation performance. It should be noted that when evaluating the effect of a specific variable, all other parameters were maintained constant. The baseline conditions were set as follows: polymer dosage = 0.5 g, temperature = 80 °C, and salinity = 0 mg/L. Furthermore, the microstructure of the gel formed under these conditions was examined using scanning electron microscopy (SEM).

Table 2. Strength grades and corresponding numerical codes [43].

Strength Grade	A	B	C	D	E	F	G	H	I	J
Numbers	1	2	3	4	5	6	7	8	9	10

4.7. Injectability Testing

The injectability of various plugging systems was evaluated by measuring the resistance factor to comprehensively assess their injection performance. Following the procedure outlined in Section 4.4, systems with different compositions (NBAP-B1, CNPA-B1, and NBNP-B1) were prepared. The influence of polymer dosage (NBAP, CNPA, and NBNP) on injectability was systematically investigated. Fifteen core samples, each with a length of 8.0 cm, diameter of 3.8 cm, and permeability of 10 mD, were selected for the experiment. The core samples were dried and subsequently loaded into a core holder. The flow apparatus was then assembled as illustrated in Figure 12. Deionized water was injected into the core at a flow rate of 0.1 mL/min until a stable pressure was attained, and the pressure difference (ΔP_w) was recorded. Subsequently, the plugging system was injected at the same flow rate until pressure stabilization, and the corresponding pressure difference (ΔP_b) was measured. The resistance factor for each system was calculated according to Equation (1).

$$F_r = \frac{\Delta P_b}{\Delta P_W} \cdot \frac{Q_W}{Q_b}. \tag{1}$$

The resistance factor F_r is dimensionless. ΔP_w and ΔP_b represent the pressure differences (in MPa) during the injection of water and the plugging system, respectively. Similarly, Q_w and Q_b denote the flow rates (in mL/min) of water and the plugging system, respectively.

Additionally, plugging systems with a fixed polymer dosage of 0.5 g were prepared. Core samples of identical dimensions but varying permeabilities (0.1, 0.5, 1, 5, 10, 50, and 100 mD) were selected to evaluate the influence of permeability on the injectability of the plugging systems using the same experimental procedure.

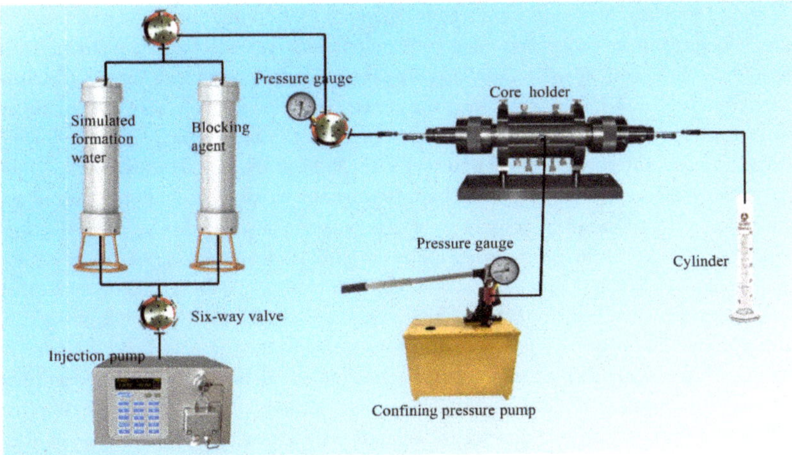

Figure 12. Schematic diagram of the fluid flooding apparatus.

4.8. Plugging Performance of NBAP-B1

The plugging performance of various systems was evaluated based on breakthrough pressure and plugging efficiency. Following the procedure described in Section 4.4, systems with different compositions (NBAP-B1, CNPA-B1, and NBNP-B1) were prepared. The effect of polymer dosage (NBAP, CNPA, and NBNP) on their injectability was also systematically investigated. Fifteen core samples, each measuring 8.0 cm in length and 3.8 cm in diameter with a uniform permeability of 10 mD, were selected for the experiment. After drying, the cores were loaded into a core holder. The flow apparatus was then connected according to the schematic shown in Figure 12. Deionized water was injected into the core at a constant flow rate of 0.1 mL/min until pressure stabilization was achieved. The injection pressure was recorded, and the initial permeability (K_a) was calculated; The plugging system was then injected at the same flow rate until a stable pressure was achieved. Following the injection, the core was transferred to an oven and maintained at 80 °C under static conditions to facilitate gel formation; Finally, deionized water was injected again at the same flow rate to measure the post-treatment permeability K_b. The plugging efficiency for each system was then calculated using Equation (2). Additionally, during the post-blocking water flooding process, the pressure at the inlet was recorded at the moment when the first water droplet appeared at the outlet. This pressure value was defined as the breakthrough pressure P_b.

$$\eta = \frac{K_a - K_b}{K_a} \times 100\% \cdot \quad (2)$$

η represents the plugging efficiency (%), while K_a and K_b denote the permeability before and after plugging (in mD), respectively. Additionally, plugging systems with a fixed polymer dosage of 0.5 g were prepared. Core samples with identical dimensions but varying permeabilities (0.1, 0.5, 1, 5, 10, 50, and 100 mD) were selected to evaluate the influence of permeability on the plugging efficiency using the same experimental procedure.

4.9. Flow and Oil Displacement Experiments

In the core flooding experiments, the change in oil displacement efficiency before and after the application of the plugging system was compared to evaluate its profile control and oil displacement performance. The NBAP-B1 system, which demonstrated the best plugging performance, was selected for this study with a consistent polymer concentration of 0.5 g. Oil displacement performance was analyzed following NBAP-B1 treatment. A heterogeneous artificial core was used, consisting of a high-permeability zone (10 mD) and a low-permeability zone (5 mD). Detailed core properties are provided in Table 3. The core was dried at 80 °C and then split along the center using a cutting machine to create an artificial fracture. The fracture was subsequently filled with sand and cemented. Following this, the core was saturated with oil and water to establish the initial oil saturation. First, vacuum the washed and dried core and apply pressure to saturate the formation water, making it reach a 100% water content state. Then, using simulated oil, slowly displace the water in the core at the reservoir temperature until no water is left. At this point, the initial water saturation and initial oil saturation of the core can be calculated by measuring the displaced water volume or by weighing. Finally, the core was mounted into a core holder, and the flow apparatus was connected according to the configuration illustrated in Figure 12. Deionized water was injected into the core at a flow rate of 0.1 mL/min until no more oil was produced at the outlet. Subsequently, the plugging system was injected at the same flow rate until the pressure stabilized. The core was then placed in an oven at 80 °C for static gelation; Finally, deionized water was injected again at the same flow rate until no additional oil was produced. The plugging efficiency of each system was then calculated using Equation (2). The brine used was NaCl-based simulated formation water with a salinity of 50,000 mg/L. The injection pressure and oil production were recorded throughout each stage of the process to evaluate the oil displacement performance of the different plugging systems.

Table 3. Core properties of oil displacement experiments.

Core Number	Diameter (cm)	Length (cm)	Permeability (mD)	Porosity (%)	Initial Oil Saturation (%)
LP-1	3.801	8.006	5/10	19.45	58.55

Author Contributions: Conceptualization, formal analysis, H.L. (Huaizhu Liu); resources, supervision, writing—review and editing, G.F.; methodology, investigation, formal analysis, funding acquisition, K.T.; writing—original draft, writing—review and editing, Z.Z.; methodology, formal analysis, D.J.; supervision, H.L. (Houyong Luo). All authors have read and agreed to the published version of the manuscript.

Funding: The researchers wish to express their gratitude for the assistance from the Shaanxi Provincial Postdoctoral Foundation (2024BSHSDZZ200) and the National Natural Science Foundation of China (52504033).

Data Availability Statement: The original contributions presented in this study are included in the article. Further inquiries can be directed to the corresponding authors.

Conflicts of Interest: Author Huaizhu Liu was employed by the company Tangshan Jiyou Ruifeng Chemical Limited Company. The remaining authors declare that the research was conducted in the absence of any commercial or financial relationships that could be construed as a potential conflict of interest.

Abbreviations

The following abbreviations are used in this manuscript:

FTIR	Fourier transform infrared spectroscopy
SEM	scanning electron microscopy
NBAP	nano bentonite/amphoteric polyacrylamide
CNAP	Amphoteric polymer
NBNP	Nano bentonite/anionic polyacrylamide
NPA	Anionic polymer
PV	Pore volume

References

1. Bai, Y.; Shang, X.; Wang, Z.; Zhao, X. Experimental study of low molecular weight polymer/nanoparticle dispersed gel for water plugging in fractures. *Colloids Surf. A Physicochem. Eng. Asp.* **2018**, *551*, 95–107. [CrossRef]
2. Hasankhani, G.M.; Madani, M.; Esmaeilzadeh, F.; Mowla, D. Experimental investigation of asphaltene-augmented gel polymer performance for water shut-off and enhancing oil recovery in fractured oil reservoirs. *J. Mol. Liq.* **2019**, *275*, 654–666. [CrossRef]
3. Nie, X.; Guo, H.; Yan, X.; Du, C.; Guo, L.; Jin, H. Analysis of water Plugging Mechanism of new gel plugging agent. In Proceedings of the IOP Conference Series: Earth and Environmental Science, Shenyang City, China, 14–15 November 2020; Volume 651, p. 32089.
4. Tong, L.; Fan, R.; Yang, S.; Zhang, Q.; Pan, Y. Experimental Study of Oilfield Water Plugging with Organic Chromium Gel. *Chem. Technol. Fuels Oils* **2022**, *57*, 991–999. [CrossRef]
5. Yan, X.; Chen, Q.; Zhu, L.; Chen, H.; Wei, D.; Chen, F.; Tang, Z.; Yang, J.; Zheng, J. High strength and self-healable gelatin/polyacrylamide double network hydrogels. *J. Mater. Chem. B* **2017**, *5*, 7683–7691. [CrossRef]
6. Helvacıoğlu, E.; Aydın, V.; Nugay, T.; Nugay, N.; Uluocak, B.G.; Şen, S. High strength poly(acrylamide)-clay hydrogels. *J. Polym. Res.* **2011**, *18*, 2341–2350. [CrossRef]
7. Fang, J.; Zhang, X.; Li, L.; Zhang, J.; Shi, X.; Hu, G. Research Progress of High-Temperature Resistant Functional Gel Materials and Their Application in Oil and Gas Drilling. *Gels* **2023**, *9*, 34. [CrossRef]
8. Yu, Z.; Li, Y.; Sha, O.; Su, Z.; Zhou, W. Synthesis and properties of amphiprotic polyacrylamide microspheres as water shutoff and profile control. *J. Appl. Polym. Sci.* **2016**, *133*, 43366. [CrossRef]
9. Cheng, C.; Liu, L. Development and Characterization of Biomimetic Profile Control and Water Plugging Polymer. *J. Appl. Polym. Sci.* **2025**, *142*, e57624. [CrossRef]
10. Cao, B.; Xie, K.; Lu, X.; Cao, W.; He, X.; Xiao, Z.; Zhang, Y.; Wang, X.; Su, C. Effect and mechanism of combined operation of profile modification and water shutoff with in-depth displacement in high-heterogeneity oil reservoirs. *Colloids Surf. A Physicochem. Eng. Asp.* **2021**, *631*, 127673. [CrossRef]
11. Gao, J.; Lu, X.; He, X.; Liu, J.; Zheng, K.; Cao, W.; Cui, T.; Sun, H. Effect Evaluation and Action Mechanism Analysis of "Profile Control + Plugging Removal" after Chemical Flooding. *Gels* **2022**, *8*, 396. [CrossRef]
12. Ji, W.; Chang, B.; Yu, H.; Li, Y.; Song, W. Effect of Polymer and Crosslinker Concentration on Static and Dynamic Gelation Behavior of Phenolic Resin Hydrogel. *Gels* **2024**, *10*, 325. [CrossRef] [PubMed]
13. Buyanov, A.L.; Gofman, I.V.; Saprykina, N.N. High-strength cellulose–polyacrylamide hydrogels: Mechanical behavior and structure depending on the type of cellulose. *J. Mech. Behav. Biomed. Mater.* **2019**, *100*, 103385. [CrossRef]
14. Zhang, L.; Yan, H.; Hao, H.; Xie, G.; Wei, A.; Wei, L.; Sun, L.; Wang, H. Chemical Fixed Point Sealing Technology in Oil Well. In Proceedings of the IOP Conference Series: Materials Science and Engineering, Shenzhen, China, 15–17 November 2019; Volume 729, p. 12023.
15. Fan, H.; Pu, W.; Du, D.; Xu, M.; Chen, L.; Xu, L.; Jin, F. Experimental study on the plugging and profile control of preformed particle gels in high temperature and salinity reservoir. *J. Appl. Polym. Sci.* **2023**, *140*, e54247. [CrossRef]
16. Zolfaghari, R.; Katbab, A.A.; Nabavizadeh, J.; Tabasi, R.Y.; Nejad, M.H. Preparation and characterization of nanocomposite hydrogels based on polyacrylamide for enhanced oil recovery applications. *J. Appl. Polym. Sci.* **2006**, *100*, 2096–2103. [CrossRef]
17. Fuqiing, Y.; Qun, Y.; Xingfeng, C. Advancements in Chemical Drive for High-Temperature and High-Salt Reservoirs in Shengli Oilfield to Improve Recovery Rate. *Pet. Geol. Recovery Effic.* **2024**, *31*, 199–208.
18. Xu, L.T. Chemical water shutoff profile research status and development trends. *IOP Conf. Ser. Earth Environ. Sci.* **2017**, *82*, 12061. [CrossRef]
19. Dai, C.; Xu, Z.; Wu, Y.; Zou, C.; Wu, X.; Wang, T.; Guo, X.; Zhao, M. Design and Study of a Novel Thermal-Resistant and Shear-Stable Amphoteric Polyacrylamide in High-Salinity Solution. *Polymers* **2017**, *9*, 296. [CrossRef]

20. Jie, C.; Hu, M.; Cheng, Y.; Li, P.; Zhang, H.; Zang, L.; Guo, J. Design of degradable hybrid dual-crosslinked polymer for high-efficiency profile control in high temperature reservoirs. *Colloids Surf. A Physicochem. Eng. Asp.* **2024**, *698*, 134530. [CrossRef]
21. Zhang, D.; Zhang, M.; Luo, Z.; Zhang, S.-X. Design, synthesis, and characterization of novel copolymer gel particles for water-plugging applications. *e-Polymers* **2024**, *24*, 20240016. [CrossRef]
22. Gu, C.; Lv, Y.; Fan, X.; Zhao, C.; Dai, C.; Zhao, G. Study on rheology and microstructure of phenolic resin cross-linked nonionic polyacrylamide (NPAM) gel for profile control and water shutoff treatments. *J. Pet. Sci. Eng.* **2018**, *169*, 546–552. [CrossRef]
23. Du, J.; Wang, Q.; Liu, P.; Xiong, G.; Chen, P.; Chen, X.; Liu, J. Nanocomposite gels for water shut-off and temporary plugging in the petroleum industry: A review. *Pet. Sci. Technol.* **2023**, *41*, 2204–2239. [CrossRef]
24. Asadizadeh, S.; Ayatollahi, S.; ZareNezhad, B. Performance evaluation of a new nanocomposite polymer gel for water shutoff in petroleum reservoirs. *J. Dispers. Sci. Technol.* **2019**, *40*, 1479–1487. [CrossRef]
25. Ran, Y.; Zhang, G.; Jiang, P.; Pei, H. Preparation Method and Performance Evaluation of a Gel Based on AM/AMPS Copolymer. *Gels* **2022**, *8*, 802. [CrossRef]
26. Yang, K.; Chen, J.; Fu, Q.; Dun, X.; Yao, C. Preparation of novel amphoteric polyacrylamide and its synergistic retention with cationic polymers. *E Polymers* **2020**, *20*, 162–170. [CrossRef]
27. Liu, D.; Shi, X.; Zhong, X.; Zhao, H.; Pei, C.; Zhu, T.; Zhang, F.; Shao, M.; Huo, G. Properties and plugging behaviors of smectite-superfine cement dispersion using as water shutoff in heavy oil reservoir. *Appl. Clay Sci.* **2017**, *147*, 160–167. [CrossRef]
28. Zhao, T.H.; Xing, J.Y.; Pu, W.F.; Dong, Z.; Yuan, C.; Peng, G.; Jin, F.; Xia, J. Synthesis and property evaluation of a novel polyacrylamide-montmorillonite composite for water shutoff and profile control in high salinity reservoirs. *Polym. Compos.* **2018**, *39*, 368–376. [CrossRef]
29. Peiyao, Q.; Shuo, W.; Ye, J.; Zhongli, T.; Jin, Q.; Junping, Z.; Xiaofei, L. Synthesis and Rheological Properties of Polyacrylamide Nanospheres and SiO_2 /PAM Composite Nanospheres. *ChemistrySelect* **2022**, *7*, e202200923. [CrossRef]
30. Cao, M.; Cai, Q.; Guo, G.; Guo, H.; Chen, Y.; Zhang, Y. The preparation and displacement performances of a hollow structure microsphere with swelling–deswelling properties for enhanced oil recovery (EOR). *Polym. Bull.* **2022**, *79*, 4323–4337. [CrossRef]
31. Dai, X.; Wang, J.; Teng, F.; Shao, Z.; Huang, X. Zr(IV)-Crosslinked Polyacrylamide/Polyanionic Cellulose Composite Hydrogels with High Strength and Unique Acid Resistance. *J. Polym. Sci. Part B Polym. Phys.* **2019**, *57*, 981–991. [CrossRef]
32. Cao, W.; Xie, K.; Wang, X.; Lu, X.; He, X.; Xu, G.; Li, X. Starch graft copolymer and polymer gel applied in Bohai oilfield for water plugging and profile control and their mechanisms. *Geosyst. Eng.* **2020**, *23*, 197–204. [CrossRef]
33. Li, X.; Yang, Y.; Wang, Q.; Gao, Y.; Zhang, W.; Liu, Z. Study on the water-tree ageing characteristics of polyethylene/organic montmorillonite and crosslinked polyethylene/organic montmorillonite nanocomposites. *High Volt.* **2023**, *8*, 262–273. [CrossRef]
34. Li, X.; Fu, M.; Liu, J.; Xiao, Q.; Tang, W.; Yang, G. Synthesis and Performance Evaluation of a Novel Heat and Salt-Resistant Gel Plugging Agent. *Polymers* **2022**, *14*, 3894. [CrossRef]
35. Lu, S.; Bo, Q.; Zhao, G.; Shaikh, A.; Dai, C. Recent advances in enhanced polymer gels for profile control and water shutoff: A review. *Front. Chem.* **2023**, *11*, 1067094. [CrossRef]
36. Pei, G. Research and application of profile control and water shutoff technology in fractured sandstone reservoir based on system analysis method. In Proceedings of the IOP Conference Series: Earth and Environmental Science, Dalian, China, 26–28 March 2021; Volume 784, p. 12037.
37. Quan, H.; Li, Z.; Huang, Z. Self-assembly properties of a temperature- and salt-tolerant amphoteric hydrophobically associating polyacrylamide. *RSC Adv.* **2016**, *6*, 49281–49288. [CrossRef]
38. Haruna, M.A.; Wen, D. Stabilization of Polymer Nanocomposites in High-Temperature and High-Salinity Brines. *ACS Omega* **2019**, *4*, 11631–11641. [CrossRef]
39. Song, Z.; Hu, J.; Liu, P.; Sun, Y. Synthesis and Performance Evaluation of Alginate-Coated Temperature-Sensitive Polymer Gel Microspheres. *Gels* **2023**, *9*, 480. [CrossRef]
40. Chen, S.; Xie, J.; Yang, Z. Effect of reactive montmorillonite with amino on the properties of polyimide/montmorillonite nanocomposite. *Polym. Bull.* **2020**, *77*, 687–699. [CrossRef]
41. Almeida, A.I.A.D.; Carvalho, L.D.O.; Lopes, R.C.F.G.; Sena, L.E.; Amparo, S.Z.D.; de Oliveira, C.P.; Calado, H.D.; Silva, G.G.; de Vasconcelos, C.K.; Viana, M.M. Enhanced polyacrylamide polymer hydrogels using nanomaterials for water shutoff: Morphology, thermal and rheological investigations at high temperatures and salinity. *J. Mol. Liq.* **2024**, *405*, 125041. [CrossRef]
42. Li, L.; Guo, J.; Kang, C. LCST-UCST Transition Property of a Novel Retarding Swelling and Thermosensitive Particle Gel. *Materials* **2023**, *16*, 2761. [CrossRef] [PubMed]
43. Pu, W.; Tian, K.; Wang, H. Synthesis of a low viscosity amphoteric polyacrylamide and its application in oilfield profile control. *J. Appl. Polym. Sci.* **2022**, *139*, e52981. [CrossRef]

gels

MDPI

Article

Experimental Investigation of Deformable Gel Particles (DGPs) for Plugging Pan-Connected Interlayer Channels in High-Water-Cut Reservoirs

Wenjing Zhao [1,2], Jing Wang [1,*], Tianjiang Wu [2], Ronald Omara Erik [1], Zhongyang Qi [1] and Huiqing Liu [1]

1 State Key Laboratory of Petroleum Resources and Prospecting, China University of Petroleum, Beijing 102249, China; zhaowenjing1_cq@petrochina.com.cn (W.Z.); ronalderikomara@yahoo.com (R.O.E.); 2024310198@student.cpu.edu.cn (Z.Q.); liuhq@cup.edu.cn (H.L.)

2 Oil and Gas Technology Research Institute of Changqing Oilfield, China National Petroleum Corporation, Xi'an 710018, China; wutianjiang_cq@petrochina.com.cn

* Correspondence: wangj@cup.edu.cn; Tel.: +86-15502917118

Abstract: Pan-connected interlayers are widely present in oil reservoirs, forming flow channels at different positions. However, conventional profile control agents struggle to plug deep interlayer channels in reservoirs, limiting the swept volume of injected water. Additionally, a clear methodology for physically simulating pan-connected reservoirs with interlayer channels and calculating interchannel flow rates remains lacking. In this study, a physical model of pan-connected interlayer reservoirs was constructed to carry out deformable gel particles (DGPs) plugging experiments on interlayer channels. A mass conservation-based flow rate calculation method for interlayer channels with iterative solution was proposed, revealing the variation law of interlayer channel flow rates during DGP injection and subsequent water flooding. Finally, oil displacement and DGP profile control experiments in pan-connected interlayer reservoirs were conducted. The study shows that during DGP injection, injected water enters the potential layer through interlayer channels in the middle and front of the water-channeling layer and bypasses back to the water-channeling layer through channels near the production well. With the increase in DGP injection volume, the flow rate of each channel increases. During subsequent water flooding, DGP breakage leads to a rapid decline in its along-path plugging capability, so water bypasses back to the water-channeling layer from the potential layer through all interlayer channels. As the DGP injection volume increases, the flow rate of each channel decreases. Large-volume DGPs can regulate interlayer channeling reservoirs in the high water cut stage. Its effectiveness mechanism involves particle migration increasing the interlayer pressure difference, which drives injected water to sweep from the water-channeling layer to the potential layer through interlayer channels, improving oil recovery by 19.74%. The flow characteristics of interlayer channels during DGP injection play a positive role in oil displacement, so the oil recovery degree in this process is greater than that in the subsequent water flooding stage under each injection volume condition. The core objective of this study is to investigate the plugging mechanism of DGPs in pan-connected interlayer channels of high-water-cut reservoirs, establish a method to quantify interlayer flow rates, and reveal how DGPs regulate flow redistribution to enhance oil recovery.

Keywords: deformable gel particles; interlayer flow channel; partially developed interlayer; physical simulation

1. Introduction

The interlayer plays a crucial role in determining the distribution and migration of oil and water, as emphasized by recent studies [1,2]. During the combined production process in multi-layer heterogeneous reservoirs, an underdeveloped interlayer can lead to the formation of interlayer channels (ICs). This, in turn, can significantly impact the efficiency of the overall reservoir development. Most studies have investigated interlaminar flow via numerical simulations and theoretical models, but these methods insufficiently capture interlayer flow dynamics [3–5]. Theoretical findings are particularly difficult to validate experimentally because recreating and observing interlayer channels under laboratory conditions poses significant challenges. As a result, experimental studies on the behavior of heterogeneous reservoirs in the presence of interlayer flow are relatively scarce.

For instance, Farshad Rezaeiakmal [6] used a visualization model to explore the influence of interlayer flow channels on foam flooding. Zhang [7] applied the Arqi method to compare combined production with layer-by-layer development, revealing that the latter was ineffective in enhancing interlayer flow in deeper reservoir sections. In another study, Lu et al. [8] used a small layer suction metering device to examine the impact of interlayer flow channels on the efficiency of water flooding, considering the variations in liquid volume within the intermediate container. However, this experiment was limited by the inconsistent sensitivity of the piston within the intermediate container, which likely compromised the accuracy of the results. In summary, existing research has not yet provided precise methods for quantitatively characterizing interlayer flow dynamics. Thus, a key challenge for future research lies in developing accurate empirical techniques for analyzing interlayer flow behavior.

The profile control in heterogeneous reservoirs becomes even more challenging during the high water cut stage due to the presence of interlayer channels. These channels not only reduce the effectiveness of profile control but also complicate water shutoff efforts. When a profile control agent is injected, subsequent water flooding can occur. However, the injected water may flow back through the interlayer channels to the dominant seepage paths or water channeling layers, thereby reducing the effectiveness of profile control. To mitigate this issue, it is essential to plug interlayer channels as effectively as possible for optimal reservoir management. But there is limited research focused on profile control strategies for heterogeneous reservoirs with interlayer channels during high water cut conditions. Currently, many studies rely on control agents designed to provide continuous medium profiles [9–11]. However, these agents have limitations, particularly in deep migration within the reservoir. Disadvantages include their susceptibility to viscous fingering in subsequent water slugs [12] and their limited effectiveness near the injection well. As a result, water can bypass the control agent and return to the water channeling layer. Consequently, more effective profile control methods are urgently needed for heterogeneous reservoirs with interlayer channels, particularly during the high water cut stage.

Deformable gel particles (DGPs) possess elastic characteristics that enable deep transport within the reservoir. Compared to other profile control agents, DGPs offer distinct advantages for deep profile control, as demonstrated in laboratory experiments [13,14] and field applications [15–19]. In theory, DGPs are better suited for reservoirs with deep interlayer channels compared to other control agents. Elsharafi and Bai [20] proposed that DGPs are particularly effective in heterogeneous reservoirs without interlayers due to their minimal impact on formation damage. Yang et al. [21] suggested using different particle positions to achieve plugging within interlayer channels. However, their evaluation was limited to friction flow as a single parameter, which may not fully account for the flow dynamics in multiple interlayer channels. In contrast, Ji et al. [22] conducted research using acid particles to block gas channeling in reservoirs without interlayers. The study revealed that DGPs effectively enhanced oil recovery in reservoirs with interconnected layers, offering insights into their potential for optimizing oil recovery in similar reservoirs. Interlayer channels vary within a reservoir, making it essential to understand the deep migration of DGPs for effective reservoir regulation [21]. Current research has mainly focused on the compatibility between DGPs and reservoirs [23,24], as well as their working mechanisms [25,26].

In this study, sandpacks were designed as physical models with multiple measuring points to capture various flow channels. By combining this design with flow calculations, we were able to identify the plugging patterns when DGPs were used. The findings of this study offer valuable insights into the impact of different factors on the recovery efficiency of interconnected reservoirs with multiple layers. These results have significant implications for the effective control of injection profiles in heterogeneous reservoirs with interlayer flow channels during the high water cut stage and provide essential guidance for future research in this area.

2. Results and Discussion

2.1. Interlayer Flow Dynamics During DGP Injection

2.1.1. Influence of Diameter Ratio of DGP to Pore Throat on Interlayer Flow Dynamics During DGP Injection

Based on the experimental results from Groups 1 to 3 in Table 1, the flow rate variations in the interlayer channels during DGP injection were calculated using the method described earlier, as shown in Figure 1. An analysis of the flow directions in Figure 1a reveals that, regardless of the diameter ratio of DGP to pore throat, the flow direction in the interlayer channel near the production well (interlayer channel-3) consistently moves from the low-permeability layer to the high-permeability layer. This behavior suggests that particle breakage is a significant transport mechanism for DGPs [27]. It can be inferred that the weaker effect of particles near the production well, due to particle breakage and percolation, leads to lower sealing strength in this region, which explains the observed flow characteristics in the interlayer channel near the production well. In contrast, the increased flow resistance in the first half of the high-permeability layer after particle injection results in a higher interlayer pressure difference, driving flow from the high-permeability layer to the low-permeability layer. As shown in Figure 1b, in an interlayer-connected reservoir with a barrier layer, the overall flow direction in the interlayer channels shifts from the high-permeability layer to the low-permeability layer, indicating that flow rates in channels 1 and 2 dominate. The particles have better mobility in the high-permeability layer when the diameter ratio of DGP to pore throat is small, leading to a smaller interlayer pressure difference. Under these conditions, the flow rates in the three interlayer channels are the lowest compared to other diameter ratios. Although small particles possess strong migration capabilities (allowing them to pass through pore throats directly), their weaker action strength results in more particles being released from the high-permeability layer.

Conversely, a larger diameter ratio causes most of particles to be retained at the injection end of the porous media, forming a filter cake. This filter cake can impede subsequent water injection. When the diameter ratio of DGP to pore throat is 2.51, the particles can deform and break, allowing them to penetrate deeper into the high-permeability layer, which increases flow resistance. As a result, the pressure differences across various parts of the porous media become larger and more uniform under these conditions. This leads to the largest interlayer pressure difference and flow rate, with the total interlayer flow rate (in absolute terms) reaching 1.26 mL/min.

Table 1. Plugging experimental parameters of interlayer pan-connected reservoir.

No.	Permeability, μm^2	Porosity	Injection Rate, mL/min	Diameter of Pore, μm	Diameter Ratio of DGP to Pore	Cumulative Injection Volume, PV
1	32.01	0.35	3	162.30	1.83	3
	0.87	0.28		29.91	9.93	
2	30.69	0.30		171.65	2.51	
	0.79	0.27		29.03	16.99	
3	30.91	0.31		169.46	3.64	
	0.82	0.28		29.04	21.24	
4	30.97	0.30	1	172.43	2.43	3
	0.79	0.29		28.01	17.39	
5	32.1	0.34	5	164.90	2.58	
	0.89	0.27		30.81	15.93	
6	31.08	0.32	3	167.25	2.52	1
	0.78	0.28		28.32	17.22	
7	29.93	0.30		169.51	2.43	5
	0.72	0.29		26.74	17.92	

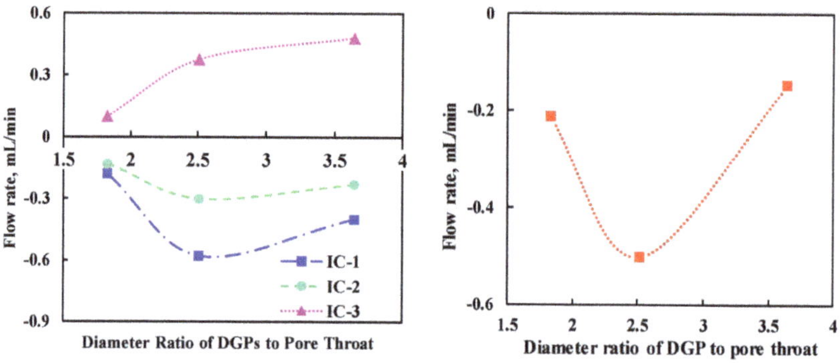

(a) (b)

Figure 1. Variation of interlayer channel flow during DGP injection under different diameter ratio of DGP to pore throat. (**a**) flow rate variation of interlayer channels. (**b**) variation of total interlayer flow rate.

2.1.2. Influence of Injection Rate on Interlayer Flow Dynamics During DGP Injection

The changes in the interlayer flow rate, calculated from the experimental results of Groups 2, 4, and 5 in Table 1, are presented in Figure 2. Since the injection rate is the primary variable in this part, the absolute flow rate alone does not adequately capture the interlayer flow behavior. Therefore, the interlayer flow rate is normalized by dividing it by the corresponding injection rate, as shown in Figure 2c. Similar to the flow direction observed during DGP injection under varying diameter ratios of DGP to pore throats, the flow pattern under different injection rates reveals that only the interlayer channel near the production well flows from the low-permeability layer to the high-permeability layer. In contrast, the other two channels exhibit flow from the high-permeability layer to the low-permeability layer due to the pressure differences. Building on the analysis of particle plugging, crushing, and migration patterns under different injection rates presented in previous research [27], it can be concluded that when DGPs are injected at a low rate (1 mL/min), shear crushing is less frequent, and particles primarily migrate through deformation within the porous media. This results in a more uniform effect in the high-permeability layer. However, this also leads to smaller interlayer pressure differences and lower interlayer flow rates during the injection process, with the total interlayer flow rate reaching only 0.35 mL/min. This has a minimal impact on the mobilization of remaining oil in the low-permeability layer during particle injection.

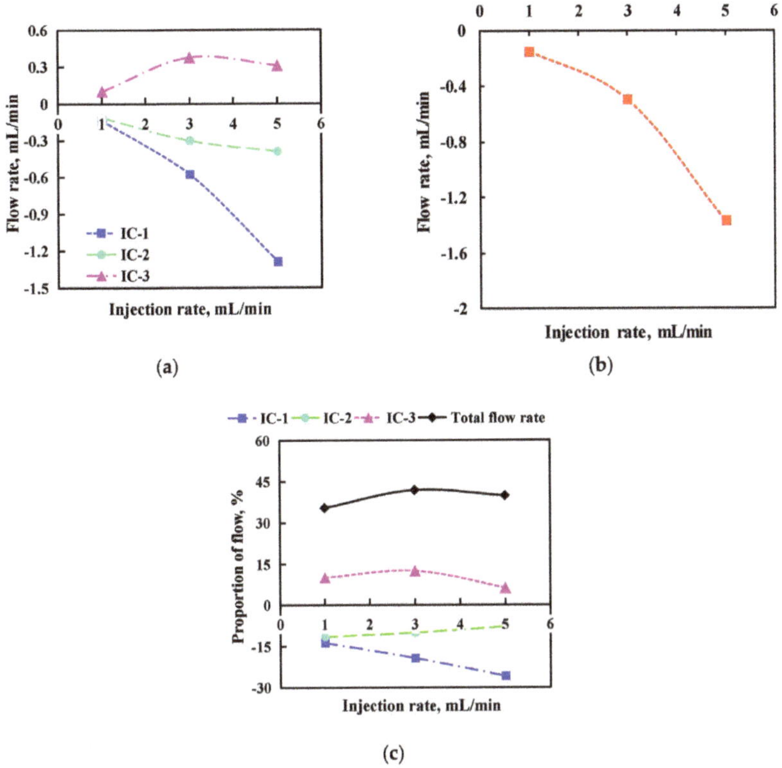

Figure 2. Variation of interlayer channel flow during DGP injection under different injection rate. (**a**) Flow rate variation of interlayer channels; (**b**) variation in total interlayer flow rate. (**c**) Proportion of flow rate of each interlayer channel.

Conversely, high-rate DGP injection results in an increase in interlayer pressure differences at the front of the porous media, which leads to higher interlayer flow rates. When the injection rate is 5 mL/min, the total flow rate increases to 1.99 mL/min, accounting for 39.83% of the total injected volume. However, high-rate injection also leads to significant particle crushing, which reduces the effectiveness of particles in deeper reservoir regions. As a result, a large volume of water flows from the high-permeability layer to the low-permeability layer through the first two interlayer channels, causing the flow rate (0.31 mL/min) in the channel near the production well to be lower than in the other two channels.

2.1.3. Influence of Injection Volume on Interlayer Flow Dynamics During DGP Injection

The study of flow rate variations in interlayer channels, influenced by the factors discussed previously, reveals that interlayer pressure difference is a key determinant of interlayer flow. This makes the injection volume of DGPs a critical factor influencing their plugging strength in deep layers. Figure 3 illustrates the changes in interlayer channel flow rates, calculated based on the experimental results from Groups 2, 6, and 7 in Table 1. The analysis shows that as the cumulative injection volume increases from 1 PV to 3 PV, the absolute interlayer flow rate rises from 0.99 mL/min to 4.99 mL/min. However, the growth rate of the interlayer flow rate gradually diminishes, and further increases in injection volume do not reverse the flow direction in interlayer channel-3. Instead, the pressure difference increases, resulting in higher flow rates in this channel. This suggests that increasing the injection volume can enhance the sweep efficiency of injected water to some extent. However, the associated costs and limitations of injection equipment with high volumes must also be considered for field applications.

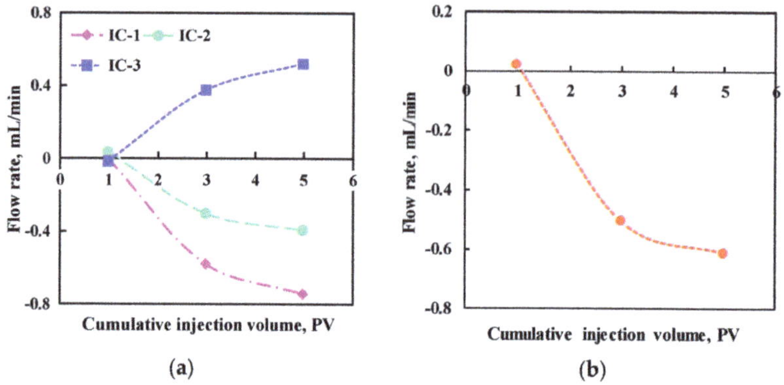

Figure 3. Variation of each interlayer channel flow during DGP injection under different injection volume. (**a**) Flow rate variation of interlayer channels. (**b**) Variation in total interlayer flow rate.

2.2. Interlayer Flow Dynamics During Subsequent Water Injection

One of the primary objectives of profile control is to block the channeling paths, enabling subsequent water to reach the potential areas and mobilize remaining oil. Therefore, this section investigates the flow behavior of injected water between layers following the plugging of channels with DGPs in interlayer pan-connected reservoirs, based on the experimental results.

2.2.1. Influence of Diameter Ratio of DGP to Pore Throat on Interlayer Flow Dynamics During Subsequent Water Injection

A filter cake forms at the entrance of the low-permeability layer after DGP injection [27]. Simultaneously, some particles invade the reservoir due to DGP breakage, resulting in a startup pressure gradient when the low-permeability layer is subsequently injected with water. The median diameter (d50) of the injected DGPs is larger than the median pore throat diameter of the high-permeability layer, which increases the plugging strength at the front end of the high-permeability layer for each diameter ratio. As a result, the subsequent water mainly enters the low-permeability layer. As shown in Figure 4, regardless of the size of the injected DGPs, water flows from the low-permeability layer to the high-permeability layer across all interlayer channels. A larger interlayer flow rate indicates a weaker plugging strength of the particles in the high-permeability layer, allowing subsequent water to enter the high-permeability layer with relatively weak seepage resistance.

It can be concluded that the plugging strength near the injection well in the high-permeability layer is highest for each diameter ratio, resulting in the smallest flow rate in the interlayer channel near the injection well. The plugging effect of particles with a diameter ratio of 2.51 is more uniform within the high-permeability layer, leading to smaller flow rates in all interlayer channels. The total flow rate in this case is 1.32 mL/min, which is less than half of the injected flow rate.

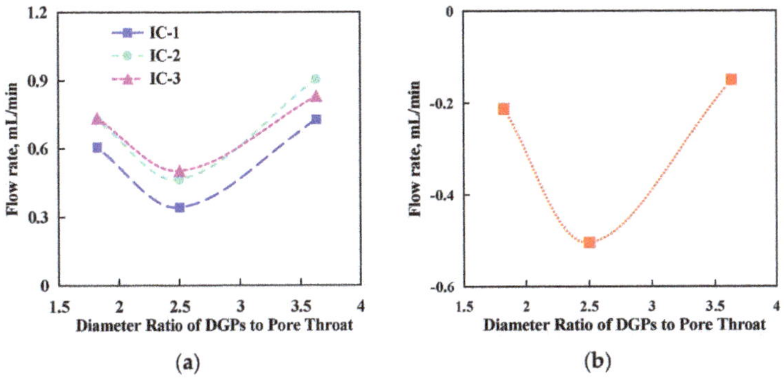

Figure 4. Relationship between interlayer channel flow rate and diameter ratio of DGP to pore throat during subsequent water injection. (**a**) Flow rate variation of interlayer channels. (**b**) Variation in total interlayer flow rate.

2.2.2. Influence of Injection Rate on Interlayer Flow Dynamics During Subsequent Water Injection

The relationship between interlayer channel flow rates and injection rates during subsequent water flooding is shown in Figure 5. Analysis reveals that, similar to the flow directions observed under different particle size-to-pore throat ratios, all interlayer channel flow rates during subsequent water injection are positive, indicating a flow direction from the low-permeability layer to the high-permeability layer. As shown in Figure 5c, the flow rates in the interlayer channels near the injection well (interlayer channel-1) and deeper within the reservoir (interlayer channel-2) increase with increasing injection rates. However, the flow rate in the interlayer channel near the production well (interlayer channel-3) accounts for 16.85% at an injection rate of 3 mL/min, slightly lower than 18.07% at 1 mL/min and 19.78% at 5 mL/min. This variation is due to the fact that at low injection rates, the driving force is insufficient, resulting in weaker deep migration ability, despite less particle crushing and migration. Conversely, at high injection rates, DGPs undergo intense shear and crushing in the front part of the high-permeability layer, weakening the sealing ability of the deep reservoir. Therefore, under both injection rates, the flow rates in the interlayer channel near the production well remain less than 3 mL/min.

2.2.3. Influence of Injection Volume on Interlayer Flow Dynamics During Subsequent Water Injection

As shown in Figure 6, with an increasing injection volume, the action intensity of DGPs along the water channeling layer also increases, resulting in more injected water entering the low-permeability layer. Additionally, the seepage resistance in the low-permeability layer is smaller than that in the water channeling layer after the action of DGPs. Specifically, the flow rates of each channel and the total flow rate decrease with increasing injection volume (Figure 6a). After injecting 5 PV of DGPs and conducting subsequent water flooding, the interlayer flow rate can be reduced to 29.6% of the injection rate. Based on the findings from the deep sealing law study, it is observed that the front part of the porous media is highly influenced by the injection volume. Consequently, the flow rate in interlayer channel-1 decreases most rapidly with increasing injection volume. However, when the injection volume is 1 PV, the action intensity in the front part is insufficient to seal interlayer channel-1, resulting in the largest flow rate in this channel. The subsequently injected water mainly bypasses this channel and returns to the water channeling layer. When the cumulative injection volume reaches 3 PV, the seepage resistance in the front part of the water channeling layer becomes sufficiently large, effectively sealing channel-1.

As a result, the flow rate in interlayer channel-1 is relatively small, and the injected water bypasses back to the potential layer through interlayer channel-2. Consequently, the flow rate in channel-2 is higher than in the other two channels. Due to the crushing effect, the residual resistance near the production well in the porous media remains small across various injection volumes, leading to the smallest flow rate change in channel-3.

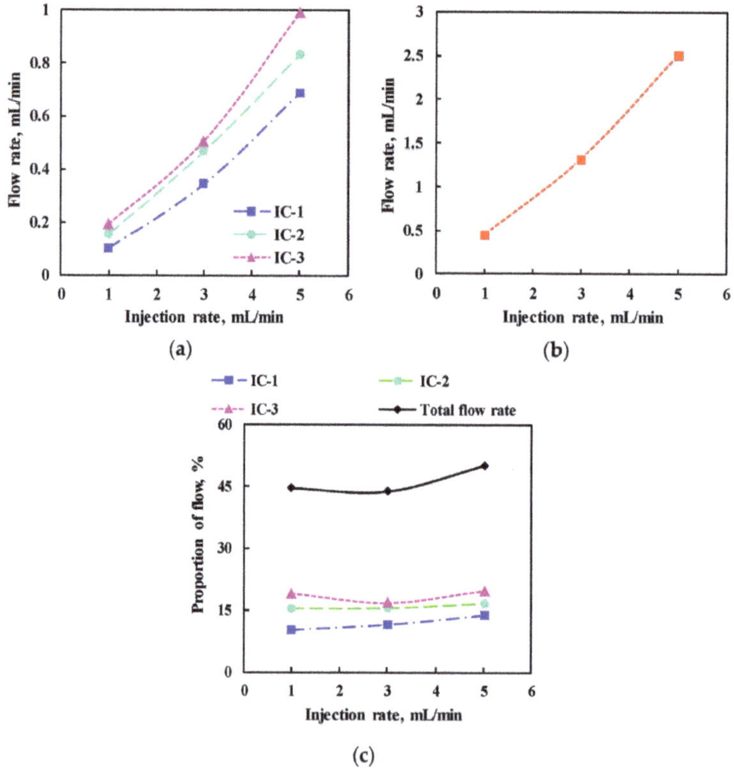

Figure 5. Relationship between interlayer channel flow rate and DGP injection rate during subsequent water injection. (**a**) Flow rate variation of interlayer channels. (**b**) Variation of total interlayer flow rate. (**c**) Proportion of flow rate of each interlayer channel.

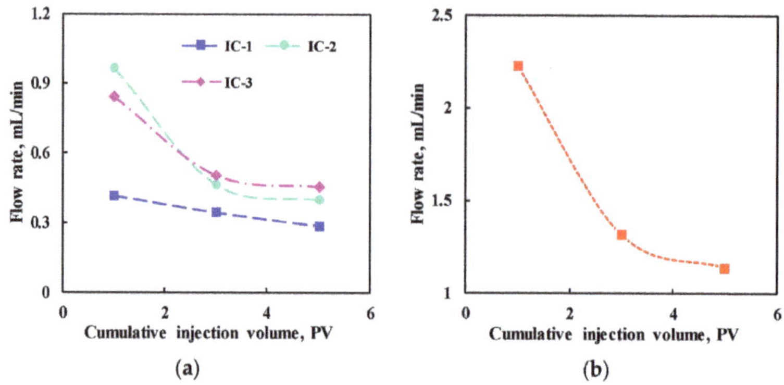

Figure 6. Relationship between flow rate of interlayer channel and DGP injection volume during subsequent water injection. (**a**) Flow rate variation of interlayer channels. (**b**) Variation in total interlayer flow rate.

2.3. Characteristics of Enhanced Oil Recovery by DGP

2.3.1. Influence of Diameter Ratio of DGP to Pore Throat on Interlayer Flow and Oil Recovery

As shown in Figure 7a,b, the improvement in oil recovery for each layer during DGP injection and subsequent water flooding is analyzed based on experiments 1 to 3 in Table 2. In both stages, the recovery degree of the high-permeability layer is generally higher than that of the low-permeability layer. However, this does not fully reflect the ability of DGPs to enhance oil recovery in the remaining oil of the high-permeability layer during the injection stage. This discrepancy is primarily due to water carrying oil from the potential layer back to the water channeling layer during the particle injection stage, where it is subsequently produced from the water channeling layer.

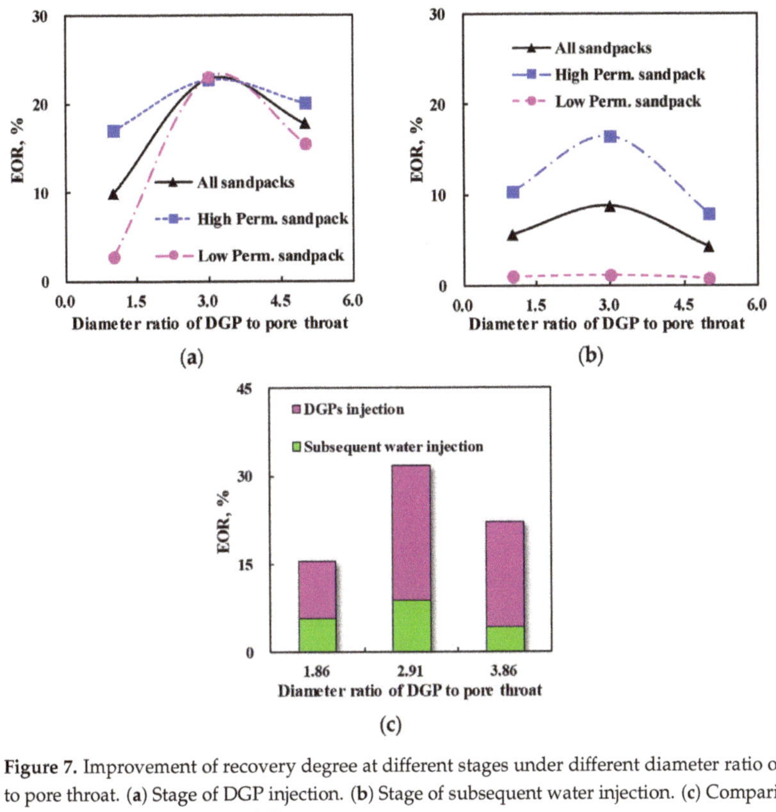

Figure 7. Improvement of recovery degree at different stages under different diameter ratio of DGP to pore throat. (**a**) Stage of DGP injection. (**b**) Stage of subsequent water injection. (**c**) Comparison of enhanced recovery degree in different stages.

As depicted in Figure 7c, the oil recovery improvement during the injection stage of DGPs is significantly higher than during the subsequent water injection stage, regardless of the particle size to pore throat ratio. For instance, when the particle size to pore throat ratio is 2.51, the oil recovery improves by 22.83% during the DGP injection stage, whereas it only improves by 8.83% during the subsequent water flooding stage. In the application and research of DGPs, scholars widely recognize their profile control and oil displacement functions. The enhanced oil recovery characteristics mentioned above are direct manifestations of these functions.

When the diameter ratio is small, the interlayer pressure difference and flow rate during DGP injection are low, resulting in minimal oil recovery improvement during this stage—only 9.85%. In contrast, when the diameter ratio is large, DGPs primarily work in the front part of the high-permeability layer, causing the subsequent water to bypass and return to the high-permeability layer through interlayer channels in the middle of the reservoir. As a result, oil recovery in this stage is the lowest, reaching only 4.35%. When the diameter ratio is 2.51, the final oil recovery in the high-permeability layer reaches 82.19%, which is notably higher than other diameter ratios but still inconsistent with realistic recovery values observed in real reservoirs. This discrepancy arises from interlayer flow, where residual oil from the low-permeability layer is displaced into the high-permeability layer under water flooding and subsequently produced at all stages. In summary, the optimal control effect is achieved when the diameter ratio is 2.51.

Table 2. Experimental parameters for improving recovery degree by DGPs in interlayer pan-connected reservoir.

No.	Permeability, μm^2	Porosity	Injection Rate, mL/min	Diameter of Pore, μm	Diameter Ratio of DGP to Pore	Initial Oil Saturation	Cumulative Injection Volume, PV
1	31.43	0.34	3	163.17	1.86	0.69	3
	0.85	0.27		30.11	10.09	0.61	
2	30.88	0.33		164.16	2.60	0.70	
	0.79	0.29		28.01	17.83	0.69	
3	29.98	0.33		161.75	3.86	0.72	
	0.83	0.26		30.32	20.58	0.60	
4	30.07	0.31	1	167.14	2.78	0.69	3
	0.77	0.27		28.66	10.36	0.58	
5	31.85	0.32	5	168.93	2.63	0.72	
	0.82	0.29		28.54	21.62	0.60	
6	28.96	0.31	3	164.03	2.57	0.71	1
	0.78	0.27		28.84	16.91	0.60	
7	31.74	0.31		171.72	2.39	0.63	5
	0.80	0.29		28.19	17.01	0.74	

2.3.2. Characteristics of Enhanced Oil Recovery Under the Influence of Injection Rate

Based on the experiments in Groups 2, 4, and 5 in Table 2, the improvement in oil recovery for each layer during the two stages with varying injection rates is presented in Figure 8. When the injection rate is 3 mL/min, the oil recovery for each layer is highest, with an overall improvement of 22.83% for the entire reservoir. In contrast, under lower injection rates, the overall improvement in oil recovery is only 17.35%.

The improvement in oil recovery during the subsequent water flooding stage is influenced by the blocking strength and penetration depth of particles in the high-permeability layer. Higher injection rates lead to shear breakage in the middle and front parts, while lower injection rates result in insufficient driving force. However, at an injection rate of 1 mL/min, more particles pass through by deformation, leading to higher blocking strength in the high-permeability layer compared to high-rate injection (5 mL/min). Additionally, due to the poor oil displacement efficiency during the injection of deformable gel particles at low injection rates, the residual oil content in the low-permeability layer remains higher. Consequently, the overall oil recovery during the subsequent water flooding stage at a low injection rate reaches 9.57%, whereas it is only 5.01% when the injection rate is 5 mL/min.

2.3.3. Characteristics of Enhanced Oil Recovery Under the Influence of Injection Volume

Based on the experimental results of Groups 2, 6, and 7 in Table 2, the characteristics of enhanced oil recovery (EOR) under different injection volumes are presented in Figure 9. The analysis indicates that, across all three injection volumes, the recovery of oil in the high-permeability layer is generally higher than that in the low-permeability layer. This is primarily due to the interlayer channels near the production well consistently facilitating the flow from the low-permeability layer to the high-permeability layer, which results in the residual oil from the low-permeability layer being mobilized into the high-permeability layer and subsequently recovered.

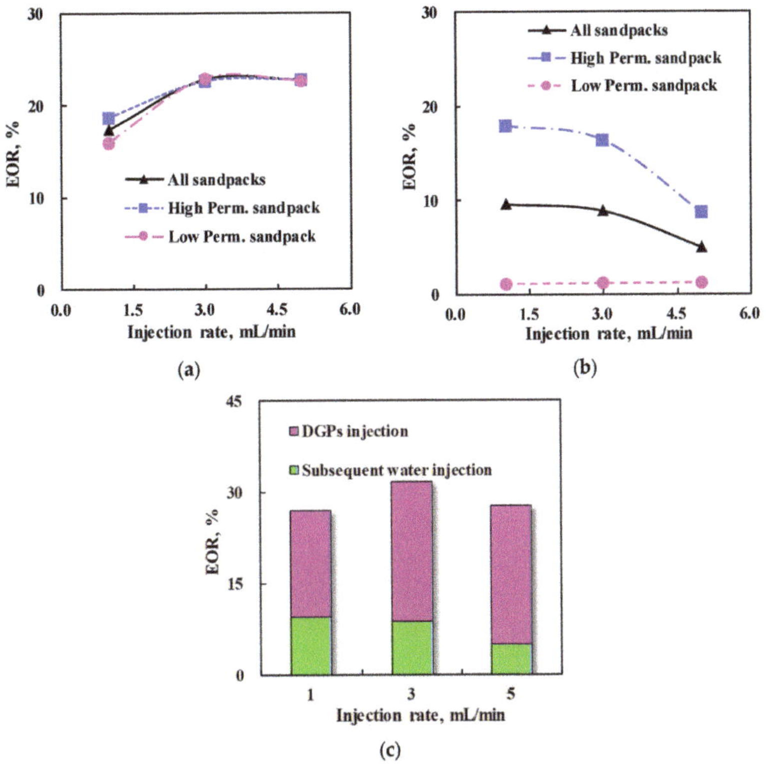

Figure 8. Improvement of recovery degree at different stages under different DGP injection rate. (**a**) Stage of DGP injection. (**b**) Stage of subsequent water injection. (**c**) Comparison of enhanced recovery degree in different stages.

During the injection of DGPs, channels 1 and 2 continuously allowed water from the suspension to enter the potential layers, significantly enhancing oil displacement. However, during subsequent water flooding, injected water consistently bypassed through all three channels, returning to the high-permeability layer. Therefore, when compared to experimental results without interlayer channels [14], the injection of DGPs led to a marked improvement in oil recovery, surpassing the enhancement observed during the subsequent water flooding stage across all injection volumes. The most significant difference was observed when the cumulative injection volume reached 5 PV. This was mainly due to two factors: first, after 5 PV of injection, less residual oil remained during subsequent water flooding, and second, the residual oil was primarily concentrated near the production well in the potential layers. However, due to the relatively weak effect of DGPs in this region and the persistent existence of the final interlayer channel, it was difficult to mobilize the remaining oil from this area.

The enhanced recovery degree of each layer after the injection of DGPs with varying diameter ratio and subsequent water flooding is shown in Figure 7. In both stages, the recovery degree of the high-permeability layer is generally higher than that of the low-permeability layer at each diameter ratio. However, this does not fully reflect the ability of DGPs to improve the recovery of residual oil in the high-permeability layer during the DGP injection stage. A significant factor contributing to this is the flow direction of DGPs during injection. As illustrated in Figure 9, the interlayer pressure difference during DGP injection drives the flow from the high-permeability layer to the low-permeability layer, with water transporting residual oil from the low-permeability layer into the high-permeability layer.

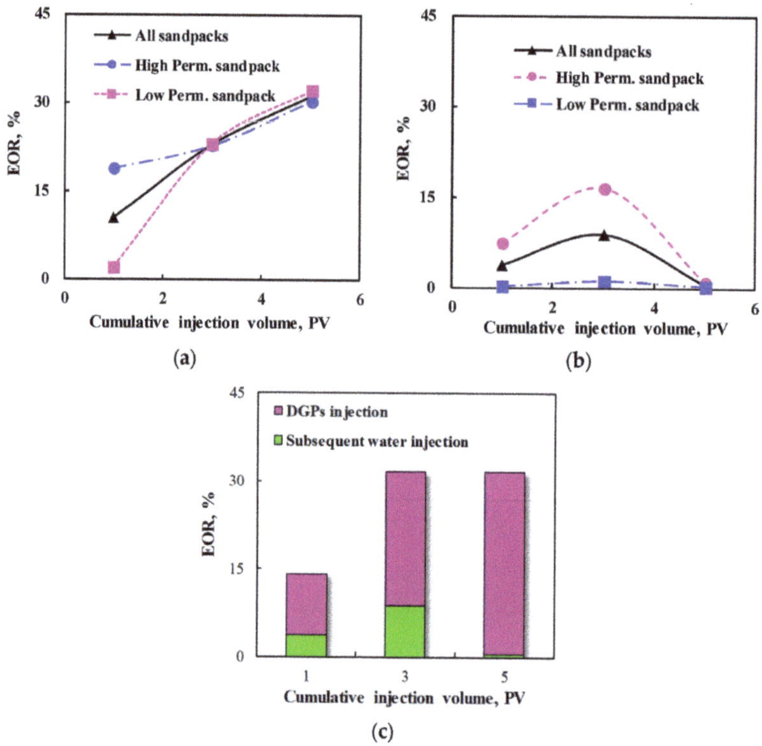

Figure 9. Improvement of recovery degree at different stages under different DGP injection volume. (**a**) Stage of DGP injection. (**b**) Stage of subsequent water injection. (**c**) Comparison of enhanced recovery degree in different stages.

From the results of the three experiments, it is evident that DGPs can effectively control interlayer pan-connected reservoirs in the high-water-cut stage. As shown in Figure 10a, interlayer channel-1 and interlayer channel-2 always flow from the high-permeability layer to the low-permeability layer, with the fluid then returning to the high-permeability layer through IC-3 during the DGP injection stage. In Figure 10b, the flow direction of all interlayer channels during the subsequent water flooding stage is from the low-permeability layer to the high-permeability layer.

Based on the recovery improvement characteristics, the control mechanism of DGPs in interlayer pan-connected reservoirs during the high-water-cut stage can be summarized as follows: during the DGP injection stage, the pressure in the water channeling layer is significantly higher than in the remaining oil enrichment layer due to deep migration. This causes the water in the particle system to enter the potential layers, with most of the fluid then flowing to the water channeling layer and being recovered from this layer after mobilizing the remaining oil from the potential layer.

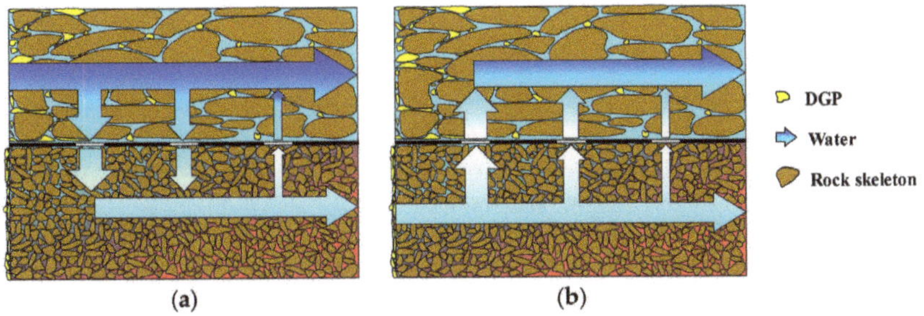

Figure 10. Comparison of interlayer channel flow direction. (**a**) Interlayer channel flow direction during DGP injection. (**b**) Interlayer channel flow direction during subsequent water flooding.

3. Conclusions

This study develops a physical model of an interlayer pan-connected reservoir, quantitatively characterizes interlayer flow, and explores the regulatory effects of deformable gel particles (DGPs) under various influencing factors. The main conclusions are as follows:

During DGP injection, the water phase in the DGP suspension passes through two interlayer channels in the middle and front parts, entering the potential layer from the water breakthrough layer and bypassing the channel near the production well to re-enter the water breakthrough layer. Experimental results indicate that interlayer flow rates are relatively high when injection volumes are large, and particle sizes and injection rates are moderate. However, the interlayer channel near the production well consistently flows from low permeability to high permeability. Under various injection conditions, the average increase in oil recovery during this stage can reach 19.74%.

During subsequent water flooding, all interlayer channels flow from the potential layer, enriched with residual oil, to the water breakthrough layer. Low injection rates limit deep migration of DGPs, while high rates intensify particle breakage. Additionally, smaller particles have weaker blocking capabilities, while larger particles primarily act in the middle and front parts of the porous media. These factors collectively result in a higher interlayer flow rate during this stage, which significantly impairs oil displacement. At this stage, the average increase in oil recovery under various injection conditions is 6.17%.

DGPs can effectively regulate widely interconnected reservoirs with interlayer barriers. Their regulatory mechanism is primarily reflected in the following: during injection, the migration of particles in the deep sections of the water breakthrough layer leads to an overall pressure increase in this layer, which, in turn, propagates the water phase to the potential layer and mobilizes the residual oil within it. Under the experimental conditions of this study, the optimal regulatory effect is achieved when DGPs with a particle size-to-pore throat ratio of 2.51 are injected at a rate of 3 mL/min for 3 pore volumes.

Despite their deep migration capability, DGPs' relatively large particle size limits their application in low-permeability reservoirs. Future studies will focus on deep profile control and displacement in interlayer channels of low-permeability reservoirs, as well as investigating reservoir damage caused by the profile control agents, aiming to optimize DGP properties for broader reservoir adaptability. A key limitation is the lack of simulation for specific reservoir environments of a particular oilfield. Future research will use real cores from a specific oilfield and conduct experiments under temperature and salinity conditions matching the reservoir to enhance field relevance.

4. Materials and Methods

4.1. Materials

Deformable gel particles (DGPs) are primarily composed of acrylamide, polyacrylamide, acrylonitrile monomer, polyvinyl alcohol, N-methylenebisacrylamide, and ammonium persulfate-sodium bisulfite, and were produced by Guangzhou Bofeng Chemical Science and Technology Co., Ltd. (Guangzhou, China). The DGPs are dry powder with a particle size range of 140–170 mesh before water absorption and expansion as shown in Figure 11.

Figure 11. Deformable gel particles (DGPs).

The change in particle size of DGPs after water absorption and expansion is shown in Figure 12a. The particle size distribution after full expansion is shown in Figure 12b. DGPs expanded 5.06 times after 12 h of water absorption, with no further size change thereafter. After that, the particle size does not change. And the particle size was measured via dynamic light scattering (Malvern Mastersizer 3000, Malvern, UK). Table 3 shows the typical properties of the DGP sample at 2000 mg/L.

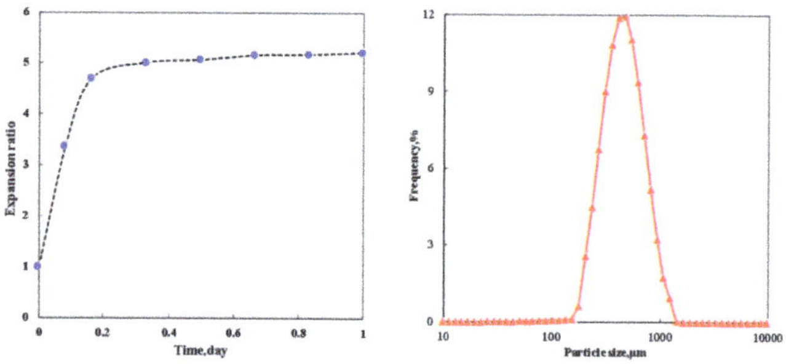

(a) (b)

Figure 12. Particle size distribution of DGPs. (a) The swollen features of DGPs, (**b**) particle size distribution after swelling.

Table 3. Typical properties of DGPs.

Properties	Values
Moisture Content (%)	<5
Apparent Bulk Density (g/mL)	0.90–0.95
Viscosity (mPa.s)	45.14
pH Value	6–8
Fully swelling Time (h)	12

Because the influence of salinity is not considered in this study, distilled water was used in the experiment. For the enhanced oil recovery (EOR) experiments, 25#-vash oil (877.6 kg/m^3,10.5 mPa.s) was used.

4.2. Physical Model

A sandpack was used to construct a physical model of porous media for the experiments. The sandpack has a length of 60 cm and a diameter of 3.8 cm. Three interfaces are located on one side of the sandpack for pressure monitoring. Pressure sensors with a measurement range of 0–5 MPa are installed at the entrance and along the three interfaces. Filters are installed at both the outlet and interfaces to prevent quartz sand from flowing out, ensuring the accuracy and uninterrupted operation of the experiment.

The sandpack dimensions are consistent with those used in the DGP flow experiments. As shown in Figure 13, each sandpack has three interfaces, which are used for pressure monitoring and interlayer channel simulation. The sandpacks are connected by a thicker pipeline with an inner diameter of 6 mm, simulating the interlayer channels. These interfaces represent the flow channels near the injection well, the reservoir depth, and the production well. A filter screen is installed at the outlet, and pressure sensors are positioned at the entrance of the high-permeability sandpack, all interfaces, and the entrance of the low-permeability sandpack. For the low-permeability sandpack, pressure sensors (0.01% F.S.) with a measurement range of 0–0.110 MPa are used along the three interfaces.

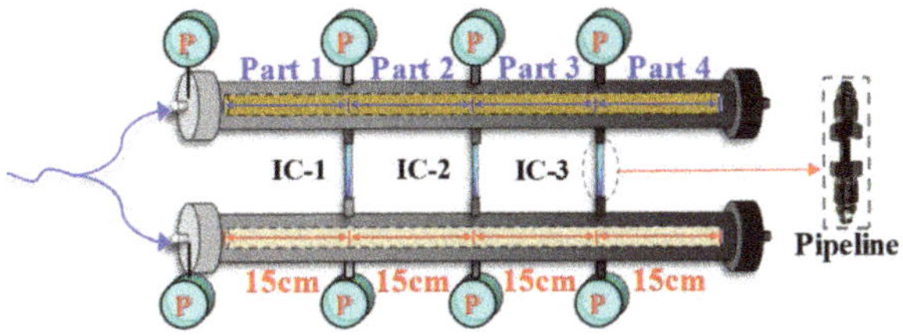

Figure 13. Physical model considering interlayer channels.

4.3. Experimental Procedures and Steps

The high- and low-permeability sandpacks were filled with quartz sand (1–10 mesh and 100–120 mesh, respectively), and the experimental setup was arranged as shown in Figure 14. To accurately characterize the particle regulation behavior in an interlayer-connected reservoir, both flow and EOR experiments were conducted simultaneously using nearly identical parameters, as outlined in Tables 1 and 2. In Tables 1 and 3: PV refers to the ratio of the injected DGP suspension to the pore volume. And the diameter ratio of DGP to pore is the ratio of the median particle size of DGPs to the pore throat diameter.

The main experimental steps are as follows:

- Equipment verification and connection: The air-tightness and connectivity of the experimental setup were thoroughly checked, focusing on the flow channel between the two sandpacks. The device was assembled as shown in Figure 14. To prevent DGP sedimentation, a magnetic stirring device was incorporated into the DGP tank.
- Physical model creation: Each sandpack was filled with the appropriate mesh-size quartz sand, as specified in Tables 1–3. The sand was pre-screened and thoroughly cleaned to ensure that smaller particles did not interfere with the experiment.
- Water saturation: The model was saturated at an injection rate of 1 mL/min. The porosity of the sandpack was determined based on the volume of injected water, and permeability was measured at an injection rate of 3 mL/min.

- Oil Saturation (EOR experiments only): Oil saturation was performed at the same rate as water saturation, and ceased once the oil-water recovery rate at the outlet stabilized post-water saturation. The oil and water saturations of the sandpacks were calculated based on the volume of the injected oil.
- Water flooding (EOR experiments only): Water injection continued until the water cut in one of the sandpacks reached 96%. The volume of oil and water produced from both sandpacks was recorded every 10 min, and pressure changes during water flooding were monitored.
- DGP injection into the model: A DGP suspension (2000 mg/L) was prepared and allowed to absorb water and fully expand for 12 h. Subsequently, the suspension was injected into the sandpacks at a rate of 3 mL/min, with varying pore volumes (PV).

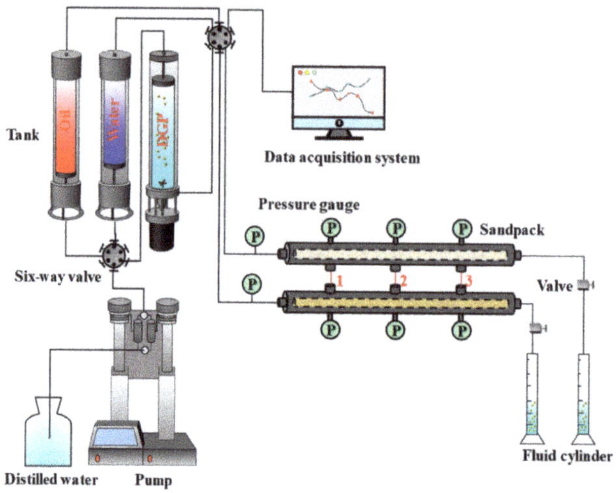

Figure 14. Physical simulation process of DGP flow experiments.

5. Calculation Method of Interlayer Channel Flow Rate

Bai [16], Elsharafi [20], and Imqam [28] proposed that when the diameter ratio of DGP to pore throat exceeds 17, or when the formation permeability falls below 100 mD, a filter cake forms at the entrance of the sandpack. Even if the DGP infiltrates into the formation, the penetration distance remains limited. Consequently, only the final three parts of the low-permeability sandpack experience water flow in the injection direction.

Experiments can obtain the pressure at each measuring point in the low-permeability sandpack. The flow rates of the two middle parts of the low-permeability tube at each moment (labeled as Q_{L2} and Q_{L3} in Figure 15) can be calculated. Taking Q_{12} as an example, substitute the pressure difference (ΔP_{12}) measured by pressure measuring points P_6 and P_7 into Darcy's formula. The calculated Q_{L2} is as follows:

$$Q_{L2} = \frac{kA\Delta P_{L2}}{\mu L} = \frac{kA(P_6 - P_7)}{\mu L} \tag{1}$$

L is the sandpack length (cm), μ is the dynamic viscosity of water (cp), A is the cross—sectional area (cm^2), and k is permeability (μm^2).

The pressure difference and permeability can be incorporated into Darcy's Law to estimate the flow rate for the central two parts of the low-permeability sandpack. The outlet flow rates for the last part of both sandpacks (part 4 in Figure 16), denoted Q_{Hout} and Q_{Lout}, can be determined experimentally by measuring the produced liquid. However, the filter cake affects the flow rate at the entrance of the low-permeability layer, preventing the calculation of Q_{Lin} based on the pressure at the entrance of the low-permeability sandpack. Assuming the flow direction from the low-permeability sandpack to the high-permeability sandpack is positive, and applying the principle of volume conservation, the flow rates for the last two flow channels can be expressed as follows:

$$Q_{C3} = Q_{L2} - Q_{Lout} \quad (2)$$

$$Q_{c2} = Q_{L1} - Q_{L2} \quad (3)$$

Furthermore, the flow rates Q_{h1} and Q_{h2} for the middle two parts of the high-permeability sandpack can be determined as follows

$$Q_{H2} = Q_{Hout} + Q_{c3} \quad (4)$$

$$Q_{H1} = Q_{H2} + Q_{H2} \quad (5)$$

Finally, Q_{Lin}, Q_{Hin}, and Q_{c1} can be iteratively solved using the following relations:

$$Q_{Hin} = Q_{c1} + Q_{H1} \quad (6)$$

$$Q_{Hin} = Q_{c1} + Q_{H1} \quad (7)$$

$$Q_{Lin} + Q_{Hin} = Q_{Lout} + Q_{Hout} = 3\text{mL/min} \quad (8)$$

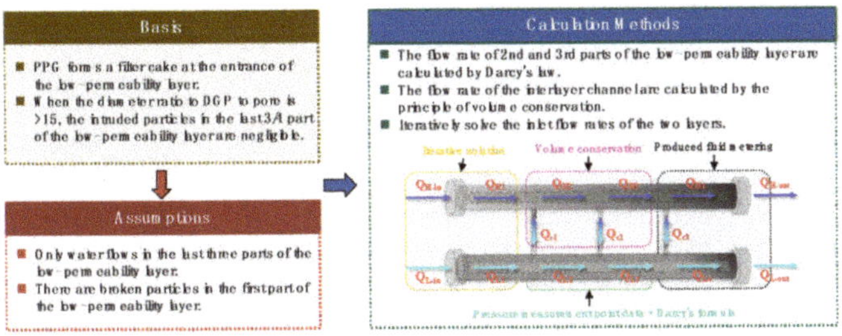

Figure 15. Schematic diagram of interlayer channel flow rate calculation method.

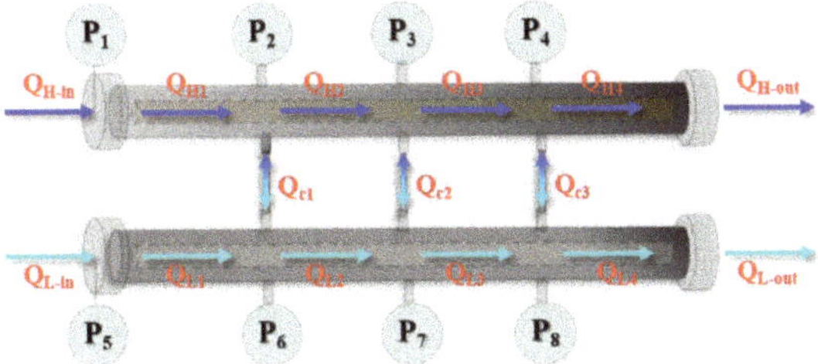

Figure 16. Schematic diagram of flow calculation for interlayer channels.

Author Contributions: Resources, W.Z.; Writing—original draft, W.Z.; Project administration, J.W. and H.L.; Funding acquisition, J.W.; Data curation, T.W.; Software, T.W. and Z.Q.; Writing—review and editing, R.O.E.; Validation, Z.Q.; Formal analysis, Z.Q. All authors have read and agreed to the published version of the manuscript.

Funding: This work is funded by National Natural Science Foundation of China (No. U24B6003, 42202292); Science Foundation of China University of Petroleum, Beijing (No.2462025YJRC010).

Conflicts of Interest: Authors Wenjing Zhao and Tianjiang Wu were employed by the company China National Petroleum Corporation. The remaining authors declare that the research was conducted in the absence of any commercial or financial relationships that could be construed as a potential conflict of interest.

References

1. Zhao, H.; Luo, X.; Xiao, Z.; Zhang, B.; Zhao, F.; Lei, Y.; Zhao, J.; Hu, C. Interlayer Features and Petroleum Geology Meaning of Donghe Sandstones in the Hadson Oilfield. *Nat. Gas Geosci.* **2014**, *25*, 824–833. [CrossRef]
2. Zhao, W.; Wang, J.; Qian, Q.; Yu, C.; Zhang, M.; Liu, H.; Huang, Y. Evolution law of dominant flow channels of water flooding in heterogeneous reservoir. *Fault-Block Oil Gas Field* **2023**, *30*, 847–857.
3. Ladipo, L.; Blunt, M.J.; King, P.R. Crossflow effects on low salinity displacement in stratified heterogeneity. *J. Pet. Sci. Eng.* **2022**, *208*, 109565–109591. [CrossRef]
4. Li, M.; Su, Y.L.; Wang, W.D.; Li, Y.J.; Dong, M.Z. Impact of cross-flow on well production in shale reservoir considering vertical variation of reservoir and fracture properties: Model and field application. *J. Pet. Sci. Eng.* **2022**, *208*, 109739–109759. [CrossRef]
5. Russell, D.G.; Prats, M. The Practical Aspects of Interlayer Crossflow. *J. Pet. Technol.* **1962**, *14*, 589–594. [CrossRef]
6. Rezaeiakmal, F.; Parsaei, R. Visualization study of polymer enhanced foam (PEF) flooding for recovery of waterflood residual oil: Effect of cross flow. *J. Pet. Sci. Eng.* **2021**, *203*, 108583–108594. [CrossRef]
7. Zhang, Q. Quantitative Characterization and Development Countermeasures of Multi-layer Heterogeneous Interlayer Channeling Flow. *Spec. Oil Gas Reserv.* **2021**, *28*, 136–141.
8. Lu, X.; Gao, J.; He, X.; Wei, W.; Su, Y.; Pei, X. Experimental study of internal channeling law of intra-layer heterogeneous reservoir: A case study of Lamadian Oilfield in Daqing. *Pet. Geol. Recovery Effic.* **2022**, *29*, 118–125.
9. Cheng, J.; Zhou, Q.; Zhou, W.; Wang, Q.; Shi, W. Plugging method of polymer gel with controllable low initial viscosity for dominant flow paths in deep reservoirs and its application. *Acta Pet. Sin.* **2020**, *41*, 969–978.
10. Seright, R.; Brattekas, B. Water shutoff and conformance improvement: An introduction. *Pet. Sci.* **2021**, *18*, 450–478. [CrossRef]
11. Sorbie, K.S.; Seright, R.S. Gel placement in heterogeneous systems with crossflow. In Proceedings of the SPE/DOE Enhanced Oil Recovery Symposium, Tulsa, OK, USA, 22–24 April 1992.
12. Seright, R.S.; Zhang, G.Y.; Akanni, O.O.; Wang, D.M. A comparison of polymer flooding with in-depth profile modification. *J. Can. Pet. Technol.* **2012**, *51*, 393–402. [CrossRef]
13. Imqam, A.; Bai, B.J.; Delshad, M. Micro-particle gel transport performance through unconsolidated sandstone and its blocking to water flow during conformance control treatments. *Fuel* **2018**, *231*, 479–488. [CrossRef]
14. Wang, J.; Zhang, H.L.; Liu, H.Q.; Zhao, W.; Liu, H.M.; Yao, C.J.; Zheng, J.P.; Shen, Y.H. Quantification of Transportation of Deformable Gel Particles in Porous Media. In Proceedings of the SPE Annual Technical Conference and Exhibition, San Antonio, TX, USA, 9–11 October 2017.
15. Bai, B.; Li, L.; Liu, Y.; Wang, Z.; Liu, H. Preformed Particle Gel for Conformance Control: Factors Affecting its Properties and Applications. In Proceedings of the SPE/DOE Symposium on Improved Oil Recovery, Tulsa, OK, USA, 17–21 April 2004.
16. Bai, B.J.; Wei, M.Z.; Liu, Y.Z. Field and lab experience with a successful preformed particle gel conformance control technology. In Proceedings of the SPE Production and Operations Symposium, Oklahoma City, OK, USA, 23–26 March 2013; pp. 23–26.
17. Kang, W.L.; Kang, X.; Lashari, Z.A.; Li, Z.; Zhou, B.B.; Yang, H.B.; Sarsenbekuly, B.; Aidarova, S. Progress of polymer gels for conformance control in oilfield. *Adv. Colloid Interface Sci.* **2021**, *289*, 102363–102379. [CrossRef] [PubMed]
18. Qiu, Y.; Wei, M.Z.; Bai, B.J. Descriptive statistical analysis for the PPG field applications in China: Screening guidelines, design considerations, and performances. *J. Pet. Sci. Eng.* **2017**, *153*, 1–11. [CrossRef]
19. Sang, Q.; Li, Y.J.; Yu, L.; Li, Z.Q.; Dong, M.Z. Enhanced oil recovery by branched-preformed particle gel injection in parallel-sandpack models. *Fuel* **2014**, *136*, 295–306. [CrossRef]
20. Elsharafi, M.O.; Bai, B.J. Effect of weak preformed particle gel on unswept oil zones/areas during conformance control treatments. *Ind. Eng. Chem. Res.* **2013**, *51*, 11547–11554. [CrossRef]
21. Yang, H.B.; Zhou, B.B.; Zhu, T.Y.; Wang, P.X.; Zhang, X.F.; Wang, T.Y.; Wu, F.P.; Zhang, L.; Kang, W.L.; Ketova, Y.A.; et al. Conformance control mechanism of low elastic polymer microspheres in porous medium. *J. Pet. Sci. Eng.* **2021**, *196*, 107708–107717. [CrossRef]
22. Ji, Z.; Zhang, Q.; Gao, Y.; Wang, J.; He, C.; Han, L.; Zhao, W. Experimental Study on Conformance Control Using Acidic Nanoparticles in a Heterogeneous Reservoir by Flue Gas Flooding. *Energies* **2023**, *16*, 315. [CrossRef]
23. Gao, Q. Compatibility between new gel particles and polymer-flooded heterogeneous oil reservoir. *Pet. Geol. Oilfield Dev. Daqing* **2022**, *41*, 112–119.
24. Yao, C.J.; Lei, G.L.; Cathles, L.M.; Steenhuis, T.S. Pore-scale investigation of micron-size polyacrylamide elastic microspheres (MPEMs) transport and retention in saturated porous media. *Environ. Sci. Technol.* **2014**, *48*, 5329–5335. [CrossRef]

25. Leng, J.; Wei, M.; Bai, B. Review of transport mechanisms and numerical simulation studies of preformed particle gel for conformance control. *J. Pet. Sci. Eng.* **2021**, *206*, 109051–109064. [CrossRef]
26. Zhao, S.; Pu, W.F.; Wei, B.; Xu, X.G. A comprehensive investigation of polymer microspheres (PMs) migration in porous media: EOR implication. *Fuel* **2019**, *235*, 249–258. [CrossRef]
27. Zhao, W.J.; Wang, J.; Qi, Z.Y.; Liu, H.Q. Study on deep transportation and plugging performance of deformable gel particles in porous media. *Pet. Sci.* **2024**, *2024*, 962–973. [CrossRef]
28. Imqam, A.; Bai, B.J.; Wei, M.Z.; Elue, H.; Muhammed, F.A. Use of hydrochloric acid to remove filter-cake damage from preformed particle gel during conformance-control treatments. *SPE Prod. Oper.* **2016**, *31*, 247–257. [CrossRef]

gels

MDPI

Article

Evaluation and Experiment of High-Strength Temperature- and Salt-Resistant Gel System

Changhua Yang [1,2], Di Xiao [1], Jun Wang [3] and Tuo Liang [1,*]

1 School of Petroleum Engineering, Xi'an Shiyou University, Xi'an 710065, China; ych569@126.com (C.Y.); xd1181163575@163.com (D.X.)
2 Engineering Research Center for Development and Control of Low-Permeability and Ultra-Low-Permeability Reservoirs in Western China, Ministry of Education, Xi'an 710065, China
3 China National Petroleum Chang Qing Oilfield Company's New Energy Division, Xi'an 710016, China; wangjun21_cq@petrochina.com.cn
* Correspondence: t_liang610@xsyu.edu.cn

Abstract: To address the issues of poor thermal stability, inadequate salt tolerance, and environmental risks in conventional gel systems for the development of high-temperature, high-salinity heterogeneous reservoirs, a triple-synergy gel system comprising anionic polyacrylamide (APAM), polyethyleneimine (PEI), and phenolic resin (SMP) was developed in this study. The optimal synthesis parameters—APAM of 180 mg/L, PEI:SMP = 3:1, salinity of 150,000 ppm, and temperature of 110 °C—were determined via response surface methodology, and a time–viscosity model was established. Compared with existing binary systems, the proposed gel exhibited a mass retention rate of 93.48% at 110 °C, a uniform porous structure (pore size of 2–8 μm), and structural stability under high salinity (150,000 ppm). Nuclear magnetic resonance displacement tests showed that the utilization efficiency of crude oil in 0.1–1 μm micropores increased to 21.32%. Parallel dual-core flooding experiments further confirmed the selective plugging capability in heterogeneous systems with a permeability contrast of 10:1: The high-permeability layer (500 mD) achieved a plugging rate of 98.7%, while the recovery factor of the low-permeability layer increased by 13.6%. This gel system provides a green and efficient profile control solution for deep, high-temperature, high-salinity reservoirs.

Keywords: response surface method; thermogravimetric analysis; microscopic visualization; nuclear magnetic analysis; parallel displacement

1. Introduction

With global oil and gas exploration extending into deeper and more geologically complex formations, high-temperature, high-salinity heterogeneous reservoirs have become a critical frontier for enhancing reserves and production [1]. However, their development is hindered by two fundamental challenges. First, elevated formation temperatures and salinities cause severe molecular chain degradation and ion-screening effects in conventional polymer gels, leading to a rapid loss of mechanical strength, dehydration, and shrinkage [2]. Second, pronounced heterogeneity—often with permeability contrasts exceeding 10—results in preferential flow through high-permeability channels [3], leaving low-permeability zones with oil recovery factors below 40%. As a result, the recovery factor of such reservoirs is typically 15–30% lower than that of homogeneous reservoirs, with annual production losses exceeding ten million tons [4]. Existing Cr/aldehyde crosslinked systems offer a degree of thermal stability, but their application is constrained by heavy-metal contamination risks and aldehyde toxicity, highlighting the urgent need for environmentally friendly, intelligent profile-control materials [5]. In recent years, organic–inorganic hybrid gels have attracted increasing attention due to their structural tunability, yet achieving long-term stability under extreme reservoir conditions remains a critical bottleneck, underscoring the necessity for innovative, sustainable solutions [6].

For profile control and water shutoff in high-temperature, high-salinity reservoirs, extensive research has focused on composite systems of polyacrylamide (APAM) with organic amine crosslinkers such as polyethyleneimine (PEI) [7]. Anionic polyacrylamide (APAM), owing to its carboxyl ionization capability, is a preferred backbone for salt resistance; however, purely electrostatically crosslinked networks tend to dissociate at elevated temperatures [8]. Polyethyleneimine (PEI), as a polyamine crosslinker, can enhance salt tolerance via charge shielding, yet its thermal stability is limited [9]. Phenolic resin (SMP) can improve the thermomechanical strength of gels but often struggles to reconcile solubility with environmental compatibility [10]. Some studies have introduced a third component into APAM/PEI systems—such as hydrophobic monomers, nanoparticles, or phenolic covalent crosslinkers—to reinforce network stability or improve mechanical performance under salt-screening conditions [11]. However, these modifications typically focus on optimizing a single performance parameter and lack a systematic, multiscale synergy design of crosslinking modes at the molecular, microscopic, and macroscopic levels. Fundamentally, a single crosslinking mechanism is insufficient to break the "strength–stability–environmental compatibility" trade-off [12,13]. To address this, the present study introduces phenolic resin (SMP) with aromatic ring structures into the established APAM/PEI system [14]. During the condensation reaction, SMP forms a covalent backbone with high thermal resistance, while the aromatic hydrophobic domains and π–π interactions create localized hydrophobic microenvironments that reduce water activity and mitigate salt-screening effects [15–18]. These features, combined with ionic coordination, give rise to a multimodal synergistic network. The novelty of this design lies in a hierarchical construction of crosslinking modes across molecular, microscopic, and macroscopic scales, enabling simultaneous optimization of network stability, toughness, and gelation controllability under harsh high-temperature and high-salinity conditions—rather than a mere additive combination of three components [19–23]. Constructing such a multi-mechanism synergistic molecular network to achieve structurally adaptive regulation under extreme reservoir environments represents a key scientific pathway to overcoming this challenge [24,25]. To validate this strategy, we first optimized the APAM/PEI/SMP composition and preparation conditions. Gel performance was then systematically characterized through rheological measurements, thermal and salt resistance tests, and self-healing/toughness evaluations. The multimodal crosslinking mechanism was elucidated using Fourier transform infrared spectroscopy (FTIR) and scanning electron microscopy (SEM). Finally, the application potential was verified via microscopic visualization, nuclear magnetic resonance (NMR) testing, core-plugging experiments, and oil recovery enhancement evaluations. The overall research workflow is illustrated in Figure 1.

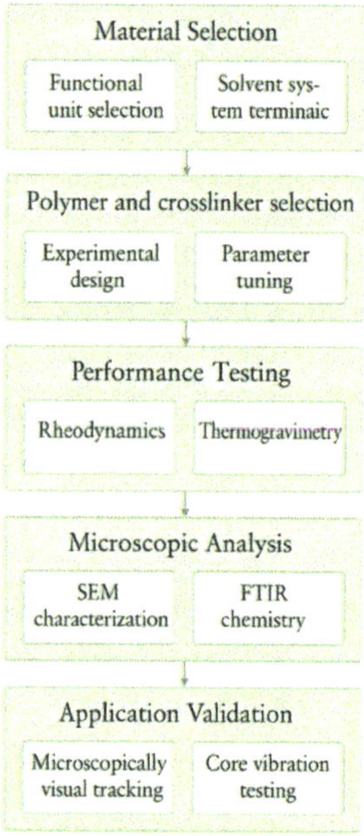

Figure 1. Research roadmap.

2. Results and Discussion

2.1. Analysis and Evaluation of Gel System

2.1.1. Two-Factor Interaction in the Response Surface

According to the designed experiments, 30 different combinations were carried out, including 16 factorial points, 8 axial points and 6 center points. Corresponding to the designed center points, 6 groups of experiments were repeated to ensure the validity of the model and reduce the estimated variance, the variance table is shown in Table 1 below. Finally, the quadratic regression polynomial model of Equation (1) was obtained.

Table 1. Model ANOVA.

Source	Sum of Squares	Degree of Freedom	Mean Square	F-Value	p-Value
Model	1201.88	14	85.85	9.13	<0.0001
A	4.08	1	4.08	3.54	0.0793
B	33.33	1	33.33	0.4341	0.5200
C	0.7500	1	0.7500	0.797	0.7815
D	21.33	1	21.33	2.27	0.1528
AB	2.25	1	2.25	0.2392	0.6318
AC	1.00	1	1.00	0.1063	0.7489
AD	9.00	1	9.00	0.9569	0.3435
BC	1.00	1	1.00	0.1063	0.7489
BD	0.2500	1	0.2500	0.0266	0.8727
CD	0.2500	1	0.2500	0.0266	0.8727
A^2	1015.05	1	1015.05	107.92	0.0023
B^2	126.30	1	126.30	13.43	<0.0001
C^2	165.76	1	165.76	17.62	0.0008
D^2	86.01	1	86.01	9.14	0.0085
Residual	141.08	15	9.14	/	/
Items not drafted	131.08	10	13.11	6.55	0.0855
Pure error	0.1280	5	0.0520	/	/
Total sum	1332.97	29	/	/	/

$$G = 92.00 + 1.67A + 0.5833B + 0.25C - 1.333D - 0.5AB - 0.75AC - 1.5AD + 0.5BC + 0.25BD + 0.25CD - 12.17A^2 - 4.27B^2 - 4.92C^2 - 3.54D^2 \quad (1)$$

It can be known from the analysis of variance table that the F of the regression model is 9.13 and $p < 0.0001$, indicating that the established regression model is extremely significant and suggesting that the fitting degree of the established quadratic multivariate regression equation is good. The misfitting term F = 6.55 and $p = 0.0855 > 0.05$, indicating that the misfitting term is not significant and proving that there are no misfitting factors in the regression model. Through the significance test of each coefficient in the formula, it can be obtained that the ranking of the influence degrees of the four factors on the gelation strength is as follows: Anionic polyacrylamide (A) > High salt resistance crosslinking agent (B) > Mineralization degree (C) > Temperature (D).

Figure 2a illustrates the effects of APAM concentration and the PEI:SMP ratio on the strength score of the gel system. In the response surface plot, the degree of curvature reflects the magnitude of influence exerted by the studied factors on the response: The more pronounced the curvature, the greater the effect of APAM concentration on gel strength. The elliptical shape of the contour lines indicates that both APAM concentration and the PEI:SMP ratio are critical factors for achieving high gel strength, with a significant interaction between them; in contrast, circular contours would suggest a less pronounced interaction. The 3D response surface opens downward, showing marked variations with changes in APAM concentration and PEI:SMP ratio, and exhibits a distinct maximum response value, confirming a strong interaction between these two parameters in influencing gel strength. When the amount of APAM is held constant, APAM concentration exerts a unidirectional effect on yield; however, when PEI:SMP is fixed, the effect of APAM concentration on strength becomes more pronounced and bidirectional. Under these conditions, the optimal APAM concentration was determined to be 180 mg/L. To validate the model's prediction, three independent replicate experiments were conducted. The results showed a gel strength score of 93.5 ± 2.3, in close agreement with the predicted value of 92.0, thereby confirming the model's reliability and predictive accuracy. As shown in Figure 2b,c, at a salinity of 150,000 ppm and a temperature of 110 °C, the interaction between salinity and APAM concentration is most pronounced. Figure 2d further reveals that the most significant interaction occurs when the PEI:SMP ratio is 3:1. Additionally, it demonstrates the influence of salinity on both APAM concentration and the PEI:SMP ratio. When the temperature is below 100 °C, poor solubility of the reactants leads to incomplete reactions and suboptimal yield. Conversely, at temperatures above 130 °C, excessive solubility causes adverse effects that outweigh the benefits, thereby compromising the stability of gel strength under otherwise optimal conditions.

Figure 2e,f present the interaction effects of PEI:SMP ratio with temperature and of salinity with temperature, respectively. When the temperature is held constant, a PEI:SMP ratio below 3:1 results in incomplete reactions and precipitation of the reactants. At temperatures above 130 °C, excessive solubility of the products leads to a marked decline in gel strength. In contrast, the interaction between salinity and temperature during the reaction process is not significant, consistent with the variance analysis results shown in Equation (1). Based on these findings, the optimal reaction temperature was determined to be 110 °C.

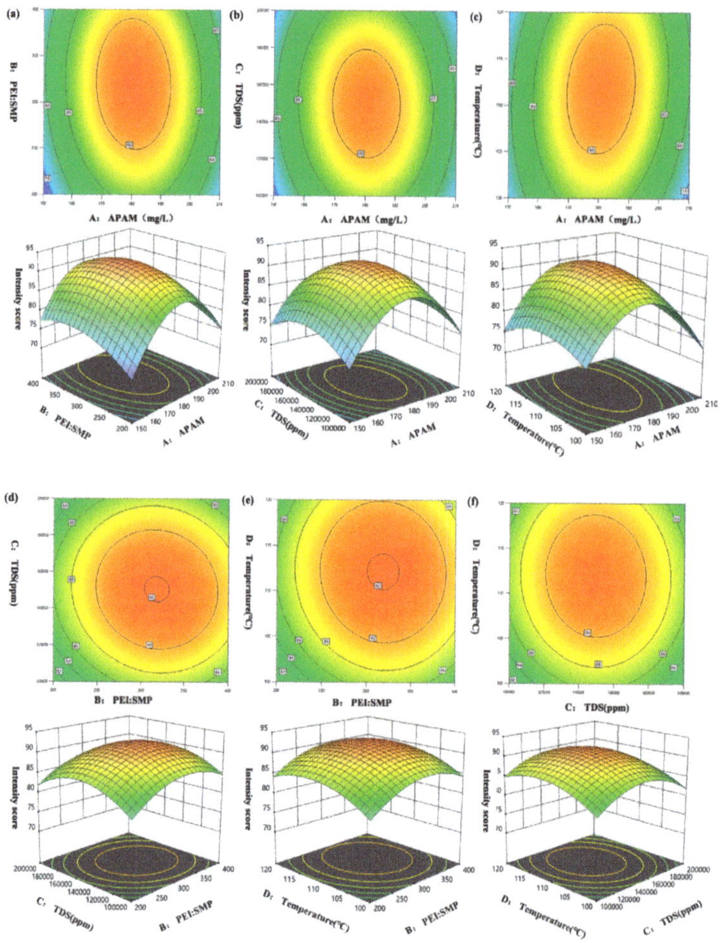

Figure 2. Response surface plot of the two-factor interaction. (**a**) The interaction between factors A and B. (**b**) The interaction between factors A and C. (**c**) The interaction between factors A and D. (**d**) The interaction between factors B and C. (**e**) The interaction between factors B and D. (**f**) The interaction between factors C and D. (The red dots in the figure represent the extreme points).

2.1.2. Time–Viscosity Relationship Fitting

As described in Section 2.1.1, to establish a direct correlation between crosslinker concentration and both system viscosity and gelation time, the time–viscosity data corresponding to different crosslinker concentrations were subjected to nonlinear surface fitting using GAUSS, COSINE, and other fitting functions. Figure 3 below shows the fitted relationship between time and viscosity.

The fitted functions yielded coefficients of determination (R^2) ≥ 0.8, indicating that the regression deviations provided a reliable explanation of the overall variance and that the model effectively captured the general trend of viscosity evolution. Under different formulation conditions, the variations of t_0 and Δt were closely related to APAM concentration and the PEI:SMP ratio: Increasing APAM concentration markedly shortened t_0, thereby accelerating gelation, whereas increasing the PEI:SMP ratio enhanced η_{max}, improving the final gel strength. A key advantage of this fitted equation is that it allows the extraction of critical parameters from a limited number of short-term experiments, enabling the prediction of gelation states under various shut-in times and thus providing a scientific basis for field operations. For example, by adjusting the APAM concentration and crosslinker ratio, t_0 can be controlled within the permissible shut-in duration on site, ensuring that the gel fully forms in the target zone without premature failure.

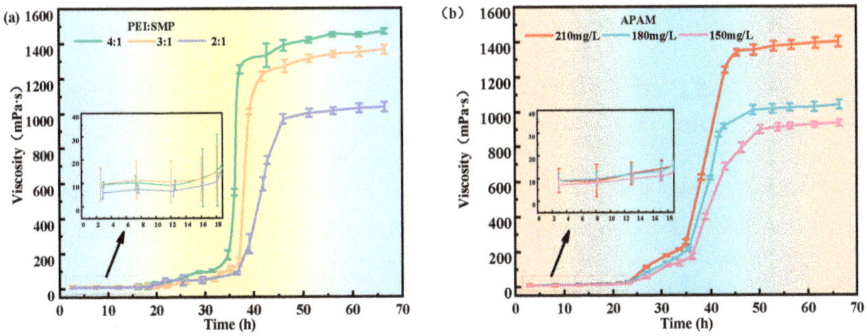

Figure 3. Viscosity–time Boltzmann fitting diagram under different main control factors. (**a**) A1: 180 mg/L APAM. (**b**) A2: PEI:SMP = 3:1.

2.2. *Test and Characterization of Gel System*

2.2.1. SEM Electron Microscope Image Analysis

After the gel system is formed under static conditions, it exhibits an irregular and complex three-dimensional network structure, as shown in Figure 4. The formation of this structure primarily arises from the interactions between the carboxyl groups of the APAM polymer chains and the protonated amine groups of the PEI crosslinker. Through a combination of crosslinking and molecular entanglement, these interactions construct a compact and interconnected network.

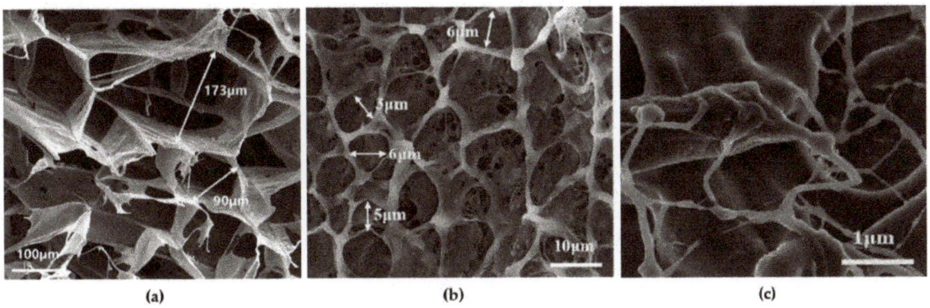

Figure 4. Scanning electron microscope images of gels of different sizes. (**a**) 100 μm. (**b**) 10 μm. (**c**) 1 μm.

Microscopic observations at different magnifications revealed that the three-dimensional network contained pores ranging from 2 μm to 8 μm in diameter. These pores impart a distinct porous morphology at the microscale, forming a complex structure capable of both encapsulating free water and binding bound water. Such structural characteristics substantially enhance the overall performance of the gel, endowing it with excellent water absorption and retention capabilities while also improving volumetric stability. To adapt the gel for high-temperature, high-salinity conditions, a structural modification strategy was implemented. The amide groups (–$CONH_2$) of APAM and the primary amine groups (–NH_2) of PEI undergo covalent condensation with the hydroxymethyl groups (–CH_2OH) of the high-temperature stabilizer SMP, thereby forming a thermally stable backbone. This approach minimizes premature chain scission and degradation of the viscous polymer solution during gelation at elevated temperatures. Concurrently, the incorporation of

SMP further increases the compactness of the gel structure. The SMP molecules, with their aromatic rings, promote strong interchain entanglement and association, resulting in tighter intermolecular interactions. As a consequence, the entire gel network exhibits higher structural stability under external stress, forming a stereoscopic lattice with SMP aromatic rings as nodal hubs. This vividly illustrates the synergistic mechanism of "electrostatic salt resistance, covalent thermal stability, and aromatic ring water locking." Even under applied mechanical forces, the three-dimensional network accommodates a certain degree of deformation due to its inherent compressible void space. This property acts as a buffering mechanism, effectively dissipating the applied stress and minimizing structural damage caused by excessive strain. Furthermore, the reduced pore spacing increases molecular shielding in space and strengthens the interlocking of the network, ensuring that, after gelation, the overall structure retains high stability while exhibiting enhanced mechanical strength.

2.2.2. Infrared Spectroscopy Analysis

As shown in Figure 5 below. Fourier transform infrared spectroscopy (FTIR) characterization revealed distinct absorption peaks in the critical wavenumber regions of the gel system, reflecting the underlying molecular interactions. A strong absorption band at 1159 cm^{-1} corresponds to the C–O stretching vibration, originating from unreacted hydroxymethyl groups ($-CH_2OH$) on the aromatic rings of SMP and the newly formed ether linkages ($-CH_2-O-CH_2-$) generated via self-crosslinking of the phenolic resin. The band at 1182 cm^{-1}, assigned to the symmetric stretching of C–N bonds in secondary amine structures ($-NH-CH_2-$), confirms the occurrence of a Mannich condensation reaction between SMP hydroxymethyl groups and the primary amines of PEI, thereby establishing covalent crosslinking. The 1395–1409 cm^{-1} range corresponds to C–N stretching vibrations, with peak broadening indicating the formation of multi-level amine structures and highlighting the complexity of the crosslinked network. An absorption band at 1435 cm^{-1}, attributed to C–H bending vibrations, increases in intensity with higher SMP content, suggesting that hydrophobic aromatic structures participate in stabilizing the three-dimensional network. A sharp peak at 1559 cm^{-1} is primarily ascribed to the asymmetric stretching of C=O groups in carboxylate structures ($-COO^-/-NH_3^+$ ion-pair complexes), indicating complete ionization of carboxyl groups and their electrostatic association with protonated amines. The notable shift in this peak to lower wavenumbers relative to free carboxyl groups further confirms the formation of ionic crosslinking. Specifically, the carboxyl groups (–COOH) of APAM are fully ionized to $-COO^-$ in simulated formation water and subsequently interact with protonated amines ($-NH_3^+$) of PEI to form stable ion-pair complexes. Additional peaks include the saturated C–H stretching vibration at 2918 cm^{-1}, associated with methylene groups ($-CH_2-$) along the APAM backbone and ortho-methyl substituents on SMP aromatic rings, and a broad N–H stretching band around 3300 cm^{-1}. The latter's breadth reflects the coexistence of protonated primary ($-NH_2$) and secondary (–NH–) amines within an extensive intermolecular hydrogen-bonding network. Overall, the spectral features align closely with the proposed triple-synergistic mechanism and molecular design strategy, providing molecular-level evidence for the gel system's structural robustness and functional adaptability in high-temperature, high-salinity reservoir environments.

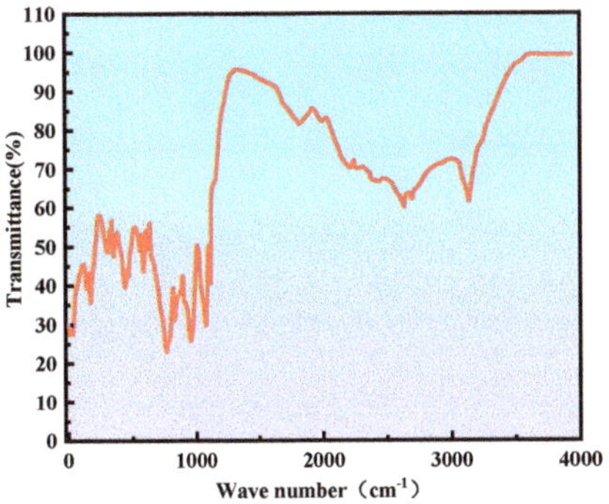

Figure 5. Infrared-spectrogram.

2.2.3. Thermogravimetric Analysis

Thermogravimetric–differential scanning calorimetry (TG–DSC) was employed over the temperature range of 35 °C to 800 °C to investigate the pyrolysis behavior of the gel system, examining changes in intermolecular interactions at different decomposition stages. This analysis also enabled the determination of the critical temperatures at which the gel network structure is disrupted, as well as the associated energy consumption during these processes. The measurements were performed on aged samples, and the resulting TG–DSC curves are presented in Figure 6.

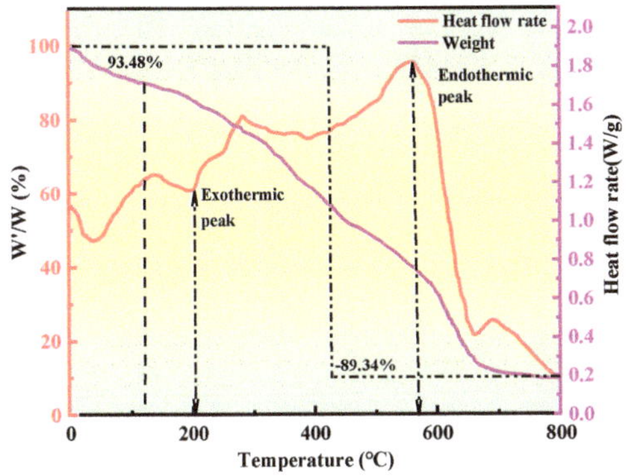

Figure 6. Thermogram-diagram.

Thermogravimetric analysis revealed that when the test temperature reached 120 °C, the gel system retained 93.48% of its initial mass, indicating that the gel used in this study is suitable for water-plugging applications under reservoir conditions up to 110 °C. Within this temperature range, minor weight loss was primarily attributed to the release of surface-bound water rather than scission of the polymer backbone. At this stage, only a fraction of the hydrophilic groups underwent thermal evaporation, while the majority of the molecular chains remained intact, without significant degradation. Upon further heating to approximately 200 °C, the gel underwent pronounced thermal decomposition, with pyrolytic depolymerization accelerating until around 500 °C, after which the degradation rate slowed. At this point, the principal hydrophilic functional groups in the gel structure—such as amide and carboxyl C=O bonds—were almost completely eliminated due to the sustained breakdown of the polymer backbone, leading to substantial thermal mass loss. Overall, the final mass loss over the entire temperature range reached 89.34%.

Differential scanning calorimetry further indicated that the gel system entered its glass transition at a relatively low temperature of approximately 82 °C, transitioning into a high-elastic state. As the temperature increased to around 190 °C, the gel began to cure, accompanied by an exothermic response, during which the rate of mass loss shifted from gradual to more rapid. Heating to approximately 200 °C initiated the melting stage, characterized by the scission of hydrophilic side chains and intensified dehydration. Upon reaching about 500 °C, the gel entered a decomposition–gasification phase, during which the progressive cleavage of the polymer backbone began to decelerate. By this stage, the gel had undergone substantial thermal dehydration and pyrolysis, demonstrating a favorable thermal resistance profile.

2.2.4. Molecular Weight Analysis

As shown in Figure 7 below. Molecular weight characterization revealed that the gel system exhibited excellent molecular weight uniformity, with a number-average molecular weight (Mn) of 135,050 g/mol, a weight-average molecular weight (Mw) of 173,435 g/mol, and a dispersity (Đ) of 1.284—significantly lower than the typical value for free-radical polymerization (Đ > 2.0). These results indicate several key features of the reaction system: (i) The PEI crosslinker precisely regulated the chain growth of APAM, effectively suppressing chain transfer side reactions; (ii) the dehydration–condensation rates of SMP hydroxymethyl groups and PEI amino groups were well-matched, leading to uniformly distributed high-molecular-weight polymers; (iii) the system overcame the conventional trade-off between gel strength and stability; and (iv) a quantitative predictive model was established linking molecular weight distribution to plugging performance. This provides an ideal plugging material for high-temperature and high-salinity reservoirs, combining high mechanical strength, slow dehydration, and rapid performance recovery.

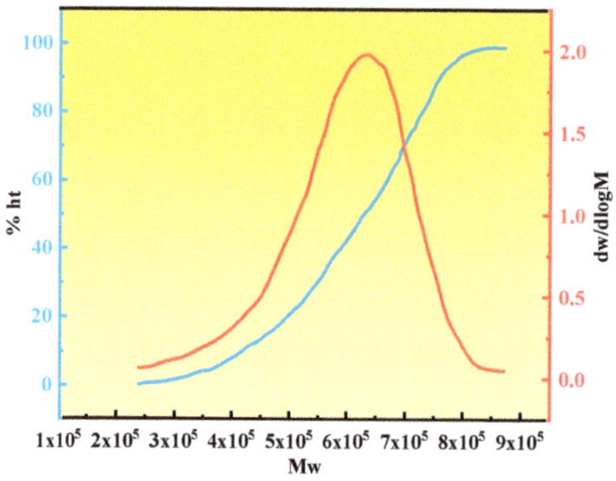

Figure 7. Molecular weight curve.

2.2.5. Rheological Curve Analysis

Gel systems with superior recovery force and impact resistance after deformation exhibit stronger self-healing and re-crosslinking capabilities following local damage. To evaluate these properties, a HAAKE RS600 rheometer was employed to perform frequency sweep tests on the gel system across different frequency ranges, measuring its storage modulus (G′) and loss modulus (G″). The tests were conducted within a frequency range of 0.01–10 Hz under a shear stress of 0.2 Pa. The corresponding scanning results are presented in Figure 8.

The rheological properties of the gel system exhibited distinct characteristics under variations in shear frequency and temperature. Based on the data in Figure 8

($G' = 4173 \pm 95$ Pa), three brine solutions with mineralization levels of 100,000 mg/L, 150,000 mg/L, and 200,000 mg/L were selected for testing within a frequency range of 0.01–10 Hz under a shear stress of 0.2 Pa. In all cases, the storage modulus (G') and loss modulus (G'') increased slightly with shear frequency, indicating stable mechanical performance of the gel across different strain rates. The overall smooth rheological curves suggest enhanced crosslinking density within the gel network and strengthened intermolecular interactions. In the temperature range of 100 °C to 120 °C, G' consistently exceeded G'', characteristic of elasticity-dominated viscoelastic behavior, confirming that the measured data fell within the viscoelastic regime. The gel existed as a viscoelastic solid, whose elasticity contributed to stable sealing of fractures and pore throats. With increasing temperature, G' decreased, primarily due to intensified molecular thermal motion that weakened crosslinking or repulsive interactions, leading to reduced rigidity and diminished elastic recovery; at elevated temperatures, such changes could cause structural loosening. Under conditions of 110 °C, pH = 1–2, and a mineralization level of 150,000 mg/L, the gel exhibited G' values in the range of 1543–4173 Pa, maintaining viscoelasticity, with elasticity as the dominant component. Gels prepared with simulated brine solutions were only slightly affected by temperature; however, mineralization had a pronounced effect on viscosity and strength before and after gelation. Possible reasons include (i) ionic interactions with polar groups of the polymer that weaken interchain attractions, (ii) ionic shielding effects forming a coating layer that hinders interchain interactions, and (iii) ion solvation effects that reduce intermolecular friction, thereby decreasing viscosity.

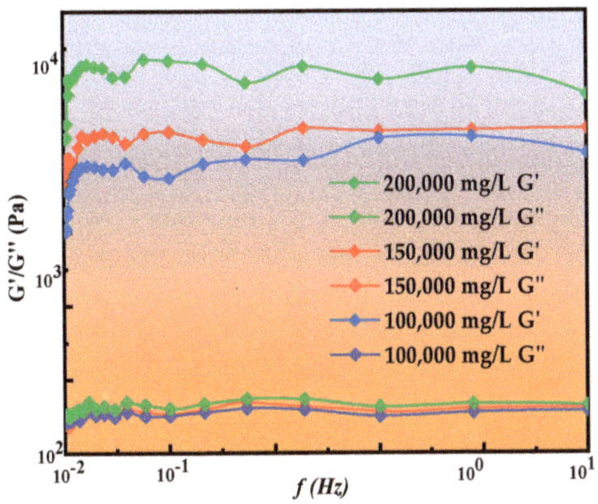

Figure 8. Gel-rheology.

2.2.6. Nuclear Magnetic Resonance Curve Analysis

Throughout the displacement process, both water flooding and gel injection were introduced from the right end of the core, with the displaced fluids produced from the left end. According to the chromatic scale, the initial oil saturation is represented by color coding, where regions with higher oil content appear green, and those with lower oil content appear blue, as shown in Figure 9a,b. The images reveal that oil saturation within the core gradually increased, with distinct color changes observed across the three displacement stages. Ultimately, the remaining oil was predominantly confined to fine pore spaces, distributed relatively uniformly, with no evident zones of residual oil enrichment.

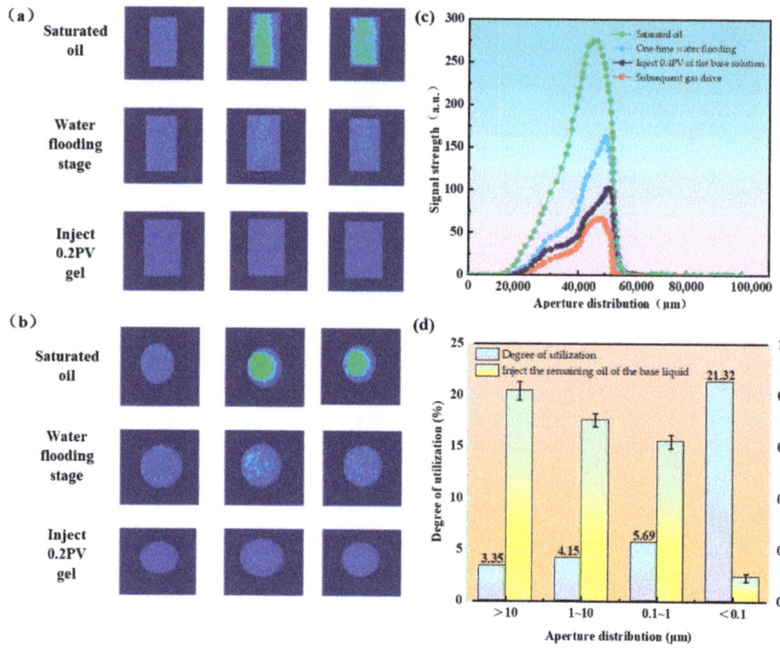

Figure 9. NMR imaging, T_2 map, and pore activation map. (**a**) Coronal plane imaging comparison before and after injection into the gel system. (**b**) Cross-sectional imaging comparison before and after injection into the gel system. (**c**) T_2 spectra of 50 mD core displacement at each stage. (**d**) The oil content distribution of the 50 mD core and the utilization degree of each pore diameter.

Following the completion of primary water flooding, 0.2 PV of the responsive gel precursor solution was injected. Upon completion, the nuclear magnetic resonance (NMR) T_2 spectrum of the core was measured. As gel injection proceeded, the signal amplitude of the T_2 spectrum further decreased. The oil saturation distribution, pore-size utilization, and contribution ratio of different pore sizes during gel injection are shown in Figure 9c,d. It can be observed that the pore structure exerted a significant influence on oil release during gel-based displacement. After the primary water flooding stage, the core retained a total of 2.20 mL of crude oil. Following gel precursor injection, the residual oil was mainly distributed across pores of various sizes: 0.82 mL in pores larger than 10 μm, 0.68 mL in 1–10 μm pores, 0.61 mL in 0.1–1 μm pores, and 0.05 mL in micropores smaller than 0.1 μm, which remained unrecovered during the primary water flooding. These results illustrate the spatial distribution of residual oil within the core and the contribution of different pore-size classes to oil displacement. During gel injection to 0.2 PV, the degree of pore utilization varied across size classes: approximately 3.35% for >10 μm pores, 4.15% for 1–10 μm pores, 5.69% for 0.1–1 μm pores, and 21.32% for micropores (<0.1 μm), representing the first mobilization of oil in the micropore fraction. The gel injection stage primarily mobilized oil from the 1–10 μm and >10 μm pore classes, while its effect on small and micropores was limited. By the end of the gel-based flooding stage, the incremental oil recovery was 2.63%, with residual oil predominantly remaining in >10 μm pores (19.63%) and 1–10 μm pores (17.11%), followed by 0.1–1 μm pores (14.35%), and only 0.97% in micropores (<0.1 μm).

2.3. Analysis of Core Oil Displacement Experimental Curves

2.3.1. Microscopic Effects

After saturating the micro-visualization model with oil and aging it in an oven at 130 °C for 12 h, the subsequent displacement experiments revealed the complexity of multi-stage fluid interactions and their influence on displacement performance. The overall injection process is illustrated in Figure 10. In stage ①, the entire micro-model was saturated with oil. In stage ②, the water-phase flooding commenced. As the water phase primarily flowed along the fracture pathways, it advanced rapidly through the larger channels. However, this preferential flow behavior soon led to pronounced channeling within the fractures. Due to the relatively large flow conduits in the fractures, the water phase moved swiftly through these zones, bypassing the smaller pore throat regions adjacent to the fractures. As a result, the simulated oil in these regions was not effectively contacted and could not be efficiently displaced.

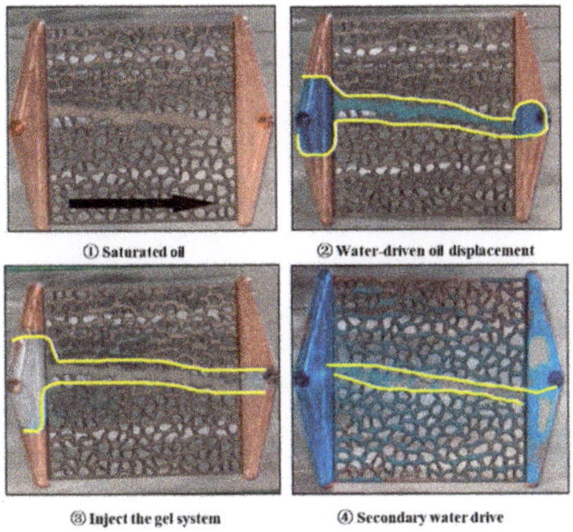

Figure 10. Microscopic visualization model of gel pore throat mobilization area (The yellow area represents the main working range of the gel).

In stage ③, the gel base fluid was injected, revealing its potential advantages in fluid displacement. The introduction of the gel base fluid significantly alleviated the flow heterogeneity observed during the water flooding in stage ②. The gel base fluid was able to penetrate the regions adjacent to the fractures and progressively extend into the small pore throat areas on both sides of the fractures, particularly near the injection inlet. This process demonstrated that the gel base fluid could, to some extent, improve the distribution of fluids within the complex pore structure, enabling the mobilization of simulated oil in the small pore throat regions flanking the fractures. Nevertheless, the flow of the gel base fluid remained influenced by the fracture-dominated pathways; although its mobilization of oil in small pore throat regions was enhanced, the overall displacement effect still did not reach a fully uniform state.

Following an 8 h aging period of the gel base fluid, stage ④ was conducted as a secondary water flooding. Owing to the flow regulation achieved by the gel system, the secondary water flooding effectively controlled the swept volume and mitigated the channeling phenomenon observed during the primary water flooding in stage ②. As a result, simulated oil in the small pore throat regions on both sides of the fractures was further mobilized. In particular, the lighter color in these regions indicated a significant reduction in oil saturation.

2.3.2. Parallel Experiment of Two Tubes

To evaluate the plugging efficiency and oil recovery of the gel under different permeability conditions, parallel double-core flooding experiments were conducted. The results, as shown in Figure 11, illustrate the plugging efficiency and recovery factor for a low-permeability core (50 mD) and a high-permeability core (500 mD) at various injection volumes. After correcting the inlet pressure, it was observed that the gel system significantly increased the plugging resistance in the low-permeability core, thereby forcing the displacement fluid to preferentially flow toward the high-resistance zones under elevated flow resistance conditions.

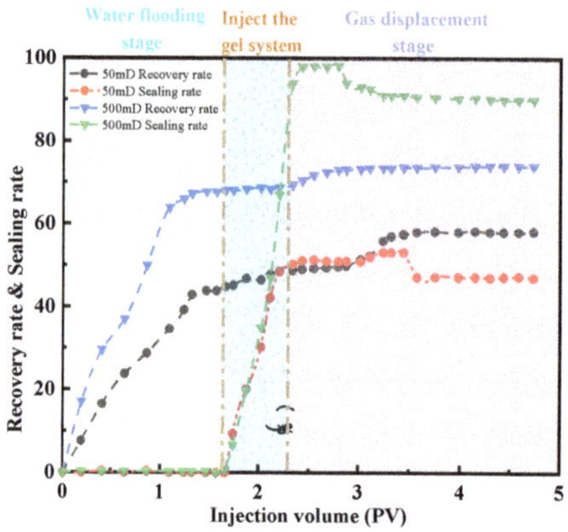

Figure 11. Block rate and recovery factor of parallel core.

For the low-permeability core, the oil recovery during the water flooding stage was 38.2%, which increased by 13.6% to reach 51.8% after gel injection. In contrast, for the high-permeability core, the recovery improved from 52.3% to 58.5%. The plugging performance of the gel system was further reflected in the increase in the plugging pressure in the low-permeability core from 0.18 MPa to 0.25 MPa, corresponding to a plugging efficiency of 38.9%. In the high-permeability core, the plugging pressure increased drastically from 0.05 MPa to 3.8 MPa, yielding a plugging efficiency of 98.7%. This indicates that the formation of a dense crosslinked network with a rigid structure created an immobile barrier, reducing the permeability from 500 mD to 6.5 mD. In the low-permeability core, only minor damage occurred, with permeability decreasing from 50 mD to 30.5 mD. From a recovery standpoint, these results elucidate the gel's displacement mechanism: The effective plugging of the high-permeability layer forced the displacement fluid to be redirected toward the low-permeability zone, leading to a pronounced recovery enhancement. The gel formed an immobile barrier in the high-permeability layer (plugging efficiency > 98%) while maintaining mobility in the low-permeability layer. Under 120 °C conditions, the gel exhibited a strength retention of over 90% after 30 days and a dehydration rate of less than 3%. The experiments confirm that under a permeability contrast of 10:1, the gel system achieved efficient plugging in the high-permeability layer and effective displacement in the low-permeability layer. The redistribution of flow paths was identified as the fundamental reason for the 13.6% increase in recovery in the low-permeability core.

3. Conclusions

(1) The study optimized the gel system's synthesis parameters using response surface methodology (APAM 180 mg/L, PEI:SMP = 3:1, salinity 150,000 ppm, 110 °C) and established the relationship between APAM concentration and system viscosity over time.

Based on this, quantitative analysis can be used to adjust the crosslinker concentration, thus controlling the progress of the crosslinking reaction. This approach enables more efficient planning of field oil well injection and shut-in waiting time for gelation, ensuring precise sealing of fractures in different reservoir sections. The gel system demonstrated a narrow molecular weight distribution (M_n = 135,050 g/mol, M_v = 173,435 g/mol, Đ = 1.284), significantly outperforming conventional free-radical polymerization systems (Đ > 2.0). This narrow dispersity promotes network uniformity and self-healing ability. The uniform molecular weight distribution ensures the nanoscale regularity of the crosslinked network, enabling the gel to exhibit rapid self-healing under shear stress (recovery rate > 98%). This study proposes a new paradigm for controlling network topology through molecular weight distribution (with plugging efficiency deviation <3% when Đ < 1.3) and establishes a quantitative predictive model for optimizing performance.

(2) The gel system exhibits excellent thermal stability, with a mass retention rate of 93.48% ± 1.02% at 120 °C, making it suitable for reservoir environments at 110 °C. Significant decomposition begins at 200 °C, and mass loss reaches 89.34% at 500 °C. Rheological performance shows that the gel predominantly behaves as a viscoelastic solid in the range of 100–120 °C, with the storage modulus (G') exceeding the loss modulus (G''). At 110 °C, G' ranges from 1543 to 4173 Pa, demonstrating strong resistance to erosion and excellent self-healing ability. Through rational formulation and structural optimization, the gel system performs exceptionally in terms of thermal and salt resistance, sealing capability, and oil displacement efficiency, offering an effective solution for the development of high-temperature, high-salinity reservoirs.

(3) Nuclear magnetic resonance imaging, micro-visualization, and parallel core displacement experiments demonstrate that the gel regulates fluid path distribution, reducing the residual oil saturation in large pores (>10 μm) to 19.63%. The underlying mechanism reveals that the gel preferentially enters high-permeability channels, forming an immobile barrier that forces subsequent displacing fluids to divert toward the low-permeability regions. This process addresses the inherent challenges of bypass flow and inefficient circulation in heterogeneous reservoirs. Through a comprehensive innovation chain involving molecular design, structural regulation, and performance validation, the gel system's triple-synergistic crosslinking mechanism, narrow molecular weight distribution network, and selective sealing ability provide technical support for the development of deep heterogeneous reservoirs, offering both engineering applicability and environmental friendliness.

4. Materials and Methods

4.1. Materials and Instruments

Anionic polyacrylamide (APAM), high-salt-resistance crosslinking agent (polyethyleneimine PEI), high-temperature stabilizer (Phenolic resin SMP), sodium chloride (NaCl), magnesium chloride ($MgCl_2$), calcium chloride ($CaCl_2$), Aladdin Biochemical Technology Co., LTD.

NDJ-9S digital viscometer, Shanghai Lichen Bangxi Instrument Technology Co., LTD. Quanta 200F field emission scanning electron microscope, FEI Company, USA EVOMA15/LS15 scanning electron microscope, German Zeiss Company. Type 101-1 Electric Heating Constant Temperature Drying Oven, Gongyi Yuhua Instrument Co., LTD. HT-PNMR12-9 nuclear magnetic resonance spectrometer, Haihuantong Science and Education Equipment Co., LTD. FT-IR Fourier transform infrared spectrometer, Bruker, Germany TGA2(LF). Thermogravimetric analyzer, Mettler Toledo Technology (China) Co., LTD. GelMax-2000 gel chromatograph, Shanghai Du Kai Biotechnology Co., LTD. HAAKE rheometer (RS600), Thermo Fisher Scientific, Germany. Hx-ii high-temperature and high-pressure displacement oven, Nantong Huaxing Petroleum Instrument Co., LTD.

4.2. Gel Preparation Process

In a beaker, 100 mL of distilled water was added to prepare a high-salinity solution, which was then placed on a magnetic stirrer platform to begin stirring. A specific mass fraction of APAM was weighed and slowly added to the solution until all solids were completely dissolved, forming a uniform transparent solution. Next, high-salt-tolerant crosslinker (PEI) and high-temperature stabilizer (SMP) were gradually added until the total monomer concentration reached the designed value. The prepared gel solution was transferred into a sealed test tube and placed in a high-temperature oven. Viscosity was measured every 4 h. Crosslinking was considered to have begun when the viscosity exceeded 1000 mPa·s and continued to rise rapidly, until the gel strength reached level I or higher. Each sample was tested only once, and the sample was discarded once opened.

4.3. Optimize the Optimal Reaction Conditions

4.3.1. The Response Surface Method Is Used to Analyze the Interaction Between the Two Factors

To optimize the gel system's strength, the Box–Behnken response surface design method was employed as an effective strategy [26]. Based on the extensive results from previous single-factor experiments, it was determined that APAM concentration, PEI concentration, salinity, and temperature have significant effects on the synthesis outcome. The optimal ranges for these factors were established: APAM concentration between 150 mg/L and 210 mg/L, PEI:SMP concentration ratio between 200 mg/L and 400 mg/L, salinity between 100,000 ppm and 200,000 ppm, and temperature between 100 °C and 120 °C. To further optimize the parameters and identify the best formulation, a 4-factor, 3-level experiment was designed to evaluate the combined effect of these 4 factors on gel strength. Each factor was set at three levels: low (−1), medium (0), and high (1). By combining the different levels of these factors, the influence of each factor on gel strength could be clarified without the need for a full-factorial experiment. The experimental goal was to measure the gel strength score, with a range of 1 to 100, where higher scores indicate better gel strength. The optimal formulation conditions were determined, and the experimental factors and levels are shown in Table 2.

Table 2. Variable factors and levels.

Level	Factor			
	A. APAM (mg/L)	**B. PEI:SMP**	**C. TDS (ppm)**	**D. Temperature (°C)**
−1	150	2:1	100,000	100
0	180	3:1	150,000	110
1	210	4:1	200,000	120

4.3.2. Time–Viscosity Relationship

Based on the results from Section 2.1.1, a numerical fitting approach was employed to establish the relationship between the two most significant interacting factors, aimed at simulating the precise profile adjustment in practical applications [13]. The data obtained in Section 2.1.1 were subjected to Boltzmann fitting, as described by Equation (2).

$$\eta(t) = \eta_0 + \frac{\eta_{max} - \eta_0}{1 + e^{(t_0 - t)/\Delta_t}} \quad (2)$$

In the equation, η represents viscosity (mPa·s), η_0 is the initial viscosity (mPa·s), η_{max} is the maximum viscosity (mPa·s), t is time (h), t_0 is the time corresponding to the curve inflection point (h), and Δ_t is the time difference for viscosity change (h). The system viscosity (η) follows an exponential relationship with time (t), and the viscosity variation curves under different independent variable factors exhibit similar patterns. This behavior can be summarized by the time–viscosity relationship in Equation (2).

4.4. Evaluation and Analysis of Gel System

4.4.1. SEM Electron Microscope

Scanning electron microscopy (SEM) imaging provides direct visualization with high magnification, making it ideal for characterizing the microstructure of test samples. The sample to be tested was placed in a Petri dish and subjected to liquid nitrogen freezing to preserve its microstructure [27]. The frozen sample was then transferred to a freeze-drying instrument and vacuum-dried for approximately 8–10 h to remove moisture. Following this, the sample underwent ethanol gradient dehydration (30–100%, each stage lasting 15–30 min) and was then dried using a critical point dryer. In this process, liquid CO_2 was used to replace the solvent in the sample (3–5 exchanges), and drying was completed under critical point conditions (31.1 °C/72.8 atm) to prevent surface tension-induced damage. After treatment, the sample was securely mounted onto the SEM copper sample holder using conductive adhesive. For SEM observation, a 3 kV accelerating voltage and a 5 mm working distance were selected, with secondary electron detection to capture surface morphology, facilitating detailed observation.

4.4.2. Infrared Spectrum

Fourier transform infrared spectroscopy (FTIR) is based on the selective absorption of infrared light at specific wavelengths by molecular structures. The gel system to be tested was freeze-dried, and a 2 mg sample was mixed with dried potassium bromide powder (approximately 1:100 mass ratio). The mixture was then ground evenly and pressed into a transparent thin film. This film was directly coated onto a potassium bromide crystal to form a uniform liquid membrane. The FTIR scan range was 4000–400 cm^{-1}, with a resolution of 0.35 cm^{-1} and 32 scans. The sample was placed in the optical path for scanning. After obtaining the spectrum, characteristic absorption peaks (e.g., C=O stretching vibrations, C-F bonds) were identified and compared with a standard spectral library to confirm the functional group structure.

4.4.3. Thermogravimetric Analysis Method and Principle

A 10 mg sample was placed in an alumina crucible and treated in a vacuum-drying oven at 40 °C for 12 h. The crucible was then placed in a thermogravimetric analyzer, with a nitrogen atmosphere set at a flow rate of 50 mL/min. The temperature range was 25–800 °C, with a heating rate of 10 °C/min. A blank crucible was run simultaneously to account for baseline drift. The weight loss curve was recorded [28], and the initial decomposition temperature (T_0), the temperature of maximum weight loss rate (T_{max}), and the residual carbon content were analyzed to assess thermal stability and decomposition behavior. The gel system's weight retention (W'/W) was evaluated. Subsequently, the glass transition temperature (Tg), crystallization temperature (Tc), and melting temperature (Tm) of the gel system were determined. The mass change (W/g) of the gel sample was observed under controlled temperature conditions, and differential scanning calorimetry (DSC) results were analyzed using the Gordon–Taylor equation.

$$\sum_{i=0}^{n} W_i \cdot A_i \cdot (T_g - T_{g,i}) = 0 \quad (3)$$

In the formula, W_i is the weight fraction of unit i, T_g is the glass transition temperature, A_i is the constant, and $T_{g,i}$ is the glass transition temperature of the homopolymer of unit i.

4.4.4. Molecular Weight Analysis Method and Principle

The test sample was dissolved in a mobile phase of DMF containing LiBr at a concentration of approximately 5 mg/mL. The solution was subjected to ultrasonic treatment for 30 min until fully dissolved. It was then filtered through a 0.22 μm PTFE membrane to remove insoluble substances and particulate impurities. The gel chromatography system was activated, and a suitable chromatographic column was selected. The mobile phase flow rate was set to 1.0 mL/min, and the column temperature was maintained at 30 °C. A calibration curve was established using narrow-distribution polystyrene standards, covering the target molecular weight range (1k–500k Da). A 20 μL sample was injected, and the elution curve was recorded using a differential refractive index detector (RID). The number-average molecular weight (M_n), weight-average molecular weight (M_w), and dispersity index ($Đ$) were calculated [29].

4.4.5. Rheological Experiment

Gel-based water shutoff systems exhibit a non-Newtonian fluid behavior, positioned between elastic solids and viscous liquids. Rheological properties, especially viscoelasticity, are commonly used to characterize gel systems. The mechanical properties of the gel structure change over time as it relaxes. The storage modulus (G') reflects the elastic nature of the gel, while the loss modulus (G'') indicates its viscosity [30]. A higher loss modulus (G'') implies greater internal friction and resistance to erosion, whereas a higher storage modulus (G') signifies better recovery and impact resistance after deformation. The gel's self-healing and re-crosslinking ability are stronger after localized damage. Under mechanical deformation, part of the energy appears elastically, corresponding to the stress response described by the storage modulus (G'), while another portion of energy is dissipated based on the molecular structure during dynamic movement, reflected by the stress response of the loss modulus (G''). During the testing process, it was confirmed that the storage modulus (G') was higher than the loss modulus (G''), ensuring that all dynamic rheological tests were conducted within the stable viscoelastic region. The storage modulus (G') and loss modulus (G'') of the gel system before and after gelation were measured using a HAAKE rheometer for viscoelastic evaluation, with a shear rate of 7.34 s^{-1}. The relationship between G', G'', and the loss angle δ is given by the following equation:

$$\tan\delta = G''/G' \tag{4}$$

In the formula, δ represents the fluid property, δ = 0–20° indicates an elastomer, δ = 20–40° indicates a viscoelastomer, and δ > 40° indicates a viscous substance.

4.4.6. Magnetic Resonance Testing

The nuclear magnetic resonance (NMR) T_2 spectrum is used to describe the pore distribution characteristics of rock cores by analyzing the distribution of fluids within the porous medium. The relaxation time reflects only the pore size but does not fully describe the pore diameter of the porous medium. Mercury intrusion porosimetry, which calculates pore size based on capillary pressure, provides a more detailed measurement. By combining NMR technology with mercury intrusion porosimetry, a permeability of 25.8 mD was selected. The NMR T_2 spectrum was compared with the pore throat distribution obtained from the mercury intrusion curve. In addition, thin-section data of the core were considered to establish a correlation between relaxation time and pore diameter [31]. The NMR testing system used in the experiment is shown in Figure 12 below.

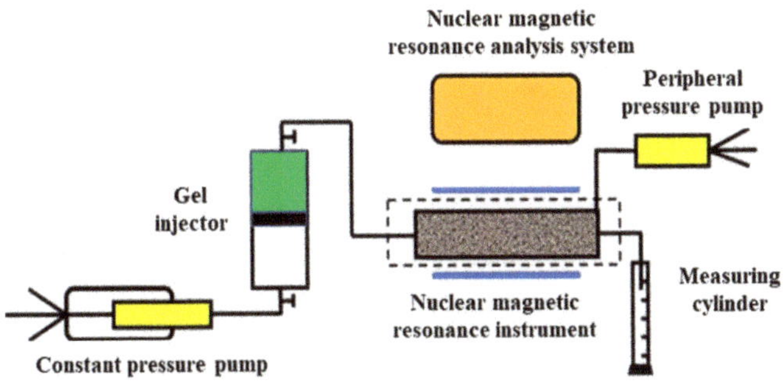

Figure 12. NMR system.

The T_2 spectrum in porous media is expressed according to the following formula:

$$T_2 = \frac{1}{T_{2S}} + \frac{1}{T_{2D}} + \frac{1}{T_{2B}} \quad (5)$$

In the formula, T_{2S} is the surface relaxation term, ms; T_{2D} is the diffusion relaxation term, ms; and T_{2B} is the body relaxation term, ms. Nuclear magnetic resonance technology is used to determine the fluid in porous media. The diffusion relaxation and bulk relaxation terms can be ignored, and the relaxation time of the fluid mainly depends on the surface relaxation term. The surface relaxation term is related to the specific surface area of the rock: The relaxation term increases as the specific surface area of the rock increases, and the T_2 relaxation time decreases. The surface relaxation term of the rock can be expressed by the following formula:

$$\frac{1}{T_{2S}} = \rho_2 \frac{S}{V} \quad (6)$$

In the formula, P_2 is the relaxation rate, μm/ms, and S/V is the specific surface area of the pore throat, 1/μm. The specific surface area is related to the pore throat radius of the porous medium:

$$\frac{S}{V} = \frac{F_S}{r} \quad (7)$$

F_S is the dimensionless shape coefficient of the pore throat, and r is the pore throat radius, μm. Therefore, T_2 can be expressed as:

$$T_2 = \frac{1}{\rho_2 F_S} r^n \quad (8)$$

The degree of micropore utilization (U) is used to quantitatively characterize the proportion of the gel that actually occupies the target pore structure:

$$U = \frac{V_{used}}{V_{total}} \times 100\% \quad (9)$$

where V_{used} is the volume of micropores occupied by the gel after sealing (measured by low-field nuclear magnetic resonance, NMR), in cm^3, and V_{total} is the total micropore volume of the target area (calculated from the total core volume and porosity), in cm^3.

4.5. Double-Tube Parallel Core Experiment

4.5.1. Gel-Sealing Mechanism and Microscopic Visualization Experiment

The mechanical initiation mechanism of the composite gel water-blocking system simulated by indoor experimental physics follows the Navier–Stokes equation of fluid dynamic motion:

$$\frac{\partial \alpha \rho}{\partial t} + \nabla \cdot \rho \alpha v = 0 \tag{10}$$

In the formula, α is the fluid volume fraction; ρ is the fluid density, kg/m^3; and v is the fluid velocity, m/s.

Due to the continuous accumulation of the continuous-phase gel during the migration process, the interaction with the fluid generates action and reaction forces. According to Newton's second law, the forces acting on the gel system mainly include gravity, the interaction force between the polymer and the crosslinking agent, and the interaction force between the gel system and the fluid. Among them, drag forces will be generated between the two mutually moving phases. Based on this, the motion equation of the gel system can be expressed by Formula (14):

$$m\frac{dv_p}{dt} = F_D + F_g + \sum F_i \tag{11}$$

In the formula, v_p is the gel migration velocity, m/s; F_D is the fluid drag force, N; F_g is gravity, N; and F_i is the the force between different gel systems, N.

The gel system follows the equation of conservation of momentum during fluid motion:

$$\frac{\partial \alpha \rho}{\partial t} + \nabla \cdot \rho \alpha \mu v = -\nabla \rho + \nabla \cdot (\mu \alpha \nabla v) + \rho \alpha g - F_D \tag{12}$$

In the formula, g is the gravitational acceleration, m^2/s; μ is the fluid viscosity, kg/(m·s); and F_D is the fluid drag force, N.

The creation of a microscopic model for a high-temperature and high-pressure visual microscopic displacement experiment involves the following steps: Design the model: Based on the experimental requirements, design the geometric shape and dimensions of the model. The commonly used model is made of transparent high-strength plastic material, allowing for a clear observation of fluid flow and displacement processes. Material selection: Use transparent materials resistant to high temperature and high pressure, such as special glass. It has good chemical corrosion resistance to adapt to the experimental environment. Mold making: Make the model mold according to the design drawings. The model shown in Figure 13 below is the model designed for the laboratory. The precision of the mold directly affects the quality of the final model, which is manufactured using 3D printing technology. Injection of material: Inject the selected transparent material into the mold to ensure uniform filling. For resin models, bubbles need to be removed in a vacuum environment after injection. Curing and processing: Subject the injected material to a curing treatment. After completion, remove the model from the mold. Carry out subsequent processing on the model and polish the surface. Install accessories: Install accessories on the model, including injection holes, sampling ports, and pressure sensor interfaces. Conduct experiments: After the model is completed, conduct high-temperature and high-pressure tests on it to verify the performance and stability of the model. Ensure that under experimental conditions, the model can work properly and provide clear visualization effects.

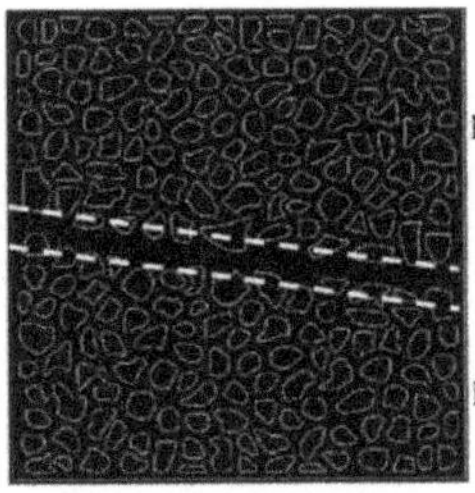

Figure 13. Microscopic model.

This model adopts an alternating configuration of small holes and slit structures, and the width of the slit structure is relatively narrow. This design is used to study the influence of alternating gaps and apertures on fluid behavior. Effective size: 30 × 30 mm. Small holes and narrow slits are alternately arranged, and the junctions of small holes and slits form specific fluid channel characteristics. Analyze the alternating effects of gaps and apertures on fluid flow and displacement behavior, as well as their manifestations under different structures.

4.5.2. Parallel Displacement Experiment

One group each of low-permeability (50 mD) and high-permeability (500 mD) standard sandstone cores (with a diameter of 2.5 cm and a length of 10 cm) was selected. After cleaning and drying, the simulated formation water (with a salinity of 150,000 PPM) was vacuum-saturated and installed in the parallel core clamps. The confining pressure was maintained at 15 MPa, and the temperature was set as the reservoir temperature (60 °C). The simulated formation water was injected at a constant flow rate (0.1 mL/min), the pressure difference at the inlet of the two cores was recorded [32], and the initial permeability was calculated:

$$K = \frac{Q\mu L}{A \cdot \Delta P} \tag{13}$$

In the formula, K is the permeability, 10^{-3} μm^2; Q is the flow rate at the core outlet, mL/min; mu is the viscosity of simulated formation water, and is taken as 8.9×10^{-4} Pa·s; L is the core length, cm; A is the cross-sectional area of the core, cm^2; and ΔP is the pressure difference at both ends of the core, MPa.

The gel was prepared and pre-crosslinked for 30 min. The gel was injected into the two cores at a rate of 0.3 mL/min. The injection was stopped when the pressure at the outlet end of the hypertonic core rose to three times the initial pressure or the flow rate of the hypotonic core decreased by 90%. The cumulative injection volume was recorded. The system was shut down and let stand for 24 h (at a constant temperature of 60 °C) to ensure that the gel was fully cross-linked to form a three-dimensional network structure. The formation water was injected again at 0.1 mL/min to monitor the changes in the flow rate at the outlets of the two cores until the pressure stabilized. The post-blockage permeability K and the water-blocking rate Γ were calculated:

$$\Gamma = \frac{K_W - K_P}{K_W} \times 100\% \tag{14}$$

In the formula, K_W and K_P are the water permeability measured before and after the injection of the gel blocking agent, respectively.

To avoid flow deviations caused by instrument system errors, a pre-pressurization test was conducted on each core channel under idle conditions before the experiment. This allowed for the calibration of the initial inlet pressure, ensuring that any subsequent pressure changes during the injection phase were attributed solely to the effects of the gel system.

Author Contributions: Conceptualization, C.Y. and D.X.; methodology, T.L.; software, J.W.; validation, C.Y.; formal analysis, D.X.; investigation, D.X.; resources, T.L.; data curation, D.X.; writing—original draft preparation, D.X.; writing—review and editing, D.X.; visualization, C.Y.; supervision, J.W.; project administration, J.W.; funding acquisition, J.W. All authors have read and agreed to the published version of the manuscript.

Funding: Xi'an Shiyou University Graduate Innovation Fund Project (YCX2512020), National Postdoctoral Researcher Program Funding (GZC20232142), Shaanxi Province Postdoctoral Research Funding Project "Research on the Seepage, Suction, Drainage and Flooding Mechanism of Small-Sized Active Sheet-like Nanofluids in Ultra-Low Permeability Reservoirs" (2023BSHTBZZ45).

Institutional Review Board Statement: Not applicable.

Informed Consent Statement: Not applicable.

Data Availability Statement: The data presented in this study are openly available in the article.

Conflicts of Interest: The authors declare no conflicts of interest.

References

1. Liu, J.; Zhong, L.; Wang, C.; Li, S.; Yuan, X.; Liu, Y.; Meng, X.; Zou, J.; Wang, Q. Investigation of a high temperature gel system for application in saline oil and gas reservoirs for profile modification. *J. Pet. Sci. Eng.* **2020**, *195*, 107852. [CrossRef]
2. Zhai, Y.; Fang, Z.; Feng, J.; Sun, C.; Deng, W.; Wen, Y. Development and performance evaluation of re-crosslinkable preformed particle gel under high temperature and high salt conditions. *Polymer* **2024**, *296*, 126824. [CrossRef]
3. Wei, Z.; Liu, J.; Zhang, L.; Yang, L.; Li, Y.; Wang, J.; Fan, H. A novel approach to investigating the influencing factors and mechanism of re-crosslinking performance of re-crosslinkable preformed particle gels. *Geoenergy Sci. Eng.* **2025**, *247*, 213679. [CrossRef]
4. Peng, Y.; Ye, J.; Li, Y.; Chen, Y.; Li, Z.; Zhang, D. Development and Performance Evaluation of Novel Self-Degradable Preformed Particulate Gels with High-Temperature and High-Salinity Resistance. *SPE J.* **2025**, *30*, 1105–1115. [CrossRef]
5. Xin, H.; Fang, B.; Yu, L.; Lu, Y.; Xu, K.; Li, K. Rheological performance of high-temperature-resistant, salt-resistant fracturing fluid gel based on organic-zirconium-crosslinked HPAM. *Gels* **2023**, *9*, 151. [CrossRef]
6. Xu, B.; Wang, Y. Profile control performance and field application of preformed particle gel in low-permeability fractured reservoir. *J. Pet. Explor. Prod. Technol.* **2021**, *11*, 477–482. [CrossRef]
7. Deng, J.; Lian, H.; Zhuang, Y.; Zhao, H.; Wang, Z.; Tian, Y.; Lin, C.; Yuan, H.; Han, M.; Lu, G.; et al. Synthesis and performance evaluation of multi-crosslinked preformed particle gels with ultra-high strength and high-temperature and high-salinity resistance for conformance control. *Fuel* **2024**, *357*, 130027. [CrossRef]
8. Guo, H.; Ge, J.; Xu, Y.; Lv, Q.; Li, Z.; Zhou, D.; Tao, Z. Preparation and Mechanism of Stability for High-Temperature and High-Salinity Gels. *SPE J.* **2022**, *27*, 3565–3578. [CrossRef]
9. Ge, J.; Guo, H.; Zhang, T.; Zhou, D.; Xu, Y.; Lü, Q. Development of temperature and salinity resistant phenolic gel and its performance regulation mechanism. *Acta Pet. Sin.* **2022**, *43*, 1145.
10. Zhou, G.; Zhang, X.; Yan, W.; Qiu, Z. Synthesis, Characteristics, and Field Applications of High-Temperature and Salt-Resistant Polymer Gel Tackifier. *Gels* **2025**, *11*, 378. [CrossRef] [PubMed]
11. Wang, Z.; Bai, B.; Zhou, E.; Pu, J.; Schuman, T.P. Experimental Evaluation of Oxidizing Breakers for a Polyacrylamide-Based Re-Crosslinkable Preformed Particle Gel. *Energy Fuels* **2019**, *33*, 5001–5010. [CrossRef]
12. Liu, J.; Zhong, L.; Cao, Z.; Hao, T.; Liu, Y.; Wu, W. High-temperature performance and cross-linking mechanism of different types of gel systems in saline environment. *J. Appl. Polym. Sci.* **2022**, *139*, 51452. [CrossRef]
13. Wu, Y.; Zhang, H.; Zhang, L.; Huang, Y.; Zhao, M.; Dai, C. Development of Novel Delayed Swelling Polymer Gel Particles with Salt Resistance for Enhanced In-Depth Permeability Control. *SPE J.* **2024**, *29*, 2060–2075. [CrossRef]
14. Shu, Z.; Qi, Y.; Luo, P. Research and performance evaluation of modified nano-silica gel plugging agent. *J. Appl. Polym. Sci.* **2023**, *140*, e53873. [CrossRef]
15. Bai, Y.; Wu, L.; Luo, P.; Li, D. Synthesis and evaluation of delayed anti-high temperature gel plugging agent. *Front. Energy Res.* **2022**, *10*, 1003473. [CrossRef]
16. Chen, J.; Qiu, H.; Djouonkep, L.D.W.; Lv, J.; Xie, B. Preparation, Evaluation and Field Application of Thermally Induced Crosslinked Polymer Gel Leakage Plugging Agent. *J. Polym. Environ.* **2024**, *32*, 5677–5688. [CrossRef]

17. Fang, J.; Sun, J.; Feng, X.; Pan, L.; Bai, Y.; Yang, J. Research and Development of a High-Temperature-Resistant, Gel-Breaking Chemical Gel Plugging Agent and Evaluation of Its Physicochemical Properties. *Gels* **2025**, *11*, 350. [CrossRef]
18. Liu, J.; Fu, H.; Luo, Z.; Chen, W.; Liu, F.; Zhao, M. Preparation and performance of pH-temperature responsive low-damage gel temporary plugging agent. *Colloids Surf. A Physicochem. Eng. Asp.* **2023**, *662*, 130990. [CrossRef]
19. Fan, H.; Pu, W.; Du, D.; Xu, M.; Chen, L.; Xu, L.; Jin, F. Experimental study on the plugging and profile control of preformed particle gels in high temperature and salinity reservoir. *J. Appl. Polym. Sci.* **2023**, *140*, 14. [CrossRef]
20. Huang, B.; Zhang, W.; Zhou, Q.; Fu, C.; He, S. Preparation and Experimental Study of a Low-Initial-Viscosity Gel Plugging Agent. *ACS Omega* **2020**, *5*, 15715–15727. [CrossRef] [PubMed]
21. Li, Y.-K.; Hou, J.-R.; Wu, W.-P.; Qu, M.; Liang, T.; Zhong, W.-X.; Wen, Y.-C.; Sun, H.-T.; Pan, Y.-N. A novel profile modification HPF-Co gel satisfied with fractured low permeability reservoirs in high temperature and high salinity. *Pet. Sci.* **2024**, *21*, 683–693. [CrossRef]
22. Liu, F.; Yin, D.; Sun, J.; Luo, X.; Huang, X. Preparation and Characterization of Temperature-Sensitive Gel Plugging Agent. *Gels* **2024**, *10*, 742. [CrossRef] [PubMed]
23. Tian, J.; Liu, J.; Liu, S.; Yuan, S.; Cai, W.; Jiang, H. A gel system with long-term high-temperature stability for deep profile modification of high-temperature oil and gas reservoirs. *J. Appl. Polym. Sci.* **2023**, *140*, e54733. [CrossRef]
24. Li, Z.; Zhao, G.; Xiang, C. Synthesis and Properties of a Gel Agent with a High Salt Resistance for Use in Weak-Gel-Type Water-Based Drilling Fluid. *Arab. J. Sci. Eng.* **2022**, *47*, 12045–12055. [CrossRef]
25. Zou, H.; Wang, Y.; Xu, Y.; Li, J.; Wu, L.; Su, G.; Yu, X.; Yang, H. Synthesis and Performance Study of Self-Degradable Gel Plugging Agents Suitable for Medium- and Low-Temperature Reservoirs. *ACS Omega* **2024**, *9*, 33702–33709. [CrossRef]
26. Wang, H.; Yang, C.; Zhang, Y.; Wang, C. Preparation and Effect of CO_2 Response Gel for Plugging Low-Permeability Reservoirs. *Gels* **2024**, *10*, 449. [CrossRef]
27. Jia, H.; Ren, Q. Evidence of the Gelation Acceleration Mechanism of HPAM Gel with Ammonium Salt at Ultralow Temperature by SEM Study. *SPE Prod. Oper.* **2016**, *31*, 238–246. [CrossRef]
28. Zhang, W.; Gao, R.; Sha, X.; Liu, G.; Wang, X.-Y.; Lin, R.-S.; Tong, Y. Strength development and thermal stability analysis of carbonation-cured red mud-based cementitious materials. *Constr. Build. Mater.* **2025**, *470*, 140724. [CrossRef]
29. Papp, D.; Carlström, G.; Nylander, T.; Sandahl, M.; Turner, C. A Complementary Multitechnique Approach to Assess the Bias in Molecular Weight Determination of Lignin by Derivatization-Free Gel Permeation Chromatography. *Anal. Chem.* **2024**, *96*, 10612–10619. [CrossRef]
30. Liu, M.; Ge, J.; Wu, Q.; Ma, A.; Li, J.; Zhang, G.; Fan, H.; Jiang, P.; Pei, H. Study on temperature-resistance and salt-tolerance aluminum gel and its rheological analysis. *Colloid Polym. Sci.* **2024**, *302*, 665–677. [CrossRef]
31. Deng, Z.; Liu, M.; Qin, J.; Sun, H.; Zhang, H.; Zhi, K.; Zhu, D. Mechanism study of water control and oil recovery improvement by polymer gels based on nuclear magnetic resonance. *J. Pet. Sci. Eng.* **2022**, *209*, 109881. [CrossRef]
32. Zhang, L.; Liu, Y.; Wang, Z.; Li, H.; Zhao, Y.; Pan, Y.; Liu, Y.; Yuan, W.; Hou, J. Evaluation of Profile Control and Oil Displacement Effect of Starch Gel and Nano-MoS_2 Combination System in High-Temperature Heterogeneous Reservoir. *Gels* **2024**, *10*, 127. [CrossRef] [PubMed]

gels

MDPI

Article

Flow Characteristics and Enhanced Oil Recovery Performance of Anionic and Zwitterionic Viscoelastic Surfactant System

Chenyue Ling [1], Yafei Liu [1,*], Xuchun Yang [1], Qi Ye [2] and Desheng Zhou [1]

[1] College of Petroleum Engineering, Xi'an Shiyou University, Xi'an 710065, China; 23212010108@stumail.xsyu.edu.cn (C.L.); 24212010117@stumail.xsyu.edu.cn (X.Y.); dshzhou@xsyu.edu (D.Z.)
[2] Oil Production Plant No.5 of Changqing Oilfield, China National Petroleum Corporation, Xi'an 710201, China; yq3_cq@petrochina.com.cn
* Correspondence: yafliu@xsyu.edu.cn

Abstract: Surfactant flooding has shown potential in enhanced oil recovery (EOR), but conventional surfactants often underperform in heterogeneous reservoirs. This study investigates the impact of a surfactant mixture, combining anionic sodium dodecyl sulfate (SDS) and zwitterionic oleylamidopropyl betaine (OAB-30), on two-phase flow behavior and its EOR potential. Six surfactant solutions with varying concentrations were first screened using an idealized dead-end shaped microchannel in combination with interfacial properties and rheological tests. The results showed that 0.2% SDS and 0.6% OAB-30 produced the highest oil recovery in the dead-end structure. Interfacial tension was reduced to 0.374 mN/m and strong viscoelastic behavior was observed using the optimized surfactant mixture. Wettability of the surface tended to be more hydrophilic after the application of the surfactant mixture as well. Subsequently, the microscale oil displacement process was examined using the optimized surfactant mixture via microfluidic devices with an idealized pore–throat network with permeability contrast and realistic pore–throat structure. The application of the optimal surfactant formula resulted in 28.46% and 49.96% improvement over conventional water flooding in a realistic pore–throat structure and idealized pore–throat network. The critical micelle concentration measurements of the mixture suggested favorable micelle formation, contributing to gel-like properties that improved sweep efficiency by lowering the mobility ratio. In heterogenous pore–throat networks, the emulsification, micellar solubilization, wettability alteration, and viscoelastic properties of the surfactant mixture favored the oil recovery process. This work provides experimental evidence and mechanistic insights for the application of viscoelastic surfactants in EOR in heterogeneous reservoirs.

Keywords: viscoelastic surfactant mixture; oil–water two-phase flow; microfluidics; enhanced oil recovery

1. Introduction

Waterflooding, attributed to its accessibility and cost-effectiveness, has become the most extensively utilized oil recovery method [1]. However, the difference in viscosity between water and crude oil has resulted in an irregular oil front diminishing sweep efficiency [2]. Moreover, relatively higher interfacial tension between oil and water leads to higher capillary resistance, which in turn results in a substantial amount of residual oil within the reservoir [3]. Given the limitations of waterflooding, chemical enhanced oil recovery [4–8], particularly surfactant flooding, is increasingly being employed to improve the oil recovery [9]. Surfactant flooding is capable of reducing the flowing resistance of crude oil and enhancing sweep efficiency by lowering the interfacial tension between oil and water and altering the wettability [10,11].

Conventional single-surfactant systems have encountered limitations in practical applications. For example, anionic surfactants such as sulfonates [12] and sulfates [13], as well as nonionic surfactants [14], exhibit inadequate salt and temperature tolerance. In high-salinity and high-temperature reservoir environments, these surfactants experience a significant decline in chemical stability. In contrast, cationic surfactants are prone to substantial adsorption losses on sandstone reservoir rock surfaces [15]. This results in the consumption of a considerable amount of active surfactant components during the displacement process. Furthermore, the dynamic interfacial behavior of single-surfactant systems is challenging to adapt to the displacement process under complex reservoir conditions [16,17]. To surmount the limitations of single-surfactant systems, researchers have proposed the application of surfactant mixtures. The primary types encompass anionic–anionic [18], anionic–cationic [19], anionic–zwitterionic [20], and anionic–nonionic [21] surfactant mixtures. Among these combinations, viscoelastic surfactant (VES) formed by combining anionic and zwitterionic surfactants exhibits strong synergistic effects, significantly enhancing solution viscosity, surface activity, and thermal/salinity tolerance, thus making it ideal for complex applications [22,23]. Some of these mixtures can form wormlike micelles and entangled structures in aqueous solutions through hydrophobic interactions, electrostatic interactions, and hydrogen bonding, resulting in viscoelastic solutions [24]. Viscoelastic surfactants possess unique rheological properties that enable the regulation of the oil displacement process at the microscopic scale. Their viscoelasticity can effectively suppress fingering during the displacement process and improve the uniformity of fluid flow in the reservoir [25]. Solid understanding of the EOR potential and oil displacement mechanism of these surfactant mixtures is necessary for a broader application of surfactants.

In recent years, numerous studies have utilized macro-scale displacement techniques and numerical simulations to assess the effectiveness of various displacement fluids [26–28]. However, research concerning the interfacial phenomena at the micro-scale is still insufficient. To gain a more comprehensive understanding of the underlying mechanisms governing the behavior of displacement fluids, investigations into the micro-scale interfacial dynamics are critical. Therefore, microfluidic approaches have been employed for advancing the understanding of these mechanisms at the microscale. It is distinguished by its ability to visualize fluid flow dynamics and replicate the pore–throat architecture of reservoir rock. These attributes render it an invaluable tool for unraveling the intricacies of how surfactant mixtures improve oil recovery efficiency [29]. Microfluidic devices are usually fabricated with a network of grooves or microchannels etched into diverse substrates, including glass [30], silicon, and polymers such as polydimethylsiloxane (PDMS) and polymethylmethacrylate (PMMA) [31,32]. Microfluidic technology has been widely applied in the field of EOR. For instance, realistic sandstone micromodels were fabricated to scrutinize the oil displacement efficacy of a novel zwitterionic Gemini surfactant [33]. A microfluidic device with an idealized pore–throat structure was applied to delineate the spontaneous imbibition phenomena occurring within distinct pore–throat configurations [34]. While previous research has predominantly focused upon the assessments of fluid properties and theoretical computations [35–37], it is still necessary to incorporate microscopic experimental investigations for a comprehensive understanding on the performance of varying surfactant mixtures. Extensive research has been conducted on surfactant mixtures including viscoelastic surfactants. Existing literature mostly demonstrated their positive effects at the macro-scale, particularly in enhancing recovery rates. However, limited attention has been paid to the micro-scale interfacial phenomena and the underlying oil displacement mechanisms. Therefore, microfluidic technology was employed in this study to investigate the micro-scale behavior of viscoelastic surfactants. Specifically, the study examined the interactions of sodium dodecyl sulfate (SDS) and oleylamidopropyl betaine (OAB-30) with various concentration ratios. Interfacial tension, viscoelasticity, viscosity, and contact angle were measured to conduct a comprehensive evaluation of the surfactant mixture's characteristics and its performance on the oil recovery. Subsequently, the oil recovery mechanism of the optimal surfactant formulation was investigated using micromodels with varying pore–throat structures.

2. Results and Discussion

2.1. *Fluid Properties of Surfactant Mixtures*

2.1.1. Interfacial Tension

Interfacial tension measurements of the six surfactant formulations revealed that the 0.2% SDS+0.2% OAB-30 solution achieved the lowest interfacial tension, closely followed by the 0.2% SDS+0.6% OAB-30 formulation. In contrast, the 0.4% SDS+0.2% OAB-30 solution exhibited the highest interfacial tension among all tested combinations. The complete interfacial tension profiles are presented in Figure 1, showing the dynamic changes during the measurement period.

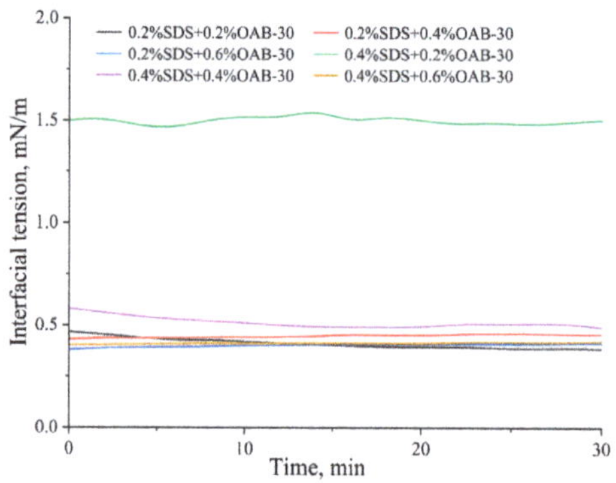

Figure 1. Dynamic interfacial tension of varying surfactant mixtures versus time.

Table 1 shows the interfacial tension data between crude oil and varying surfactant mixtures measured in this study.

Table 1. Interfacial tension between crude oil and varying surfactant mixtures.

Solution Composition	Interfacial Tension (mN/m)	
	Mean Value	Standard Deviation
0.2% SDS+0.2% OAB-30	0.376	0.020
0.2% SDS+0.4% OAB-30	0.444	0.011
0.2% SDS+0.6% OAB-30	0.395	0.023
0.4% SDS+0.2% OAB-30	1.539	0.076
0.4% SDS+0.4% OAB-30	0.521	0.016
0.4% SDS+0.6% OAB-30	0.430	0.021

2.1.2. Viscosity

Viscosity represents a fluid's internal frictional resistance to flow deformation that reflects the hindrance of molecular interactions to fluid motion. As shown in Figure 2, at a temperature of 30 °C and a shear rate of 170 rpm, the blended oil (crude oil + liquid paraffin) exhibited a viscosity of 55.8 mPa·s. The results showed that surfactant solution with 0.2% SDS+0.6% OAB-30 displayed the highest viscosity with a 1:4 ratio to the crude oil viscosity.

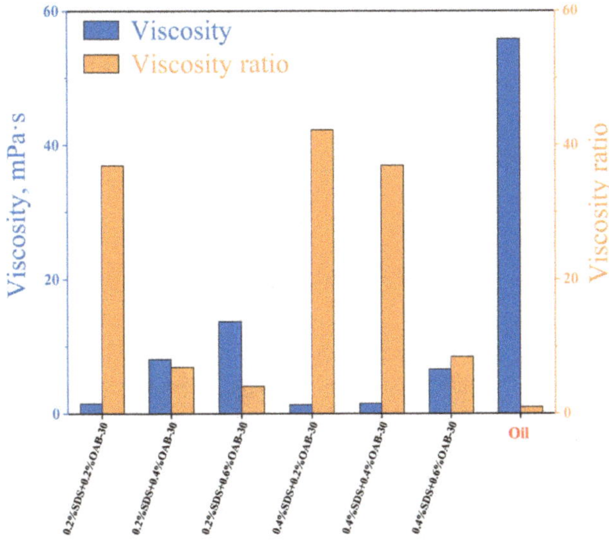

Figure 2. Viscosity of crude oil and surfactant mixtures with different concentrations and viscosity ratio between oil and the surfactant solutions.

2.1.3. Contact Angle

The initial contact angle measured after crude oil aging was 64.6 ± 2.38°. As shown in Figures 3 and 4, all surfactant mixtures resulted in wettability alteration towards a more hydrophilic state. The maximum wettability alteration occurred when SDS and OAB-30 were mixed at a 1:1 concentration ratio. The 0.2% SDS+0.6% OAB-30 formulation (1:3 ratio) showed the least effective wettability alteration among all tested solutions.

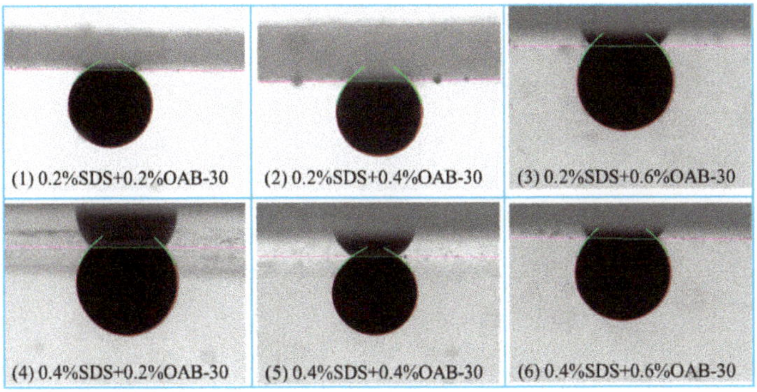

Figure 3. Images of contact angle measurement using different surfactant mixtures.

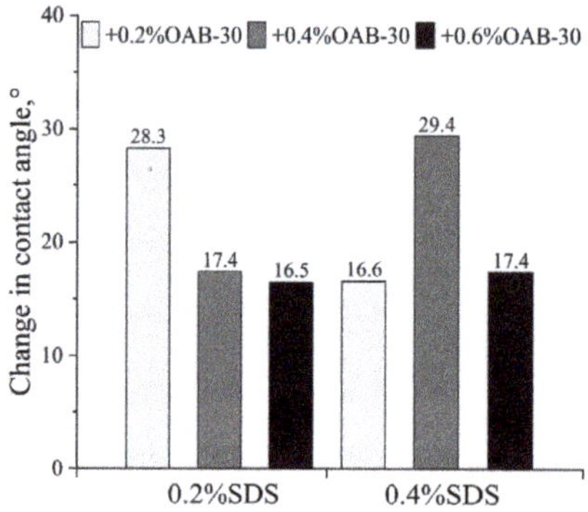

Figure 4. The change in the contact angle for different surfactant mixtures.

2.1.4. Viscoelasticity

Rheological measurements were conducted to characterize the viscoelastic properties of each formulation. Analyses of the six sets of angular frequency versus storage modulus (G′) and loss modulus (G″) curves are presented in Figure 5. At a fixed concentration of SDS, an increase in the concentration of OAB-30 results in a decrease in the cross-over angular frequency and an increase in the relaxation time and an enhancement in elasticity. Conversely, at a fixed concentration of OAB-30, an increase in the concentration of SDS results in an increase in the cross-over angular frequency, a decrease in the relaxation time and a relatively greater enhancement in viscosity. The 0.2% SDS+0.6% OAB-30 and 0.4% SDS+0.6% OAB-30 solutions exhibit a plateau region with a relatively high plateau modulus, which indicates a stronger elastic network. The viscoelastic properties of both groups are relatively good overall, with the 0.2% SDS+0.6% OAB-30 group exhibiting better elasticity at lower SDS concentrations. The enhanced viscoelasticity of the formulated solution indicates improved structural integrity and increased flow resistance. These characteristics are essential for maintaining effective mobility control during oil displacement processes.

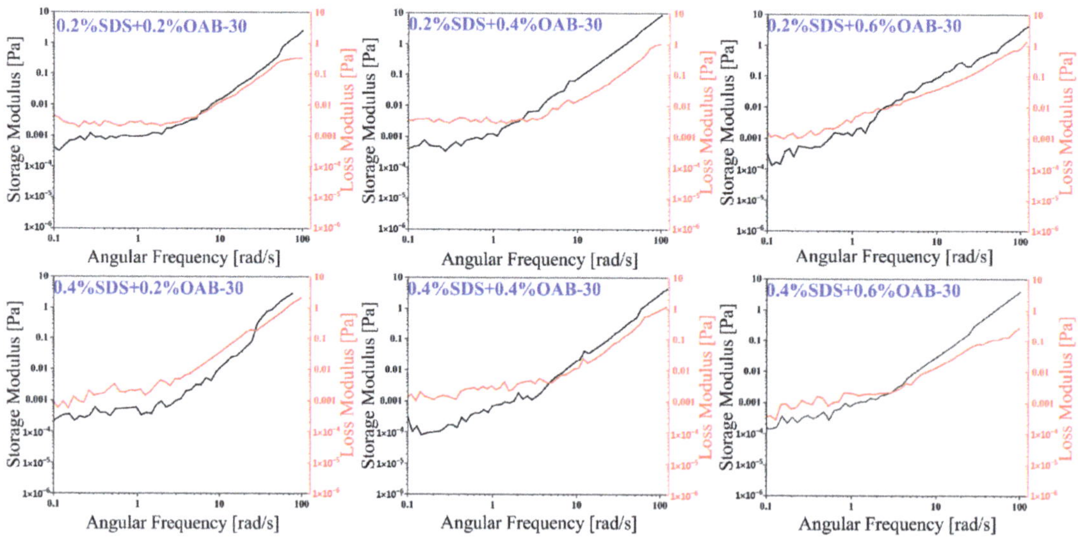

Figure 5. Storage modulus (G′) and loss modulus (G″) versus angular frequency for different surfactant mixtures.

2.2. Screening of the Optimal Surfactant Formula

Figure 6 presents the results of displacement experiments in dead-end pores utilizing six distinct surfactant formulations. Quantitative analysis was performed using Image J software (NIH, Win64) to determine the respective oil recovery efficiencies. The final recovery rates were calculated by averaging results from three central L-shaped structures for each formulation.

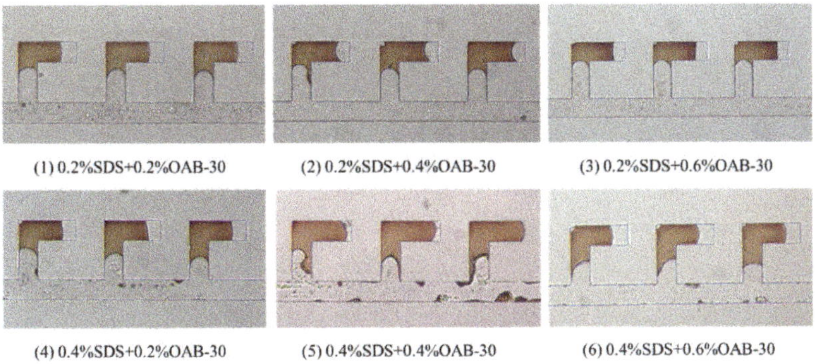

Figure 6. Final image after displacement process in the L-shaped dead-end pore structure using different surfactant mixtures.

As illustrated in Figure 7, displacement tests in the dead-end pores revealed that the optimal performance was achieved by the 0.2% SDS+0.6% OAB-30 solution, which demonstrated a superior recovery rate of 37.10%.

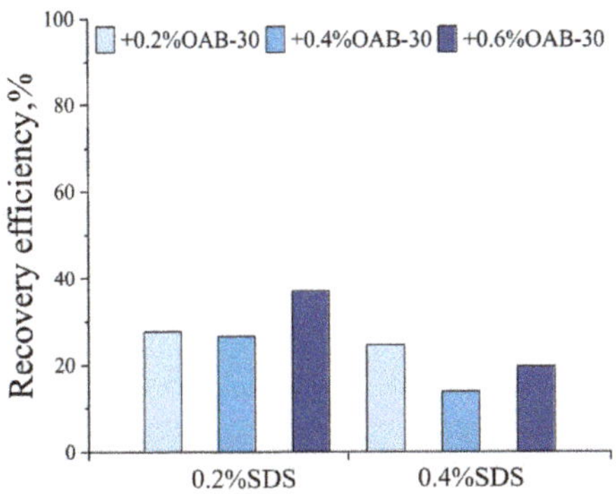

Figure 7. Oil recovery efficiency in L-shaped dead-end pore structure for different surfactant solutions.

Fluid property measurements revealed that the 0.2% SDS+0.6% OAB-30 solution exhibited the highest viscosity with a favorable 1:4 viscosity ratio to crude oil. This elevated viscosity ratio could potentially suppress viscous fingering and promote stable displacement fronts with uniform advancement. In addition, this system achieved relatively lower interfacial tension, favorable viscoelastic properties, and wettability alteration. The relatively high viscoelasticity improves the rheological properties of the displacement fluid, enabling a more uniform flow field within the pores. This could potentially reduce fluid fingering and enhance sweep efficiency. Additionally, even with limited changes in wettability, the adhesive force between oil droplets and pore walls can be reduced, thereby improving oil mobility. The 0.2% SDS+0.6% OAB-30 surfactant mixture resulted in efficient oil displacement through synergistic effects: IFT reduction, wettability modification and viscoelastic flow control. These synergistic effects collectively contributed to its optimal oil recovery performance. Consequently, this formulation was selected as the optimized surfactant formula for subsequent displacement experiments.

2.3. Microscale Displacement Experiments Using Two Different Pore–Throat Structures

In the pore–throat structure emulating realistic reservoir rock, water flooding achieved a recovery factor of 34.18% as shown in Figure 8. Subsequent surfactant flooding significantly enhanced the oil recovery up to 62.64%, demonstrating a remarkable 28.46% improvement over conventional water flooding.

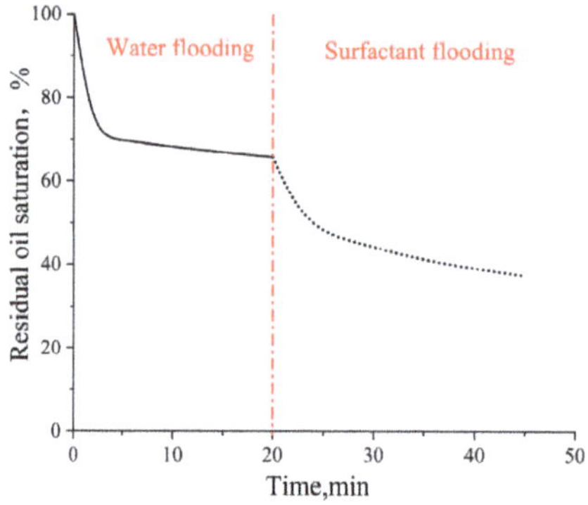

Figure 8. Residual oil saturation change in the pore–throat structure emulating realistic reservoir rock. Solid line represents the waterflooding stage and dashed line represents the surfactant flooding stage.

For the idealized pore–throat network with permeability contrast, conventional formation water flooding resulted in residual oil saturations of 42.11% in the high-permeability (High-perm) zone, 97.36% in the medium-permeability (Medium-perm) zone, and 98.82% in the low-permeability (Low-perm) zone as shown in Figure 9. Subsequent surfactant flooding using the optimized surfactant solution significantly reduced these values to 17.57%, 35.93%, and 44.76%, respectively, indicating effective oil mobilization from both medium and low permeability zones. The overall final oil recovery rate was enhanced by 49.96% compared to that achieved by formation water flooding.

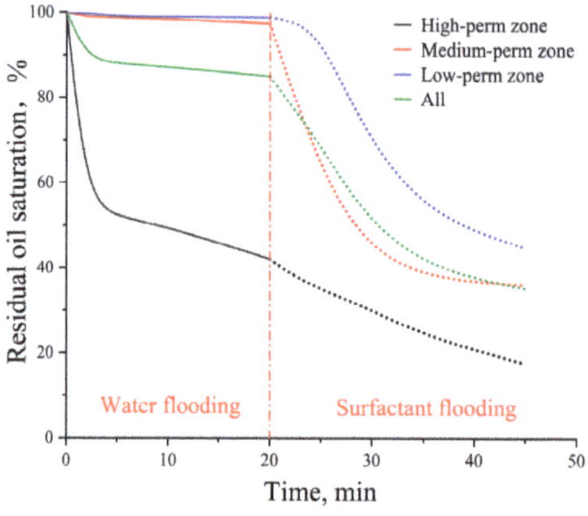

Figure 9. Residual oil saturation change in the idealized pore–throat network with permeability contrast. Solid lines represent the waterflooding stage and dashed lines represent the surfactant flooding stage.

When comparing the recovery rates in the two structures, structural variability leads to different degrees of recovery enhancement. In the pore–throat structure emulating realistic reservoir rock, the pore–throat topology is complicated, possessing tortuous channels and differently shaped dead-end pores. On the other hand, for the idealized pore–throat network with permeability contrast, the pore–throat space is regular and there is no dead-end space. Therefore, the enhancement over conventional waterflooding by the surfactant mixture is lower in the pore–throat structure emulating realistic reservoir rock.

The experimental results have demonstrated the superior performance of anionic and zwitterionic surfactant mixtures in enhanced oil recovery. Table 2 shows various anionic–zwitterionic surfactant mixtures and their respective increases in oil recovery efficiency in the published work. Overall, the combination of anionic and zwitterionic surfactants demonstrates excellent effectiveness in enhancing oil recovery for various systems.

Table 2. Different anionic–zwitterionic surfactant formulations and the increase in oil recovery efficiency.

No.	Solution Composition	Increase in Oil Recovery Efficiency, %
1	Sodium alkyl glucosyl hydroxypropyl sulfonate + octadecyl betaine [38]	20
2	Sodium dodecyl benzene sulfonate + cocoamidopropyl sulfonate betaine [39]	18.5
3	Sodium alkylphenol polyoxyethylene ether carboxylate + dodecyl dimethyl hydroxypropyl sulfobetaine [40]	14
4	Nonylphenol polyoxyethylene ether carboxylate + oleic amide propyl betaine [41]	14.8

2.4. Oil Displacement Mechanism of the Optimized Surfactant Mixture

In an oil–water two-phase system, the addition of surfactants leads to the formation of mixed adsorption layers at the interface with the lipophilic groups penetrating into the oil phase and the hydrophilic groups in the aqueous phase. This configuration transforms the originally high-energy oil–water interface into a low-energy surfactant adsorption layer, thus effectively mitigating interphase repulsion forces and achieving significant reduction in interfacial tension.

The combination of SDS and OAB-30 facilitates the formation of micellar networks, leading to the development of gel-like substances. This endows the system with superior viscoelastic properties, thereby enhancing the ability to mobilize and transport crude oil. The critical micelle concentration (CMC) values for individual surfactants were determined through the surface tension measurements, and the results are presented in Figure 10. The CMCs for SDS and OAB-30 are 1.73 and 0.11 mmol/L respectively which are marked with red stars in Figure 10.

The CMC of the surfactant mixture was determined using the surface tension method as well. As shown in Figure 11, the measured CMC for the surfactant mixture is 0.014 mmol/L which is marked with the red star.

According to the Rubingh theory, for two different surfactants forming mixed micelles in an aqueous phase, the interaction parameter β^M in the mixed micelles can be calculated using the following equations:

$$\frac{\left(X_1^M\right)^2 \ln\left(\alpha_1 C_{12}^M / X_1^M C_1^M\right)}{\left(1 - X_1^M\right)^2 \ln\left[(1-\alpha_1) C_{12}^M / \left(1 - X_1^M\right) C_2^M\right]} = 1 \quad (1)$$

$$\beta^M = \frac{\ln\left(\alpha_1 C_{12}^M / X_1^M C_1^M\right)}{\left(1 - X_1^M\right)^2} \quad (2)$$

where $X_1{}^M$ represents the molar fraction of surfactant 1 in the mixed micelles, and β^M is a parameter that characterizes the nature and extent of the interaction between the two different surfactants in the mixed micelles; α_1 is the molar fraction of SDS in the solution with a value of 0.34. $C_{12}{}^M$ is the CMC of the 0.2% SDS+0.6% OAB-30 surfactant mixture. $C_1{}^M$ and $C_2{}^M$ are the CMCs of SDS and OAB-30, respectively. By iteratively solving Equation (1) to obtain the value of $X_1{}^M$, the interaction parameter β^M can be evaluated using Equation (2). A positive β^M value indicates antagonistic interactions in the mixed system, while a negative β^M value suggests synergistic interactions [42]. The calculated value of β^M is −13.2, indicating that the combination of the two surfactants exhibits a synergistic effect.

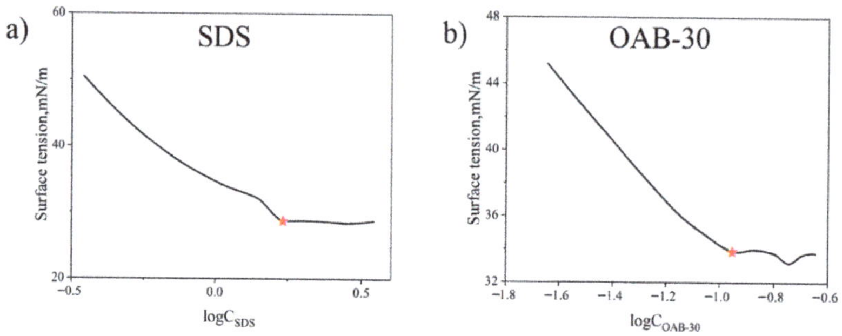

Figure 10. Measured surface tension versus varying surfactant concentrations for (**a**) SDS and (**b**) OAB-30.

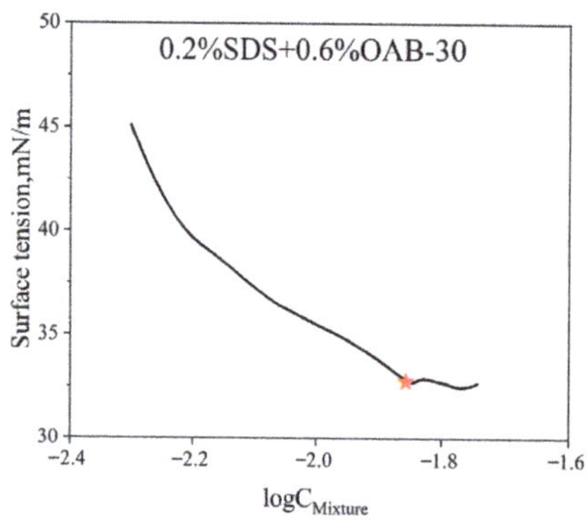

Figure 11. Measured surface tension versus log surfactant mixture concentrations of the surfactant mixture.

Figure 12 reveals that in the pore–throat structure emulating realistic reservoir rock, the substantial heterogeneity in flow resistance across different pore networks leads to the development of preferential flow paths through well-connected macropores during conventional waterflooding. This results in significant bypassing of oil trapped in poorly connected dead-end pores and narrow throat channels. Substantial residual oil saturation remained unrecoverable through standard water injection methods. Upon switching to surfactant solution flooding, a remarkable improvement in displacement efficiency was observed. The swept area expands substantially, and previously unrecoverable oil by waterflooding was rapidly mobilized and displaced. This pronounced enhancement in recovery performance directly results from the surfactant mixture's ability to reduce interfacial tension, alter wettability, and improve mobility control. Therefore, the capillary trapping that constrains waterflood effectiveness in heterogeneous pore structures was overcome. The visual evidence confirms the superior oil mobilization capability of the viscoelastic surfactant solutions in complex porous media.

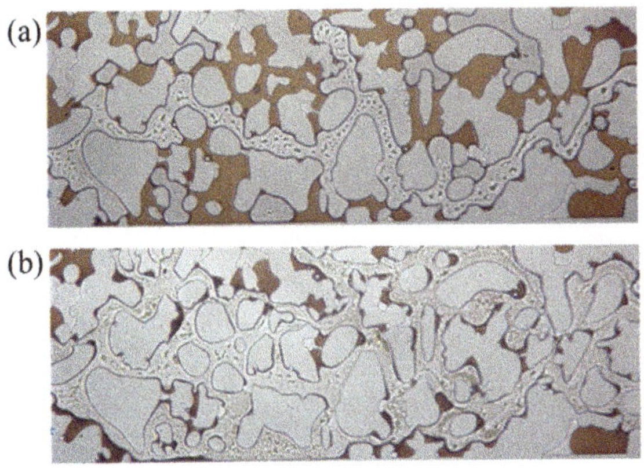

Figure 12. Sweep comparison (**a**) before and (**b**) after surfactant flooding.

The observations from Figure 13 indicate that the regions circled by blue dotted lines demonstrate the surfactant system's ability to displace fluids from dead-end structures. The regions circled by red dotted lines highlight that during waterflooding, the aqueous phase encountered significant flow resistance at constricted pore throats. This resistance could not be overcome regardless of the injection duration. Upon transitioning to surfactant flooding, the reduction in interfacial tension induces changes in flow dynamics, thereby lowering the mobilization pressure for crude oil. This process is accompanied by distinct emulsification and solubilization phenomena, where crude oil was dispersed into discrete droplets and partially solubilized in the aqueous phase. This dual mechanism enables the gradual formation of an oil bank, effectively accessing and producing trapped hydrocarbons from dead-end structures and narrow throats. The superior performance of the surfactant system is quantitatively confirmed by the significantly reduced residual oil saturation compared to waterflooding. Real-time visualization captured three key recovery processes: the continuous mobilization of emulsified droplets through flow channels, the complete displacement of oil in pore bodies following surfactant mixture flooding, and the advancement of the oil bank through pore–throat networks. These observations elucidate how viscoelastic surfactant solution overcame the capillary trapping mechanisms that limit the efficiency of waterflooding in heterogeneous porous media.

Figure 14 illustrates the unique oil displacement mechanism of the optimized viscoelastic surfactant mixture exhibiting gel-like properties. In the regions circled by red lines, the oil phase adhering to the surface was deformed under the influence of viscous and elastic forces and detached as droplets under shear stress. This process results in the mobilization and removal of the oil phase from the surface. In the regions circled by green lines, the small oil droplets formed after shearing collectively prompted the bulk oil within the narrow throats as they pass through. This phenomenon highlights the synergistic role of viscoelasticity in overcoming capillary trapping and improving interfacial flow dynamics.

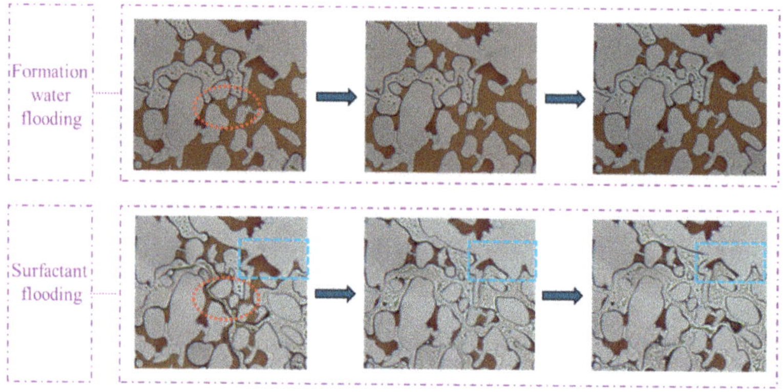

Figure 13. Images of the flooding experiment using formation water and surfactant in the pore–throat structure emulating realistic reservoir rock.

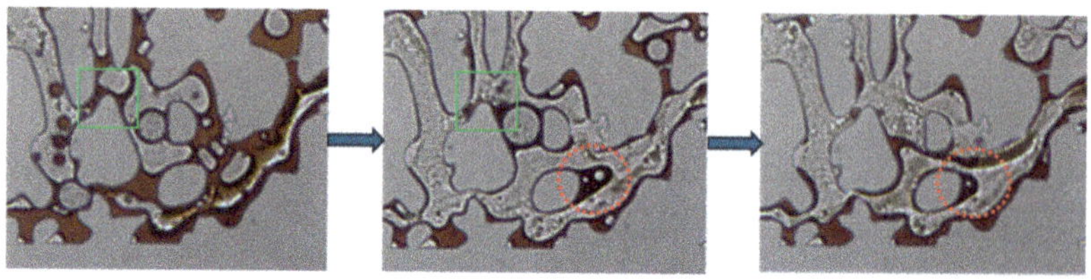

Figure 14. The viscoelastic phenomena during the surfactant flooding process in the pore–throat structure emulating realistic reservoir rock.

Owing to the planar heterogeneity of the structure, the fingering phenomenon of the injected fluid occurred during the waterflooding process, resulting in significant differences in recovery efficiency across various regions. As shown in Figure 15, formation water readily established dominant flow channels in the high-permeability zone. In contrast, the medium and low permeability zones exhibit higher capillary forces and greater flow resistance leading to the entrapment of a substantial amount of crude oil in these regions. Upon switching to surfactant solution flooding, the reduction in oil–water interfacial tension diminished capillary forces and decreased the resistance to oil flow. Consequently, crude oil in the medium permeability zone was mobilized followed by that in the low permeability zone after a period of displacement. The surfactant solution emulsified crude oil into smaller droplets, thus reducing the flow resistance of the oil.

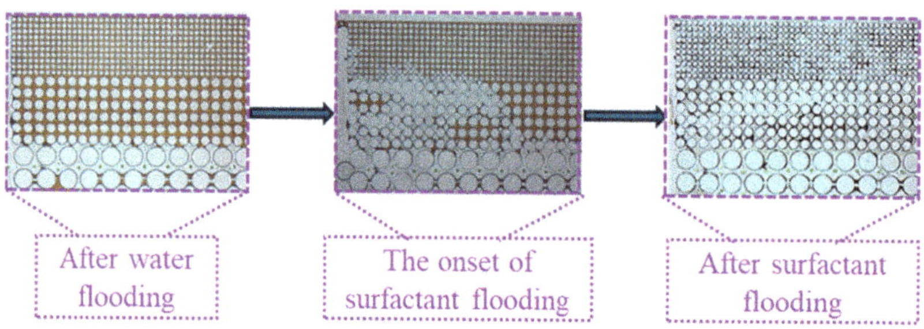

Figure 15. Mobility variation in the idealized pore–throat network with permeability contrast.

Figure 16 shows that in an idealized pore–throat network with permeability contrast, the blue-lined area exhibits distinct emulsification and dissolution phenomena. In regions where surfactant solutions cannot flush directly, a noticeable emulsification and solubilization phenomenon occurs at the oil–water interface. Under the influence of surfactants, crude oil was either dissolved in water or emulsified into small droplets, which was then carried away by the displacement fluid. The red-lined area shows that crude oil adhering to a solid surface was significantly stretched and fragmented into small droplets due to the solution's viscoelastic properties and the combined effects of elastic and viscous forces. These droplets were then carried away by the displacement fluid.

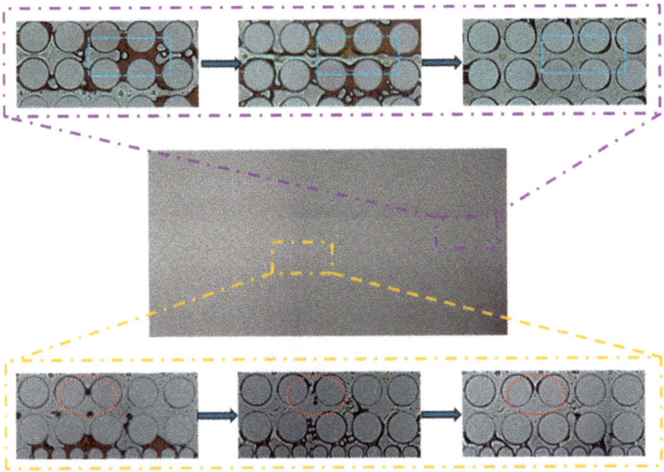

Figure 16. Images of the flooding experiment using the optimized surfactant formula in the idealized pore–throat network with permeability contrast.

3. Conclusions

This study evaluated various combinations of anionic (SDS) and zwitterionic (OAB-30) surfactants in displacement tests, identifying a 0.2% SDS+0.6% OAB-30 mixture as the most effective formulation. This solution exhibited the highest viscosity, reduction interfacial tension, and favorable viscoelastic properties, thus resulting in superior oil displacement efficiency in dead-end pores. When applied to a micromodel simulating realistic reservoir rock, the optimized surfactant solution enhanced oil recovery by 28.46% over conventional water flooding. It effectively displaced crude oil from dead-end pores and throats. In a heterogeneous pore–throat network with permeability contrast, the surfactant increased oil recovery by 49.96% compared to water flooding. The gel-like properties of the solution improved sweep efficiency and mobility control, therefore mitigating the fingering in high-permeability zones, confirming its potential for enhanced recovery in complex reservoir systems.

4. Materials and Methods

4.1. Materials

Oil was prepared by blending crude oil and high-viscosity liquid paraffin (Shanghai Macklin Biochemical Co., Ltd., Shanghai, China) at a 1:1 volume ratio. Simulated formation water (0.03% KCl, 0.07% $MgCl_2$, 0.1% $CaCl_2$, and 0.8% NaCl), sodium dodecyl sulfate (SDS, Tianjin Damao Chemical Reagent Factory, Tianjin, China), and oleic amide propyl betaine (OAB-30, 30%, Shandong Usolf Chemical Technology Co., Ltd., Linyi, China) were used in this study. Crude oil used in this work consists of 2.09% asphaltene, 22.78% resin, 4.27% aromatic hydrocarbons, and 70.86% saturated hydrocarbons. The chemical structures of the surfactants are shown in Figure 17. Six surfactant mixtures with varying compositions were prepared and the details are presented in Table 3.

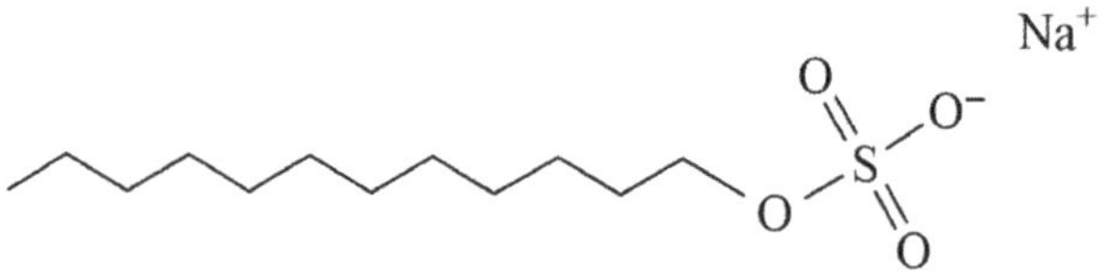

(a)

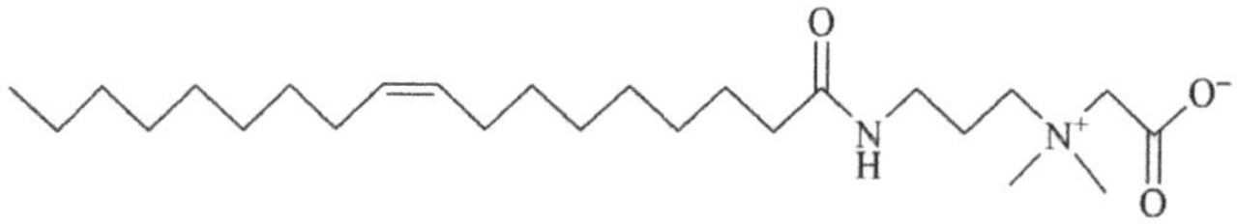

(b)

Figure 17. Chemical structure of the examined surfactants. (**a**) SDS; (**b**) OAB-30.

Table 3. Composition of surfactant solutions.

No.	Solution Composition	SDS (mmol/L)	OAB-30 (mmol/L)
1	0.2% SDS+0.2% OAB-30	6.94	4.52
2	0.2% SDS+0.4% OAB-30	6.94	9.04
3	0.2% SDS+0.6% OAB-30	6.94	13.55
4	0.4% SDS+0.2% OAB-30	13.87	4.52
5	0.4% SDS+0.4% OAB-30	13.87	9.04
6	0.4% SDS+0.6% OAB-30	13.87	13.55

4.2. Microfluidic Experiment Setup

The microfluidic system consists of a microinjection pump, an inverted microscope, a high-speed camera, and a computer. Microfluidic devices were fabricated using poly-dimethylsiloxane (PDMS) by mixing pre-polymer and curing agent (10:1 w/w) (SYLGARD 184 Silicone Elastomer Kit, Dow Corning, Midland, MI, USA) and curing on a silicon mold with a designed pattern at 65 °C. After demolding, fluidic ports were punched, followed by oxygen plasma bonding to a glass slide. The schematic of the experimental setup is shown in Figure 18. A microfluidic device was secured under the microscope equipped with the high-speed camera, which was connected to the computer. The microinjection pump was used to deliver fluids into the device at predetermined flow rates. The entire process was monitored and recorded in real-time via high-speed camera. Microfluidic devices with different pore–throat structures were used in this work.

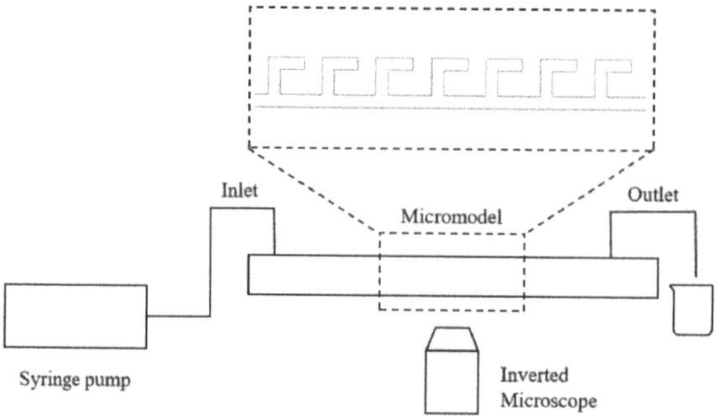

Figure 18. Schematic of the experimental setup.

4.3. Experimental Procedures

4.3.1. Screening of Surfactant Formulations Using Microfluidic Device with Dead-End Structure

The screening experiment was conducted using parallel L-shaped dead-end microfluidic structures, with the microfluidic device configuration illustrated in Figure 19. The experimental procedures are as follows: (1) prior to surfactant flooding, the fabricated microfluidic devices were initially saturated with simulated formation water, followed by oil injection until stable oil saturation was achieved, thereby simulating initial reservoir conditions; (2) subsequently, surfactant solutions were injected at a constant flow rate of 0.5 μL/min until a stable residual oil saturation was achieved. Images were taken periodically to calculate the oil saturation change and final oil recovery efficiency. The procedures were repeated for all six surfactant formulations. Three central structures (circled by red dotted lines in Figure 19) were selected for performance evaluation and the oil displacement efficiency in these structures was calculated and the average was taken as the oil displacement efficiency. The entire process was monitored using a microscope and high-speed camera to analyze fluid dynamics and displacement patterns within the microchannels.

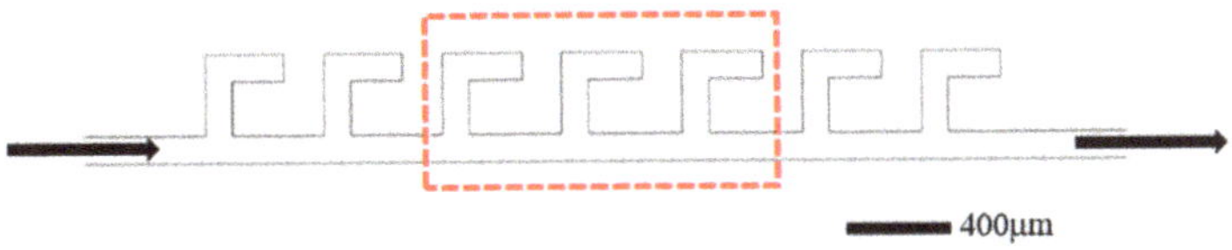

Figure 19. Schematic of the L-shaped dead-end pore structure. The arrows represent the flow direction.

The ImageJ software (NIH) was employed to process and analyze the images obtained during the experimental process. As illustrated in Figure 20, after binarization using the software, the white regions represent the crude oil. The area of the white regions was calculated using the software that was then used to represent the oil saturation given the micromodel depth was constant. Each image was measured three times, and the results exhibited a measurement error less than 3%.

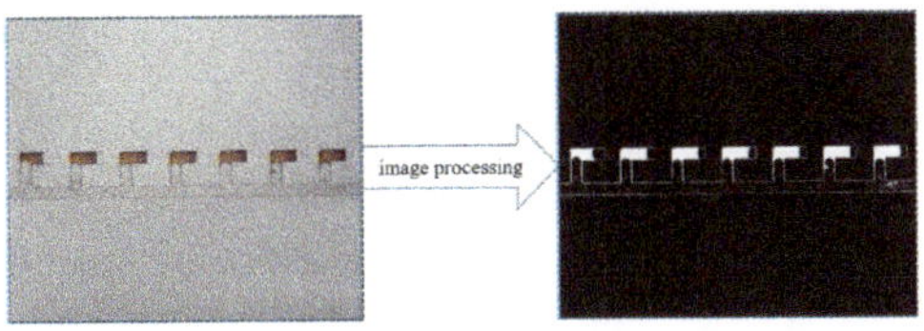

Figure 20. Demonstration of the image processing procedure.

4.3.2. Microscopic Displacement Experiments with Microfluidic Devices with Different Pore–Throat Structures

After screening surfactants using the device with L-shaped dead-end pores, microscopic displacement experiments were conducted using two distinct pore–throat configurations: pore–throat structure emulating realistic reservoir rock and idealized pore–throat network with permeability contrast as shown in Figures 21 and 22. The experimental procedure involved initial saturation of the device with formation water and subsequent oil injection to establish irreducible water saturation and initial oil saturation, thus simulating the reservoir conditions. Formation water flooding was first performed. Upon stabilization of residual oil saturation, the injection fluid was switched to the optimized surfactant mixture solution with a constant injection rate of 0.5 μL/min. The same procedures were applied to both microfluidic devices.

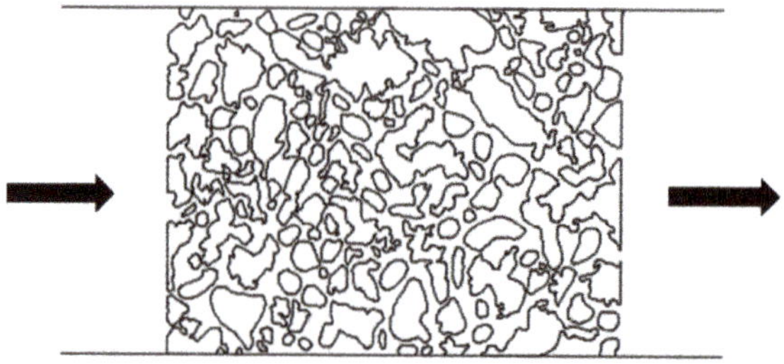

Figure 21. Schematic of the pore–throat structure emulating realistic reservoir rock. The arrows represent the flow direction.

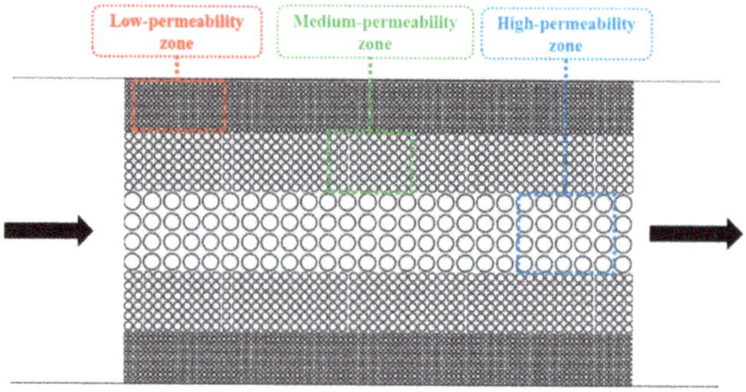

Figure 22. Schematic of the idealized pore–throat network with permeability contrast. The arrows represent the flow direction.

4.3.3. Determination of the Interfacial Tension

The interfacial tension between surfactant solutions and crude oil was measured using the spinning drop method (CNGTX, Shengwei Technology Co., Ltd., Beijing, China). Prior to testing, the centrifuge tube was thoroughly cleaned with organic solvent and deionized water to eliminate contaminant residues, followed by air drying. To minimize interference from impurities, the tubes were rinsed 2–3 times with small amounts of the surfactant solution. A droplet of crude oil was injected into the center of the solution using a microsyringe. Rotation speed was set at 5000 rpm and temperature was set at 30 °C. The measurement duration was 30 min, with automatic data acquisition every 2 min. Upon initiating the test procedure, the system continuously recorded dynamic interfacial tension variations over time. The stabilized segment of the resulting curve was selected as the final measurement value. To ensure statistical reliability, each sample was measured three times to minimize experimental errors.

4.3.4. Determination of the Contact Angle

The contact angle was measured using the captive bubble method incorporating digital image analysis and contour fitting algorithms for enhanced measurement precision (KRÜSS DSA100, KRÜSS GmbH, Hamburg, Germany). The experimental procedures were as follows: (1) glass slides were aged in crude oil for 4 days to simulate reservoir conditions; (2) residual crude oil was thoroughly removed from the slides using anhydrous kerosene prior to contact angle measurement; (3) the sample cell was secured on the stage, and distilled water was added to fully submerge the substrate. The aged glass slide was then positioned within the cell; (4) a specially designed inverted needle was vertically inserted into the water, and carefully adjusted to dispense an oil droplet beneath the slide; the diameter of the oil droplets was approximately 3–5 mm and capture time was within 15 s.

(5) The camera's focus, lighting, and position were optimized to ensure clear visualization of the submerged oil droplet's contour; (6) the droplet image was captured and analyzed to determine the initial contact angle. Following the initial measurement, the glass slide was subjected to an additional 4-day aging process in the surfactant solution, after which the contact angle was measured again to evaluate wettability alteration.

4.3.5. Determination of the Viscosity of Surfactant Mixtures

The viscosity measurements in this study were conducted using a rotational viscometer (ROTAVISC me-vi, IKA, Staufen, Germany). Prior to testing, the instrument was carefully leveled and preheated to a stabilized temperature of 30 °C using the integrated temperature control system. With the rotational speed set to 170 rpm, viscosity readings were recorded only after achieving stable measurement values. This standardized procedure was repeated for all fluid samples under identical experimental conditions to ensure measurement consistency and reliability.

4.3.6. Determination of the Viscoelasticity of Surfactant Mixtures

The viscoelastic properties of six surfactant solutions were characterized using a rheometer (Anton Paar MCR 102, Anton Paar GmbH, Graz, Austria). Steady-state shear tests were first conducted by applying controlled shear rates to obtain strain versus storage modulus (G′) and loss modulus (G″) curves. The optimal strain value was determined by identifying the plateau region where both moduli remained stable. Subsequent dynamic oscillatory tests were performed at this stabilized strain level across a frequency range of 0.1–100 rad/s. The resulting angular frequency versus G′ and G″ curves were recorded to fully characterize the viscoelastic behavior of each polymer solution under dynamic conditions.

Author Contributions: Methodology, C.L.; formal analysis, X.Y.; investigation, C.L. and X.Y.; resources, Q.Y. and D.Z.; data curation, C.L.; writing—original draft preparation, C.L.; writing—review and editing, Y.L.; visualization, Q.Y.; funding acquisition, Y.L. and D.Z. All authors have read and agreed to the published version of the manuscript.

Funding: This work was funded by the National Natural Science Foundation of China (Grant No. 52174028 and U23B2089).

Institutional Review Board Statement: Not applicable.

Informed Consent Statement: Not applicable.

Data Availability Statement: The authors confirm that the data are provided within the article.

Conflicts of Interest: The author Qi Ye was employed by the Changqing Oilfield Company. The company had no role in the design of the study; in the collection, analyses, or interpretations of the data; in the writing of the manuscript, or in the decision to publish the results The remaining authors declared no conflicts of interest.

References

1. Khurshid, I.; Al-Attar, H.; Alraeesi, A. Modeling cementation in porous media during waterflooding: Asphaltene deposition, formation dissolution and their cementation. *J. Pet. Sci. Eng.* **2018**, *161*, 359–367. [CrossRef]
2. Tsuji, T.; Jiang, F.; Christensen, K.T. Characterization of immiscible fluid displacement processes with various capillary numbers and viscosity ratios in 3D natural sandstone. *Adv. Water Resour.* **2016**, *95*, 3–15. [CrossRef]
3. Singh, K.; Jung, M.; Brinkmann, M.; Seemann, R. Capillary-dominated fluid displacement in porous media. *Annu. Rev. Fluid Mech.* **2019**, *51*, 429–449. [CrossRef]
4. Pi, Y.; Su, Z.; Cao, R.; Li, B.; Liu, J.; Fan, X.; Zhao, M. Experimental study on enhanced oil recovery of ppg/asp heterogeneous system after polymer flooding. *Gels* **2023**, *9*, 427. [CrossRef] [PubMed]

5. Wang, Z.; Hu, R.; Ren, G.; Li, G.; Liu, S.; Xu, Z.; Sun, D. Polyetheramine as an alternative alkali for alkali/surfactant/polymer flooding. *Colloids Surf. A Physicochem. Eng. Aspects* **2019**, *581*, 12. [CrossRef]
6. Al-Yaari, A.; Ching, D.L.C.; Sakidin, H.; Muthuvalu, M.S.; Zafar, M.; Haruna, A.; Merican, Z.M.A.; Yunus, R.B.; Al-Dhawi, B.N.S.; Jagaba, A.H. The effects of nanofluid thermophysical properties on enhanced oil recovery in a heterogenous porous media. *Case Stud. Chem. Environ. Eng.* **2024**, *9*, 15. [CrossRef]
7. Guo, Y.; Zhang, L.; Zhu, G.; Yao, J.; Sun, H.; Song, W.; Yang, Y.; Zhao, J. A pore-scale investigation of residual oil distributions and enhanced oil recovery methods. *Energies* **2019**, *12*, 3732. [CrossRef]
8. Ning, T.; Xi, M.; Hu, B.; Wang, L.; Huang, C.; Su, J. Effect of viscosity action and capillarity on pore-scale oil–water flowing behaviors in a low-permeability sandstone waterflood. *Energies* **2021**, *14*, 8200. [CrossRef]
9. Demirbas, A.; Alsulami, H.E.; Hassanein, W.S. Utilization of surfactant flooding processes for enhanced oil recovery (EOR). *Pet. Sci. Technol.* **2015**, *33*, 1331–1339. [CrossRef]
10. Hirasaki, G.J.; Miller, C.A.; Puerto, M. Recent advances in surfactant EOR. *SPE J.* **2011**, *16*, 889–907. [CrossRef]
11. Howe, A.M.; Clarke, A.; Mitchell, J.; Staniland, J.; Hawkes, L.; Whalan, C. Visualising surfactant enhanced oil recovery. *Colloids Surf. A Physicochem. Eng. Asp.* **2015**, *480*, 449–461. [CrossRef]
12. Yang, H.; Britton, C.; Liyanage, P.J.; Solairaj, S.; Kim, D.H.; Nguyen, Q.; Upali, W.; Pope, G.A. *Low-Cost, High-Performance Chemicals for Enhanced Oil Recovery*; SPE Improved Oil Recovery Symposium: Tulsa, OK, USA, 2010; p. 129978.
13. Solairaj, S.; Britton, C.; Kim, D.H.; Weerasooriya, U.; Pope, G.A. *Measurement and Analysis of Surfactant Retention*; SPE Improved Oil Recovery Symposium: Tulsa, OK, USA, 2012; p. 154247.
14. Zhao, G.; Khin, C.C.; Chen, S.B.; Chen, B.H. Nonionic surfactant and temperature effects on the viscosity of hydrophobically modified hydroxyethyl cellulose solutions. *J. Phys. Chem. B* **2005**, *109*, 14198–14204. [CrossRef]
15. Ma, K.; Cui, L.; Dong, Y.; Wang, T.; Da, C.; Hirasaki, G.J.; Biswal, S.L. Adsorption of cationic and anionic surfactants on natural and synthetic carbonate materials. *J. Colloid Interface Sci.* **2013**, *408*, 164–172. [CrossRef]
16. Massarweh, O.; Abushaikha, A.S. The use of surfactants in enhanced oil recovery: A review of recent advances. *Energy Rep.* **2020**, *6*, 3150–3178. [CrossRef]
17. Kamal, M.S.; Hussein, I.A.; Sultan, A.S. Review on surfactant flooding: Phase behavior, retention, IFT, and field applications. *Energy Fuels* **2017**, *31*, 7701–7720. [CrossRef]
18. Xue, J.; Liu, T.; Liu, Y. Interaction of Novel Anionic Gemini Surfactants with Dodecyl Sodium Sulfate in Aqueous NaCl Medium and the Performance of Their Mixtures. *J. Dispers. Sci. Technol.* **2012**, *33*, 599–604. [CrossRef]
19. Li, Y.; Puerto, M.; Bao, X.; Zhang, W.; Jin, J.; Su, Z.; Shen, S.; Hirasaki, G.; Miller, C. Synergism and performance for systems containing binary mixtures of anionic/cationic surfactants for enhanced oil recovery. *J. Surfactants Deterg.* **2017**, *20*, 21–34. [CrossRef]
20. Torquato, L.M.; Tyagi, G.; Sharratt, W.N.; Ahmad, Z.; Mahmoudi, N.; Gummel, J.; Robles, E.S.J.; Cabral, J.T. Concentration Dependent Asymmetric Synergy in SDS–DDAO Mixed Surfactant Micelles. *Langmuir* **2024**, *40*, 7433–7443. [CrossRef] [PubMed]
21. Li, L.; Liu, Y.; Jin, X.; Dong, Y.; Dai, C. Compound Anionic–Nonionic Surfactant System with Excellent Temperature and Salt Resistance for Enhanced Oil Recovery. *Energy Fuels* **2023**, *38*, 1047–1055. [CrossRef]
22. Haque, M.E.; Das, A.R.; Rakshit, A.K.; Moulik, S.P. Properties of mixed micelles of binary surfactant combinations. *Langmuir* **1996**, *12*, 4084–4089. [CrossRef]
23. Misselyn-Bauduin, A.M.; Thibaut, A.; Grandjean, J.; Broze, G.; Jérôme, R. Mixed micelles of anionic–nonionic and anionic–zwitterionic surfactants analyzed by pulsed field gradient NMR. *Langmuir* **2000**, *16*, 4430–4435. [CrossRef]
24. Li, K.X.; Jing, X.Q.; He, S.; Ren, H.; Wei, B. Laboratory study displacement efficiency of viscoelastic surfactant solution in enhanced oil recovery. *Energy Fuels* **2016**, *30*, 4467–4474. [CrossRef]
25. Hu, R.; Tang, S.; Mpelwa, M.; Jiang, Z.; Feng, S. Research progress of viscoelastic surfactants for enhanced oil recovery. *Energy Explor. Exploit.* **2021**, *39*, 1324–1348. [CrossRef]
26. Deng, R.; Dong, J.; Dang, L. Numerical simulation and evaluation of residual oil saturation in waterflooded reservoirs. *Fuel* **2025**, *384*, 134018. [CrossRef]
27. Yin, B.; Ding, T.; Wang, S.; Wang, Z.; Sun, B.; Zhang, W.; Zhang, X. Deformation and migration characteristics of bubbles moving in gas-liquid countercurrent flow in annulus. *Pet. Explor. Dev. Online* **2025**, *52*, 471–484. [CrossRef]
28. Yu, H.; Zhao, Z.; Taleghani, A.D.; Lian, Z.; Zhang, Q. Modeling thermal-induced wellhead growth through the lifecycle of a well. *Geoenergy Sci. Eng.* **2024**, *241*, 15. [CrossRef]
29. Xu, F.; Jin, Y.; Fan, Y. A study on enhancing oil recovery efficiency through bubble displacement based on microfluidic technology. *Can. J. Chem. Eng.* **2025**, *103*, 1450–1460. [CrossRef]
30. Fani, M.; Pourafshary, P.; Mostaghimi, P.; Mosavat, N. Application of microfluidics in chemical enhanced oil recovery: A review. *Fuel* **2022**, *315*, 123225. [CrossRef]
31. Kumar, G.; Mani, E.; Sangwai, J.S. Microfluidic investigation of surfactant-assisted functional silica nanofluids in low-salinity seawater for enhanced oil recovery using reservoir-on-a-chip. *Energy Fuels* **2023**, *37*, 10329–10343. [CrossRef]

32. Liu, Y.; Block, E.; Squier, J.; Oakey, J. Investigating low salinity waterflooding via glass micromodels with triangular pore-throat architectures. *Fuel* **2021**, *283*, 119264. [CrossRef]
33. Pattanayak, P.; Singh, S.K.; Gulati, M.; Vishwas, S.; Kapoor, B.; Chellappan, D.K.; Anand, K.; Gupta, G.; Jha, N.K.; Gupta, P.K.; et al. Microfluidic chips: Recent advances, critical strategies in design, applications and future perspectives. *Microfluid. Nanofluid* **2021**, *25*, 99. [CrossRef]
34. Liu, Y.; Liu, Z.; Wang, Y.; Hu, Z.; Zhu, Z.; Jie, Y.; Zhang, Y. Effect of Different Surfactants and Nanoparticles on Pore-Scale Oil Recovery Process Using Heterogeneous Micromodel. *Int. J. Energy Res.* **2024**, *2024*, 5319748. [CrossRef]
35. Dong, Z.; Zheng, Y.; Zhao, J. Synthesis, physico-chemical properties and enhanced oil recovery flooding evaluation of novel zwitterionic gemini surfactants. *J. Surfactants Deterg.* **2014**, *17*, 1213–1222. [CrossRef]
36. Liu, Y.; He, F.; Hu, Z.; Zhu, Z.; Ling, C. Microfluidic Study on the Effect of Single Pore–Throat Geometry on Spontaneous Imbibition. *Langmuir* **2024**, *40*, 19209–19219. [CrossRef]
37. Abdel-Rahem, R. Synergism in mixed anionic–amphoteric surfactant solutions: Influence of anionic surfactant chain length. *Tenside Surfactants Deterg.* **2009**, *46*, 298–305. [CrossRef]
38. Wu, H.R.; Tan, R.; Hong, S.P.; Zhou, Q.; Liu, B.Y.; Chang, J.W.; Luan, T.-F.; Kang, N.; Hou, J.R. Synergistic anionic/zwitterionic mixed surfactant system with high emulsification efficiency for enhanced oil recovery in low permeability reservoirs. *Pet. Sci.* **2024**, *21*, 936–950. [CrossRef]
39. Ma, H.; Ranimol, S.; Cameron, A. Synergetic system of zwitterionic/anionic surfactants with ultralow interfacial tension and high salt resistance for enhancing heavy oil recovery. *Energy Fuels* **2022**, *36*, 8216–8223. [CrossRef]
40. Xiao, L. Enhance oil recovery in fracture-cave carbonate reservoirs using zwitterion-anionic composite surfactant system. *Energies* **2025**, *18*, 383.
41. Pei, H.; Liu, Y.; Shan, J.; Zhao, J.; Zhang, J.; Wu, Y.; Zhang, G. Enhancement of in-situ emulsification performance through the synergistic effects of zwitterionic and nonionic-anionic surfactants to improve heavy oil recovery. *J. Mol. Liq.* **2024**, *410*, 11. [CrossRef]
42. Boucher, E.A.; Grinchuk, T.M.; Zettlemoyer, A.C. Surface activity of sodium salts of alpha-sulfo fatty esters: The air-water interface. *J. Am. Oil Chem. Soc.* **1968**, *45*, 49–52. [CrossRef]

gels

MDPI

Article

Research on the Energy Transfer Law of Polymer Gel Profile Control Flooding in Low-Permeability Oil Reservoirs

Chen Wang [1,2,3], Yongquan Deng [1,2], Yunlong Liu [4,5], Gaocheng Li [1,2], Ping Yi [4,5], Bo Ma [4,5] and Hui Gao [1,2,3,*]

1 School of Petroleum Engineering, Xi'an Shiyou University, Xi'an 710065, China; cwangxsyu@163.com (C.W.); 15066439673@163.com (Y.D.); 15318794631@163.com (G.L.)
2 Engineering Research Center of Development and Management for Low to Ultra-Low Permeability Oil & Gas Reservoirs in West China, Ministry of Education, Xi'an 710065, China
3 Key Laboratory of Exploration and Development of Complex and Difficult-to-Produce Oil & Gas Reservoirs, Ministry of Education, Xi'an 710065, China
4 Oil and Gas Technology Research Institute, PetroChina Changqing Oilfield Company, Xi'an 710018, China; 13236507857@163.com (Y.L.); 13776738856@163.com (P.Y.); 19992605070@163.com (B.M.)
5 National Engineering Laboratory for Exploration and Development of Low Permeability Oil and Gas Fields, Xi'an 710018, China
* Correspondence: ghtopsun1@163.com or gh@xsyu.edu.cn

Abstract: To investigate the energy conduction behavior of polymer gel profile control and flooding in low-permeability reservoirs, a parallel dual-tube displacement experiment was conducted to simulate reservoirs with different permeability ratios. Injection schemes included constant rates from 0.40 to 1.20 mL/min and dynamic injection from 1.20 to 0.40 mL/min. Pressure monitoring and shunt analysis were used to evaluate profile control and recovery performance. The results show that polymer gel preferentially enters high-permeability layers, transmitting pressure more rapidly than in low-permeability zones. At 1.20 mL/min, pressure onset at 90 cm in the high-permeability layer occurs earlier than in the low-permeability layer. Higher injection rates accelerate pressure buildup. At 0.80 mL/min, permeability contrast is minimized, achieving a 22.96% recovery rate in low-permeability layers. The combination effect of 1.2–0.4 mL/min is the best in dynamic injection, with the difference in shunt ratio of 9.6% and the recovery rate of low permeability layer increased to 31.23%. Polymer gel improves oil recovery by blocking high-permeability channels, expanding the swept volume, and utilizing viscoelastic properties.

Keywords: low-permeability oil reservoirs; polymer gel; profile control flooding; energy transfer law; injection methods; enhanced oil recovery

1. Introduction

The proven reserves of low-permeability oil fields account for more than 60% of the domestic oil reserves, making them a crucial pillar of China's petroleum industry [1–3]. However, low-permeability reservoirs face significant development challenges due to their complex physical properties and unique seepage characteristics. In recent years, the continuous development of these reservoirs has become an essential task. With the growing energy demand in China and the increasingly severe domestic energy security situation, the energy supply is facing enormous challenges. Meanwhile, the proportion of unconventional oil and gas production within the overall oil and gas output has been rising year by year. As an important unconventional oil and gas resource, low-permeability reservoirs cannot be overlooked [4–8].

Water injection, as an effective method for enhancing formation energy and crude oil production, is influenced by various factors such as water injection rates and injection pressure, which directly impact oil displacement efficiency [9,10]. To improve the development efficiency of low-permeability oil reservoirs, technologies like water injection and deep profile control flooding must be adopted. However, during the later stages of water injection development, particularly when the oil field enters a high water cut period, the effectiveness of water injection often falls short of expectations. For wells with a high water cut, deep profile modification technology is a commonly used solution. However, due to the unclear causal relationship between injection and production, the effectiveness of deep profile modification is often unsatisfactory. By optimizing and adjusting the displacement rate, it is hoped that the reservoir's seepage structure can be improved, thereby enhancing the overall profile control flooding effect across the development interval or the entire reservoir.

Shen et al. developed a green and environmentally friendly bio-based profile control flooding agent, Bio Nano30, using non-covalent supramolecular interactions, specifically designed for ultra-low permeability oil reservoirs. Physical simulation experiments revealed the mechanism of profile control flooding with Bio Nano30. Both laboratory experiments and field applications demonstrated excellent results, including effective removal of wellbore blockages in oil wells, pressure reduction and enhanced injection in water wells, and successful profile modification and oil displacement in well groups [11]. Zhou et al. optimized the formulas of three deep profile control flooding gel systems for Block H of the Daqing Oilfield, a low-permeability reservoir, through orthogonal experiments. The profile control flooding effects of each system were further evaluated through core experiments [12]. Li et al. conducted compatibility experiments on profile control agents (PCAs) and dominant flow channels, which are key targets for deep profile modification agents in low-permeability reservoirs. They investigated three commonly used PCAs and three levels of dominant flow channels based on their permeability. Considering the permeability differences between layers, they proposed a systematic criterion for optimizing slug flow structures and engineering parameters through principal component analysis. This approach, based on the principles of "blocking, controlling, and displacing," provides a valuable reference for PCA selection and field implementation parameters [13]. Zhao et al. employed polymer microspheres to enhance deep profile modification in low-permeability oil reservoirs. The plugging performance and deformability of the polymer microspheres improved with swelling time, making them effective pore channel plugging agents with strong deep profile modification capabilities in low-permeability cores [14]. Zou et al. introduced a new type of in situ polymer microsphere profile modification agent to resolve the issue of balancing injectability and profile modification effectiveness in low-permeability oil reservoirs. The in situ polymer microsphere (ISPM) emulsion is injected into the formation, where it undergoes polymerization to form microspheres for profile modification. Dual-layer heterogeneous core displacement experiments showed that ISPM can expand the swept volume of subsequent injected fluids, resulting in increased oil production and improved water control [15]. Abdullah et al. studied oil extraction in two-phase incompressible fluid in a two-dimensional rectangular porous uniform region filled with oil and without capillary pressure. By comparing three kinds of nanoparticles—SiO_2, Al_2O_3, and CuO—the oil displacement process of nano-fluid and the influence of inlet temperature are simulated. The results show that adding nanoparticles into the base fluid can improve the oil recovery by more than 20% [16].

Li et al. selected three commonly used deep profile control agents (PCAs): nano-microsphere suspensions, polyethylene glycol single-phase gel particles (PEG), and cross-linked gel swelling particles (CBG-SP), and investigated the compatibility of these agents with fracture channels through a series of experiments. The experimental results demonstrated that nano-microspheres with particle sizes of 100 nm and 300 nm exhibit good injectability and deep migration ability, maintaining a relatively high core sealing rate [17]. Yin et al. combined the characteristics and evolution mechanisms of well pattern fractures, established a core model of outyield plate fractures considering fracture direction, and conducted core displacement experiments using the HPAM/Cr3+ gel deep conditioning system. They analyzed the mechanism of enhancing oil recovery by integrating deep profile control with circulating water injection. The results indicated that combining deep profile control with circulating water injection can be carried out simultaneously to further enhance recovery rates [18]. Xu et al. prepared preformed granular gel and optimized its synthesis conditions, investigating its temperature and salt resistance as well as its sealing performance. Field application tests were also conducted. The results revealed that PPGs possess good temperature and salt resistance. PPG with larger particle sizes exhibited greater sealing strength but a shorter effective period. Field tests showed that multiple rounds of profile control using PPGs of different particle sizes could achieve deep profile control [19]. By selecting two different types of reservoirs, Pryazhnikov et al. conducted oil displacement experiments with silica nanosol. The results show that the oil displacement efficiency increases with the increase in nanoparticle concentration and the decrease in nanoparticle size [20]. Yin et al. optimized the polymer/chromium ion deep profile control system, considering the characteristics of fractured and low-permeability reservoirs in the Chaoyanggou Oilfield, through viscosity evaluation, fluidity experiments, and oil displacement tests. The experimental results demonstrated that both high-molecular-weight main agent/low-concentration systems and low-molecular-weight main agent/high-concentration systems meet the required gel strength [21].

The Jurassic reservoirs of the Changqing Oilfield exhibit low permeability. After prolonged water flooding, dominant flow channels have become well-developed, causing injected water to rapidly advance along high-permeability layers. At the same time, the widespread development of edge and bottom water has led to a rapid rise in water cut in oil wells, resulting in significant oil production loss and a decline in the efficiency of conventional water injection development. Profile control flooding has emerged as one of the most promising enhanced oil recovery techniques, with successful industrial applications in oil fields such as Daqing and Shengli, achieving recovery rate increases of over 10% [22–24]. Previous research has highlighted that viscoelasticity is a critical property in profile control flooding. Polymer gel solutions improve the water-to-oil mobility ratio by enhancing the viscosity of the aqueous phase, thereby increasing the macroscopic swept volume. The elastic effects of these gels also improve microscopic oil displacement efficiency [25–28]. As a result, viscoelastic profile control flooding demonstrates superior oil displacement capabilities. Pilot tests in the Changqing Jurassic reservoirs have yielded positive results, stabilizing oil production and controlling water production. However, field observations have shown that, after polymer gel injection based on geological allocation, some wells exhibit both declining fluid levels and liquid production. This has created a pressing need to optimize the injection and production scheme for profile control flooding to boost formation energy around the production wells and achieve balanced liquid production. Due to the strong heterogeneity of low-permeability reservoirs, conventional injection methods and parameters for the main oil layers are no longer effective. Presently, there are few studies or reports addressing the adjustment of injection and production parameters during profile control flooding in such reservoirs.

This study is based on laboratory parallel double-tube displacement experiments. By utilizing an online real-time pressure monitoring system, the energy transfer dynamics during the profile control flooding process are quantitatively assessed. Additionally, the study investigates various combinations of injection rates to offer essential technical support for the efficient development of heterogeneous reservoirs.

2. Results and Discussion

2.1. Pressure Conduction Characteristics

After restoring the saturated crude oil in the sand-filled pipe model to its original oil–water distribution, water flooding was conducted at an injection rate of 0.50 mL/min. Figure 1 illustrates the pressure variation curves during the water flooding process for both the high-permeability and low-permeability models. Due to differences in permeability across the models and the injection method combining injection and separate extraction, the initial injection pressure of the low-permeability model is relatively high. The pressure in the high-permeability model rises sharply, eventually aligning with the inlet pressure of the low-permeability model. Following this, the pressure in both models decreases initially, then increases as the injection volume rises, eventually stabilizing [29–31]. When the injection volume reaches 1.40 PV, the pressure at each measurement point in the high-permeability sand-filled pipe reaches its lowest point and begins to stabilize. During the stage of oil pressure decline in the high-permeability sand-filled pipe, the pressure drop at each monitoring point remains relatively consistent. Once the injection volume hits 1.50 PV, the pressure at all measurement points in both the high-permeability and low-permeability sand-filled pipes stabilizes. The pressure in the low-permeability sand-filled pipe drops significantly at the 70 cm and 90 cm positions, closer to the outlet, during the early stage of water flooding. However, when the injection volume reaches 0.50 PV, the pressure at these points tends to stabilize. From the pressure curve, it can be observed that the pressure values remain generally low throughout the water flooding process. The maximum pressure in the high-permeability sand-filled pipe is 0.10 MPa, while the minimum pressure in the low-permeability sand-filled pipe is 0.10 MPa. The water flooding process ends when the cumulative moisture content at the outlet ends of both sand-filled pipes exceeds 90%. This phase simulates the situation on production sites where a large amount of water is produced at the extraction end after long-term water flooding.

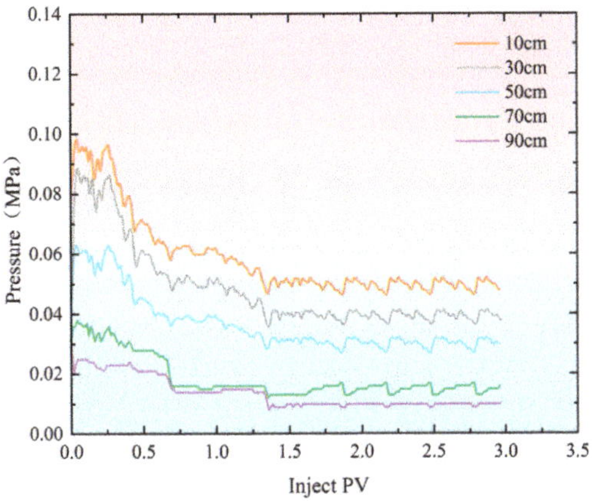

(a)

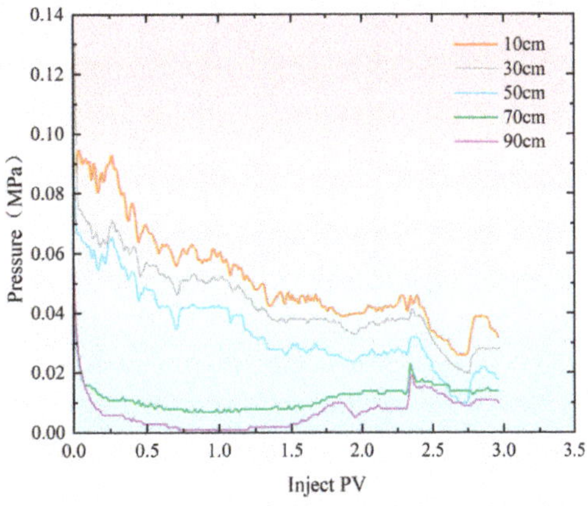

(b)

Figure 1. Pressure variation curve during the water flooding stage. (**a**) Hypertonic model; (**b**) hypotonic model.

After water flooding, a viscoelastic polymer gel solution is injected. The increased viscosity of the displacing medium leads to higher seepage resistance. Maintaining a constant injection rate inevitably causes the pressure to rise [32,33]. By monitoring pressure changes at different locations in high-permeability and low-permeability models, the specific flow direction of the gel at various injection volumes can be determined, and the pressure transfer characteristics during profile control flooding can be investigated.

During the profile control and flooding phase, injection rate significantly impacts pressure changes within the sand-filled pipe. As shown in the pressure change curve in Figure 2, within the injection rate range of 0.40 mL/min to 1.20 mL/min, polymer gel preferentially enters the high-permeability sand-filled pipe. The pressure transmission rate in the high-permeability pipe is significantly faster than in the low-permeability pipe, with all monitoring points showing pressure fluctuations, particularly near the inlet. At an injection rate of 0.40 mL/min (Figure 2a,b), the injection volume reaches 0.25 PV, and pressure in both sand-filled pipes begins to rise simultaneously at 10 cm. As injection progresses, the pressure at 90 cm in the high-permeability sand-filled pipe noticeably increases at 1.75 PV. However, the low-permeability sand-filled pipe does not experience pressure buildup until the injection volume reaches 3.10 PV at 90 cm. After injecting 2.70 PV and 3.10 PV, the pressures in both the high-permeability and low-permeability sand-filled pipes gradually stabilized. The maximum pressure at 10 cm in both pipes reached 6.50 MPa. When the injection rate is increased to 0.80 mL/min (Figure 2c,d), the pressure at the inlet end (10 cm) shows a noticeable increase after injecting 0.50 PV, indicating gel blockage at this point. However, the pressure in the low-permeability pipe does not begin to rise until the injection volume reaches 0.90 PV. At the outlet end (90 cm), the initial injection volumes for the high-permeability and low-permeability pipes are 1.50 PV and 2.80 PV, respectively. When the injection rate reaches 1.20 mL/min (Figure 2e,f), pressure in the high-permeability sand-filled pipe begins to rise at 10 cm with an injection volume of only 0.10 PV. In contrast, the low-permeability pipe only experiences pressure change when the injection volume reaches 0.20 PV, further confirming that the viscoelastic self-regulating agent preferentially enters the high-permeability channel. Compared to the initial pressure at 90 cm, the high-permeability sand-filled pipe exhibits an initial pressure of 1.49 PV, while the low-permeability sand-filled pipe shows 2.13 PV. This further demonstrates that pressure transmission is faster in the high-permeability pipe than in the low-permeability pipe.

Figure 3 shows the distributary ratio characteristics during the profile control and flooding phase at injection rates ranging from 0.40 to 1.20 mL/min. The distributary ratios in high- and low-permeability sand-filled pipes show both similarities and differences. At 0.40 mL/min (Figure 3a), the initial shunt rate of high-permeability pipes is 67.36%, roughly double that of low-permeability pipes (32.64%). As gel injection continues, high-permeability channels are blocked, reducing the diversion rate in high-permeability pipes and increasing it in low-permeability pipes, improving water injection balance [34–36]. When the injection volume reaches 2.40 PV, the diversion rates stabilize at 53.60% for high-permeability pipes and 46.40% for low-permeability pipes. At 0.80 mL/min (Figure 3b), the shunt rates change more rapidly. Initially, high-permeability pipes have a shunt rate of 88.09%, while low-permeability pipes have 11.90%. With continued injection, high-permeability channels are blocked, and the shunt ratio changes. By 2.50 PV, the diversion rates stabilize at 52.30% for high-permeability pipes and 47.70% for low-permeability pipes. At 1.20 mL/min (Figure 3c), the shunt rates initially differ significantly, with high-permeability pipes at 72.49% and low-permeability pipes at 27.51%. As gel injection progresses, high-permeability channels are blocked, and the shunt curves converge. However, the diversion rate curves reverse in the early mid-stage, suggesting that the gel's blocking effect is limited. Ultimately, the diversion rates stabilize at 61.10% for high-permeability pipes and 38.90% for low-permeability pipes.

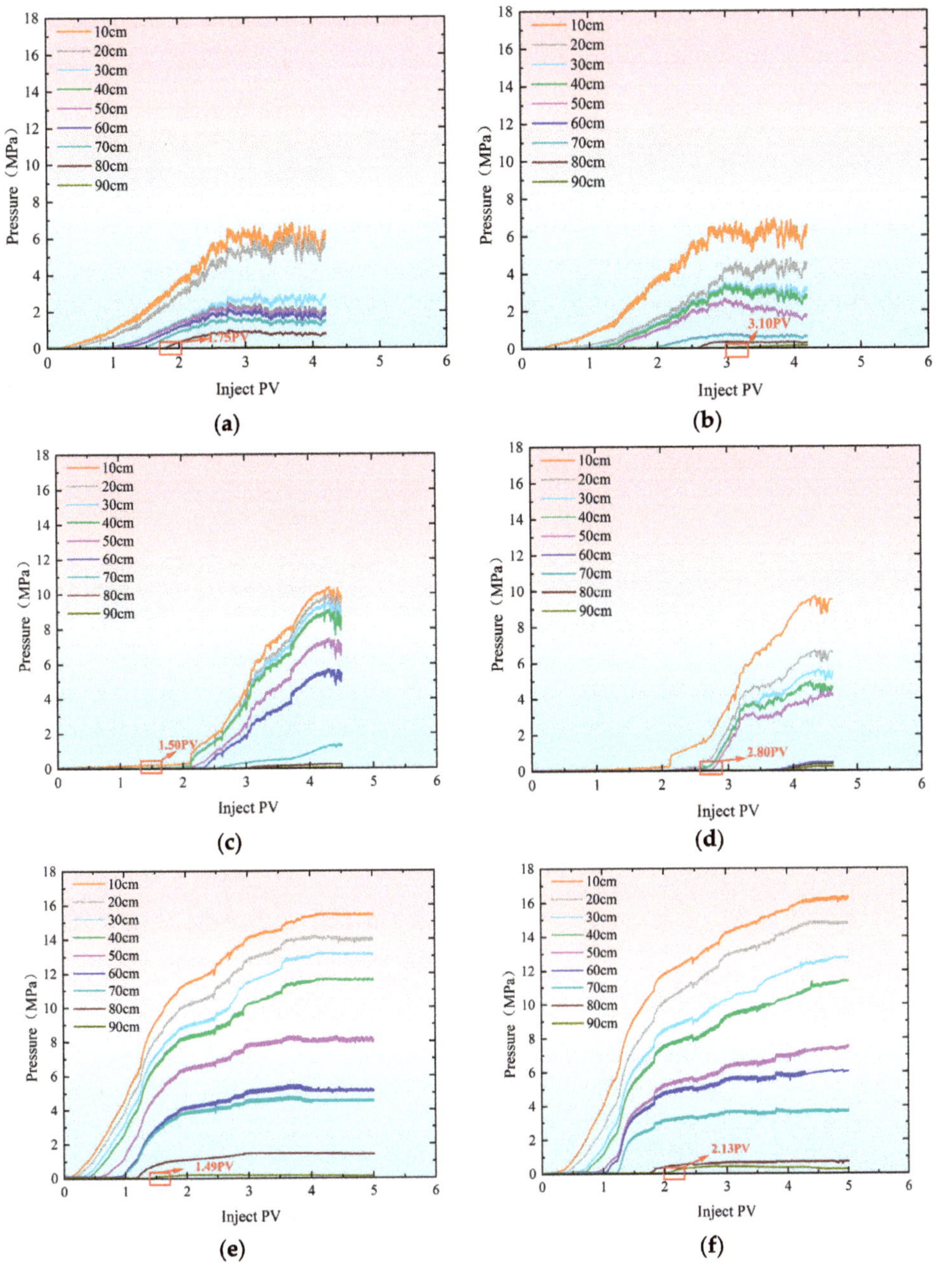

Figure 2. Pressure change curve at injection rate of 0.40 mL/min–1.20 mL during profile control and flooding. (**a**) Hypertonic model (injection rate: 0.40 mL/min); (**b**) hypotonic model (injection rate: 0.40 mL/min). (**c**) Hypertonic model (injection rate: 0.80 mL/min); (**d**) hypotonic model (injection rate: 0.80 mL/min). (**e**) Hypertonic model (injection rate: 1.20 mL/min); (**f**) hypotonic model (injection rate: 1.20 mL/min).

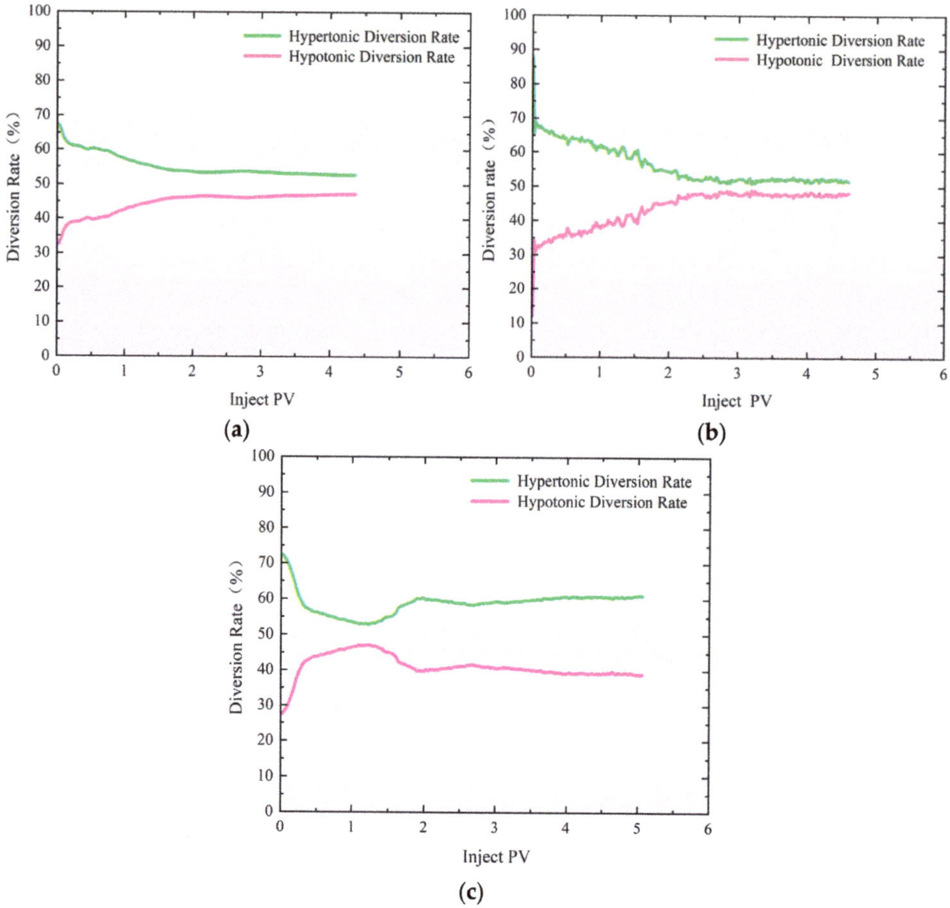

Figure 3. Special evidence of profile control and flooding shunt rate curve with injection rate of 0.40 mL/min–1.20 mL/min. (**a**) Injection rate: 0.40 mL/min; (**b**) injection rate: 0.80 mL/min; (**c**) injection rate: 1.20 mL/min.

Due to the permeability grade differences, when the polymer gel solution is injected into a heterogeneous reservoir, it tends to flow more into the high-permeability regions, resulting in an increase in seepage resistance and a rapid rise in pressure. Consequently, the suction pressure difference in the low-permeability zones increases, causing the gel to start flowing towards these areas. As a result, the pressure rise in the low-permeability zones lags behind that in the high-permeability zones [37–40]. Additionally, the injection volume required for the outlet of the high-permeability region to begin pressurizing is less than that required for the low-permeability region, indicating that pressure conduction in the high-permeability model, dominated by the dominant channels, occurs more rapidly than in the low-permeability matrix.

During the experiment, pressure fluctuations were observed at each monitoring point, particularly near the injection end. This phenomenon can be attributed to the viscoelastic properties of the system. When the shear force is large, the physical associations are disrupted, leading to a decrease in viscosity and a drop in pressure. Conversely, when the shear force is small or removed, the association effect is restored, the structural viscosity is re-established, and the pressure increases.

During the constant rate profile control flooding process, differences in the split flow rates between the high-permeability and low-permeability models were observed at different stages. To reduce these differences and improve the recovery rate, the flow rate was adjusted during the displacement process, and a dynamic profile control flooding method was employed for the experiment.

Figure 4 illustrates the pressure change curves during the profile control flooding process with different multi-stage flow rates in the high-permeability and low-permeability models. As the injection volume of polymer gel increased, the pressure rose rapidly. The initial injection rate was 1.20 mL/min, which was reduced to 0.80 mL/min when the injection volume reached 2.10 PV. Figure 5 shows that at this point, the difference in split flow rates between the high-permeability and low-permeability models increased. To stabilize the split flow rate, the injection rate was adjusted to 0.80 mL/min, resulting in a significant drop in pressure at each monitoring point. As the injection volume continued to rise, the pressure gradually increased. When the injection volume reached 3.15 PV, the difference in split flow rates still increased, prompting a further reduction in flow rate to stabilize the split flow rate. From the split flow rate curve, it is evident that after the flow rate was reduced to 0.40 mL/min, the split flow rate gradually stabilized. When the injection volume reached 3.91 PV, the flow rate was further reduced to 0.10 mL/min in an attempt to reduce the split flow rate difference. However, the experimental data indicated that the difference could not be minimized. As a result, the displacement process was stopped when the injection volume reached 4.25 PV. Ultimately, the split flow rate in the high-permeability sand-packed tube was 70.40%, while the split flow rate in the low-permeability sand-packed tube was 32.60%.

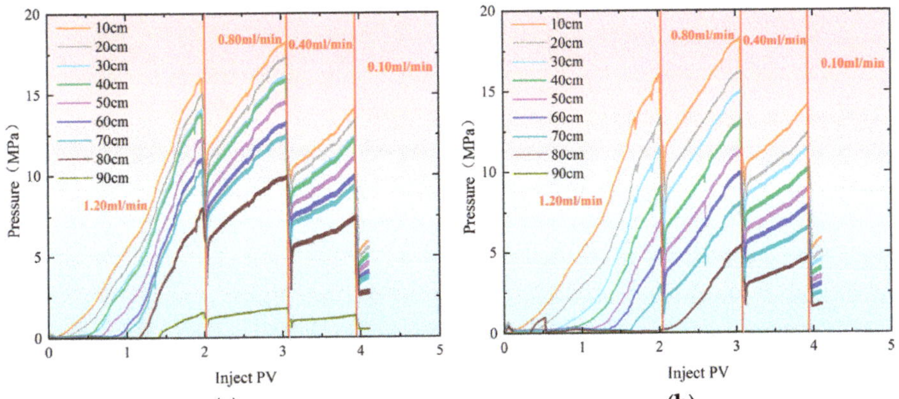

Figure 4. Pressure change curves during profile control flooding at injection rates of 1.20–0.80–0.40–0.10 mL/min. (**a**) Hypertonic model (injection rate: 1.20–0.80–0.40–0.10 mL/min); (**b**) hypotonic model (injection rate: 1.20–0.80–0.40–0.10 mL/min).

As shown in Figure 6, the pressure increases gradually as the injection volume rises during profile control flooding. The initial injection rate was set at 0.80 mL/min, which was increased to 1.20 mL/min once the injection volume reached 2.10 PV. From Figure 7, it is evident that when the injection volume reached 2.10 PV, the split flow rate had already stabilized. As a result, the injection rate was adjusted to 1.20 mL/min to observe how the recovery rate changed with the variation in split flow rate. As the injection volume continued to increase, the pressure also rose steadily. By the time the injection volume reached 4.87 PV, the final split flow rate in the high-permeability model was 80.90%, while in the low-permeability model, it was 19.10%.

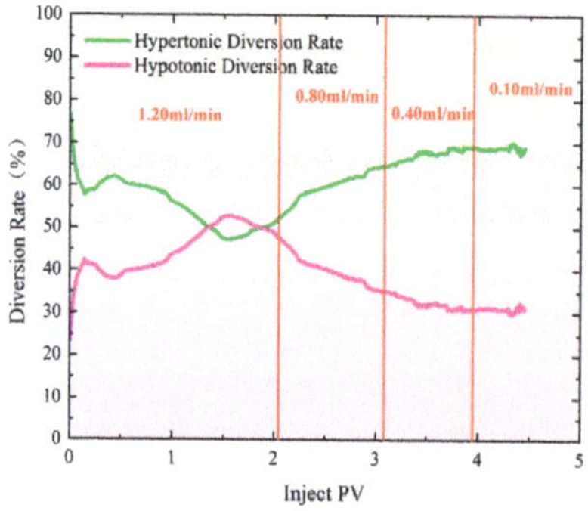

Figure 5. Characteristics of split flow rate curves during profile control flooding at injection rates of 1.20–0.80–0.40–0.10 mL/min.

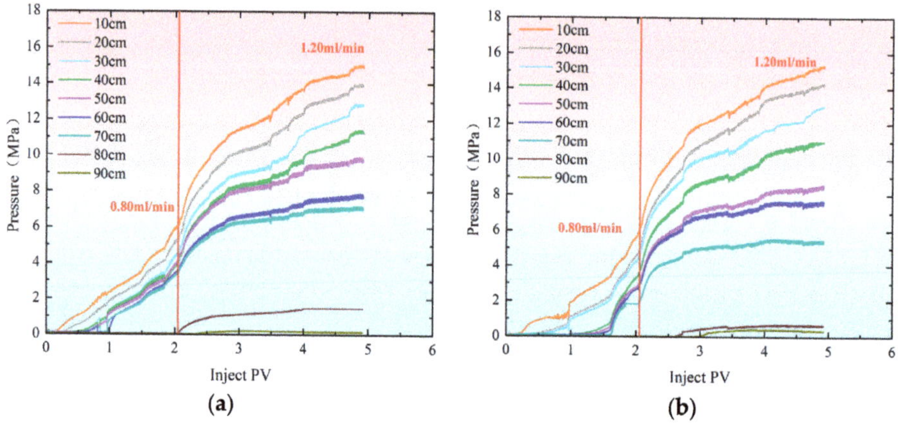

Figure 6. Pressure change curves during profile control flooding at injection rates of 0.80–1.20 mL/min. (**a**) Hypertonic model (injection rate: 0.80–1.20 mL/min); (**b**) hypotonic model (injection rate: 0.80–1.20 mL/min).

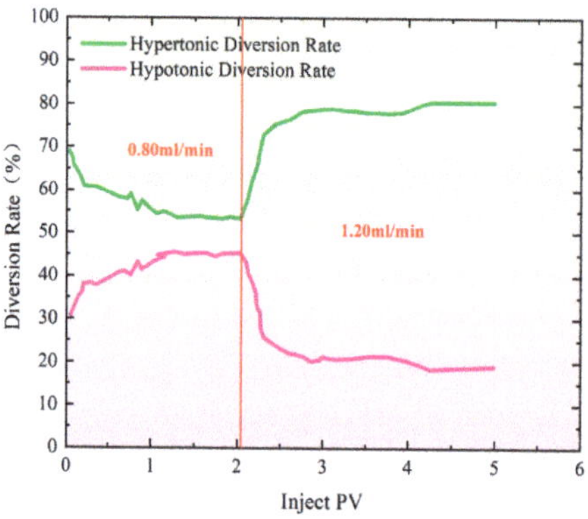

Figure 7. Characteristics of split flow rate curves during profile control flooding at injection rates of 0.80–1.20 mL/min.

It can be observed from Figure 8 that as the injection volume of the polymer gel increases, the pressure rises rapidly. Initially, the injection rate was set at 1.20 mL/min. When the injection reached 1.50 PV, the injection was halted, and the flow rate was reduced to 0.40 mL/min. Figure 9 shows that when the injection volume reached 1.50 PV, the separation rate of the fracture hyperpermeability model approached 50% of that of the matrix hypopermeability model. Therefore, to maintain a better separation rate, the injection rate was reduced at this point. Afterward, the separation rate gradually diverged and eventually stabilized. In the end, the hyperpermeability separation rate was 54.80%, while the low permeability diversion rate was 45.20%.

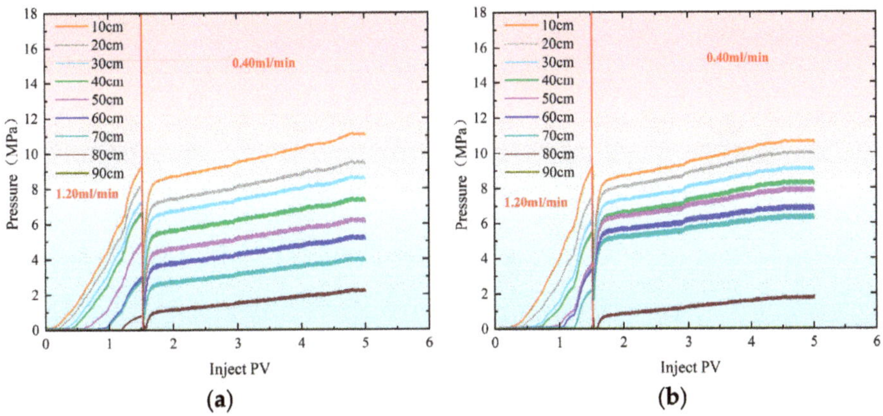

Figure 8. Pressure change curves during profile control flooding at injection rates of 1.20–0.40 mL/min. (**a**) Hypertonic model (injection rate: 1.20–0.40 mL/min); (**b**) hypotonic model (injection rate: 1.20 mL/min).

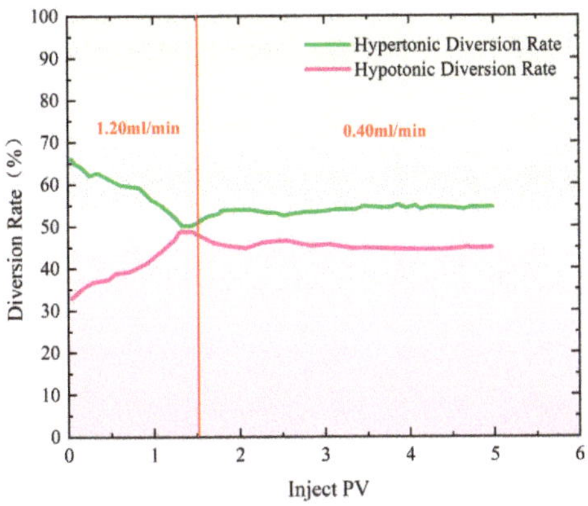

Figure 9. Characteristics of split flow rate curves during profile control flooding at injection rates of 1.20–0.40 mL/min.

2.2. *Law of Energy Conduction*

By comparing the pressure transfer characteristics under different conditions, the energy transfer law during viscoelastic profile control flooding in heterogeneous reservoirs was clarified. The stabilized pressure at each monitoring point was statistically analyzed, the pressure gradient between adjacent locations was calculated, and the pressure gradient distribution in different position segments was plotted.

As shown in Figure 10, the pressure gradient in the hypertonic model, with an injection rate of 0.80 mL/min, was highest in the 60–70 cm range, indicating the greatest accumulation of gel and the best sealing effect in this region. The pressure gradually decreased as it extended to both sides, forming a unimodal pattern. In the low-permeability model, the highest pressure gradient occurred in the 50–60 cm section, followed by the 10–20 cm section, with a 40 cm gap between them. These correspond to the main and secondary wave peaks, respectively, suggesting that the sealing effect follows a stepped pattern and is more effective.

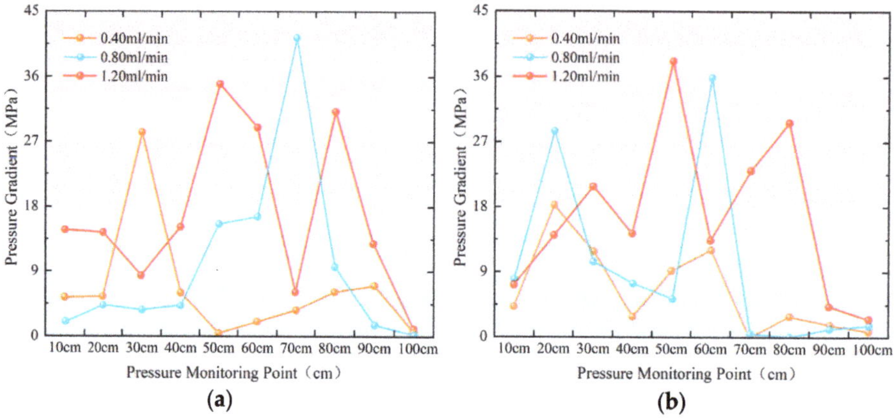

Figure 10. Shows the line graph of the average pressure gradient distribution at each position of the sand-filling pipe under different injection speeds. (**a**) Hypertonic model; (**b**) hypotonic model.

The pressurization time at the production end under different conditions provides a relatively intuitive reflection of the pressure conduction rate in the sand-filling pipe. As shown in Figure 11, when the injection speed is increased under similar conditions in permeability ratio, the amount of polymer gel required for the end pressure of production in both high-permeability and low-permeability models decreases gradually, and the change in low permeability is more obvious. Compared with the injection speed of 0.40 mL/min, the amount of gel required for the end pressure of production decreases by 0.30 PV at 0.80 mL/min and by 0.97 PV at 1.20 mL/min, which indicates that the increase in injection speed makes the pressure low. Within a certain range, the energy conduction efficiency of the formation is positively correlated with the injection speed, meaning that a higher injection rate at the injection end leads to a faster pressure increase at the production end.

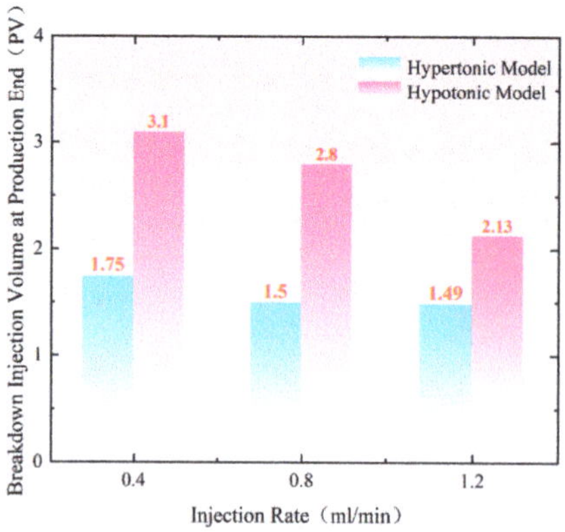

Figure 11. Shows the distribution of pressure buildup time at the production end under different constant injection speeds.

As shown in Figure 12, at an injection speed of 0.80 mL/min, the difference in the final diversion rate between the hypertonic and hypotonic models is the smallest, at only 4.60%. At this point, the instantaneous flow rates entering both models are nearly identical, indicating that the hypertonic channel is effectively blocked and the reservoir's heterogeneity is significantly improved. When the injection rate is 1.20 mL/min, the high pressure gradient allows better utilization of the hypotonic layer in the early stage of injection. However, as the amount of retained gel increases, the suction pressure difference gradually grows. In the later stage, profile reversal begins, and the gel enters the hypertonic layer. The final diversion rate difference increases to 22.24%, and the effect deteriorates.

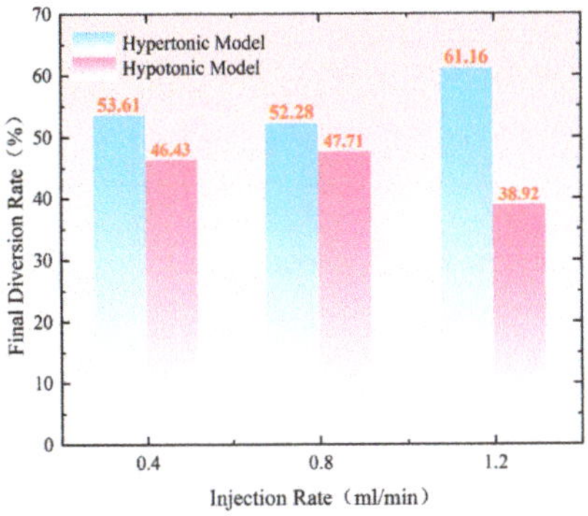

Figure 12. Comparison of the final diversion rate under different constant injection speeds.

As shown in Figure 13, the difference in the final diversion rate between the hypertonic and hypotonic models at an injection rate of 1.20–0.40 mL/min is 9.60%. This relatively small difference indicates that the hypertonic channels are effectively blocked, significantly improving the heterogeneity of the reservoir. When the injection speeds were 1.20–0.80–0.40–0.10 mL/min, the difference in the final diversion rate between the hypertonic and hypotonic models increased to 37.80%. At an injection speed of 0.80–1.20 mL/min, the difference in the diversion rate rose to 61.72%. If the difference in the diversion rate is too large, most of the fluid will flow through the hypertonic crack model, resulting in poor overall effectiveness.

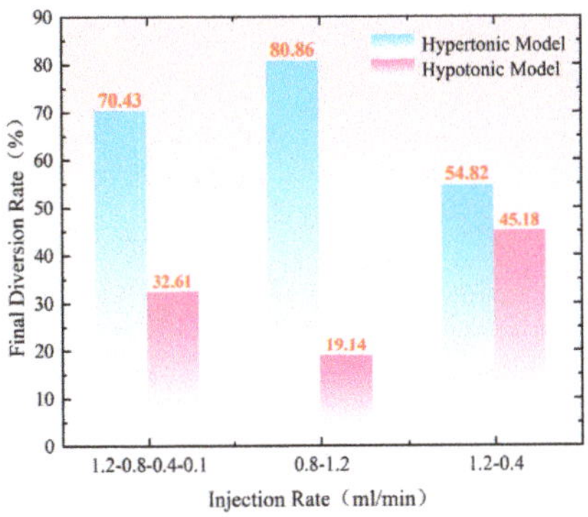

Figure 13. Comparison of final diversion rates under multi–pole flow rates.

2.3. Recovery Rate Characteristics

Before conducting the static profile control flooding experiments, water flooding experiments were initially performed. During the water flooding process, a combined injection and separate production method was employed. The water flooding was halted once the cumulative water cut at the outlet of both the high-permeability model and the matrix model exceeded 90%. Following this, profile control flooding experiments were conducted with different constant injection rates. The results demonstrated that the recovery rate improved to varying extents after the profile control flooding. The experimental data is presented in Table 1.

Table 1. Recovery rate data for water flooding and profile control flooding.

Core Number	Water Flooding Flow Velocity (mL/min)	Profile Control Flooding Rate (mL/min)	Permeability (mD)	Water Flood Recovery Rate (%)	Profile Control Flooding Recovery Rate (%)	Total Recovery Rate (%)
1-1	0.50	0.40	107.30	77.78	9.45	87.23
1-2			54.20	70.13	14.60	84.73
2-1	0.50	0.80	101.00	76.83	15.27	92.10
2-2			47.90	68.52	22.96	91.48
3-1	0.50	1.20	100.60	75.19	9.69	84.88
3-2			47.05	65.71	17.25	82.96

The injection rate for water flooding in all three experimental groups was set at 0.5 mL/min. As shown in Figure 14, the recovery rate during the water flooding stage was approximately 76% for the high-permeability model and around 68% for the low-permeability model, demonstrating good repeatability of the experiments. When comparing the changes in recovery rate after profile control flooding, the increase in recovery rate for both the high-permeability and low-permeability models was smallest at an injection rate of 0.40 mL/min. In contrast, at an injection rate of 1.20 mL/min, the recovery rate for the low-permeability model increased by 2.65%, reaching 17.25%. The most significant increase in recovery rate for the low-permeability model occurred at an injection rate of 0.8 mL/min, where it rose by 8.36% compared to 0.4 mL/min, reaching 22.96%. This improvement is largely attributed to the optimal split flow rate achieved at this injection rate. At 0.80 mL/min, the difference in the final split flow rate between the high-permeability and low-permeability models was smallest, at only 4.57%, with nearly identical instantaneous flow rates entering both models. This indicates that the high-permeability channels were effectively blocked, leading to a significant improvement in the reservoir's heterogeneity. At an injection rate of 1.20 mL/min, the higher pressure gradient allowed better utilization of the low-permeability layer during the early stage of injection. However, as the amount of retained gel increased, the suction pressure difference gradually escalated, resulting in profile reversal in the later stages, with the gel entering the high-permeability layer. This led to a final difference in split flow rate of 22.24%, resulting in a less effective outcome.

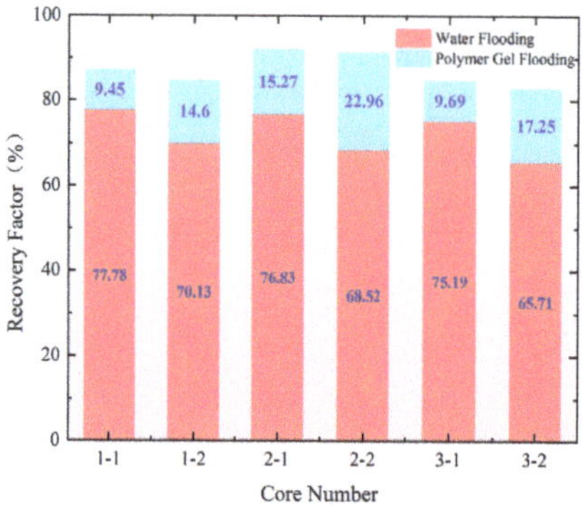

Figure 14. Comparison of the extraction degrees at each stage of the static displacement experiment.

Through dynamic profile control flooding experiments, the changes in recovery rate under multi-stage flow rates were explored, and the optimal injection parameter experimental data are shown in Table 2.

Table 2. Comparison of recovery rates at multiple flow rates.

Core Number	Water Flooding Flow Velocity (mL/min)	Profile Control Flooding Rate (mL/min)	Permeability (mD)	Water Flooding Recovery Rate (%)	Profile Control Flooding Recovery Rate (%)	Total Recovery Rate (%)
3-1	0.50	1.20–0.80–0.04–0.10	101.00	65.14	15.16	80.30
3-2			48.00	64.53	14.06	78.59
3-3	0.50	0.80–1.20	106.90	65.12	10.38	75.5
3-4			49.50	64.58	17.81	82.39
3-5	1.20–0.40	1.20–0.40	103.80	65.78	19.49	85.27
3-6			49.40	64.61	31.23	95.84

The injection rate for water flooding in all three experimental groups was set at 0.50 mL/min. As shown in Figure 15, the recovery rate during the water flooding stage was approximately 65.50% for the high-permeability model and around 64.50% for the low-permeability model, indicating a good repeatability of the experiments. When comparing the changes in recovery rate after profile control flooding, the smallest increase in recovery rate for both the high-permeability and low-permeability models occurred at injection rates of 1.20, 0.80, 0.40, and 0.10 mL/min. In contrast, at an injection rate of 0.80–1.20 mL/min, the recovery rate for the low-permeability model increased by 3.75%, reaching 17.81%. The most significant improvement in recovery rate for the low-permeability model was observed at an injection rate of 1.20–0.40 mL/min, where it increased by 17.17% compared to the 1.20–0.80–0.40–0.10 mL/min injection rate, reaching 31.23%. This change is closely related to the optimal split flow rate achieved at this injection rate. Specifically, at an injection rate of 1.20–0.40 mL/min, the difference in the final split flow rate between the high-permeability and low-permeability models was the smallest at only 9.64%, with nearly identical instantaneous flow rates entering both models. This indicates that the high-permeability channels were effectively blocked, leading to a significant improvement in the reservoir's heterogeneity.

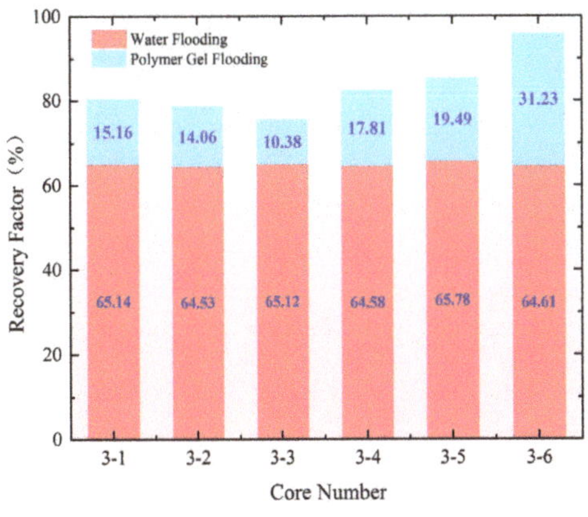

Figure 15. Comparison of the extraction degrees at each stage of the dynamic displacement experiment.

2.4. Microscopic Mechanism of Profile Control Flooding

Profile control flooding technology aims to enhance reservoir heterogeneity and improve crude oil recovery rates, with its microscopic mechanisms being particularly critical [41,42]. In heterogeneous reservoirs, profile control agents play a vital role by preferentially entering larger pores with higher permeability. These agents form high-strength plugging barriers through processes such as adsorption, bridging, and cross-linking, forcing the injected fluid to alter its flow direction. As a result, the fluid gradually moves into low-permeability areas, expanding the swept volume and effectively mobilizing oil from low-permeability reservoirs [43–45]. The ratio of G′/G″ of polymer gel is 5.81, and the elastic modulus is greater than the viscous modulus, so the gel can better resist shear deformation. Moreover, due to its viscoelasticity, the residual oil flows at the pore throat by relying on the extrusion swelling effect, which significantly improves the oil recovery (Figures 16 and 17) [46–49].

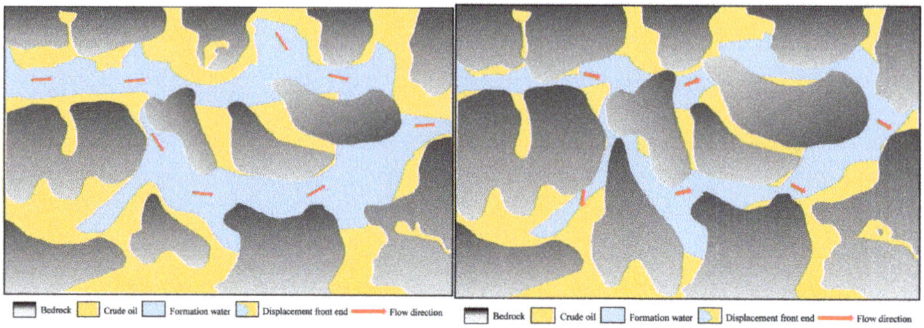

(a) (b)

Figure 16. Microscopic mechanism of water flooding. (**a**) Water flooding state of the hypertonic model; (**b**) water flooding state of the low-permeability model.

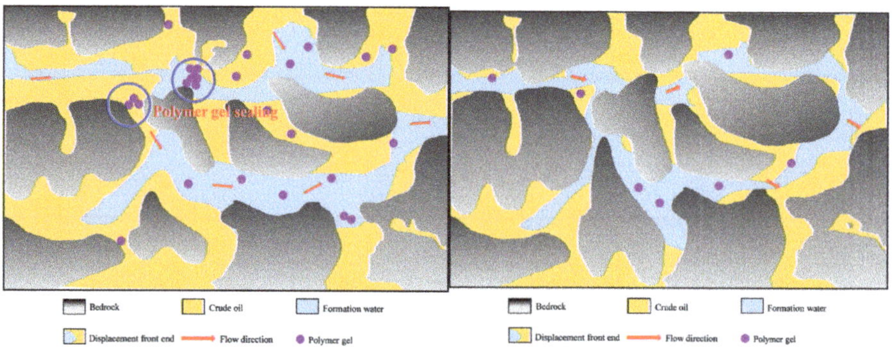

(a) (b)

Figure 17. Microscopic mechanism of polymer gel profile control flooding. (**a**) Profile control flooding state of high-permeability model; (**b**) profile control flooding state of low-permeability model.

3. Conclusions

Based on sand-packed tube physical model experiments, a study on polymer gel profile control flooding was conducted to address the challenges of combined injection and separate production in high-permeability and low-permeability channels. The research explored the laws of formation energy transfer and the microscopic mechanisms during the water injection stage of polymer gel profile control flooding under different injection and production conditions. The following conclusions were drawn:

(1) After polymer gel is injected into the heterogeneous reservoir, it preferentially enters the high-permeability layer, causing an increase in seepage resistance and a rapid rise

in pressure. As the suction pressure differential increases, the gel gradually shifts towards the low-permeability layer, resulting in a delayed pressure response in the low-permeability layer. The pressure transfer speed in high-permeability channels is significantly faster than in low-permeability channels.

(2) The formation energy transfer efficiency increases with the injection rate. The higher the injection rate, the faster the pressure increases at the production end, leading to a more significant reduction in the amount of polymer gel required to buildup pressure in the low-permeability reservoir layer.

(3) Initially using a high flow rate to promote rapid pressure transfer to the low-permeability area, followed by a reduction in flow rate to maintain a stable split flow rate, can effectively block the high-permeability channels and significantly improve the recovery rate.

(4) When the difference in split flow rate between the high-permeability and low-permeability reservoir layers is minimized, the heterogeneity of the reservoir is most effectively improved, resulting in the largest increase in recovery rate.

(5) By blocking the high-permeability channels, polymer gel causes the fluid to gradually shift towards the low-permeability reservoir, thereby expanding the swept volume. Moreover, due to its viscoelasticity, the polymer gel utilizes the extrusion swelling effect to overcome capillary forces, causing the residual oil in the pore throats to flow and thus increasing the crude oil recovery rate.

4. Materials and Methods

4.1. Materials and Instruments

The quartz sand ground from the Yan 9 outcrop of Hujianshan Oilfield has a particle size of 40–60 mesh (0.3–0.5 mm). The degassed crude oil of Hujianshan Oilfield has a viscosity of 1.50 mPa·s at 50 °C. In terms of water types, the simulated formation water is a $CaCl_2$ water type with a mineralization degree of 52,197 mg/L; the simulated injected water is a Na_2SO_4 water type with a mineralization degree of 1470 mg/L. Polymer gel comprised 1000 mg/L of the mixture (this was produced by Xi'an Changqing Chemical Industry Group Co., Ltd. Xi'an, China. It was analytically pure, with the elastic modulus of the system greater than the viscosity modulus, and the viscosity remaining at low shear stress and decreasing at high shear, showing good viscoelastic characteristics. It has good thermal stability, and thermogravimetric analysis shows that the fastest range of weight loss rate is 320–400 °C. Its salt resistance is outstanding; the viscosity is 6.1 mPa·s in saline water with a salinity of 50,000 mg/L, and the viscosity retention rate was about 75%).

Figure 18 shows a schematic diagram of the indoor parallel double-tube displacement experimental device, which consists of four key components: (1) Heterogeneous reservoir simulation system: Two parallel sand-filling pipes are used to simulate the reservoir matrix and high-permeability bands, respectively, with combined injection and separate production (sand-filling pipe, length × diameter = 1000 mm × 50 mm, 9 pressure monitoring points, set at equal intervals of 100 mm). (2) Fluid injection system: 1 L intermediate container, 2PB-104IV type constant flow pump; (3) Online real-time pressure monitoring system: A total of 18 pressure monitoring points, which can record the pressure changes at different parts of the sand-filling pipe from the injection end to the production end in real time. (4) Oil–water metering system: Liquid collection bottle.

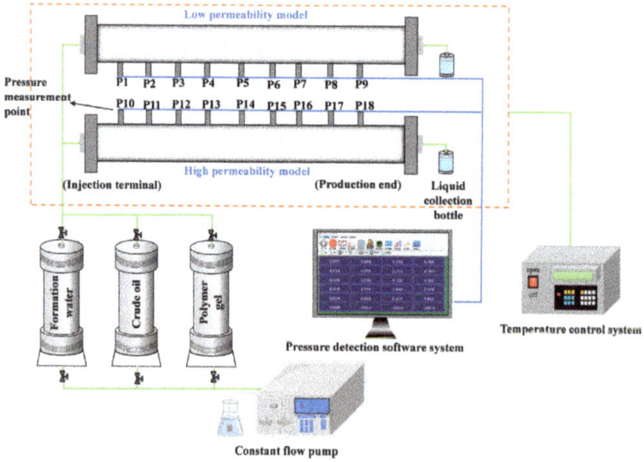

Figure 18. Schematic diagram of the indoor parallel double-tube displacement experimental device.

4.2. Experimental Principle

Convert the injection speed based on the on-site injection linear speed [50], and use Equation (1) to approximate the injection speed for the large-scale object model replacement experiment. This will allow for the setting of the experimental parameters, as outlined in Tables 3 and 4.

$$Q = \phi A V_a = \frac{\pi \phi D^2 V_a \times 100}{4 \times 24 \times 60} \quad (1)$$

Table 3. Comparison between on-site injection linear velocity and experimental injection velocity.

Method	On-Site Injection Linear Velocity (m/d)	Experimental Injection Rate (ml/min)
Water flooding	0.013	0.50
Profile control and flooding	0.01	0.40
	0.02	0.80
	0.03	1.20

Table 4. Design of displacement experiment parameters.

Experiment Number	Experimental Content	Core Number	Penetration Rate (mD)	The Penetration Rate Is Extremely Poor	Porosity (%)	Polymer Gel Concentration (mg/L)	Injection Speed (mL/min) Water Flooding	Profile Control Flooding
1	Static Displacement	1-1	107.30	2.00	37.20	1000	0.50	0.40
		1-2	54.20		23.90			
2		2-1	101.00	2.10	34.50			0.80
		2-2	47.90		26.50			
3		3-1	100.60	2.10	33.40			1.20
		3-2	47.10		29.20			
4	Dynamic Displacement	4-1	101.00	2.10	29.20			1.20–0.80–0.40–0.10
		4-2	48.00		26.50			
5		5-1	106.90	2.20	29.30			0.80–1.20
		5-2	49.50		26.60			
6		6-1	103.80	2.10	29.50			1.20–0.40
		6-2	49.40		26.40			

Of these, Q is the injection speed of the experiment, ml/min; ϕ is porosity, %; A is the cross-sectional area of the core, cm^2; V_a is the linear velocity at the time of injection in the mine, m/d; and D represents the diameter of the core, in cm.

The diversion rate is defined as the ratio of the liquid production from high-permeability (or low-permeability) cores to the total liquid production. During the displacement process, the liquid output from both high and low permeability models is measured in real-time, and their respective diversion rates are calculated using Equations (2) and (3), as shown in Sections [51–54].

$$D_h = \frac{V_h}{V_h + V_l} \times 100\% \tag{2}$$

$$D_l = \frac{V_l}{V_h + V_l} \times 100\% \tag{3}$$

Of these, D_h represents the diversion rate of the hypertonic model, %; D_l represents the diversion rate of the hypotonic model, in %; V_h represents the liquid production volume of the hypertonic model, in ml; V_l represents the liquid production volume of the hypotonic model, in mL.

4.3. Experimental Procedures

Based on parallel double-tube displacement experiments in the laboratory, combined with an online real-time pressure monitoring system, the energy transfer law during the profile control flooding process was quantitatively evaluated. The specific steps of the physical flow simulation experiments for profile control flooding in the sand-packed tube are as follows:

(1) Considering the characteristics of the reservoir, such as strong heterogeneity and significant oil–water separation in the Yan'an Formation, a mixture of quartz sand and clay was prepared for sand packing in tubes with different permeabilities, in accordance with the experimental requirements.
(2) The sand-packed tubes were filled with the prepared sand mixture, and the weight of each tube was recorded.
(3) The intelligent pressure monitoring system was installed, and the airtightness of the device connections was thoroughly checked.
(4) The two sand-packed tubes were saturated with formation water at a flow rate of 0.5 mL/min. The permeability was measured using the liquid, and after complete saturation, the sand-packed tubes were weighed to calculate their pore volume.
(5) The two sand-packed tubes were then saturated with crude oil at a flow rate of 0.5 mL/min. Saturation was halted when the liquid at the outlet reached 100% crude oil.
(6) The sand-packed tubes were connected in parallel, and water flooding experiments were conducted at a flow rate of 0.5 mL/min. The displacement was stopped when the cumulative water cut of the two sand-packed tubes reached 90%. During the water flooding process, pressure changes at different positions of the sand-packed tubes were monitored in real time. Additionally, the water cut and oil displacement efficiency of the two sand-packed tubes were recorded every half hour.
(7) After completing the water flooding experiment, the viscoelastic self-regulating agent displacement experiment was carried out at the set experimental flow rate. Throughout the experiment, pressure changes at various positions of the sand-packed tubes were monitored and recorded in real time, with the water cut and oil displacement efficiency of the tubes also recorded every half hour.
(8) The displacement was stopped when the injected volume reached five times the total pore volume of the two sand-packed tubes, or when the water cut at the outlet of the sand-packed tubes remained unchanged.
(9) After completing the displacement experiment, the packing material in the sand-packed tubes was removed, the tube walls were cleaned, and after complete drying, the permeability contrast of the sand-packed tubes was reassessed. Alternatively, adjustments were made to the injection rate of the profile control flooding agent, or the timing of rate changes during dynamic profile control flooding. Steps 2 to 8 were then repeated as necessary.

Author Contributions: Conceptualization, C.W. and H.G.; methodology, C.W.; software, Y.D.; validation, Y.D., Y.L. and G.L.; formal analysis, P.Y.; investigation, B.M.; resources, H.G.; data curation, Y.D.; writing—original draft preparation, Y.D.; writing—review and editing, C.W.; visualization, G.L.; supervision, Y.L.; project administration, C.W.; funding acquisition, C.W. All authors have read and agreed to the published version of the manuscript.

Funding: This research was supported by the National Natural Science Foundation of China (No.52374041), the National Natural Science Foundation of China (No.52174030), and the Key Research and Development Projects of Shaanxi Province (No.2023-YBGY-306).

Institutional Review Board Statement: Not applicable.

Informed Consent Statement: Not applicable.

Data Availability Statement: All data are included in the manuscript.

Acknowledgments: Thanks to all authors for their contributions.

Conflicts of Interest: Authors Yunlong Liu, Ping Yi, and Bo Ma were employed by the company PetroChina Changqing Oilfield. The remaining authors declare that the research was conducted in the absence of any commercial or financial relationships that could be construed as potential conflicts of interest.

Abbreviations

The following abbreviations are used in this manuscript:

PCAs	Profile Control Agents
PEG	Polyethylene Glycol Single-phase Gel Particles
CBG-SP	Cross-linked Gel Swelling Particles

References

1. Li, X.; Yang, Z.; Li, S.; Huang, W.; Zhan, J.; Lin, W. Reservoir characteristics and effective development technology in typical low-permeability to ultralow-permeability reservoirs of China National Petroleum Corporation. *Energy Explor. Exploit.* **2021**, *39*, 1713–1726. [CrossRef]
2. Cao, J.; Hao, M.; Chen, Y.; Li, B.; Liu, Y.; Xu, J. Advancing PetroChina's development strategies for low-permeability oil reservoirs. *Processes* **2024**, *12*, 351. [CrossRef]
3. Wang, J.; Feng, L.; Steve, M.; Tang, X.; Gail, T.E.; Mikael, H. China's unconventional oil: A review of its resources and outlook for long-term production. *Energy* **2015**, *82*, 31–42. [CrossRef]
4. Muther, T.; Qureshi, H.A.; Syed, F.I.; Aziz, H.; Siyal, A.; Dahaghi, A.K.; Negahban, S. Unconventional hydrocarbon resources: Geological statistics, petrophysical characterization, and field development strategies. *J. Pet. Explor. Prod. Technol.* **2022**, *12*, 1463–1488. [CrossRef]
5. Ding, M.; Li, Q.; Yuan, Y.; Wang, Y.; Zhao, N.; Han, Y. Permeability and heterogeneity adaptability of surfactant-alternating-gas foam for recovering oil from low-permeability reservoirs. *Pet. Sci.* **2022**, *19*, 1185–1197. [CrossRef]
6. He, S.; Fang, F.; Song, K.; Li, X.; Zhang, J.; Cao, J.; Wang, Y.; Huang, X.; Bian, C. Research progress on the method of determining gas–water relative permeability curve in unconventional reservoirs. *J. Pet. Explor. Prod. Technol.* **2025**, *15*, 36. [CrossRef]
7. Shi, X.; Wei, J.; Bo, H.; Zheng, Q.; Yi, F.; Yin, Y.; Chen, Y.; Dong, M.; Zhang, D.; Li, J. A novel model for oil recovery estimate in heterogeneous low-permeability and tight reservoirs with pseudo threshold pressure gradient. *Energy Rep.* **2021**, *7*, 1416–1423. [CrossRef]
8. Gao, H.; Luo, K.; Wang, C.; Li, T.; Cheng, Z.; Dou, L.; Zhao, K.; Zhang, N.; Liu, Y. Impact of dissolution and precipitation on pore structure in CO_2 sequestration within tight sandstone reservoirs. *Pet. Sci.* **2025**, *22*, 868–883. [CrossRef]
9. Wei, J.; Zhou, X.; Zhou, J.; Li, J.; Wang, A. Recovery efficiency of tight oil reservoirs with different injection fluids: An experimental investigation of oil-water distribution feature. *J. Pet. Explor. Prod. Technol.* **2021**, *195*, 107678. [CrossRef]
10. AlHammadi, M.; Mahzari, P.; Sohrabi, M. Fundamental investigation of underlying mechanisms behind improved oil recovery by low salinity water injection in carbonate rocks. *Fuel* **2018**, *220*, 345–357. [CrossRef]
11. Shen, Y.; Wu, J.; Shi, F.; He, Y.; Li, H.; Zhou, W.; Zhao, M.; Wang, C.; Yan, Y.; Lu, M. Study on Green Nanofluid Profile Control and Displacement of Oil in a Low Permeability Reservoir. *ACS Omega* **2023**, *8*, 38926–38932. [CrossRef] [PubMed]

12. Zhou, Z.; Zhao, J.; Zhou, T.; Huang, Y. Study on in-depth profile control system of low-permeability reservoir in block H of Daqing oil field. *J. Pet. Explor. Prod. Technol.* **2017**, *157*, 1192–1196. [CrossRef]
13. Li, J.; Wang, Z.; Yang, H.; Yang, H.; Wu, T.; Wu, H.; Wang, H.; Yu, F.; Jiang, H. Compatibility evaluation of in-depth profile control agents in dominant channels of low-permeability reservoirs. *J. Pet. Explor. Prod. Technol.* **2020**, *194*, 107529. [CrossRef]
14. Zhao, F.; Zhang, H.; Wu, Y.; Wang, D.; Zhang, Y. Preparation and Performance Evaluation of Polymeric Microspheres Used for Profile Control of Low-Permeability Reservoirs. *J. Chem.* **2020**, *2020*, 5279608. [CrossRef]
15. Zou, J.; Yue, X.; Dong, J.; Shao, M.; Wang, L.; Gu, J.; An, W. Novel in-depth profile control agent based on in-situ polymeric microspheres in low permeability reservoir. *J. Dispers. Sci. Technol.* **2019**, *441*, 1254–1264. [CrossRef]
16. Abdullah, A.; Chuan, L.D.C.; Hamzah, S.; Sundaram, M.M.; Mudasar, Z.; Yousif, A.; Hezam, A.A.S.; Roil, M.B. Thermophysical Properties of Nanofluid in Two-Phase Fluid Flow through a Porous Rectangular Medium for Enhanced Oil Recovery. *Nanomaterials* **2022**, *12*, 1011. [CrossRef]
17. Li, X.; Wang, J.; Ni, J.; Fu, L.; Xu, A.; Zhao, L. Compatibility Evaluation of In-Depth Profile Control Agents for Low-Permeability Fractured Reservoirs. *Gels* **2022**, *8*, 575. [CrossRef]
18. Yin, D.; Zhou, W. Mechanism of Enhanced Oil Recovery for In-Depth Profile Control and Cyclic Waterflooding in Fractured Low-Permeability Reservoirs. *Geofluids* **2021**, *2021*, 6615495. [CrossRef]
19. Xu, B.; Wang, Y. Profile control performance and field application of preformed particle gel in low-permeability fractured reservoir. *J. Pet. Explor. Prod. Technol.* **2020**, *11*, 477–482. [CrossRef]
20. Pryazhnikov, M.I.; Zhigarev, V.; Minakov, A.V.; Nemtsev, I.V. Spontaneous imbibition experiments for enhanced oil recovery with silica nanosols. *Capillarity* **2024**, *10*, 73–86. [CrossRef]
21. Yin, D.; Song, S.; Xu, Q.; Liu, K. Experimental Research on the Optimization and Evaluation of the Polymer/Chromium-Ion Deep Profile Control System for the Fractured Low-Permeability Reservoirs. *Processes* **2020**, *8*, 1021. [CrossRef]
22. Druetta, P.; Picchioni, F. Polymer and nanoparticles flooding as a new method for Enhanced Oil Recovery. *J. Pet. Sci. Eng.* **2019**, *177*, 479–495. [CrossRef]
23. Tavakkoli, O.; Kamyab, H.; Shariati, M.; Mohamed, A.M.; Junin, R. Effect of nanoparticles on the performance of polymer/surfactant flooding for enhanced oil recovery: A review. *Fuel* **2022**, *312*, 122867. [CrossRef]
24. Ambaliya, M.; Bera, A. A Perspective Review on the Current Status and Development of Polymer Flooding in Enhanced Oil Recovery Using Polymeric Nanofluids. *Ind. Eng. Chem. Res.* **2023**, *62*, 2444–2459. [CrossRef]
25. Zeynalli, M.; Mushtaq, M.; Al-Shalabi, E.W.; Alfazazi, U.; Hassan, A.M.; AlAmeri, W. A comprehensive review of viscoelastic polymer flooding in sandstone and carbonate rocks. *Sci. Rep.* **2024**, *13*, 17679. [CrossRef] [PubMed]
26. Kamal, M.S.; Sultan, A.S.; Al-Mubaiyedh, U.A.; Hussein, I.A. Review on Polymer Flooding: Rheology, Adsorption, Stability, and Field Applications of Various Polymer Systems. *Polym. Rev.* **2015**, *55*, 491–530. [CrossRef]
27. Sharafi, M.S.; Jamialahmadi, M.; Hoseinpour, S.A. Modeling of Viscoelastic Polymer Flooding in Core-Scale for Prediction of Oil Recovery Using Numerical Approach. *J. Mol. Liq.* **2018**, *250*, 295–306. [CrossRef]
28. Mahajan, S.; Yadav, H.; Rellegadla, S.; Agrawal, A. Polymers for Enhanced Oil Recovery: Fundamentals and Selection Criteria Revisited. *Appl. Microbiol. Biotechnol.* **2021**, *105*, 8073–8090. [CrossRef]
29. Alizadeh, A.H.; Piri, M. Three-Phase Flow in Porous Media: A Review of Experimental Studies on Relative Permeability. *Rev. Geophys.* **2014**, *52*, 468–521. [CrossRef]
30. Feng, S.-J.; Zhang, X.; Zheng, Q.-T.; Chen, H.-X.; Zhao, Y.; Yang, C.-B.-X. Modeling the Spreading and Remediation Efficiency of Slow-Release Oxidants in a Fractured and Contaminated Low-Permeability Stratum. *Chemosphere* **2023**, *337*, 139271. [CrossRef]
31. Yin, F.; Gao, Y.; Zhang, H.; Sun, B.; Chen, Y.; Gao, D.; Zhao, X. Comprehensive Evaluation of Gas Production Efficiency and Reservoir Stability of Horizontal Well with Different Depressurization Methods in Low Permeability Hydrate Reservoir. *Energy* **2022**, *239*, 122422. [CrossRef]
32. Zhang, D.; Wang, Y.; Ye, P.; Li, S.; Wei, J.; Zhong, L.; Zhou, R. Oil Recovery Mechanism of Polymer Gel Injection Between Injection Wells and Production Wells to Block the Dominant Channel of Water Flow. *Gels* **2025**, *11*, 337. [CrossRef] [PubMed]
33. Xiangyu, W.; Chenwei, L.; Jiacheng, N.; Chenru, Z.; Chenmei, H.; Long, L.; Mingzhong, L. Microscopic Seepage Mechanism of Nanofluids Assistant CO_2 Aerosol Foam System Flooding after Water Flooding. *Fuel* **2025**, *384*, 134049. [CrossRef]
34. Abdelgawad, K.Z. Polymer Induced Permeability Reduction: The Influence of Polymer Retention and Porous Medium Properties. *J. Pet. Sci. Eng.* **2022**, *217*, 110821. [CrossRef]
35. Wang, Y.; Zhang, L.; Liu, M.; Yu, X. Comparative Study on the Mechanical Properties of Solidified Silty-Fine Sand Reinforced by Permeable Polymer and Traditional Grouting Materials. *Constr. Build. Mater.* **2024**, *419*, 135485. [CrossRef]
36. Song, W.; Liu, X.; Zheng, T.; Yang, J. A Review of Recharge and Clogging in Sandstone Aquifer. *Geothermics* **2020**, *87*, 101857. [CrossRef]
37. Mirzaie Yegane, M.; Boukany, P.E.; Zitha, P. Fundamentals and Recent Progress in the Flow of Water-Soluble Polymers in Porous Media for Enhanced Oil Recovery. *Energies* **2022**, *15*, 8575. [CrossRef]

38. Zhang, Y.; Yang, C. Study on the Imbibition Production Mechanism and Main Controlling Factors of Low-Permeability Reservoir. *RSC Adv.* **2024**, *14*, 31183–31192.
39. Ning, T.; Xi, M.; Hu, B.; Wang, L.; Huang, C.; Su, J. Effect of Viscosity Action and Capillarity on Pore-Scale Oil–Water Flowing Behaviors in a Low-Permeability Sandstone Waterflood. *Energies* **2021**, *14*, 8200. [CrossRef]
40. Al-shajalee, F.; Arif, M.; Myers, M.; Tadé, M.O.; Wood, C.; Saeedi, A. Rock/Fluid/Polymer Interaction Mechanisms: Implications for Water Shut-off Treatment. *Energy Fuels* **2021**, *35*, 12809–12827. [CrossRef]
41. Xu, Z.; Li, S.; Li, B.; Chen, D.; Liu, Z.; Li, Z. A Review of Development Methods and EOR Technologies for Carbonate Reservoirs. *Pet. Sci.* **2020**, *17*, 990–1013. [CrossRef]
42. Sun, J.; Sun, L.; Bao, B. Recovery Mechanism of Supercritical CO_2 Miscible Flooding in 30 Nm Nanomatrices with Multi-Scale and Fracture Heterogeneity. *Fuel* **2025**, *392*, 134835. [CrossRef]
43. Cao, B.; Xie, K.; Lu, X.; Cao, W.; He, X.; Xiao, Z.; Zhang, Y.; Wang, X.; Su, C. Effect and Mechanism of Combined Operation of Profile Modification and Water Shutoff with In-Depth Displacement in High-Heterogeneity Oil Reservoirs. *Colloids Surf. A* **2021**, *631*, 127673. [CrossRef]
44. Wang, F.; Yang, H.; Jiang, H.; Kang, X.; Hou, X.; Wang, T.; Zhou, B.; Sarsenbekuly, B.; Kang, W. Formation Mechanism and Location Distribution of Blockage during Polymer Flooding. *J. Pet. Sci. Eng.* **2020**, *194*, 107503. [CrossRef]
45. Mohamed, A.I.A.; Khishvand, M.; Piri, M. Entrapment and Mobilization Dynamics during the Flow of Viscoelastic Fluids in Natural Porous Media: A Micro-Scale Experimental Investigation. *Phys. Fluids* **2023**, *35*, 047119. [CrossRef]
46. Shao, W.; Yang, J.; Wang, H.; Chang, J.; Wu, H.; Hou, J. Recent research progress on imbibition system of nanoparticle-surfactant dispersions. *Capillarity* **2023**, *8*, 34–44. [CrossRef]
47. Deng, R.; Dong, J.; Dang, L. Numerical simulation and evaluation of residual oil saturation in waterflooded reservoirs. *Fuel* **2025**, *384*, 134018. [CrossRef]
48. Yin, B.; Zhang, C.; Ding, T.; Chen, C.; Feng, K.; Wang, Z.; Sun, B. An experimental and numerical study of gas-liquid two-phase flow moving upward vertically in larger annulus. *Eng. Appl. Comp. Fluid* **2025**, *19*, 2476605. [CrossRef]
49. Yin, B.; Ding, T.; Wang, S.; Wang, Z.; Sun, B.; Zhang, W.; Zhang, X. Deformation and migration characteristics of bubbles moving in gas-liquid countercurrent flow in annulus. *Pet. Explor. Dev.* **2025**, *52*, 471–484. [CrossRef]
50. You, Q.; Wen, Q.; Fang, J.; Guo, M.; Zhang, Q.; Dai, C. Experimental Study on Lateral Flooding for Enhanced Oil Recovery in Bottom-Water Reservoir with High Water Cut. *J. Pet. Sci. Eng.* **2019**, *174*, 747–756. [CrossRef]
51. Wang, K.; Wang, Z.; Liu, K.; Cheng, L.; Bai, Y.; Jin, G. Optimizing Flood Diversion Siting and Its Control Strategy of Detention Basins: A Case Study of the Yangtze River, China. *J. Hydrol.* **2021**, *597*, 126201. [CrossRef]
52. Yan, Q.; Jing, Z.; Xia, Z.; Zhang, S. Flood Control Capacity of the Three Gorges Project for Different Frequency Floods. *Environ. Eng. Sci.* **2021**, *38*, 1195–1205. [CrossRef]
53. An, Y.; Ren, F.; Liu, X.; Hu, J.; Fang, T.; Wang, B.; Luo, M. Migration and Plugging Laws of Plugging-Selection Balls in Fine Profile Control Process of Layered Waterflooding Wells. *Sci. Rep.* **2025**, *15*, 1776. [CrossRef] [PubMed]
54. Liu, P.; Zhang, X.; Wu, Y.; Li, X. Enhanced Oil Recovery by Air-Foam Flooding System in Tight Oil Reservoirs: Study on the Profile-Controlling Mechanisms. *J. Pet. Sci. Eng.* **2017**, *150*, 208–216. [CrossRef]

gels

MDPI

Article

Dynamic Simulation of Nano-Gel Microspheres for Plugging Preferential Flow Channels and Enhancing Oil Recovery in Waterflooded Reservoirs

Long Ren [1,2,*], Cong Zhao [1,2,*], Jian Sun [1,2], Cheng Jing [1,2], Haitao Bai [1,2], Qingqing Li [1,2] and Xin Ma [1,2]

1 School of Petroleum Engineering, Xi'an Shiyou University, Xi'an 710065, China; xjkelsj@163.com (J.S.); jich.0704@163.com (C.J.)

2 Shaanxi Key Laboratory of Advanced Stimulation Technology for Oi & Gas Reservoirs, Xi'an 710065, China

* Correspondence: renlong@xsyu.edu.cn (L.R.); 18791954836@163.com (C.Z.)

Abstract: This study addresses the unclear mechanisms by which preferential flow channels (PFCs), formed during long-term waterflooding, affect nano-gel microsphere (NGM) flooding efficiency, utilizing CMG reservoir numerical simulation software. A dynamic evolution model of PFCs was established by coupling CROCKTAB (stress–porosity hysteresis) and CROCKTABW (water saturation-driven permeability evolution), and the deep flooding mechanism of NGMs (based on their gel properties such as swelling, elastic deformation, and adsorption, and characterized by a "plugging-migration-replugging" process) was integrated. The results demonstrate that neglecting PFCs overestimates recovery by 8.7%, while NGMs reduce permeability by 33% (from 12 to 8 mD) in high-conductivity zones via "bridge-plug-filter cake" structures, diverting flow to low-permeability layers (+33% permeability, from 4.5 to 6 mD). Field application in a Chang 6 tight reservoir (permeability variation coefficient 0.82) confirms a >10-year effective period with 0.84% incremental recovery (from 7.31% to 8.15%) and favorable economics (ROI ≈ 10:1), providing a theoretical and engineering framework for gel-based conformance control in analogous reservoirs.

Keywords: nano-gel microspheres (NGMs); preferential flow channels (PFCs); plugging mechanism; numerical simulation; enhanced oil recovery (EOR)

1. Introduction

In oilfield development, particularly during the late stages of waterflooding, the formation of preferential flow channels (PFCs) between injection and production wells presents a central challenge to enhancing reservoir recovery [1,2]. As reported by Dai et al. [3] and others, prolonged water injection and reservoir development exacerbate heterogeneity and facilitate the gradual formation of high-conductivity flow channels. These PFCs cause injected water to bypass substantial volumes of residual oil, significantly reducing waterflood efficiency. This triggers the characteristic contradiction of a sudden decline in oil production accompanied by a surge in water production—a phenomenon widely documented in mature oilfields globally [4–6].

In recent years, nano-gel microsphere (NGM) modulation drive and dissection technology has shown significant potential for application. Feng et al. [7] experimentally demonstrated, through physical simulations, that nano-elastic microspheres effectively block high-conductivity channels via physical retention and elastic deformation. Combining NMR technology, they observed a transport characteristic in porous media where microspheres migrate through large pores but are retained in small pores, thereby enabling water

https://doi.org/10.3390/gels11070536

cut control. Using environmental scanning electron microscopy (ESEM), Zhang et al. [8] revealed significant swelling characteristics in polymer-based NGMs, with swelling ratios reaching up to 15-times. Through microscopic displacement experiments, they observed the formation of a composite blocking structure—comprising bridge plugging and filter cake mechanisms—within pore throats, attributed to mechanical retention. These findings verify the formation of a composite migration–plugging–filter cake blocking structure within pore throats and corroborate the dynamic regulation mechanism characterized by sequential migration, plugging, breakthrough, and replugging [9–12].

In the field of numerical simulation, scholars are continuously improving relevant models. For instance, Su et al. [13] established a multi-physics coupled model considering microsphere expansion, migration, and plugging characteristics. By incorporating pore structure parameters and fluid–solid interaction characteristics, they achieved the quantitative characterization of microsphere transport behavior. However, key technical bottlenecks persist in existing studies. Firstly, numerical models often fail to accurately characterize the formation mechanism and dynamic evolution process of PFCs. Secondly, the coupled simulation methodology for the plugging mechanism of nanospheres in high-permeability channels and the resulting fluid diversion effect remains underdeveloped.

Building upon this foundation, this study integrates reservoir engineering theory with CMG numerical simulation technology. Through the coupling of the key parameters of CROCKTAB (stress–porosity hysteresis) and CROCKTABW (water saturation-driven permeability evolution), a multi-field coupling model of PFCs and nanospheres is constructed. This model enables the systematic analysis of the evolution of reservoir properties and their influence on development indicators. The research findings aim to provide a technical solution possessing both theoretical value and engineering applicability for enhancing recovery in long-term waterflooded reservoirs.

2. Results and Discussion

2.1. Numerical Simulation of PFCs in Waterflooding Development

To mitigate computational expense and convergence challenges, a one-injection and one-production well-pair model was adopted. The grid dimensions are $150 \times 150 \times 2$, with planar grid spacing of $1 \text{ m} \times 1 \text{ m}$ and vertical spacing of 5 m. The injector–producer distance is 150 m. Key model parameters are detailed in Table 1.

Table 1. Physical properties of the reservoir simulation model.

Reservoir Properties	Value
Depth to oil layer midpoint (m)	1210
Reference pressure (MPa)	9.62
Crude oil formation volume factor (rm^3/sm^3)	1.206
Crude oil compressibility (MPa^{-1})	9.816×10^{-5}
Crude oil density (kg/m^3)	841
Crude oil viscosity (mPa·s)	2.62
Water compressibility (MPa^{-1})	5.0×10^{-5}
Water viscosity (mPa·s)	0.5
Rock compressibility (MPa^{-1})	7.135×10^{-5}

2.1.1. Coupled Simulation Setup for PFC Evolution

CROCKTAB (stress–porosity hysteresis) and CROCKTABW (water saturation-driven permeability evolution) keywords were implemented to simulate PFC formation during waterflooding. CROCKTAB defines pressure-dependent rock compressibility/expansion factors, while CROCKTABW governs pore volume variation as a function of water saturation differentials under varying pressures. The parameterization schemes are illustrated in Figure 1.

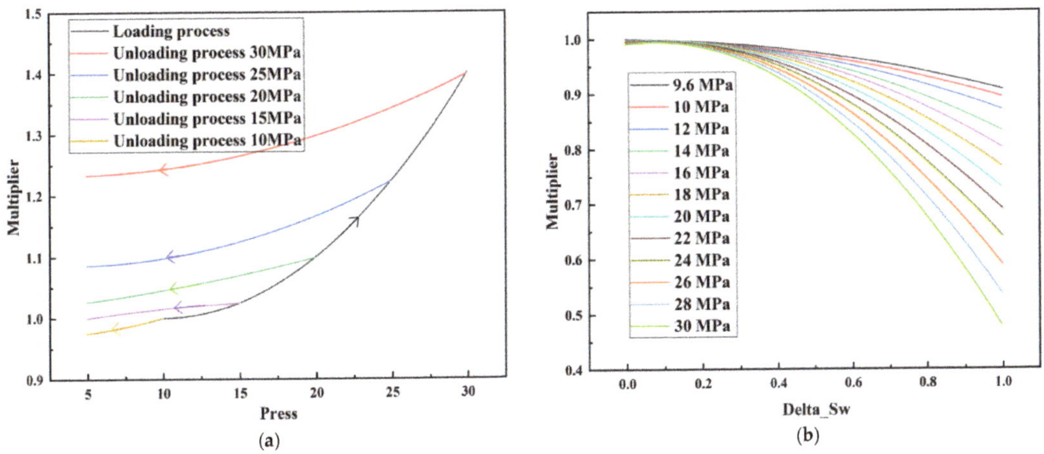

Figure 1. Multiplier effects of pressure (CROCKTAB) and water saturation differential (CROCKTABW). (**a**) CROCKTAB parameterization; (**b**) CROCKTABW parameterization.

(1) CROCKTAB parameters (see Figure 1a) reveal nonlinear porosity growth during loading (black curve), with pronounced hysteresis during unloading. The elastic recovery capacity decays significantly under high-pressure unloading, confirming irreversible deformation from prolonged water injection.
(2) CROCKTABW parameters (see Figure 1b) demonstrate a positive correlation between pore volume expansion and water saturation increments. High-stress conditions (e.g., 30 MPa) intensify pore structure adjustment during aqueous-phase seepage, quantitatively supporting PFC evolution.

2.1.2. Dynamic Evolution of Reservoir Properties During Waterflooding

Integrated CROCKTAB and CROCKTABW simulations reveal distinct property evolution in high-permeability and low-permeability layers (Tables 2–4).

Table 2 documents reservoir property evolution during waterflooding from initial development to the 95% water-cut stage. The high-permeability layer exhibits progressive porosity enhancement (initial: from 0.12 to 0.16) and permeability increases (from 8 to 12 mD), demonstrating a linear permeability enhancement trend along the injector–producer alignment. This trend arises from prolonged aqueous-phase scouring that enlarges pore throats and diminishes rock compaction effects, consistent with the effective stress–porosity coupling captured by the CROCKTAB and CROCKTABW models. Concurrently, oil saturation declines sharply from 0.45 to 0.30, forming a distinct low oil-saturation zone between wells. This spatial distribution directly manifests injected-water bypassing due to PFCs, confirming the reservoir heterogeneity challenge characterized by high-permeability channeling.

Table 2. Property evolution in high-permeability layer during waterflooding stages.

	Initial Development Stage	Waterless Oil Recovery Stage	Water Breakthrough Stage	High Water Cut Stage	
Porosity			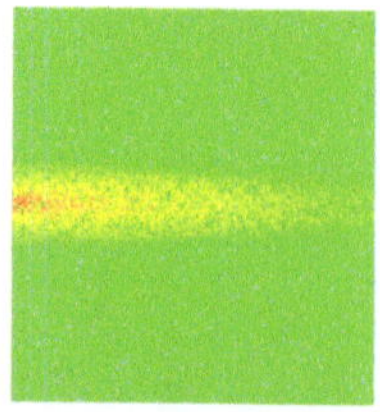	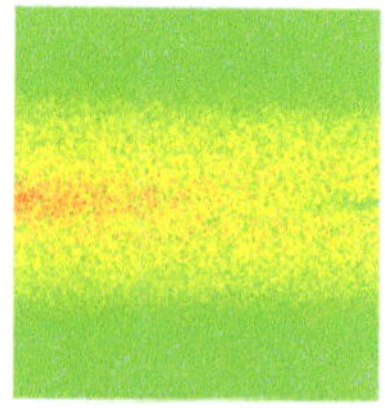	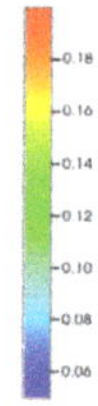
Permeability	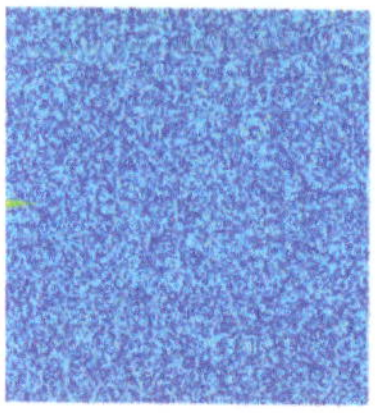	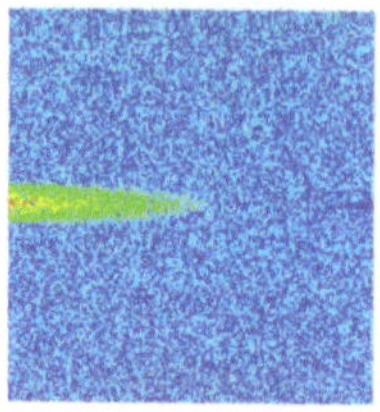	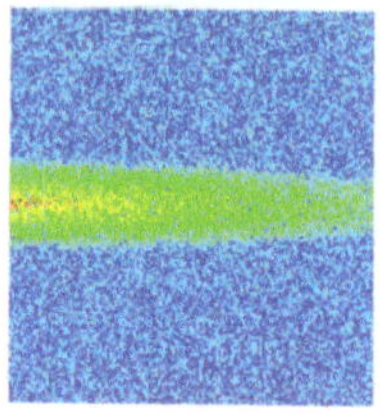	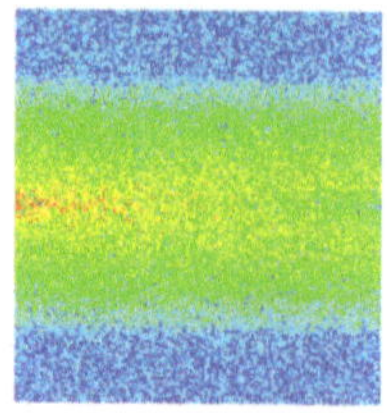	
Oil saturation				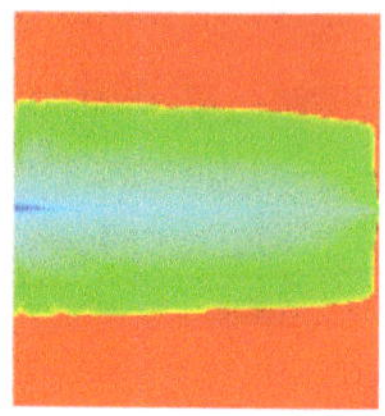	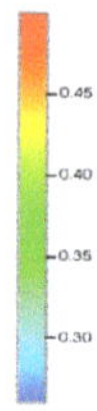

Table 3. Property evolution in low-permeability layer during waterflooding stages.

	Initial Development Stage	Waterless Oil Recovery Stage	Water Breakthrough Stage	High Water Cut Stage	
Porosity			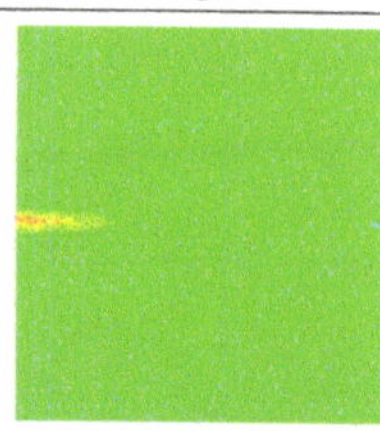	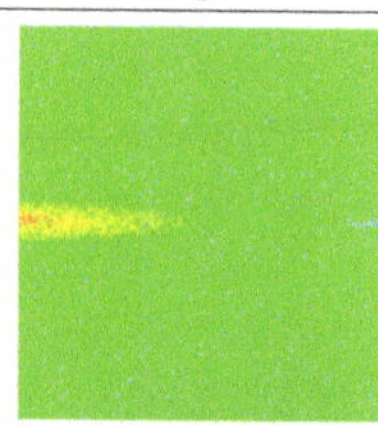	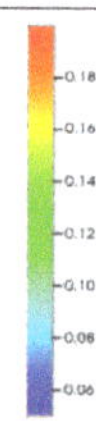
Permeability	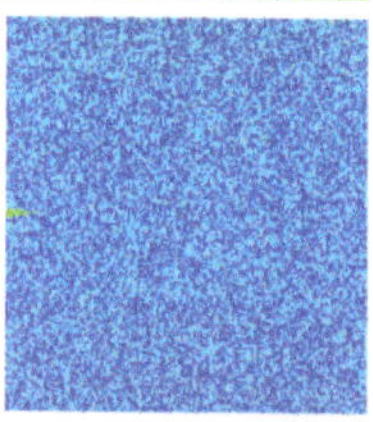	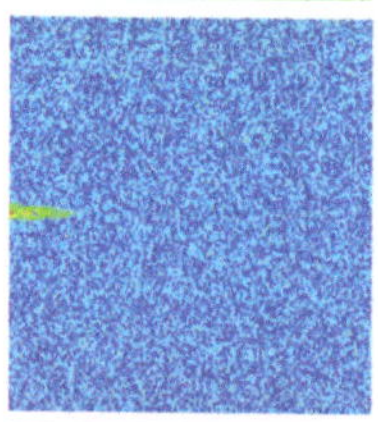	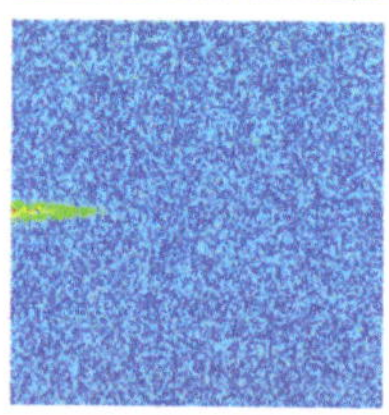	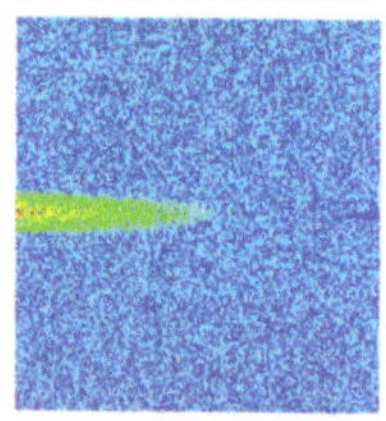	
Oil saturation					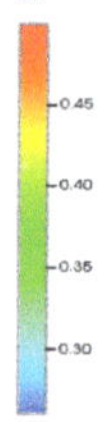

Table 4. Longitudinal profile evolution along injector–producer connection line during waterflooding stages.

	Initial Development Stage	Waterless Oil Recovery Stage	Water Breakthrough Stage	High Water Cut Stage
Porosity		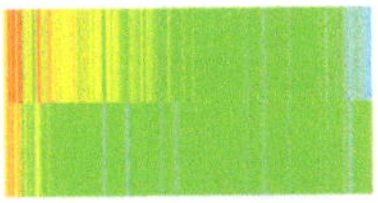		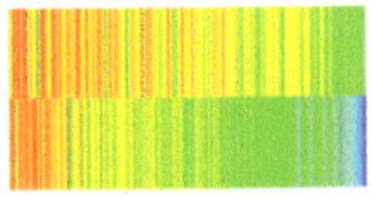
Permeability		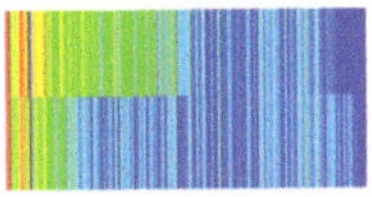	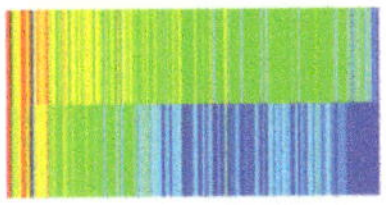	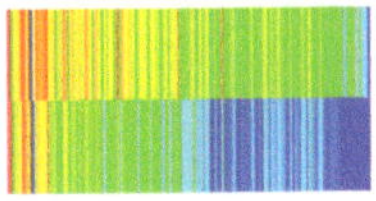
Oil saturation				

Table 3 reveals distinct property evolution in the low-permeability layer compared to the high-permeability layer (Table 2). Porosity increased modestly (from 0.08 to 0.10) and permeability rose to 4.5 mD from 3 mD—only one-third the magnitude of the high-permeability layer's enhancement (12 mD from 8 mD). This muted response stems from injected-water channeling in high-permeability zones, which impedes energy replenishment and delays pore-pressure transmission in low-permeability strata, empirically validating the "high-permeability layer shielding effect." Oil saturation stabilizes at 0.35 post-decline—significantly exceeding the high-permeability layer's 0.30 (Table 2)—confirming residual oil enrichment during waterflooding. Limited flow capacity prevents effective exploitation of this resource, establishing the targetable reservoir volume for subsequent NGM flooding through selective fluid diversion.

Table 4 reveals that the longitudinal profile along the injector–producer connection line exhibits a distinct strip-shaped high-permeability zone in the high-permeability layer, accompanied by a corresponding funnel-shaped low oil-saturation zone. This pattern confirms the formation of a highly connected PFC between the wells, consistent with literature findings that long-term water injection exacerbates reservoir heterogeneity.

To visualize the temporal evolution of porosity and permeability governed by the CROCKTAB and CROCKTABW keywords, specific grid cells were selected for monitoring: five in the high-permeability layer ((15, 75, 1), (45, 75, 1), (75, 75, 1), (105, 75, 1), (135, 75, 1)) and five corresponding grid cells in the low-permeability layer ((15, 75, 2), (45, 75, 2), (75, 75, 2), (105, 75, 2), (135, 75, 2)). The resulting property changes are presented in Figure 2.

Figure 2a,b demonstrates that permeability in the high-permeability layer grids exhibits non-linear growth with water injection time. Notably, permeability near the injection well (e.g., grid (15, 75, 1)) increases from 8 mD to 12 mD within 300 days. In contrast, the grid farthest from the injector (135, 75, 1) requires 600 days to achieve the same increase, confirming the "pressure conduction timeliness", as injected water preferentially channels through high-permeability pathways. This observation aligns with the effective stress–permeability hysteresis relationship defined by the CROCKTAB keyword. A transient 20% permeability decrease occurred initially in grid (135, 75, 1) due to the pore pressure near the producer falling below the formation pressure, inducing elastic rock expansion. Subsequent pressure recovery from water injection reversed this trend, allowing the permeability to rebound and continue growing.

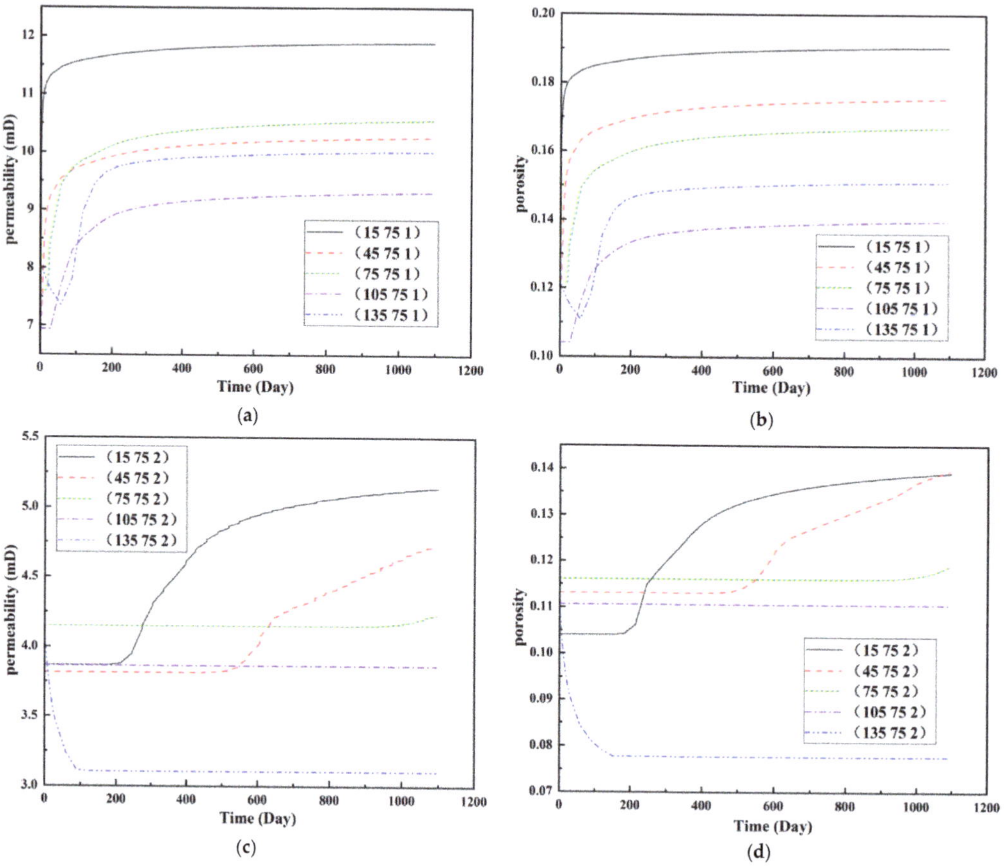

Figure 2. Temporal evolution of porosity and permeability in grid cells of different layers. (**a**) Permeability in high-permeability layer; (**b**) Porosity in high-permeability layer; (**c**) Permeability in low-permeability layer; (**d**) Porosity in low-permeability layer.

Porosity in the high-permeability layer also increased from 0.12 to 0.16, displaying a gradient growth trend along the injector–producer direction. The porosity growth rate at grid (15, 75, 1) reached 0.01 per 100 days, while grid (105, 75, 1), situated at the edge of the primary seepage channel, exhibited a significantly lower rate of only 0.005 per 100 days. This verifies the mechanism whereby "the intensity of water flow scouring governs the pore-throat expansion rate". The porosity evolution in grid (135, 75, 1) closely mirrored its permeability trend.

Figure 2c,d reveals that permeability in the low-permeability layer increased only modestly from 3 mD to 4.5 mD, representing just 50% of the amplitude observed in the high-permeability layer. Proximity to the injection well enabled grid (15, 75, 2) to initiate a slow permeability increase after 600 days of injection. Conversely, grid (135, 75, 2), persistently located in a low-pressure zone, maintained permeability below 4 mD, reflecting the "seepage shielding effect" imposed by the high-permeability layer. Porosity remained constrained between 0.08 and 0.10, substantially lower than in the high-permeability layer. Grid (105, 75, 2), largely unaffected by direct injected water flow, showed negligible porosity change. This confirms the "physical property inertia" characteristic of the low-permeability layer during waterflooding, providing the physical basis for subsequent residual oil enrichment targeted by NGM flooding.

2.1.3. Impact of PFCs on Waterflooding Performance

To quantify the macroscopic impact of PFCs on waterflooding efficiency, a 3-year simulation was conducted. The comparative results between the model incorporating PFC evolution (dominant channel model) and the base model (neglecting PFCs) are presented in Figure 3.

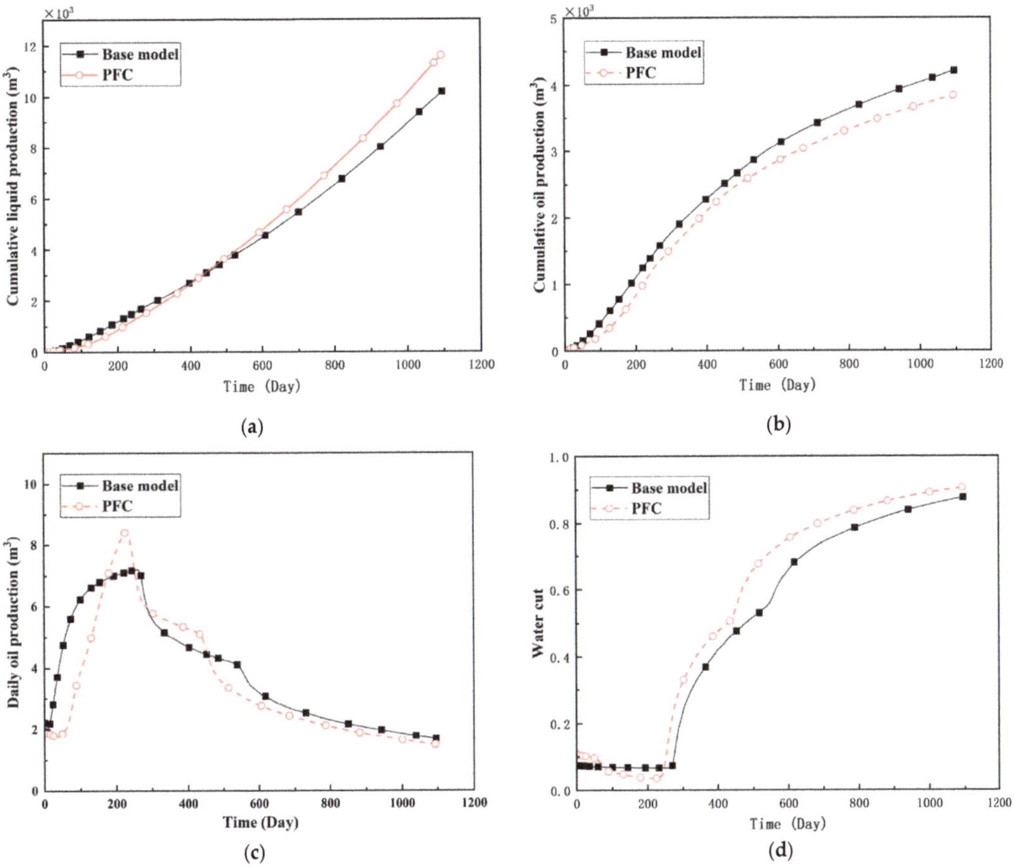

Figure 3. Impact of PFCs on waterflooding performance. (**a**) Cumulative liquid production; (**b**) Cumulative oil production; (**c**) Daily oil production; (**d**) Water cut.

The emergence and evolution of PFCs, rigorously captured through coupled geomechanical–fluid flow modeling (CROCKTAB and CROCKTABW), drive fundamentally distinct flooding regimes, with profound implications for recovery efficiency. Figure 3 provides conclusive evidence through four interlinked responses. (1) Early-stage flow divergence vs. late-stage channelization (see Figure 3a): Initial suppression of cumulative liquid production (<500 days) transitions to a 13.85% surplus (11,596 m^3 vs. 10,185 m^3) post-PFC maturation. This inversion confirms progressive flow conductivity enhancement as injected water increasingly bypasses low-permeability strata through established high-k pathways. (2) Sweep efficiency deterioration (see Figure 3b): Cumulative oil production decreases by 8.7% (3842 m^3 vs. 4207 m^3) due to volumetric sweep reduction, where PFCs sequester >35% of movable oil in unswept low-k zones. (3) Accelerated production decline (see Figure 3c): Post-600 days, daily oil production stabilizes at 15% lower (2.5 m^3/d) under the "short-circuit effect"—a direct consequence of rapid fluid migration through high-k conduits, while the base model overestimates recovery by neglecting the PFC-induced shielding of low-k contributions. (4) Premature water breakthrough (see Figure 3d): Water cut rises 30 days earlier and plateaus 4% higher (92% vs. 88%), spatially correlating with the low oil-saturation corridor in Figure 2. This behavior validates CROCKTABW's critical feedback mechanism: incremental water saturation → permeability amplification → accelerated channelization.

2.2. Simulation and Application of NGM Flooding After Waterflooding

2.2.1. Mechanism of NGM Profile Control

The numerical simulation technique for NGM profile control is based on three synergistic mechanisms:

(1) NGMs block PFCs through physical retention and elastic deformation, redirecting injected water into low-permeability zones to mobilize residual oil [7].
(2) NGM injection increases the displacing-phase viscosity, enabling viscous forces to surpass capillary thresholds and mobilize trapped crude oil from pore throats [14–16].
(3) NGMs' adsorption reduces oil-wetting sites on rock surfaces, shifting wettability toward water-wet states (validated by ESEM studies [8]). This decreases contact angles, reducing capillary forces that trap residual oil.

Consequently, oil-phase relative permeability K_{ro} increases since weakened oil–rock adhesion improves oil mobility, while water-phase relative permeability K_{rw} decreases as adsorbed NGMs constrict pore throats and increase flow resistance (Figure 4).

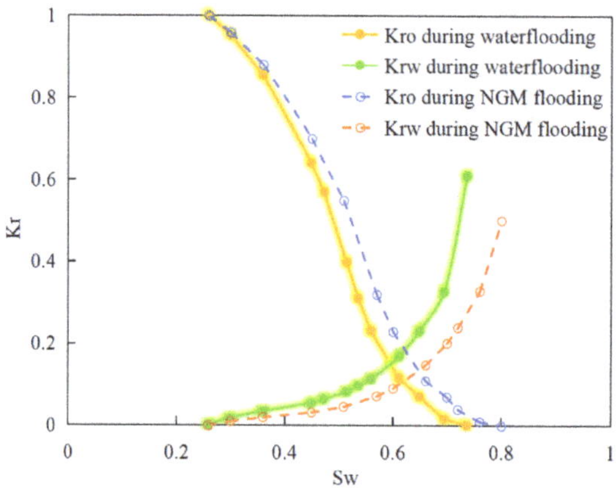

Figure 4. Influence of NGM flooding on oil-water relative permeability curves.

Figure 4 illustrates alterations in the phase-relative-permeability curves before and after NGM displacement. The yellow and green solid lines represent these parameters during waterflooding, whereas blue and orange dashed lines denote oil-phase and water-phase relative permeability during NGM displacement after waterflooding, explicitly showing the K_{ro} increase and K_{rw} decrease described above.

Comparing the relative permeability curves of waterflooding and NGM flooding shows different features. In the early stage (S_w 0~0.3), K_{ro} values are very low and decrease slowly, while K_{rw} values remain low with minimal increases. This indicates that at low water saturation, water exists primarily in small pores or in a dispersed state with limited flow capacity, while oil dominates the pore channels. Consequently, low S_w increases have a minimal impact on oil-phase permeability. In the middle stage (S_w 0.3~0.6), increasing S_w causes more significant Krw increases as water occupies the oil-phase pore channels, restricting oil flow. In the late stage (S_w > 0.6), the initial Krw rises faster than the NGM K_{rw}, indicating that NGMs alter water-phase viscosity and increase flow resistance. The NGM-phase relative permeability curve resembles the oil–water curve, but post-NGM flooding, oil-phase relative permeability end values increase and residual oil saturation decreases. The rightward shift of the iso-saturation point (from 0.59 to 0.65) further confirms enhanced oil mobility. The water-phase relative permeability at residual oil saturation decreased by 0.109 (from 0.61 to 0.501). Post-nano-fluid flooding-phase permeability curve changes resulted from nanoparticle adsorption-induced wettability reversal in the porous medium. The reduction in oil-wetting sites due to NGM adsorption shifts rock wettability toward water-wet states, decreasing the contact angle and capillary trapping forces. At high water saturation stages (Sw = 0.65), this elevates the oil-phase relative permeability (Kro) by 0.07 (from 0.07 to 0.14) and suppresses the water-phase relative permeability (Krw) by 0.09 (from 0.23 to 0.14).

Collectively, these mechanisms mobilize crude oil trapped within capillaries from pore throats, displacing residual oil and reducing phase relative permeability residual oil saturation. The calculation formula for the reduction value of residual oil saturation by NGMs is as follows:

$$\Delta S_{\text{or}} = \alpha \cdot C_{\text{ads}} (S_{\text{ori}} - S_{\text{orE}}) \quad (1)$$

where ΔS_{or} denotes the reduction value of residual oil saturation; α is the calibrated coefficient from coreflood experiments; C_{ads} denotes the NGM adsorption concentration; S_{ori} represents the initial residual oil saturation in the waterflooding stage; and S_{orE} indicates the water-driven residual oil saturation in the EOR stage.

The dynamic water-phase endpoint permeability is scaled by the NGM adsorption concentration C_{ads}, as follows:

$$K_{\text{rw}}^{\text{end}} = K_{\text{rwi}}^{\text{end}} (1 - \beta \cdot C_{\text{ads}}) \quad (2)$$

where K_{rw}^{end} represents the dynamic water-phase endpoint permeability in NGM flooding; K_{rwi}^{end} represents the initial water-phase endpoint permeability in the waterflooding stage; and β is calibrated coefficient from coreflood experiments.

Relative permeability curves are dynamically adjusted using CMG's ADSCOMP and ADSTABLE keyword, linking C_{ads} to grid-block properties in the dynamic effective permeability calculation formula. This method integrates geomechanical effects (CROCKTAB and CROCKTABW) with chemical flooding responses (ADSCOMP and ADSTABLE), enabling the full coupling of wettability dynamics.

NGMs' properties critically enable profile control; significant swelling (up to 15×) allows 3~6× in-situ expansion, forming essential "bridge-plug and filter-cake" structures for PFC blockage. Concurrently, pore-surface adsorption reduces effective pore radii, while elastic deformation facilitates pore-throat migration and structural adaptation. These mechanisms collectively enhance PFC plugging and oil recovery.

2.2.2. Simulation of NGM Flooding Effects

Following the establishment of PFCs in the high-permeability layer at waterflooding termination (Section 2.1.2), a 1% NGM suspension was injected. The NGMs comprised environmentally friendly polyacrylamide/cellulose composite gels, nano-modified for enhanced thermal stability (>90 °C) and salinity tolerance (>20,000 ppm). Based on these stability thresholds, NGMs were assumed to maintain stable gel properties throughout the 10-year simulation period, though degradation effects will be incorporated in future studies as experimental data becomes available. This formulation renders them suitable for low-permeability/fractured reservoirs. Key property evolution patterns are quantified in Tables 5–7.

Table 5. Characteristics of dynamic changes in physical properties of relatively high permeability layers during NGM flooding after waterflooding.

	Early Stage of NGMs	Middle Stage of NGMs	Late Stage of NGMs	Final Stage of NGMs
Porosity	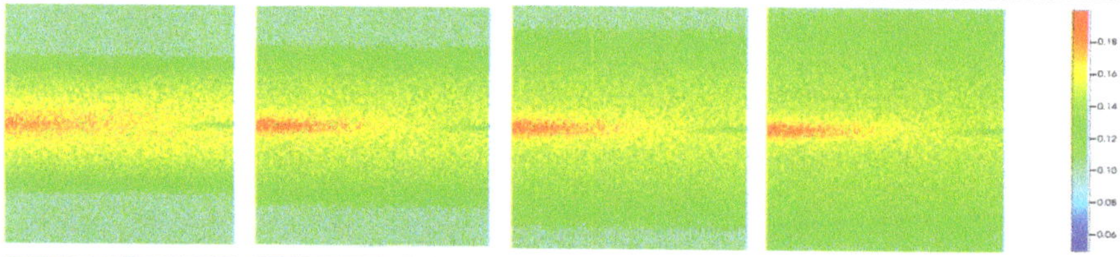			
Permeability	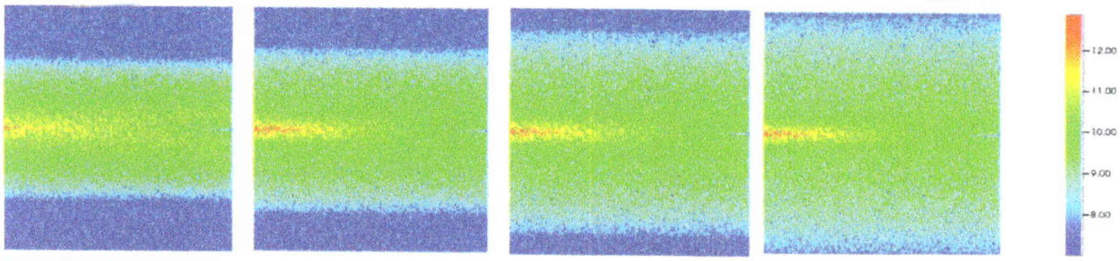			
Oil saturation	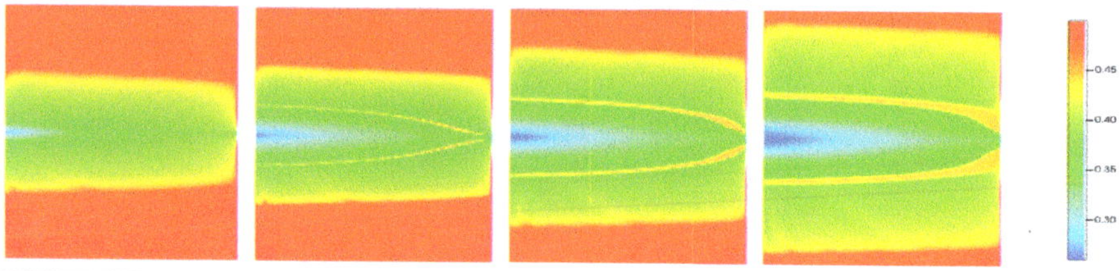			

High-permeability layer dynamics (Table 5): Permeability decreased abruptly from 12 mD to 8 mD post-injection, followed by gradual recovery to 10 mD over 2000 days. This reduction arose from microsphere-induced pore-throat blockage via a "bridge-plug and filter-cake" mechanism, elevating the displacement pressure—consistent with the inverse pressure–permeability correlation in the CROCKTABW model (Section 2.2). Partial permeability recovery after microsphere breakthrough reflected "dynamic plugging-rebreakthrough" behavior, though values remained below initial waterflooding levels. Porosity concurrently declined from 0.16 to 0.14, confirming that NGM retention reduced the effective pore-throat volume. Oil saturation rebounded from 0.30 to 0.35, indicating fluid diversion to low-permeability zones and secondary displacement of residual oil.

Low-permeability layer response (Table 6): Permeability increased by 33% (from 4.5 mD to 6.0 mD) after a 500-day lag—attributable to preferential NGM plugging of high-conductivity pathways. This delayed enhancement aligns with the "block-before-diversion" fluid-flow principle. Near-injector grids (e.g., (15, 75, 2)) exhibited a permeability growth rate of 0.003 mD/100 days, doubling the waterflooding rate (0.0015 mD/100 days), validating the diversion efficacy. Oil saturation declined from 0.35 to 0.28, demonstrating residual oil mobilization. The 40% acceleration in oil-saturation decline relative to waterflooding confirmed NGMs' enhanced recovery via seepage-field reconfiguration.

Longitudinal profile evolution (Table 7): The injector–producer connection line transitioned from a "strip-shaped high-permeability zone" (Section 4.2, Table 4) to a "dumbbell-shaped low-permeability zone". Permeability decreased by 30% in central high-k regions while increasing by 20% at low-k peripheries, with the seepage-direction deflection reaching 45°. This reconstructed flow field elevated the inter-well pressure gradient from 0.02 MPa/m to 0.035 MPa/m, improving driving-energy distribution uniformity.

Table 6. Characteristics of dynamic changes in physical properties of relatively low permeability layers during NGM flooding after waterflooding.

	Early Stage of NGMs	Middle Stage of NGMs	Late Stage of NGMs	Final Stage of NGMs	
Porosity	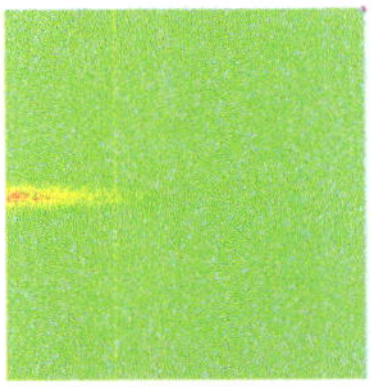		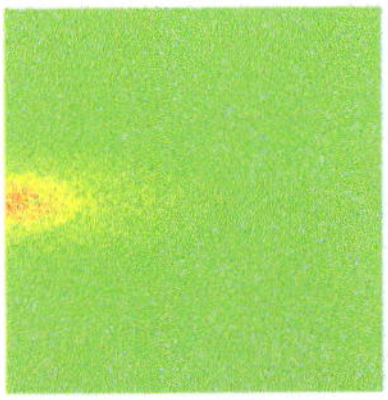	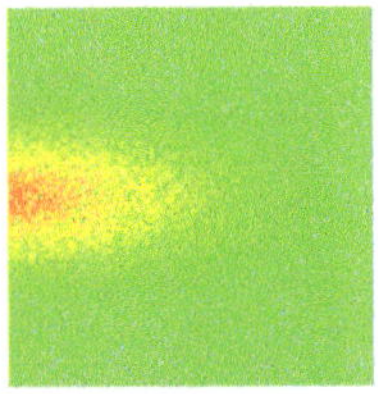	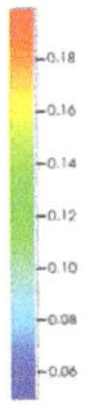
Permeability	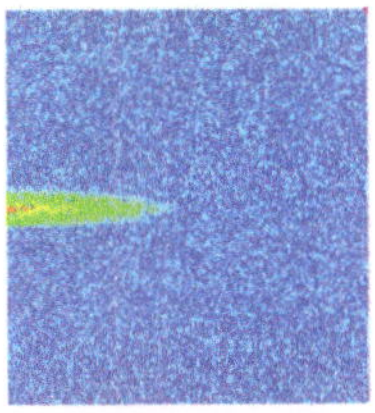	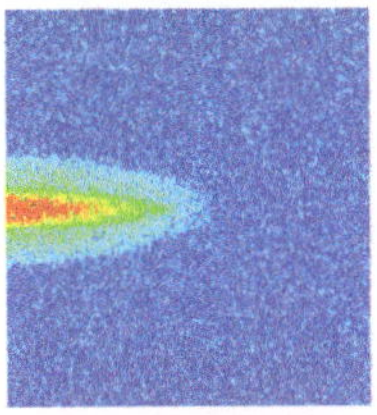	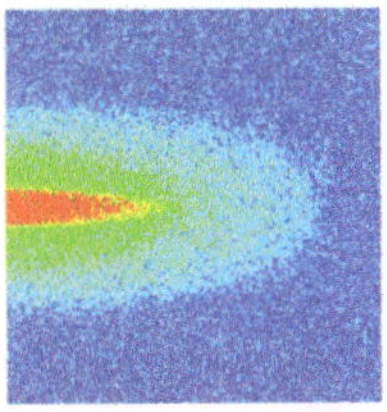	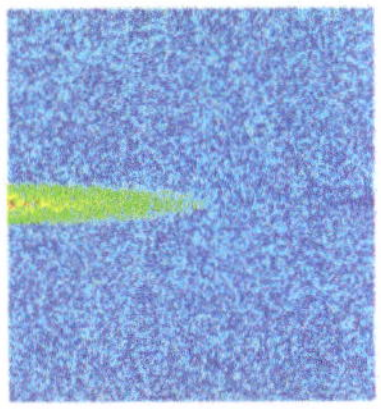	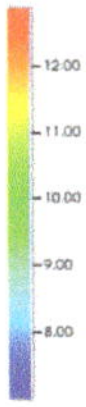
Oil saturation					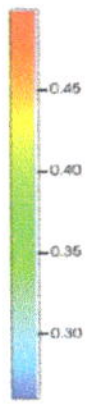

Table 7. Characteristics of dynamic changes in physical properties of longitudinal profiles between injection and production well during NGM flooding after waterflooding.

	Early Stage of NGMs	Middle Stage of NGMs	Late Stage of NGMs	Final Stage of NGMs
Porosity		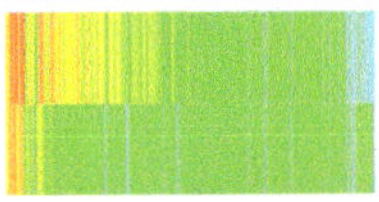	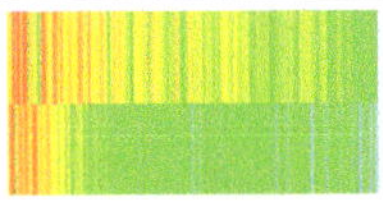	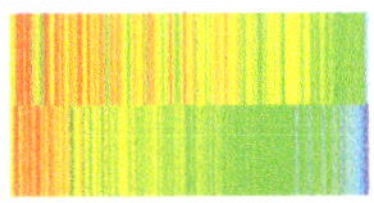
Permeability	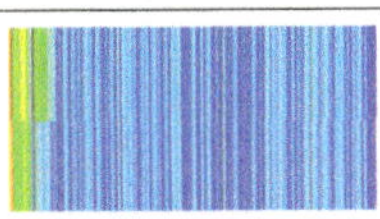	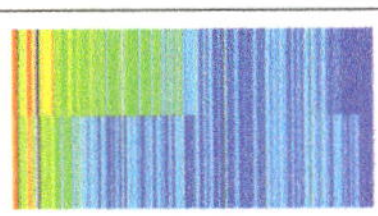	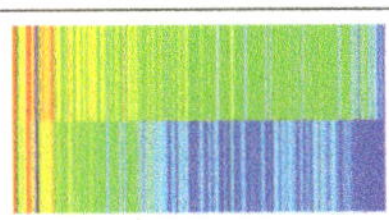	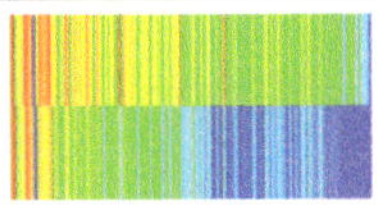
Oil saturation		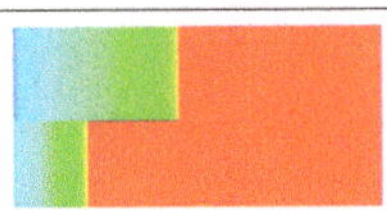		

To visualize the plugging effect of NGMs on large pore throats and PFCs, five grids in the high-permeability layer ((15, 75, 1), (45, 75, 1), (75, 75, 1), (105, 75, 1), (135, 75, 1)) and five grids in the low-permeability layer ((15, 75, 2), (45, 75, 2), (75, 75, 2), (105, 75, 2), (135, 75, 2)) were selected to monitor changes in permeability and porosity. The resulting property evolution curves during the entire waterflooding and NGM flooding cycle are shown in Figure 5.

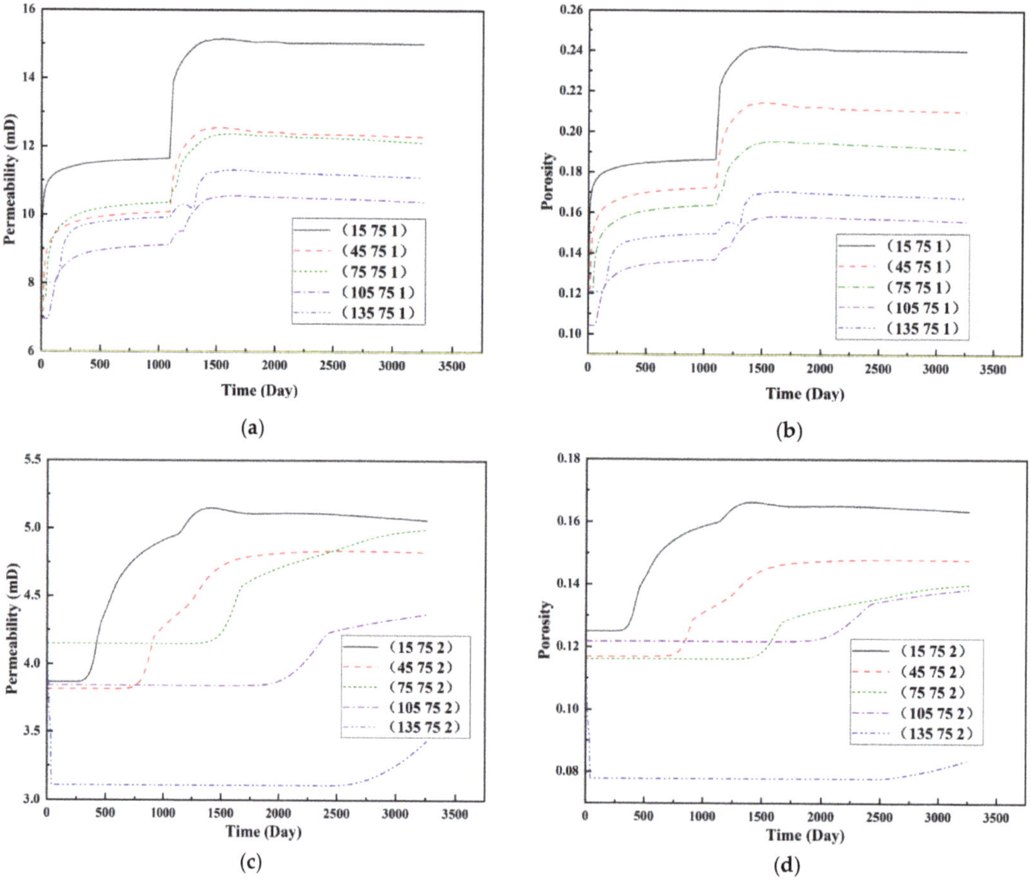

Figure 5. Evolution of porosity and permeability after NGM flooding. (**a**) Permeability of high-permeability layer grids; (**b**) Porosity of high-permeability layer grids; (**c**) Permeability of low-permeability layer grids; (**d**) Porosity of low-permeability layer grids.

Comprehensive analysis of Figure 5a,b reveals that following NGM injection, porosity and permeability trends in high-permeability grids exhibit an initial rapid increase followed by a gradual decline. This behavior arises because NGM blockage of large pore throats elevates the displacement pressure, temporarily enhancing the porosity and permeability. Subsequent NGM breakthrough reduces the pore pressure, causing corresponding property declines. Locally, NGM-driven changes align with trends observed during pure waterflooding, with grids near injector and producer wells showing accelerated pressure responses. Critically, porosity fluctuations in grid (135, 75, 1) are attributable not to near-wellbore pressure-induced rock deformation, but to NGM plugging and breakthrough dynamics within pore throats.

Analysis of Figure 5c,d indicates that grids (15, 75, 2) and (45, 75, 2) in the low-permeability layer display similar NGM-induced trends. However, reduced initial water influx into this layer mitigates property fluctuations compared to high-permeability zones. Furthermore, while NGMs block major flow channels, inferior reservoir properties promote radial propagation of injected fluids and pressure. Consequently, grids (75, 75, 2), (105, 75, 2), and (135, 75, 2) exhibit delayed pressure responses without distinct decline phases.

2.2.3. Macro-Scale Impact of NGMs on Oil Recovery

To quantitatively evaluate the macro-scale impact of NGMs on reservoir development efficiency, Figure 6 presents comparative simulation results of production dynamics for two development scenarios: continuous waterflooding versus NGM flooding initiated after high water-cut (90%) in production wells.

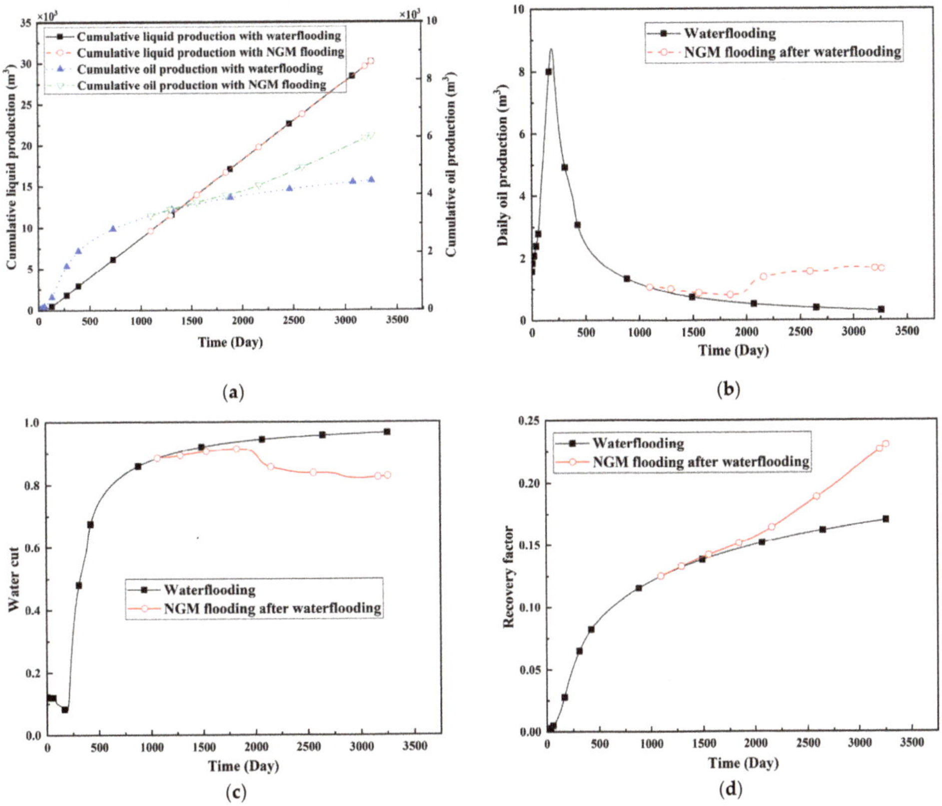

Figure 6. Macro-scale impacts of NGMs on production performance. (**a**) Cumulative liquid and oil production; (**b**) Daily oil production; (**c**) Water cut; (**d**) Recovery factor.

Comparative analysis of development schemes reveals that NGMs significantly enhance recovery efficiency. Under equivalent liquid production volumes, NGM implementation increases cumulative oil production by 7000 m^3 and elevates ultimate recovery by 0.84 percentage points. This improvement stems from NGMs' "bridge-plug-filter cake" mechanism obstructing high-permeability channels, forcing fluid diversion into previously under-swept low-permeability zones.

The modest early-stage oil production increase reflects initial PFC plugging and residual oil mobilization in waterflooded high-permeability regions. A critical inflection point occurs at 2096 days (1000 days post-NGM injection), where cumulative oil production curves diverge sharply and daily production surges. This correlates with permeability enhancement in low-permeability layers (Section 2.2.2, Table 6), confirming the "plug-before-divert" mechanism. Radial flow propagation—induced by PFC blockage and exacerbated by inferior reservoir properties in low-permeability zones—delays pressure transmission to peripheral grids (e.g., (75,75,2), (105,75,2), and (135,75,2)), explaining their attenuated property responses.

2.3. Field Application: Case Study of the Chang 6 Tight Oil Reservoir

The Jiyuan Oilfield, primarily comprising lithologic and structural–lithologic composite reservoirs, hosts the Chang 6 tight oil reservoir. This reservoir exhibits pronounced lateral and vertical heterogeneity (with a permeability variation coefficient of 0.82), characterized by directional permeability anisotropy from natural fractures (K_x/K_y = 8), tight lithology and complex oil–water relationships. A numerical simulation model, incorporating a three-injector and six-producer well pattern based on actual geological parameters, was constructed to represent this challenging reservoir. The pronounced heterogeneity accelerates the PFCs during waterflooding, manifested by a rapid water cut rise to 92%, which severely constrained the contribution of liquid production to recovery. To investigate the impact of NGMs on reservoir performance and recovery factor under these complex conditions, NGMs were introduced after the initial waterflooding period. The simulation spanned a total of 20 years. The corresponding results are presented in Figures 7 and 8.

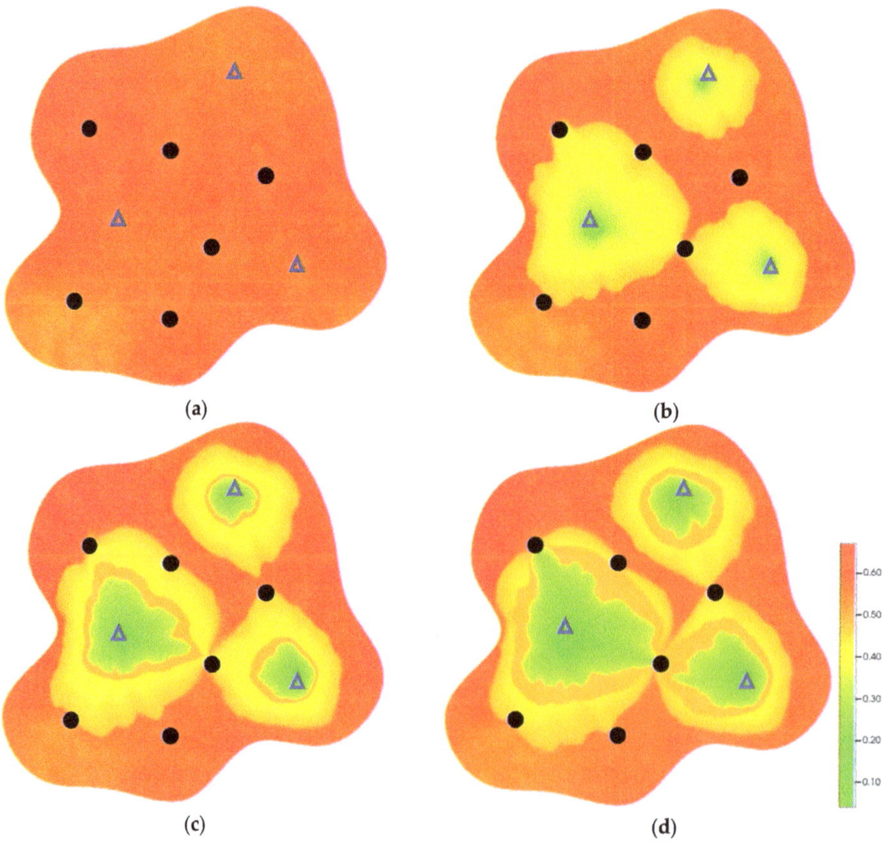

Figure 7. Full-cycle evolution of oil saturation fields under waterflooding and NGM flooding in Chang 6 Reservoir (Triangles represent injector wells, circles represent producer wells). (**a**) Initial stage; (**b**) End of waterflooding; (**c**) 6 years post-NGM injection; (**d**) 12 years post-NGM injection.

Comparison of the oil saturation fields across different time periods reveals that during the initial 8 years of waterflooding, injected water reached the production wells, establishing PFCs between the injectors and producers. Following NGM injection, an oil bank forms around the injection well and subsequently propagates towards the production wells. Furthermore, the residual oil saturation following NGM flooding is demonstrably lower than that observed after conventional waterflooding.

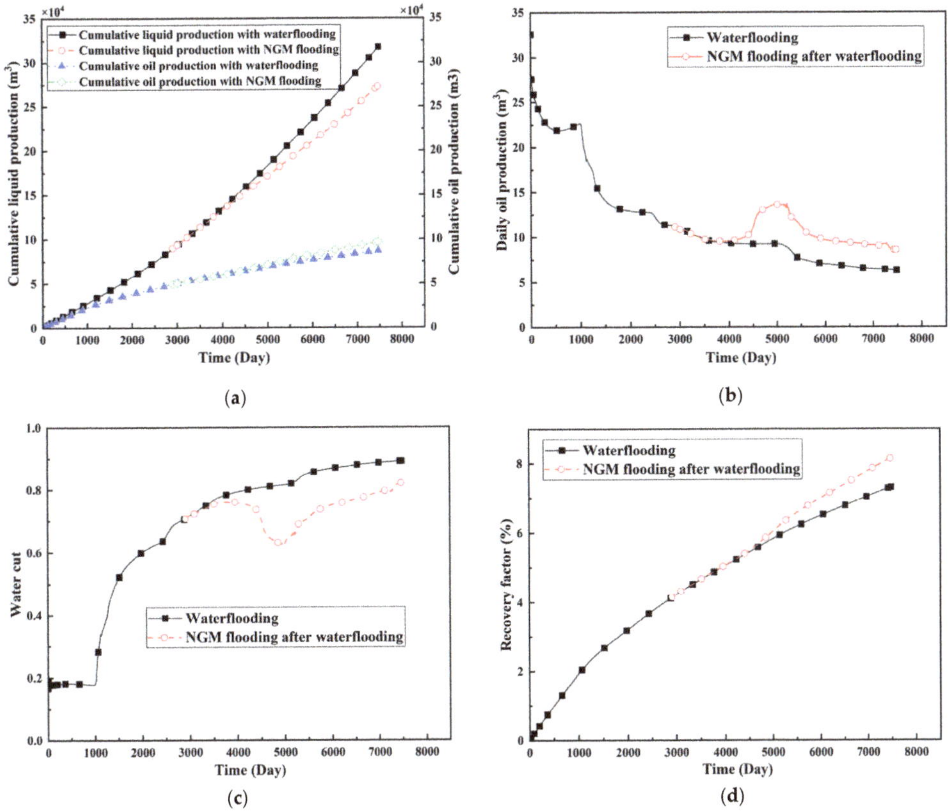

Figure 8. Production dynamics of Chang 6 Reservoir under different development schemes. (**a**) Cumulative liquid and oil production; (**b**) Daily oil production; (**c**) Water cut; (**d**) Recovery factor.

Analysis of production parameters for the different development schemes indicates that cumulative liquid production was higher for the scenario without NGMs compared to that with NGMs, whereas cumulative oil production exhibited the opposite trend. This demonstrates that NGMs effectively block the PFCs formed during long-term waterflooding development, thereby diverting injected water to displace residual oil from smaller pores. NGM flooding exhibits a significant response lag; the cumulative oil production curves diverge markedly at approximately 4000 days, coinciding with an increase in daily oil production. Following the divergence point, daily oil production under NGM flooding rapidly increased to 17 m^3/d before gradually declining to 8 m^3/d; however, it consistently exceeded the production rate achieved without NGMs. Simultaneously, the results indicate that the effective period of incremental oil production attributable to NGMs exceeded ten years, accompanied by a significant reduction in water cut for the simulated reservoir block. At the conclusion of the 20-year simulation period, the recovery factor reached 7.31% without NGMs and 8.15% with NGMs, representing an increase of 0.84 percentage points. At a conservative oil price of 70/bbl USD, preliminary economic assessment confirms a favorable ROI $\approx$ 10:1 for NGM flooding in the Chang 6 reservoir, where incremental revenue from the 0.84% recovery gain offsets chemical and operational costs, with key economic drivers being the NGM unit price, heterogeneity severity, and treatment lifetime.

3. Conclusions

This study establishes an integrated fluid–solid coupling model for PFCs and NGMs using CMG numerical simulation, systematically revealing the synergistic mechanisms of NGM plugging and EOR. Key conclusions are drawn as follows:

(1) Dynamic quantification of PFC evolution was achieved through coupled CROCKTAB (stress–porosity hysteresis) and CROCKTABW (water saturation-driven permeability evolution) modeling, revealing that conventional simulations overestimate recovery by 8.7% due to unaccounted permeability growth (from 8 to 12 mD, +50%) in high-conductivity layers during long-term waterflooding.

(2) NGMs enable effective conformance control via synergistic "bridge-plug-filter cake" structures leveraging gel properties (15× swelling ratio, elasticity, adsorption), reducing thief-zone permeability by 33% (from 12 to 8 mD) while enhancing low-permeability layer flow capacity by 33% (from 4.5 to 6 mD) and decreasing residual oil saturation (from 0.35 to 0.28).

(3) Field validation in the Chang 6 tight reservoir (permeability variation coefficient 0.82) confirmed a >10-year sustained performance: 0.84% incremental recovery (from 7.31% to 8.15%) with 8% water-cut reduction and favorable economics (ROI ≈ 10:1 at 70/bbl USD), demonstrating engineering viability for heterogeneous reservoirs.

(4) This work establishes a transferable framework for gel-based EOR, with future research prioritizing adaptive NGM injection (size, concentration optimization) and integration with auxiliary techniques (e.g., surfactant, thermal methods) for complex pore systems in ultra-deep or high-temperature reservoirs.

4. Mechanisms and Methods

4.1. Formation and Numerical Characterization of PFCs

4.1.1. Formation Mechanism of PFCs

During prolonged water injection, injected water leaches and transports mudstone components, weakening interparticle cementation strength. This process promotes gradual pore-throat enlargement, reduces flow path tortuosity, and ultimately increases reservoir permeability by up to 50% in affected zones. Concurrently, crude oil adsorption capacity decreases while polar component adsorption increases. These changes shift rock wettability toward hydrophilicity, thereby enhancing fluid mobility (particularly aqueous phase). These coupled mechanisms establish high-connectivity preferential flow channels (PFCs) between injector and producer wells [17]. PFC development substantially exacerbates reservoir heterogeneity, causing injected water to bypass significant residual oil volumes. This constitutes the primary constraint on recovery enhancement during late-stage waterflooding.

4.1.2. Numerical Modeling of PFC Evolution

Samier et al. [18,19] established that dynamic reservoir stress field alterations (e.g., rock compaction, fracture reactivation) during water injection constitute the primary driver of permeability evolution, which cannot be accurately captured by one-way coupling models. Therefore, this study employs the coupled CROCKTAB and CROCKTABW keywords to construct a multi-field coupling model. CROCKTAB defines complete loading and unloading paths, accounting for elastic and partial plastic rebound during pressure variations. Conversely, CROCKTABW defines a single loading path where rock compression and expansion occur without elastic rebound during pressure changes.

The selection of this dual-keyword coupling approach is grounded in three critical requirements:

(1) Physical fidelity requirement: Conventional one-way coupling models fail to capture the hysteresis effects in stress–permeability relationships during cyclic water injection. The dual-keyword approach dynamically couples rock deformation with aqueous-phase pressure evolution—essential for simulating irreversible PFC formation in long-term waterflooding scenarios.
(2) Field validation imperative: The CROCKTABW keyword explicitly links rock compaction to water injection operations (dominant in the Chang 6 reservoir), resolving the "static table limitation" of conventional geomechanical models. Field data confirm that the permeability changes correlate with water saturation increments.
(3) Numerical efficiency advantage: The coupled framework achieves accurate PFC characterization with <5% additional computation time versus decoupled approaches.

The CROCKTAB keyword in the CMG core defines the hysteresis relationship (loading and unloading path) between rock porosity and permeability as a function of effective stress (overlying formation stress–pore pressure). However the model is essentially a static table of rock mechanic ontogenetic relationships, which does not explicitly couple the hydrodynamic source of pore pressure variation with rock deformation. In real reservoirs, pressure changes are realized by fluid injection (pressurization) or extraction (depressurization), and fluid flow, phase distribution, and pressure propagation are the fundamental driving forces that drive changes in the effective stresses of compaction and trigger compaction and rebound. Therefore, in order to more realistically simulate the fluid–rock interaction in the process of water-driven development, the keyword CROCKTABW is also chosen in this paper. The core objectives are:

(1) To establish a coupling mechanism dominated by water-phase pressure: The water-phase fluid pressure is explicitly used as a direct input variable for calculating the effective stress change of rock grid cells. This allows the compaction or rebound response of the rock to be directly and dynamically bound to the flow, injection, or extraction processes of the water phase and the pressure field evolution in the model.
(2) Realize true fluid–solid coupling (partial coupling): With CROCKTABW, the change of rock properties no longer depends only on a preset, static pressure field or stress path, but responds in real time to local water-phase pressure changes generated by the water-driven process.
(3) Focus on water-driven dominated deformation processes: The model is explicitly instructed to focus on scenarios where the water phase is the main driver of pore pressure changes and rock deformation. This is especially applicable to oilfields where water injection is the main development method, and the main controlling factor of the rock mechanical response is the spatial and temporal distribution of water-phase pressure.
(4) Reflecting engineering reality: in oilfield sites, water injection operations directly affect reservoir pressure, which in turn changes the effective stress and leads to rock deformation. The choice of CROCKTABW is to reproduce the dynamic change of rock properties directly triggered by the waterflooding operation in the numerical model, so that the simulation results are more in line with the physical reality and engineering observations.

This choice resolves three critical gaps: hysteresis effects in stress–permeability relationships, waterflooding-specific deformation triggers, and computational efficiency for field-scale simulations.

4.1.3. Characterization of PFC Dynamics via Coupled Simulation

Most geologic models are derived from initial logging curves obtained during the early development stage of oilfields. After prolonged waterflooding, quantitatively characterizing the preferential flow channels (PFCs) formed between injection and production wells under continuous aqueous-phase scouring constitutes a critical unresolved challenge. It is established that PFCs exhibit elevated flow diversion rates, high water saturation, and increased porosity and permeability relative to low-permeability strata. Consequently, porosity and permeability can be modeled as functions of water saturation [20].

(1) Pressure-induced temporal evolution of reservoir properties (CROCKTAB)

Permeability is governed by fluid porosity, which is influenced by pressure-dependent rock compaction during reservoir development. As pressure propagates through the fluid medium, permeability and porosity variations can be expressed as functions of pressure:

$$\phi(P) = \phi_{\mathrm{i}} \cdot m \tag{3}$$

$$K(P) = K_{\mathrm{i}} \cdot m \tag{4}$$

where ϕ_{i} denotes the initial porosity; K_{i} denotes the initial permeability; ϕ(P) is the porosity at pressure P; K(P) denotes the permeability at pressure P; and m is the multiplier at pressure P.

(2) Temporal variation of reservoir properties due to water saturation (CROCKTABW):

$$\phi(\Delta S_{\mathrm{w}}) = \phi_{\mathrm{i}} \cdot (1+n) \tag{5}$$

$$K(\Delta S_{\mathrm{w}}) = K_{\mathrm{i}} \cdot (1+n) \tag{6}$$

$$\Delta S_{\mathrm{w}} = S_{\mathrm{w}} - S_{\mathrm{wi}} \tag{7}$$

where S_{w} is the dynamic water saturation; S_{wi} is the initial water saturation; ΔS_{w} is the incremental water saturation during displacement; $\phi(\Delta S_{\mathrm{w}})$ is the dynamic porosity; $K(\Delta S_{\mathrm{w}})$ is the dynamic permeability; and n is the multiplier determined by ΔS_{w}.

When a grid cell is subject to both CROCKTAB and CROCKTABW, the resultant porosity and permeability changes are:

$$\phi(\Delta S_{\mathrm{w}}, P) = \phi_{\mathrm{i}} \cdot \max_{m \geq 1 \,\cap\, n \leq 1} (m, 1+n) \tag{8}$$

$$K(\Delta S_{\mathrm{w}}, P) = K_{\mathrm{i}} \cdot \max_{m \geq 1 \,\cap\, n \leq 1} (m, 1+n) \tag{9}$$

4.2. Mechanisms of NGMs for Flooding and Profile Control

Wang et al. [21] demonstrated that conventional polymers exhibit peak oil-enhancement periods of only 0.5–1.5 years and effective durations of approximately 1.5–4.5 years, whereas NGMs achieve peak enhancement periods of 1.0–4.0 years and effective durations exceeding 8.0 years. Lin et al. [22] identified that microsphere migration involves surface adsorption, mechanical capture, hydrodynamic retention, and interparticle interactions. Their proposed multistage sub-nanosphere deep profile control technology for submarine fractured reservoirs significantly extends breakthrough times and mitigates gas channeling damage.

4.2.1. Plugging Mechanism of NGMs

Lin et al. [23] utilized mercury intrusion porosimetry (constant-rate and constant-pressure methods) to establish that prolonged waterflooding enlarges pore throats, with larger throats dominating fluid seepage pathways. Jia et al. [24] further concluded that primary pore-throat radii expand after waterflooding, enhancing their seepage contribution, while pore-throat distributions remain largely unchanged. This confirms that pore-throat characteristics—rather than void space properties—govern reservoir-scale seepage behavior. Hua et al. [25] employed nuclear pore membrane filtration, sandpack displacement, core flooding, micromodel visualization, and capillary flow experiments to demonstrate that water-dispersed NGMs expand 3–6-fold in size, effectively plugging nuclear pore membranes. These systems selectively obstruct high-permeability zones and displace crude oil from low-permeability regions in parallel sandpacks. Cai et al. [26] derived an inverse relationship between the pore-throat ratio and permeability:

$$K = \frac{\phi R_{pt}^2 r_p^2}{2(1 + R_{pt}^4)(1 + R_{pt}^2)\tau^2} \quad (10)$$

where K denotes the permeability; ϕ is the porosity; R_{pt} denotes the pore throat ratio; r_p denotes the pore-throat radius; and τ denotes the pore tortuosity.

4.2.2. Fluid Diversion Mechanism

Prior studies have confirmed that partial or complete pore-throat blockage in high-conductivity channels elevates viscosity or reduces permeability, thereby increasing the pressure gradient and diverting flow toward low-permeability strata. Li et al. [27] simulated formation fractures using single-sandpack glass capillaries, revealing that diminished deep-reservoir capillary forces reduce aqueous-phase seepage velocity. Under such conditions, particle Brownian motion becomes significant, further suppressing the water-phase permeability and enabling the gradual conversion of injected water into formation energy. This facilitates deep-reservoir energy buildup and fluid diversion. Li et al. [28] and Wang et al. [29] mathematically established that NGM deposition significantly increases the pore-wall surface area while reducing the capillary channel volume and elevating the specific surface area. They identified nanoscale polymer microsphere adsorption on pore surfaces as the fundamental permeability-alteration mechanism. Jiang et al. [30] experimentally observed solid–liquid interfacial adsorption during microsphere transport through pore channels; wall-adhered microspheres reduce effective pore radii, enabling flow steering under incomplete blockage. Wang et al. [31] combined experiments with SEM imaging to demonstrate that adsorbed polymer microspheres reduce pore-throat dimensions while increasing the specific surface area, thereby lowering high-permeability-layer permeability and inducing fluid diversion.

4.2.3. Numerical Modeling of NGM Flooding Mechanisms

Key parameters (concentration, size, adsorption coefficients) were empirically optimized through laboratory core-flooding and validated in field applications, ensuring reliability without additional sensitivity analysis. In ongoing experimental and simulation studies aimed at optimizing NGM concentration and properties, preliminary findings indicate that a 1% concentration offers a favorable balance between performance and cost, though this is subject to reservoir-specific conditions [32–35]. Relevant research findings reveal concentration-dependent tradeoffs. Specifically, a 1% concentration strikes a balance between plugging efficiency (achieving permeability reduction of over 30%) and injectivity (keeping the pressure rise under 20%). The size range of nano-gel microspheres is determined to be 100~500 nm via pore-throat matching using mercury intrusion porosimetry. Among them, smaller microspheres (100~200 nm) can enhance access to low-permeability zones, but require higher concentrations to achieve equivalent PFC blockage. Injection strategies with adaptive concentration adjustments could further enhance NGM flooding efficiency. These insights have been incorporated into our numerical model to guide more accurate simulations of NGM flooding mechanisms.

The resistance coefficient F_r and residual resistance coefficient F_{rr} serve as critical indices for quantifying elastic microsphere plugging performance [36,37]. Zhou et al. [38]

injected diverse fluids into formations and computed these coefficients using Hall plot slope variations across injection stages to assess permeability reduction in high-permeability zones. For the current simulations, we assume that no significant degradation of NGMs occurs within the 10-year timeframe based on their demonstrated thermal stability (>90 °C) and salinity tolerance (>20,000 ppm), though degradation kinetics will be incorporated in future models as the experimental data mature. These coefficients are derivable via Hall curve analysis during three key phases: the initial waterflooding, NGM injection, and post-flushing stages.

Based on the above theory, the absolute permeability within NGM-influenced regions is calculated as:

$$\overline{K} = \frac{K(\Delta S_{\mathrm{w}}, P)}{R_{\mathrm{K}}} \tag{11}$$

where $R_{\mathrm{K}} = 1 + (F_{\mathrm{rr}} - 1)\cdot\frac{AD}{ADMAXT}$ and $F_{\mathrm{rr}} = \frac{\lambda_{\mathrm{w}}}{\lambda'_{\mathrm{w}}} = \frac{k_{\mathrm{w}}/\mu_{\mathrm{w}}}{k'_{\mathrm{w}}/\mu'_{\mathrm{w}}}$ represent the resistance and residual resistance coefficients, respectively; AD is the cumulative polymer adsorption per unit rock volume; $ADMAXT$ is the maximum polymer adsorption per unit rock volume; $\overline{K}$ is the absolute permeability in NGM-affected zones; λ_{w} and λ'_{w} denote the pre- and post-NGM injection aqueous-phase mobility; k_{w} and k'_{w} are the pre- and post-NGM effective aqueous-phase permeabilities; and μ_{w} and μ'_{w} are the pre- and post-NGM aqueous-phase viscosities.

Then, the absolute permeability within NGM-influenced regions is calculated as:

$$\overline{K} = \frac{K(\Delta S_{\mathrm{w}}, P)}{1 + \left(\frac{k_{\mathrm{w}}/\mu_{\mathrm{w}}}{k'_{\mathrm{w}}/\mu'_{\mathrm{w}}} - 1\right)\cdot\frac{AD}{ADMAXT}} \tag{12}$$

Thus, the effective permeability for any fluid phase 'a' in NGM-swept zones is:

$$k_{\mathrm{a}} = \overline{K}\cdot k_{\mathrm{ra}} = \frac{K(\Delta S_{\mathrm{w}}, P)\cdot k_{\mathrm{ra}}}{1 + \left(\frac{k_{\mathrm{w}}/\mu_{\mathrm{w}}}{k'_{\mathrm{w}}/\mu'_{\mathrm{w}}} - 1\right)\cdot\frac{AD}{ADMAXT}} \tag{13}$$

where K_{a} is the phase-effective permeability and K_{ra} is the phase-relative permeability.

Incorporating Equation (8) for dual CROCKTAB and CROCKTABW effects yields the dynamic effective permeability:

$$k_{\mathrm{a}} = \overline{K}\cdot k_{\mathrm{ra}} = \frac{k_{\mathrm{i}} k_{\mathrm{ra}} \max\limits_{\mathrm{m}\geq 1 \cap \mathrm{n}\leq 1}(m, 1+n)}{1 + \left(\frac{k_{\mathrm{w}}/\mu_{\mathrm{w}}}{k'_{\mathrm{w}}/\mu'_{\mathrm{w}}} - 1\right)\cdot\frac{AD}{ADMAXT}} \tag{14}$$

Author Contributions: Methodology, L.R. and C.Z.; validation, J.S., C.J. and H.B.; investigation, Q.L. and X.M.; writing—original draft preparation, L.R.; writing—review and editing, L.R. and C.Z.; supervision, L.R. and J.S. All authors have read and agreed to the published version of the manuscript.

Funding: This research was supported by the National Natural Science Foundation of China (No. 52304036; No. 51704235), Natural Science Basic Research Plan in Shaanxi Province of China (No. 2023-JC-YB-344) and the Youth Innovation Team of Shaanxi Universities (Reserve Geological-Engineering Integration Innovation Team for Efficient Development of Challenging Oil and Gas Resources).

Institutional Review Board Statement: Not applicable.

Informed Consent Statement: Not applicable.

Data Availability Statement: The raw data supporting the conclusions of this article will be made available by the authors on request.

Conflicts of Interest: The authors declare no conflicts of interest.

References

1. van der Linden, J.H.; Tordesillas, A.; Narsilio, G.A. Preferential flow pathways in a deforming granular material: Self-organization into functional groups for optimized global transport. *Sci. Rep.* **2019**, *9*, 18231. [CrossRef] [PubMed]
2. Zhao, P.; Zhu, H.Y.; Li, G.S.; Chen, Z.; Chen, S.J.; Shangguan, S.T.; Qi, X.F. Large-scale physical simulation of injection and production of hot dry rock in Gonghe Basin, Qinghai Province, China, Petroleum Exploration and Development. *Pet. Explor. Dev.* **2024**, *51*, 741–752. [CrossRef]
3. Dai, C.L.; You, Q.; Zhao, M.W.; Zhang, G.; Zhang, F.L. *Principles of Enhanced Oil Recovery*; Springer Nature: Singapore, 2010. [CrossRef]
4. Zhang, X.; Zhu, H.H.; Zhang, T.S.; Meng, Q.C.; Wang, H.T.; Zhang, J.X.; Yong, J.J.; Li, H.J.; Wang, X.X. Instantaneous resistivity vector difference used to simulate the seepage of polymer flooding in an oilfield. *J. Pet. Sci. Eng.* **2020**, *191*, 106922. [CrossRef]
5. Ma, X.; Zhou, Y.H.; Yi, P.; Zhou, S.; Wang, Y.; Yang, D.Z. Design, preparation and properties of new polyacrylamide based composite nano-microspheres with like "ball in ball" structure, Colloids and surfaces. *Colloids Surf. A Physicochem. Eng. Asp.* **2022**, *654*, 130037. [CrossRef]
6. Zhang, H.J.; Li, M.T.; Cai, C.; Zhang, S.; Liu, W.D.; Ma, Z.; Ding, B.; Zeng, J. Minireview on Plugging Agents for High-Conductivity Channels in Reservoirs. *Energy Fuels* **2024**, *38*, 19139–19155. [CrossRef]
7. Feng, X.; Zhang, H.; Liu, H.; Hong, J.; Liu, J.; Yang, Y.; Liu, Z.; Abdullah, M.; Yang, H.; Yu, H. Investigation into the Flow Mechanism of Nano-Elastic Microspheres during Water Invasion. *Processes* **2023**, *11*, 3342. [CrossRef]
8. Zhang, F.M.; Hou, B.F.; Wang, S.; Chen, H.; Yang, J.B.; Fan, H.M.; Gui, L. Study on the Profile Control and Oil Displacement Mechanism of a Polymer Nano-microsphere for Oilfield. *Energy Fuels* **2023**, *37*, 2692–2701. [CrossRef]
9. Ge, L.; Chen, X.; Wang, G.; Zhang, G.; Li, J.; Liu, Y.; Xiao, L.; Wen, Y.; Yuan, W.; Qu, M.; et al. Analysis of the Distribution Pattern of Remaining Oil and Development Potential after Weak Gel Flooding in the Offshore LD Oilfield. *Gels* **2024**, *10*, 236. [CrossRef]
10. Li, X.; Liu, S.; Zhang, J.; Han, S.; Zhao, L.; Xu, A.; Wang, J.; Zhou, F.; Li, M. High-Temperature-Resistant Profile Control System Formed by Hydrolyzed Polyacrylamide and Water-Soluble Phenol-Formaldehyde Resin. *Gels* **2024**, *10*, 413. [CrossRef]
11. Wang, Z.; Jiang, J.; Huang, W.; Gan, Y.; Bai, Y. Construction of a 'Simple, Fast and Accurate' Evaluation Method for Profile Control and Plugging Effect of Gel Plugging Agent Based on Simulations. *Gels* **2025**, *11*, 115. [CrossRef]
12. Zhang, L.; Liu, Y.; Wang, Z.; Li, H.; Zhao, Y.; Pan, Y.; Liu, Y.; Yuan, W.; Hou, J. Evaluation of Profile Control and Oil Displacement Effect of Starch Gel and Nano-MoS2 Combination System in High-Temperature Heterogeneous Reservoir. *Gels* **2024**, *10*, 127. [CrossRef] [PubMed]
13. Su, Y.-C.; Li, Y.-L.; Wang, L.-L.; He, Y.-F. Experimental and Pilot Tests of Deep Profile Control by Injecting Small Slug-Size Nano-Microsphere in Offshore Oil Fields. In Proceedings of the Offshore Technology Conference, Houston, TX, USA, 6–9 May 2019. [CrossRef]
14. Gao, Y.; Chen, Z.; Liang, X.; Li, Y.; Shen, S.; Li, D.; Huang, Z. Study on Permeability Enhancement and Heat Transfer of Oil Sands Reservoir Based on Hydrophobic Nanofluids. *Energies* **2025**, *18*, 927. [CrossRef]
15. Liang, L.H.; Meng, S.W.; Tao, J.P.; Gong, C.P.; Shen, M.; Jin, X.; Fu, H.T. Displacement Mechanisms of Modified Carbon Black Nanofluid as a Novel Flooding System in Ultralow-Permeability Reservoirs. *SPE J.* **2025**, *30*, 2909–2922. [CrossRef]
16. Ju, Y.T.; Luo, H.M.; Zhang, X.N.; Zhang, J. Microscopic regulation and displacement mechanism of polymer–microsphere plug combination for enhanced oil recovery in sandstone reservoirs. *Can. J. Chem. Eng.* **2025**, *4*, 25713. [CrossRef]
17. Li, Y.T.; Zhang, B.; Wang, L.; Wu, Y.; Wang, H.X.; Peng, Z.H. Identification of dominant seepage channels in fractured rock masses of underground water–sealed oil storage: A case study. *Bull. Eng. Geol. Environ.* **2022**, *81*, 357. [CrossRef]
18. Samier, P.; Onaisi, A.; Fontaine, G. Coupled Analysis of Geomechanics and Fluid Flow in Reservoir Simulation. In Proceedings of the SPE Reservoir Simulation Symposium, Houston, TX, USA, 3–6 February 2003; Society of Petroleum Engineers: Richardson, TX, USA, 2003. [CrossRef]
19. Samier, P.; Onaisi, A.; de Gennaro, S. A Practical Iterative Scheme for Coupling Geomechanics with Reservoir Simulation. *SPE Res. Eval. Eng.* **2008**, *11*, 892–901. [CrossRef]
20. Torskaya, T.; Jin, G.; Torres-Verdin, C. Pore-Level Analysis of the Relationship Between Porosity, Irreducible Water Saturation, and Permeability of Clastic Rocks. In Proceedings of the SPE Annual Technical Conference and Exhibition, Anaheim, CA, USA, 11–14 November 2007; Society of Petroleum Engineers: Richardson, TX, USA, 2007. [CrossRef]
21. Wang, M.C.; Zhu, W.Y.; Wang, G.F.; Wang, M. Simulation study of nanopolymer microspheres in a medium-seepage high-water content oilfield. *J. Southwest Pet. Univ.* **2010**, *32*, 105–108. Available online: https://kns.cnki.net/kcms2/article/abstract?v=a4fp6zKrpgYjE4gJvKmIewGQrvXhS0lR4IfXVjMAK_QQg4sVY55WaenlBowUP3FLMP-xWG3f_S0vajYbjW84YNIHD8AITS4WanM2GvWtJdpkosuG2JEmsRs1XkcDh2v51d3mMzARSlkOr4le9k2HukqgGgIhWZlxqSdl43LsjCenaTi1QN3Yow==&uniplatform=NZKPT&language=CHS (accessed on 12 November 2010).
22. Lin, R.Y.; Luo, P.Y.; Sun, Y.; Pan, Y.; Sun, L. Experimental Study on the Optimization of Multi-level Nano–Microsphere Deep Profile Control in the Process of Gas Injection in Fracture-Type Buried-Hill Reservoirs. *ACS Omega* **2021**, *6*, 24185–24195. [CrossRef]

23. Lin, Y.B.; Zhang, J.; Lin, X.G.; Zhou, H.T. Characteristics of pore structure in the late high water-bearing reservoir of Lamaidean Oilfield. *Pet. Explor. Dev.* **2008**, *2*, 215–219. Available online: https://kns.cnki.net/kcms2/article/abstract?v=a4fp6zKrpgZpMCX2xc3SQtqOLq1DIskWbS8nJ_EMc4ubQiFOutjrRZhCRK99QvaPnMbFncasvNPVnp9lM6qeON5ErZoQAre8xHKcFOw0NAZB9qZL6yCC5gH9uRVjZDHvdAbAWuHx1IcuRsMzv_jU2c27N26OBglmaNuDy-wKmRMq2645PBgE-Q==&uniplatform=NZKPT&language=CHS (accessed on 16 May 2008). [CrossRef]
24. Jia, X.F.; Lei, G.L.; Yin, J.H.; Yao, C.J.; Wang, C.; Li, W.Z. Theoretical study on the relationship between pore-throat scale elastic modulation drive microspheres and reservoir matching. *Oil Drill. Technol.* **2011**, *39*, 87–89. Available online: https://kns.cnki.net/kcms2/article/abstract?v=a4fp6zKrpgY1HvXmvDxLzpyH5zuISvc4gUXzKhKoSGeNitB_GsfuF5akHt23MzSwKx-P0zGnZ4GISXSmTsphIIqU8xfMT92hVZrPCBK6YmJWLKjP1vfHtypiInH9N8tC0Rjq2-6LaGBD2_7IP5XsmD1Wx6aYHBscQRtbwRklR07mdXTVGH8NYQ==&uniplatform=NZKPT&language=CHS (accessed on 17 October 2011).
25. Hua, Z.; Lin, M.Q.; Dong, Z.X.; Li, M.Y.; Zhang, G.Q.; Yang, J. Study of deep profile control and oil displacement technologies with nanoscale polymer microspheres. *J. Colloid Interface Sci.* **2014**, *424*, 67–74. [CrossRef] [PubMed]
26. Cai, J.C.; Zhang, Z.E.; Wei, W.; Guo, D.M.; Li, S.; Zhao, P.Q. Thc critical factors for permeability-formation factor relation in reservoir rocks: Pore–throat ratio, tortuosity and connectivity. *Energy* **2019**, *188*, 116051. [CrossRef]
27. Li, Z.B.; Wang, M. Research and application evaluation of deep conditioning and driving technology of nano-microspheres in low-permeability oilfields. *Xinjiang Oil Gas* **2020**, *16*, 61–64. Available online: https://kns.cnki.net/kcms2/article/abstract?v=a4fp6zKrpgbzwlWSQ9hOqkZT7_jIT3B7BkfnP28TuQa3ZJCCShdYK2Zak3Nc6FpMtARYW5kdIjwJgFsePtO0yxMEKt9TgDRrOHxZHPx99lHVKM6wWNhgCiOqjZtW_erisjuZApheb1v9s5ID1MpEfQ1tYVNTUdcZi5Ix2LLBmk5FR2BujNf0Kw==&uniplatform=NZKPT&language=CHS (accessed on 8 June 2020).
28. Li, J.B.; Niu, L.W.; Lu, X.G. Migration characteristics and deep profile control mechanism of polymer microspheres in porous media. *Energy Sci. Eng.* **2019**, *7*, 2026–2045. [CrossRef]
29. Wang, G.Z.; Shen, H.; Gao, Y.; Xiang, Y.Q.; Yang, Z.H.; Lv, Q.C.; Dong, Z.X.; Lin, M.Q. Study on plugging, migration, and EOR capability of dual crosslinked terpolymer microspheres. *J. Mol. Liq.* **2025**, *424*, 127101. [CrossRef]
30. Jiang, X.; Chen, J.B.; Zhang, D.J.; Cao, Y.; Sun, C. Static Adsorption of Nano-spheres on Mineral Surface in Liquid Phase. *Oilfield Chem.* **2022**, *39*, 281–287. [CrossRef]
31. Wang, J.; Zhou, F.J.; Li, J.J.; Yang, K.; Zhang, L.F.; Fan, F. Evaluation of the oil/water selective plugging performance of nano-polymer microspheres in fractured carbonate reservoirs. *J. Zhejiang Univ. Sci. A* **2019**, *20*, 714–726. [CrossRef]
32. Firozjaii, A.; Saghafi, H. Review on chemical enhanced oil recovery using polymer flooding: Fundamentals, experimental and numerical simulation. *Petroleum* **2020**, *6*, 115–122. [CrossRef]
33. Gao, F.; Liu, B.; Liu, Y.; Xing, L.; Zhang, Y. Preparation and Enhanced Oil Recovery Mechanisms of Janus-SiO_2-Reinforced Polymer Gel Microspheres. *Gels* **2025**, *11*, 506. [CrossRef]
34. Wang, Z.; Lin, M.; Jin, S.; Yang, Z.; Dong, Z.; Zhang, J. Combined Flooding Systems with polymer microspheres and Nonionic Surfactant for Enhanced Water Sweep and Oil Displacement Efficiency in Heterogeneous Reservoirs. *J. Dispers. Sci. Technol.* **2020**, *41*, 267–276. [CrossRef]
35. Shagymgereyeva, S.; Sarsenbekuly, B.; Kang, W.; Yang, H.; Turtabayev, S. Advances of polymer gel microspheres and Its Applications for Enhanced Oil Recovery. *Colloids Surf. B Biointerfaces* **2024**, *233*, 113622. [CrossRef] [PubMed]
36. Wen, Y.; Hou, J. Experimental Study on Multi-Dimensional Visualization Simulation of Gas and Gel Foam Flooding in Fractured-Vuggy Reservoirs. *Gels* **2023**, *9*, 722. [CrossRef] [PubMed]
37. Zhao, S.; Pu, W.F.; Wei, B.; Xu, X.G. A comprehensive investigation of polymer microspheres (PMs) migration in porous media: EOR implication. *Fuel* **2019**, *235*, 249–258. [CrossRef]
38. Zhou, H.Y.; Liu, B.; Sun, Q.; Zhai, C.C.; Yue, B.L. Evaluation of multi-rounds driving effect of nano-gel microspheres in offshore fluvial-phase thick oil field. *Xinjiang Oil Gas* **2021**, *17*, 60–64. Available online: https://kns.cnki.net/kcms2/article/abstract?v=a4fp6zKrpgamI9YmNPg-Zeq63sm_F3loNMRQ8mdtkw5jhDsELBknfp3DA_t1FXSrIG0kzwVrdkNA2eztELNubY-CNq3PQsbdBhYj7_Sr_SbHUQ6C3ji1isW0hGVlmoqQuirKsgASkyV1m3TS0kkLIKi0kqligu4AsQJMqtwKvh_fxScH52aImw==&uniplatform=NZKPT&language=CHS (accessed on 31 March 2021).

gels

Article

Construction of a 'Simple, Fast and Accurate' Evaluation Method for Profile Control and Plugging Effect of Gel Plugging Agent Based on Simulations

Zengbao Wang [1,2,*], Junjie Jiang [1], Weian Huang [1,2], Yuwei Gan [1] and Yingrui Bai [1,2]

[1] School of Petroleum Engineering, China University of Petroleum (East China), Qingdao 266580, China; jjunjie2000@163.com (J.J.); masterhuang1997@163.com (W.H.); hh498082@163.com (Y.G.); smart-byron@163.com (Y.B.)

[2] Key Laboratory of Unconventional Oil & Gas Development, China University of Petroleum (East China), Ministry of Education, Qingdao 266580, China

* Correspondence: wangzengbao@upc.edu.cn

Abstract: At present, the evaluation perspective of the gel plugging agent assessment method is incomprehensive, due to which the experimental results deviate from the field data. By analyzing the current indoor evaluation methods and the factors controlling the sealing capability of gel plugging agents, an experimental device and method for evaluating the blocking effect of oilfield gel plugging agents has been designed. In contrast to traditional assessment methods, the proposed approach offers advantages such as simple operation, rapid experimentation, and accurate results. The experimental results show that gels selected using conventional methods are inconsistent with the results of plugging displacement tests. This discrepancy can be attributed to the fact that these methods focus solely on cohesive strength while neglecting adhesive strength. Considering that the evaluation perspective of conventional methods is relatively limited, an evaluation method for the sealing effect of the plugging agent was developed. This method comprehensively incorporates factors such as cohesion strength, adhesion capability, shear resistance, and the long-term anti-dehydration performance of the gel. The evaluation results of the method were consistent with the results of the plugging displacement experiments. The newly constructed method defines Γ as the comprehensive evaluation parameter for the gel. A new experimental system with a comprehensive evaluation index (Γ) of 8.97 Pa^2 was selected. After the profile control of the system, the effluent ratio of the high and low permeability layers reached 1:9, and its erosion resistance was greater than 20 PV. Meanwhile, the profile control effect was also stable. Through verification based on field data, the injection pressure of the system optimized by the proposed method was found to be 2.5 times higher than that of the original system. Meanwhile, the plugging validity period was >2 times of the original system. The test results were consistent with the plugging capability evaluation index. In summary, the performance evaluation method of the designed gel plugging agent was reasonable in principle and the results were accurate and reliable. Therefore, it is considered to be of guiding significance for the selection of efficient profile control plugging agents in oilfields.

Keywords: gel plugging agent; evaluation method; experimental design; cohesive strength; adhesive strength; degree of damage

1. Introduction

With the continuous water injection development of oilfields, the heterogeneity of the formation has aggravated [1]. Meanwhile, some production wells have reached a high water-cut stage, and the development benefits are getting worse and worse [2]. The early reaching of high water cuts in production wells is mainly caused by water channeling [3]. Generally, the fluid in the reservoir will flow out preferentially along the channel with the minimum flow resistance so that the injected water will form a water channeling passage in the high-permeability layer [4], which will form the phenomenon of "repeated flushing of the high-permeability channel and non-entry of the low-permeability channel". In addition to altering the water injection pattern to address water channeling, profile modification and water plugging are also widely adopted methods [5]. The profile-controlling and water-plugging agent is a special material used to adjust the water absorption profile and realize the diversion of liquid flow to deal with water channeling [4,6,7] (see Figure 1). Among many profile-controlling and water-plugging agents, gel plugging agents are widely used in oilfields due to their adjustable strength, good injection performance, and simple construction [8]. For example, Li et al. [9] developed a high-elastic expansion particle as well as the phenolic gel system, which was used in nine wells of the Jilin oilfield, with a cumulative reduction in water loss of 56,900 m^3 and a cumulative oil increase of 3200 tons. The overall profile-controlling and plugging effect was obvious. Li et al. [10] research on polymer gel plugging agents, primarily based on acrylic resins (MMA). The gel system exhibits a plugging strength of 6×10^4 mPa·s and a plugging rate of 99.79%, demonstrating exceptional performance in water control and profile modification.

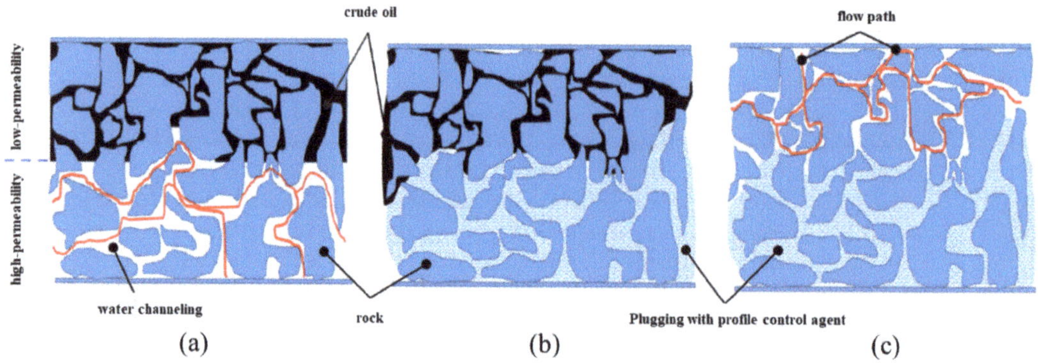

Figure 1. Schematic of the regulation of plugging agent to improve water flooding flow field. (**a**) Occurrence of water channeling; (**b**) Utilization of profile control agent; (**c**) Turning of water flow.

In field applications, the profile-controlling effect of gel is affected by many factors. For example, the gel is sheared by the ground equipment and the migration of formation [11], resulting in the degradation of some polymers. Under strong water flooding, some of the gel is discharged without bonding with the reservoir [12]. These factors will reduce the plugging effect of the gel and affect the effectiveness of the on-site process implementation. Therefore, it is required that the gel system should have good strength, shear resistance, and adhesion capacity. However, how to properly characterize, evaluate, and screen the gel system is still not studied in detail in the literature.

At present, the most commonly used performance evaluation methods of gel plugging agent include breakthrough vacuum degree method [13], transition pressure method [14], cylinder immersion depth method [15], void resistance factor method [16], visual code method [17], rheological parameter method [18], rotary viscometer method [19], falling ball viscosity method [20], capillary viscometer method [21], flow length evaluation method [22], and gel strength rank (GSR) index method [23]. In the breakthrough vacuum degree method, the maximum degree of vacuum (BV) of air breaking through the gel using negative pressure is measured to characterize the strength. The transition pressure method compares the gel strength by recording the pressure at which the gel overcomes the intermolecular binding from the flow state. The cylinder immersion depth method involves recording the depth of the cylinder sinking into the gel within the specified time, which is represented as the gel strength index. The strength index is inversely proportional to the gel strength. By comparing the same volume of gel and polymer solution through the same length and permeability of the sand filling tube time, the method of void resistance factor method is used to reflect the relative strength of colloid and polymer solution. In the visual code method, the sample tube is inverted to observe the shape of the gel in the tube to intuitively evaluate the gel strength. This method is simple, convenient, and applicable to a wide range of conditions, and is the most commonly used method. In the rheological parameter method, the storage modulus of the gel system was measured using a rheometer to compare the strength of different gel formulations. The rotary viscometer method is used for characterizing the strength of gel by measuring the ease of flow of gel at a certain temperature and shear rate using a rotary viscometer. Similarly to the cylinder immersion depth method, the falling ball viscosity method is suitable for high strength gels. The capillary viscometer method is used for indirectly measuring the strength of gel by measuring the time required for inhaling a unit volume of gel in a capillary. By comparing the length of the gel after vertical and horizontal placements, the flow length evaluation method defines a criterion for evaluating the colloidal strength. The GSR index method involves measuring the strength of a gel by determining the time it takes to pass through a length of capillary viscometer.

Among them, the first four methods mainly characterize the capability of the gel plugging agent to resist shear breaking, while the latter seven methods mainly show the adhesive capability of the gel plugging agent. However, the effectiveness of a gel plugging agent needs to have both good strength and adhesion capability [24]. The single evaluation method is not used in the actual application in oilfields and lacks scientific basis and practicability [25,26]. Although the plugging capability of a gel plugging agent, evaluated by a core tube displacement test, can be closely combined with the actual construction site, the operation is complicated and the experimental time is long, which are not suitable for the optimization of the formulation and the screening of the batch gel plugging agent [27].

With this backdrop, considering the strength, adhesion, and anti-shearing capability of the plugging gel, an experimental device and a characterization method for evaluating the plugging effect of the gel plugging agent are designed, which are simple to operate, fast in experiment and accurate in evaluation. A more realistic comprehensive evaluation index (defined herein as $\Gamma = \frac{\delta \times \eta}{\lambda}$) is established to characterize the plugging effect of the gel plugging agent on the reservoir. Finally, the rationality, accuracy, and reliability of the experimental method are established through heterogeneous reservoir plugging displacement experiments and field tests. It could provide some effective theoretical basis for field evaluation and optimization of plugging agents.

2. Results and Discussion

2.1. Screening Using Conventional Test Methods

In the study of indoor synthetic gel, the visual code method [14] and the breakthrough vacuum method are generally used to quickly evaluate the gelling performance of the gel [28–31]. The former can directly determine the gelling time and gelling strength semi-quantitatively by observing the gelling state of the water shutoff agent after inversion, whereas the latter only needs to use simple instruments such as a vacuum pump and suction filter bottle to reflect the strength of the gel. During the breakthrough vacuum method, when the degree of vacuum increases to the maximum and air breaks through the gel, the BV reading is measured to do further calculations. Both methods have the characteristics of simple operation and are fast and convenient. Based on the visual code method, three gel systems with the same gel strengths (see Table 1) were selected for comparative evaluation, and the breakthrough vacuum of the gel system was determined.

Table 1. Comparison of different gel systems for evaluation.

Gel System	Composition of Gel System	The Gel Strength Level Obtained by Visual Code Method	Breakthrough Vacuum Degree, MPa	Remark
Gel system 1	3000 mg/L HPAM + 0.3% phenolic resin + 0.3% g agent + 0.15% crosslinking accelerator	H	−0.094	The system used before the mine
Gel system 2	3000 mg/L HPAM + 0.3% phenolic resin + 0.15% Resorcinol + 0.3% hexamethylenetetramine + 0.15% crosslinking accelerator	H	−0.066	The present system of ore field
Gel system 3	3000 mg/L HPAM + 0.3% amino resin + 0.15% crosslinking accelerator	H *	−0.063	Comparison system

* description: visual code method gel strength level H—when the sample bottle is vertically inverted, only the surface of the gel is slightly deformed (Figure 2).

According to the results for the three gel systems (as shown in Table 1), the gel-forming strength of the three gel systems obtained by the visual code method was consistent. However, Gel system 1 exhibited the highest breakthrough vacuum strength, and therefore, the most preferable result was Gel system 1. However, in the field application, it was found that the sealing effect of the preferred Gel system 1 was not ideal. Due to the reason that the adhesion effect between the gel and the reservoir pore throat was not considered, the plugging period was short and it was easy to be destroyed by a subsequent water injection. The reservoir profile control capability was not up to the expectations, and the optimization results of the method were inconsistent with the field application results.

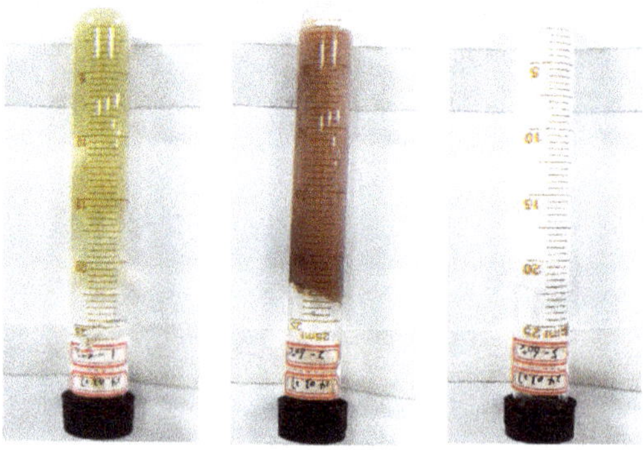

(a) (b) (c)

Figure 2. Gelation morphologies of the gels. (**a**) Gel system 1; (**b**) Gel system 2; (**c**) Gel system 3.

2.2. Comparison of Actual Plugging Capability of Gel System

Based on the results of the field feedback, the sand-filled pipe or core was used to simulate the application of the field plugging agent and explore the causes of the experimental results. The permeability of the sand-filled pipe filled with 20–80-mesh quartz sand was calculated to be 8.5 μm^2 using Equation (1). The plugging capabilities of the different gel plugging agents under this permeability were tested. The experimental results of the plugging performance are presented in Table 2, and the displacement pressure curve after plugging is shown in Figure 3.

Table 2. Test results for the plugging performance of the gel systems.

Gel System	Permeability Before Plugging × 10^3, μm^2	Permeability After Plugging × 10^3, μm^2	Permeability After Displacement of 16 PV × 10^3, μm^2	Plugging Rate, %	Plugging Rate After Displacement of 16 PV, %	Breakthrough Pressure Gradient, MPa/m
Gel system 1	8493.6	2.23	1.98	99.97	99.98	14.77
Gel system 2	8495.8	0.90	1.45	99.99	99.98	36.57
Gel system 3	8542.1	6.41	46.3	99.92	99.45	4.70

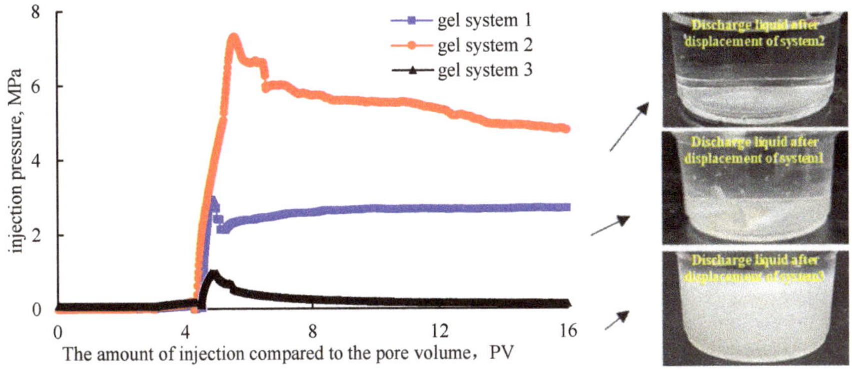

Figure 3. Displacement pressure curve and produced fluid of the gel system after profile control.

It can be seen from the results presented in Table 2 that the maximum plugging rate from the three gel systems to the sand-filled pipe and the plugging rate after displacement scouring at 16 PV were all higher than 99%, with little difference in the numerical values. Therefore, the plugging effect from various gel systems could not be effectively distinguished from each other. From the comparison of the results presented in Figure 3, it can be seen that the plugging capacity (maximum breakthrough pressure of 7 MPa) and the erosion resistance (pressure of water injection displacement of 16 PV: 4.8 MPa) of Gel system 2 were obviously better than those of Gel system 1 and Gel system 3. The breakthrough pressure gradient reached the value of 36.57 MPa/m, which was 2.5 times higher than that of Gel system 1 and 7.8 times higher than that of Gel system 3, respectively. The plugging effect was obviously different for the three gel systems. It can be seen from the picture of the discharged liquid after displacement in Figure 3 that, after Gel system 2 was flushed by displacement, almost no gel was washed out from the sand-filling pipe. However, part of Gel system 1 was scoured, sheared, and washed out of the sand-filled tube. Gel system 3 was almost all washed out of the sand-filling pipe, showing the weakest plugging effect. The test results did not agree with the results of the visual code method + breakthrough vacuum method; however, they were basically consistent with the application in the minefield.

The sand body in the sand-filling pipe was taken out to observe the morphology of the gel/sand mixture. Moreover, the adhesive properties of the gel system were also analyzed. The images of the three types of gel adhesion to sand particles are shown in Figure 4. The sand/gel mixture formed by Gel system 1 was pale yellow. Meanwhile, the residual amount of the colloid is minimal, indicating that most of the gel was displaced early in the displacement experiment due to inadequate adhesive performance. The residual colloid detached from the quartz sand and formed a layered structure due to gravity segregation, further reflecting poor adhesion. The gel's hanging length is 2.4 cm, with high cohesive strength. On the other hand, the adhesion performance was poor. The sand/gel mixture formed by Gel system 2 was dark brown. The quartz sand and the colloid basically combined into one and were uniformly dispersed. The hanging length was 3.2 cm, which had good adhesion. The sand/gel mixture formed by Gel system 3 was white. The colloid and sand particles were evenly distributed, though the content was small. Only a small amount of mixture could be clamped, and the strength was poor. The results showed that Gel system 2 had excellent adhesion capability compared with the other systems and was more effective in resisting erosion due to water flooding.

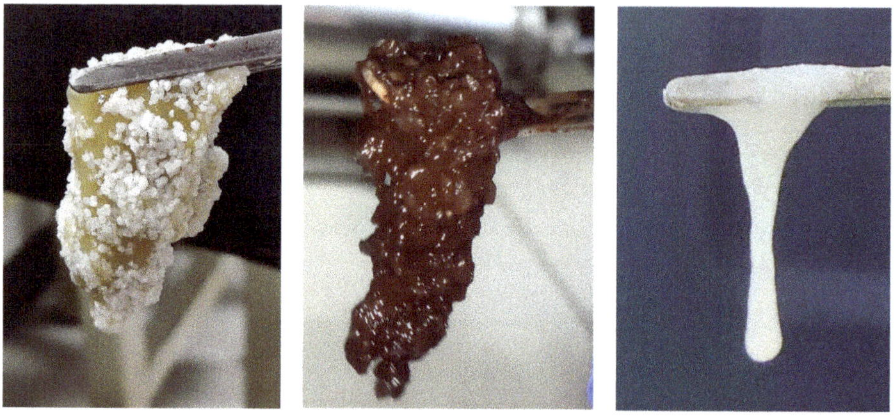

(a) (b) (c)

Figure 4. The morphologic diagram of the gel/sand mixtures. (**a**) Gel system 1; (**b**) Gel system 2; (**c**) Gel system 3.

In order to find out the reason why the results about the plugging capabilities of the three gel systems were inconsistent, the three gel systems were placed in a constant-temperature oven for continuous aging for 2 d. The micro-morphology of the gel was analyzed using a freeze scanning electron microscope. It can be seen from the experimental results (Figure 5) that after Gel system 1 formed the gel, the gel became closely connected with itself, forming a cemented structure with high solid strength. However, the surface of the gel had point-shaped or rod-shaped protrusions (Within the ellipse in Figure 5a), so there was a gap when the gel system adhered to the rock surface, which reduced the adhesion area and weakened the adhesion performance. Gel system 2 had a dense grid texture, and each crosslinking formed a whole, resulting in higher cohesion strength. The surface had no obvious protrusions. The overall flatness and roughness were low (Within the ellipse in Figure 5b). The adhesion area increased, and the formed adhesion force became high. The whole Gel system 3 had crosslinking. However, the gel network structure of this system was sparse. The mesh spacing was large. Part of the connection was weak and had the tendency to fracture (Within the ellipse in Figure 5c). The mesh density was obviously lower than the other two types, while the strength was obviously lower. The larger hole-like structure significantly reduced the adhesion area, resulting in poor adhesion performance.

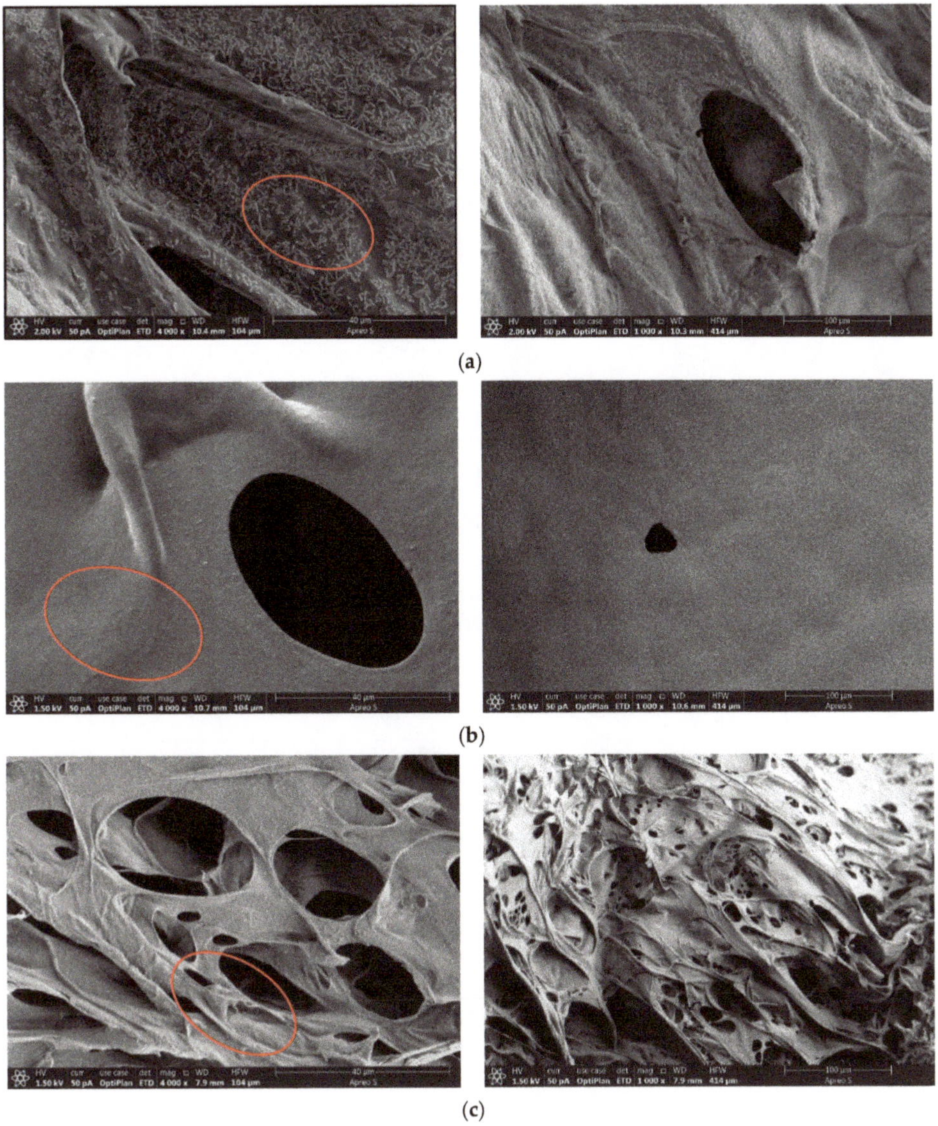

Figure 5. Cryo-SEM images of the three gel systems. (**a**) Cryo-SEM image of Gel system 1; (**b**) Cryo-SEM image of Gel system 2; (**c**) Cryo-SEM image of Gel system 3.

Based on the above experimental analysis, the skeleton density and the degree of networking of the gel system constituted the main factors affecting the strength of the gel system. The micro-surface roughness of materials affects their adhesion capability by influencing the adhesion area. Because of its excellent microstructure, Gel system 2 had the characteristics of high strength, high adhesion, and high plugging compared to Gel systems 1 and 3.

2.3. Optimization of the Plugging Capacity of Gel Plugging Agent by Evaluation Index Method

Although the visual code method and the breakthrough vacuum method are simple and easy to use, they only pay attention to the strength of the gel body and neglect the adhesion energy and erosion resistance of the gel to the reservoir pore throat, which eventually leads to the deviation from the results obtained for actual applications. Using a sand-filling pipe or core to simulate the reservoir to conduct profile control plugging displacement test of the gel plugging agent, the displacement test is the closest evaluation method to field conditions. However, this method is complicated and not suitable for the mass selection of gel systems. The morphological observation of the gel/sand mixture and SEM analysis can be operated easily and visually, which can not only effectively compare the strength of the gel adhering to the rock wall, but can also reflect the compatibility of gel and sand. This method can only be used for qualitative comparison and lacks quantitative criterion, due to which it cannot be used as the optimal selection standard. Therefore, the current evaluation method for screening gel systems is not perfect. It is necessary to establish an evaluation method for the gel plugging agent that is simple in operation, quick in experimentation, and accurate in its evaluation of the gel plugging agent.

Considering the strength and adhesion strength of the gel for plugging, the evaluation index method of gel plugging capacity is established to make up for the deficiency of the current indices for evaluating the plugging capacities of gels. The evaluation indices for the plugging capability of the three gel systems were tested and calculated. The strength of gel plugging agents was tested with stainless steel porous mesh with 80 mesh and a thickness of 2.0 mm. The adhesive strength test of the gel plugging agent adopted the reservoir rock slice with a permeability of about 5 μm^2. The corresponding experimental results are presented in Table 3.

Table 3. Evaluation index of plugging capability of gel plugging agent in different gel systems.

Gel System	Body Strength δ, KPa	Adhesion Strength η, N/m^2			Degree of Injury λ, Dimensionless	Plugging Capacity Evaluation Index Γ, Pa^2
		Adhesive Force, mN	Core Slice Diameter, cm	η, N/m^2		
Gel system 1	90	190	2.5	387.3	6.4	5.45
Gel system 2	66	140	2.5	285.4	2.1	8.97
Gel system 3	63	60	2.5	122.3	2.3	3.35

Based upon the comparison of the results presented in Table 3, it can be seen that the cohesion strength and adhesion strength of Gel system 1 were 90 KPa and 190 mN, respectively, both of which were the best. However, the gel system was easy to be damaged by shear, with a degree of damage (λ) of 387 N/m^2, which was the weakest among the studied gel systems. Gel system 2 and Gel system 3 had similar bulk strength and degree of damage; however, Gel system 2 had better adhesion capability. Compared with Gel system 1, Gel system 2 had weak cohesion strength and adhesion capacity. However, the introduction of a resorcinol crosslinking agent forms a higher-strength interpenetrating network structure [32], which improves the resistance to shear damage of the gel system. According to the evaluation index of the plugging capability of the gel plugging agent in different gel systems, the gel system suitable for reservoir profile control and plugging is Gel system 2.

2.4. Comparison of Profile Control Capability of the Gel Plugging Agent

Using a sand-filled pipe or core to simulate the reservoir for the profile controlling of the gel plugging agent, the plugging displacement test is the most suitable testing method. However, this method is complex in operation and not suitable for screening gel systems in large quantities. Therefore, the method can be used as the final verification method for the indoor plugging control of the selected gel systems. According to Equation (1), the permeability of sand-filling pipe filled with 20–80-mesh and 80–120-mesh quartz sand was 8.5 μm^2 and 1.3 μm^2, respectively. The high and low permeability reservoirs of heterogeneous reservoirs were simulated in parallel to evaluate the plugging and profile control capability of Gel system 1 (prefield system) and Gel system 2 (optimized field system). The variation curve of the injection flow rate of double-pipe parallel heterogeneous profile control injection for displacement after plugging with the gel plugging agent is shown in Figures 6 and 7.

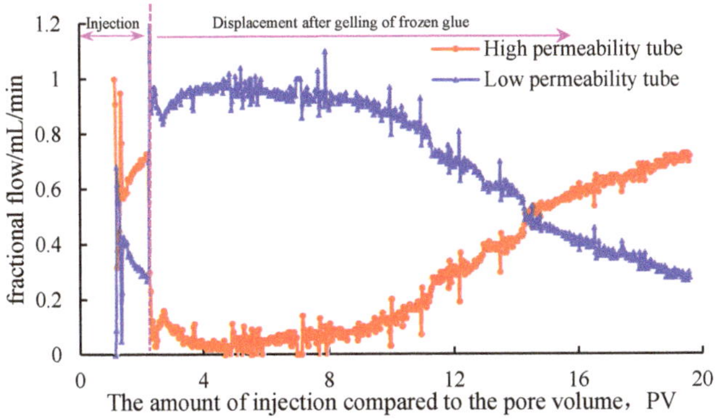

Figure 6. Variation in injection flow rate of Gel system 1 double tube parallel heterogeneous profile control.

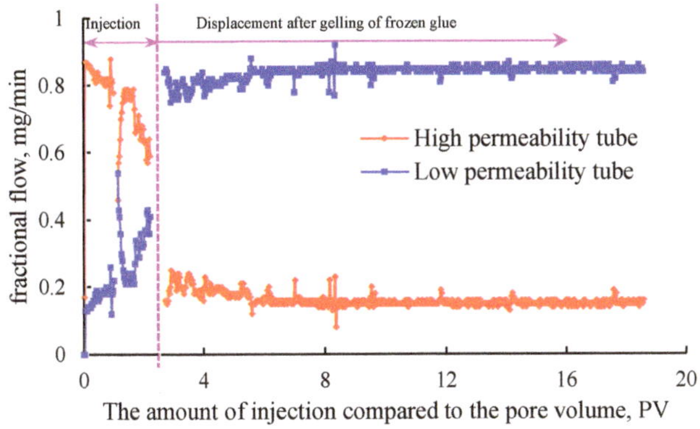

Figure 7. Variation in injection flow rate of 2 double-tube parallel heterogeneous profile control in gel system.

It can be seen from Figures 6 and 7 that in the gel injection stage, both Gel systems 1 and 2 entered the high permeability tube in large quantities. After the gel system was crosslinked into the gel, the water injection displacement erosion was carried out. In the initial stage, both Gel systems 1 and 2 could reverse the injection flow rate of the high and low permeability pipes, and the liquid ratio of the high and low permeability pipes reached 1:9, respectively, indicating that good profile control was achieved.

It can be seen from Figure 6 that with the water flooding and flushing, when the high and low permeability sand-filling pipes of Gel system 1 were injected and displaced to 8 PV, the difference in the partial flows gradually decreased, and the partial flows of the high and low permeability pipes began to turn over for the second time at 14.3 PV, indicating that the profile control effect was invalid at this time. At 20 PV, the flow ratio of the high and low permeability pipes reached 7:3, and the profile-controlling effect was completely lost. The experimental results also explained the problems that the plugging effect of Gel system 1 was not ideal and the effective period of the plugging was short. Compared with the results presented in Figure 7, it can be seen that the high and low permeability sand-filling pipe of Gel system 2 for plugging control had good profile control and diversion effect, and the comprehensive plugging control effect was superior to that of Gel system 1.

By comparing Figures 6 and 7, it can be seen that Gel system 2 had better profile control and plugging effect than Gel system 1, especially in long-term profile control and plugging capability. The experimental results were consistent with the evaluation index of the plugging capability, showing the feasibility and effectiveness of the method.

2.5. Field Tests

Two water injection wells (Ngs (6(1)) and Ngs (6(2))) at the same horizon and similar reservoir properties were selected to carry out the profile control before and after the optimization of the gel system. Figure 8 shows the changes in the pressure curve for the two wells before and after the injection of the profile-controlling agent.

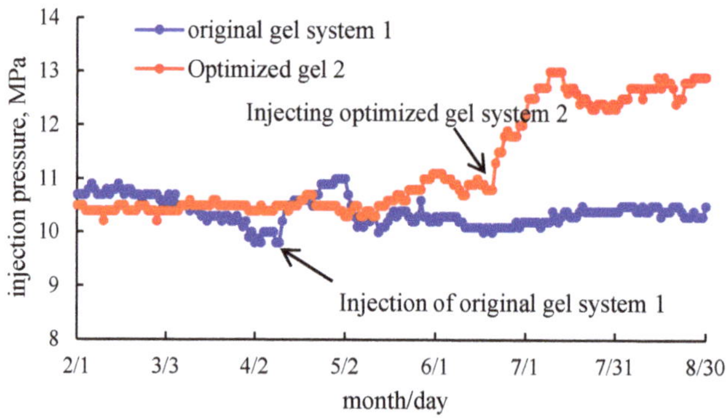

Figure 8. Comparison curve of field test before and after optimization.

It can be seen from Figure 8 that the pressure of the water injection well injected with the original Gel system 1 increased from 9.8 MPa to 11 MPa since 9 April, indicating that the profile-controlling agent had achieved the profile-controlling effect. However, after 28 days of water injection, the pressure decreased rapidly after the breakthrough, and the final pressure was maintained at the same level as the injection pressure before profile control, indicating that the profile-controlling effect was lost. The pressure of water injection wells injected with the optimized Gel system 2 began to increase from 20 June (up to 13 MPa), and the subsequent pressure decreased slightly though it remained at about 12.5 MPa, indicating that the injection of the profile-controlling agent achieved a good profile-controlling effect and maintained the subsequent stable plugging effect. Gel system 2 was 2.5 times the plugging rate of Gel system 1, while the effective plugging time was more than 60 days. The field test results were consistent with the conclusions obtained by comparing the plugging capabilities.

3. Conclusions

1. In the present work, gel system 1 (3000 mg/L HPAM + 0.3% phenolic resin + 0.3% organic chromium crosslinking agent + 0.15% crosslinking accelerator) was screened from three different systems using the conventional breakthrough vacuum method and visual code method. Based on field tests and profile-controlling displacement plugging experiments, the screened gel was easy to separate from the reservoir, resulting in poor plugging capacity. The evaluation effect (strength code: H, degree of breakthrough vacuum: 0.094 MPa, indicating favorable gel performance) was different from the actual plugging capacity (after plugging, the permeability was 2.23×10^3 μm^2, and the breakthrough pressure gradient was 14.77 MPa/m, indicating poor gel performance). The results showed that the adhesive capability and erosion resistance of the gel to the reservoir pore throat were neglected in the conventional method, and the evaluation angle was found not to be comprehensive enough. The plugging capacity test of the gel plugging agent is cumbersome and not suitable for on-site rapid detection.
2. After analyzing the current laboratory evaluation procedure and summarizing the controlling factors of the plugging capacity of the gel plugging agent, a new evaluation method for the plugging effect of the gel plugging agent was established. Based on this, the gel plugging capability evaluation index method (with the correlation of $\Gamma = \frac{\delta \times \eta}{\lambda}$) was constructed, which comprehensively considered the gel strength, gel adhesion capability, shear resistance, and long-term anti-dehydration performance. The experimental results showed that the evaluation results of the proposed method were consistent with the experimental results of profile control and displacement plugging.
3. A new gel system (3000 mg/L HPAM + 0.3% phenolic resin + 0.15% Resorcinol + 0.3% hexamethylenetetramine + 0.15% crosslinking accelerator) was selected by the newly established method, and its comprehensive evaluation index was found to be 8.97 Pa^2. Compared with the old system, the failure of profile control gradually occurred when 8 PV was displaced. The new system could achieve a 1:9 ratio of high and low permeability tubes, and the effective period was more than 20 PV. Moreover, the profile control and the plugging capability were obviously improved.
4. A field in the Shengli oilfield was selected for the verification of the experimental results. The gel system optimized by the evaluation method achieved an injection pressure of up to 13 MPa, with a plugging duration exceeding 60 days. The injection pressure of the new system was 2.5 times that before the optimization. The effective plugging period was prolonged to more than two times that of the original system. The field test results were consistent with the plugging capacity evaluation index. To sum up, the newly established method is more scientific and practical, and can provide a reference for screening and evaluating batches of gel plugging agent systems under actual reservoir conditions.

4. Materials and Methods

4.1. Materials

Resorcinol, hexamethylenetetramine, and ammonium chloride (as the crosslinking accelerator) were analytically pure, and obtained from Sinopharm Chemical Reagent Co., Ltd., Shanghai, China. Phenolic resin ($(C_6H_6O)_n$.$(CH_2O)_n$; molecular weight of 12,000; alkali number: 20 mg KOH/g), urea–formaldehyde resin ($(C_2H_6N_2O_2)$; molecular weight of 15,000; alkali number: 50 mg KOH/g), organic chromium crosslinking agent ($Cr(C_4H_6O_4)_3$), and partially hydrolyzed polyacrylamide (HPAM; molecular weight of 22 million; degree of hydrolysis of 22–24%) were obtained from the construction site of a block in Shengli

Oilfield, Sinopec, Dongying, China. The simulated formation water with a salinity of 20,262 mg/L is prepared in the laboratory, and the specific formula is as follows: Table 4.

Table 4. Formula table of simulated formation water.

Ionic Types	Ca^{2+}	Na^{+}	Mg^{2+}	Cl^{-}	SO_4^{2-}	HCO_3^{-}
ion concentration, mg/L	1022	6422	259	12,326	31	202

4.2. *The Traditional Test Method of Gel Performance*

4.2.1. Visual Code Method [33]

The sample tube was turned upside down to observe the form of a gel in the tube to visually evaluate the strength of the gel. The gel-forming time was determined and the gel strength was adjudged according to the strength code (Table 5) [17]. This method is simple, convenient, and applicable to a wide range of conditions, but the test results are easily affected by the observation of the test container and the tester.

Table 5. Code standards for gel strength.

Grade	Strength Classification Standard
A	No continuous gel formation was detected: the viscosity of the gelling system was the same as that of the same concentration of polymer solution without a crosslinking agent, but sometimes there might be some unconnected gel blocks with great viscosity in the system.
B	Highly flowable gel: The viscosity of the gel system is slightly higher than that of a polymer solution of the same concentration without a crosslinking agent.
C	Flowing gel: When the sample bottle is inverted, most of the gel flows to the cap.
D	Medium flow gel: Only a small portion (10~15%) when the vial is inverted vertically. The gel does not easily flow to the cap (typically described as long tongue type gel).
E	Difficult to flow gel: When the sample bottle is vertically inverted, the gel flows very slowly to the cap or a large part (>15%) does not flow to the cap.
F	Highly Deformed No Flow Gel: Gel does not flow to the cap when the vial is inverted vertically.
G	Moderate deformation non-flowing gel: When the sample bottle is vertically inverted, the gel is deformed downward to about half of the position.
H	Slight deformation does not flow gel: When the sample bottle is vertically inverted, only the surface of the gel is slightly deformed.
I	Rigid gel: When the sample bottle is inverted vertically, the surface of the gel does not deform.

4.2.2. Vacuum Breakthrough Method

The specific experimental device is as shown in Figure 9. The colorimetric tube filled with the formed gel was connected according to the schematic shown in Figure 9. When the vacuum pump was activated and the air broke through the gel, the vacuum on the vacuum gauge increased to the maximum reading (the breakthrough vacuum (BV)). The pressure gauge showed 0 MPa before use. Each sample (or condition) was measured in parallel 3 times, and the average value was used for the analysis (atmospheric pressure at the time of measurement was 0.1 MPa). The larger the BV value, the higher the strength.

4.2.3. Adhesion Sand Test

The adhesion experiment of the gel and sand was carried out to observe the adhesion capability of the system. In the plugging capacity test of the gel plugging agent, the sand-filling pipe was injected with clean water. The advection pump was closed after 16 PV of water was stably displaced. The sand-filling tube was removed, and the mixture of quartz sand and gel was taken out at the 10 mL outlet. The mixture was clamped with tweezers to observe the adhesion of gel and sand.

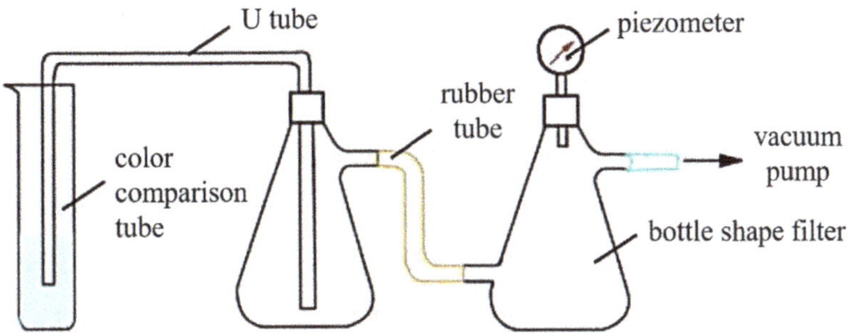

Figure 9. Schematic of the setup used to measure the strength using the breakthrough vacuum method.

4.2.4. Cryo-SEM Test of Gel

The microstructure of the gel was observed using scanning electron microscopy (SEM). SU8010 scanning electron microscope used in this experiment was provided by Hitachi High-Tech Analytical Science Co., Ltd., Ibaraki Japan. [34]. First, the gel sample was pretreated. Liquid nitrogen was used to attain a low temperature of −160 °C. The pressure around the sample was reduced to sublimate the frozen water in the gel so that the water would not crystallize during the solidification process to form glassy water. This is important to avoid the increase in the volume of water due to expansion, which can damage the original structure of the gel in the aqueous solution [35]. The dried polymer mesh was carefully placed on the conductive tape for conductive spraying, and finally, the microtopography analysis was performed using a scanning electron microscope. The Adhesion Sand Test and Cryo-SEM test of the gel can only qualitatively analyze the adhesion ability, and cannot be quantitatively tested.

4.3. Experimental Method of Profile Control Plugging Displacement of Gel Plugging Agent

Using the custom-made displacement device, the plugging capability of the gel plugging agent was analyzed to simulate the plugging effect of the gel system under reservoir conditions [36–39]. The profile-controlling capability of the gel plugging agent was tested to verify the feasibility of the comprehensive evaluation index of the plugging effect. The flow diagram of the unit is shown in Figure 10. A 20 cm long and 2.5 cm diameter sand-filling pipe was filled with quartz sand to simulate the sandstone reservoir. Sand-filled tubes filled with 20–80-mesh quartz sand simulated the high-permeability reservoirs (4000~5000 mD), whereas sand-filled tubes filled with 80–120-mesh quartz sand simulated the low-permeability reservoirs (400~500 mD).

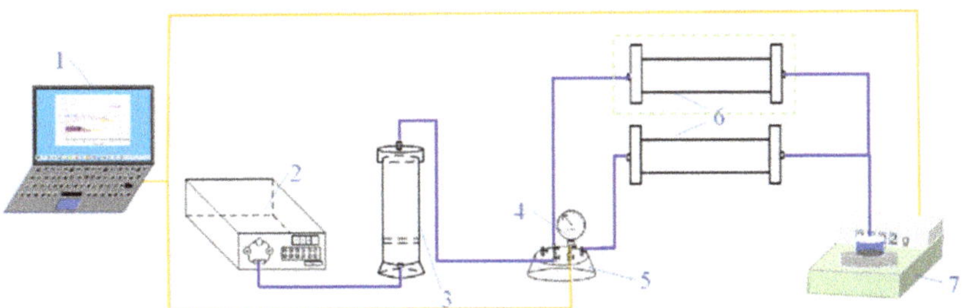

Figure 10. Schematic of the gel plugging agent's profile-controlling plugging displacement experimental device. 1—data acquisition system; 2—constant-flux pump; 3—intermediate container; 4—pressure sensor; 5—six-way valve; 6—sand-filling pipe; 7—electronic balance.

4.3.1. Plugging Capacity Test of the Gel Plugging Agent

First, a sand-filling pipe was filled with 20–80-mesh quartz sand. The dry weight was determined, and the experimental device was connected according to Figure 10. Clean water was injected at 1 mL/min until the pressure at the injection end stabilized [40]. The pressure was recorded, and the sand-filling pipe was removed to measure the wet weight. The permeability [41] (Kw) and pore volume (PV) were calculated. The sand-filling tube was reconnected, and after injecting 1 PV gel system at the speed of 1 mL/min, the sand-filling tube was removed. Both the ends of the tube were sealed and placed in the thermostat that was set to the gel-forming temperature. After the gel system was gelled, the sand-filling pipe was taken out, and placed into the displacement device, where clean water was injected at the flow rate of 1 mL/min. The amount of injected clean water and the injection pressure were recorded. Finally, the plugging rate and the breakthrough pressure gradient were calculated.

Permeability:

$$K_{\mathrm{w}} = \frac{Q\mu_{\mathrm{w}}\Delta L}{10A\Delta P} \tag{1}$$

where K_{w} is the water phase permeability (μm^2), Q is the liquid phase flow (cm^3/s), μ_{w} is the fluid viscosity (mPa·s), ΔL is the core length (cm), A is the core's cross-sectional area (cm^2), and ΔP is the pressure difference between the injection and outlet ends (MPa).

Plugging rate [23]:

$$\gamma = \frac{k_1 - k_2}{k_1} \times 100 \tag{2}$$

where γ is the core plugging rate (%), k_1 is the initial water phase permeability (μm^2), and k_2 is the permeability of the water phase after plugging (μm^2).

Breakthrough Pressure Gradient [19]:

$$p_{\mathrm{m}} = \frac{p_{\mathrm{max}}}{L} \tag{3}$$

where p_{m} is the breakthrough pressure gradient (MPa/cm), p_{max} is the maximum breakthrough pressure (MPa), and L is the length of the sand-filling tube (cm).

4.3.2. Profile-Controlling Capacity Test of Gel Plugging Agent

The sand-filling pipe was replaced with two parallel sand-filling pipes filled with quartz sand of 20–80 meshes and 80–120 meshes, respectively. The other experimental parameters and methods were the same as the plugging capacity test of the gel plugging agent.

4.4. Construction of the Evaluation Method for Plugging Effect of Gel Plugging Agent

4.4.1. Body Strength Test Method

Besides good cohesion strength, the shear resistance of gel plugging agents in reservoir throats is also one of the important parameters that help realize a good plugging effect. By simulating the process of the gel plugging agent breaking through the reservoir pore throat, an experimental device for testing the cohesion strength of the gel plugging agent is designed. The schematic of the device is shown in Figure 11. The cohesion strength of the gel plugging agent is characterized using the pressure difference when the gel plugging agent breaks through the simulated pore throat.

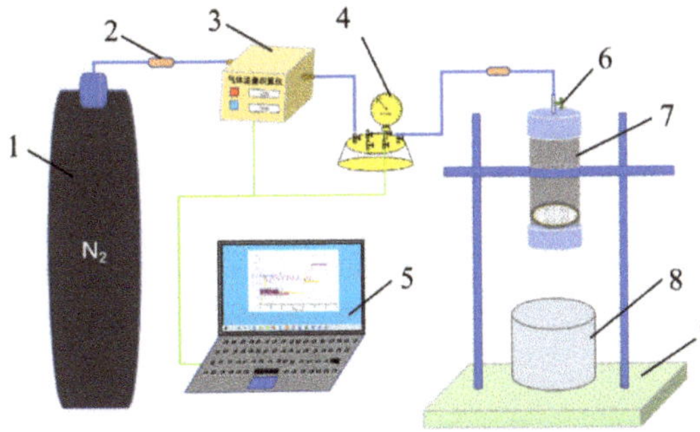

Figure 11. Schematic of the experimental device for cohesion strength test of the gel plugging agent. 1—nitrogen cylinder; 2—check valve; 3—gas flow totalizer; 4—pressure sensor; 5—gas flow control and pressure acquisition system; 6—valve; 7—gel strength test container; 8—sample collection container; 9—bracket table.

The experimental device for testing the strength of the gel plugging agent mainly consisted of a pressure supply system (represented by 1, 2, 3, and 4 in Figure 11), a data acquisition system (represented by 5 in Figure 11), and a simulation system for the gel plugging agent to break through the pore throat (represented by 6 and 7 in Figure 11). In order to avoid the influence of pressure difference on the experimental results, the gas flow integrating instrument in the pressure supply system (represented by 3 in Figure 11) was designed to control the rate of gas output, and the rate of increase in pressure in the gel strength test vessel (represented by 7 in Figure 11).

The gel strength test container is key to the device used for evaluating the strength of the gel plugging agent. The schematic of the container is shown in Figure 12. The vessel cavity (represented by 4 in Figure 12) was made up of transparent plexiglass, thus meeting the pressure requirements, which can visually observe the gel plugging agent breakthrough process. The porous mesh (represented by 5 in Figure 12) was a regularly arranged mesh. Different mesh sizes can be used to simulate different reservoir permeabilities. The materials can be stainless steel or organic glass. The porous mesh can also be replaced by the cemented core sheet or the core sheet cut and processed by the on-site reservoir.

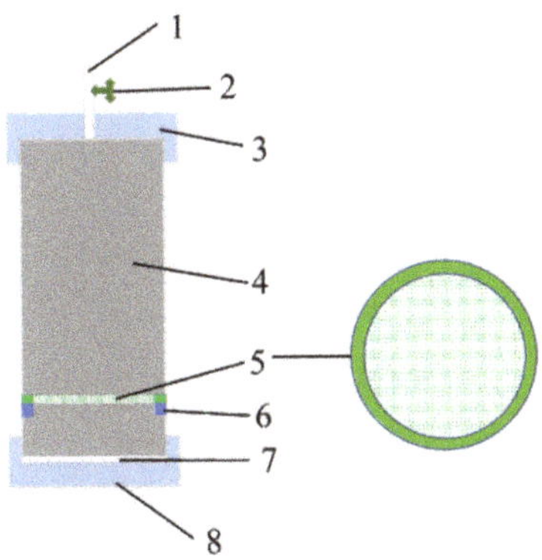

Figure 12. Schematic of the gel strength test container. 1—opening pipeline joint; 2—valve; 3—container top cover; 4—container cavity; 5—porous mesh; 6—partition plate in the vessel; 7—liner; 8—container lower lid.

The inner lining (represented by 7 in Figure 12) corresponding to the lower cover of the container was made up of polytetrafluoroethylene. During the experiment, a layer of Vaseline was applied to reduce the adhesion between the gel and the lower cover and reduce the experimental error.

Since the transfer of a gel plugging agent to any strength-measuring device will cause some damage to its performance, the measured value is usually not representative of the real gel strength [19,42]. Therefore, the gel plugging agent is designed to first gel and crosslink in the container used to measure the gel strength, and then, the strength test is performed. The experimental device used for determining gel strength is shown in Figure 11. The detailed experimental procedures are outlined as follows:

1. The solution of the gel system was poured into the container used to measure the strength of the gel. The liquid level was 2 cm higher than the porous mesh.
2. The container was sealed and placed in the thermostat box, which was set to the gelling temperature.
3. After the gel system was gelled, the container was taken out. The upper cover valve was opened, emptied, and fixed on the bracket table. The experimental device was connected as shown in Figure 11.
4. The lower cover of the container was opened. The gas flow rate was set to be 20 mL/min. The pressure reading was recorded. The maximum pressure value was collected when the gel broke through, which was the body strength of the gel plugging agent.

4.4.2. Adhesion Strength Test Method

The gel plugging agent has good adhesion to the pore throat of the reservoir, which can improve erosion resistance. Adhesion is another necessary condition for achieving a good plugging effect. The schematic of the test device used for evaluating adhesion between the gel plugging agent and the reservoir pore throat wall is shown in Figure 13. The core slice (represented by 4 in Figure 13) adopted the standard cylindrical slice (a diameter of 2.54 cm) of stratum core, with a thickness of 5–10 mm. If the formation core was not available, it could be replaced by other rock slices with similar material and permeability.

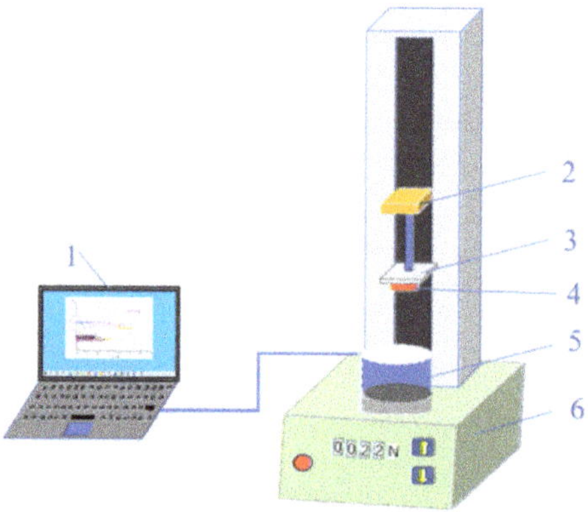

Figure 13. Schematic of the test device used for adhesive strength test of gel plugging agent. 1—data acquisition system; 2—upper and lower lifting platform; 3—plate of lifting table; 4—core slice; 5—transparent sample cup; 6—precision tensile testing machine.

The adhesive strength of the gel plugging agent was characterized using the maximum tensile force between the pore throat wall and the gel plugging agent per unit area. The adhesive strength is given by Equation (4).

$$\eta = \frac{F}{S} \quad (4)$$

where η is the adhesive strength of the gel plugging agent to reservoir pore throat (N/m^2), F is the maximum tensile force when the gel plugging agent is peeled off from the core sheet (N), and S is the surface area under the core (m^2).

The adhesive strength of the gel plugging agent was measured using the experimental device shown in Figure 13. The detailed experimental procedures are outlined as follows:

1. The gel system solution was poured into a transparent sample cup to a height of 4–6 cm. The cup was sealed and placed in a constant-temperature box that was set to the gelling temperature.
2. After the gel system was gelled, the sample cup was taken out. The cover of the sample cup was removed and placed on the sample table of the precision tensile testing machine.
3. The cutting flat core piece was bonded to the lifting platform plate, and the experimental device was connected as shown in Figure 13.
4. The lifting platform was controlled to drop lower until the lower end of the core slice was in close contact with the gel and maintained for a period of 5–10 min.
5. The lifting speed of the lifting platform was set to 2 mm/min. The value of the tension was collected when it started to rise. The maximum tension value (F) was collected, and the adhesion strength of the gel plugging agent was calculated by Equation (4).

4.4.3. Method for Measuring the Degree of Damage

Dehydration of a gel plugging agent significantly reduces its plugging capacity, which is an important reason for shortening the effective period of profile control and water plugging in oilfields [43]. The process of migration and shearing will increase the dehydration of the gel. The rate of dehydration of the gel was adopted to represent the degree of damage to the gel. The detailed experimental procedures are outlined as follows:

1. After the strength test of the gel body, the gel sheared by the porous mesh was taken. The mass of the gel was m_1 (g). The gel was placed in a suitable cover screw glass test tube and sealed.
2. The sealed gel was placed in the constant-temperature box for 48 h, which was set to the reservoir temperature.
3. The screw glass test tube was removed. The amount of water from the gel was noted down, and the mass was observed as m_2 (g). The gel damage index is calculated using the expression $\lambda = \frac{m_2}{m_1} \times 100$.

4.5. Construction of the Comprehensive Evaluation Index of Plugging Effect

The plugging capability of the gel plugging agent is positively correlated with the cohesion strength and adhesion strength of the gel plugging agent. The plugging capability is negatively correlated with the degree of loss. Therefore, the comprehensive evaluation index (Γ) is defined to evaluate the plugging effect of the gel plugging agent on the reservoir. The comprehensive evaluation index is given by Equation (5).

$$\Gamma = \frac{\delta \times \eta}{\lambda} \quad (5)$$

Author Contributions: Methodology, Z.W.; validation, Z.W. and Y.B.; investigation, W.H.; writing—original draft preparation, Z.W. and J.J.; writing—review and editing, Z.W. and J.J.; supervision, Z.W. and Y.G. All authors have read and agreed to the published version of the manuscript.

Funding: This research was funded by the National Natural Science Foundation of China (No. 52374026).

Institutional Review Board Statement: Not applicable.

Informed Consent Statement: Not applicable.

Data Availability Statement: The raw data supporting the conclusions of this article will be made available by the authors on request.

Conflicts of Interest: The authors declare no conflicts of interest.

Abbreviations

The following abbreviations are used in this manuscript:

GSR	gel strength rank index method
BV	breakthrough vacuum
HPAM	partially hydrolyzed polyacrylamide
PV	pore volume

References

1. Deng, R.; Dong, J.; Dang, L. Numerical Simulation and Evaluation of Residual Oil Saturation in Waterflooded Reservoirs. *Fuel* **2025**, *384*, 134018. [CrossRef]
2. Wang, Z.; Zhai, D.; Zhao, X. The Evaluation Method of Residual Oil Distribution Based on Three-Distribution Visible Model. *Res. Explor. Lab.* **2016**, *35*, 13–16.
3. Yin, H.; Zhang, G.; Wu, Q.; Cui, F.; Yan, B.; Yin, S.; Soltanian, M.R.; Thanh, H.V.; Dai, Z. Unraveling Overlying Rock Fracturing Evolvement for Mining Water Inflow Channel Prediction: A Spatiotemporal Analysis Using ConvLSTM Image Reconstruction. *IEEE Trans. Geosci. Remote Sens.* **2024**, *62*, 4510417. [CrossRef]
4. Pu, W.; Wen, C.; Liu, R.; Jin, F.; Wang, C.; Liao, Z. Evaluation of a Novel Profile Control Agent for Enhancing an Oil-Recovery Application. *J. Appl. Polym. Sci.* **2016**, *133*, 43756. [CrossRef]
5. Singh, V.; Ruwali, N.; Pandey, R.K.; Vaferi, B.; Wood, D.A. Applying a Deep-Learning Surrogate Model to Simulate and Compare Achievable Oil Recovery by Different Waterflood Scenarios. *Pet. Sci. Technol.* **2024**, *42*, 4405–4423. [CrossRef]
6. Zhu, D.; Bai, B.; Hou, J. Polymer Gel Systems for Water Management in High-Temperature Petroleum Reservoirs: A Chemical Review. *Energy Fuels* **2017**, *31*, 13063–13087. [CrossRef]
7. Lu, S.; Bo, Q.; Zhao, G.; Shaikh, A.; Dai, C. Recent Advances in Enhanced Polymer Gels for Profile Control and Water Shutoff: A Review. *Front. Chem.* **2023**, *11*, 1067094. [CrossRef]
8. Cao, B.; Xie, K.; Lu, X.; Cao, W.; He, X.; Xiao, Z.; Zhang, Y.; Wang, X.; Su, C. Effect and Mechanism of Combined Operation of Profile Modification and Water Shutoff with In-Depth Displacement in High-Heterogeneity Oil Reservoirs. *Colloids Surf. A Physicochem. Eng. Asp.* **2021**, *631*, 127673. [CrossRef]
9. Li, Y.; Xiao, C.; Wang, L. Research and Application of Profile Control Technology for Improving the Recovery Factor of Medium and High Permeability Reservoirs. In Proceedings of the International Field Exploration and Development Conference 2023, Wuhan, China, 20–22 September 2023; pp. 453–459.
10. Li, Y.; Shi, H.; Chen, H.; Wang, L.; Wang, S. Water analysis and chemical plugging technology of oil wells in oilfields of western water injection development. *Oil-Gasfield Surf. Eng.* **2010**, *29*, 23–25.
11. Wang, Y.; Seright, R.; Huang, B.; Fu, C.; Zeng, Y. Describing Flow Behavior of Polymer Gel in Fracture Using Rheology. *J. Pet. Sci. Eng.* **2017**, *157*, 1209–1219. [CrossRef]
12. Sun, Z.; Li, Z.; Qu, K.; Zhang, Z.; Niu, Y.; Xu, W.; Ren, C. A Review on Recent Advances in Gel Adhesion and Their Potential Applications. *J. Mol. Liq.* **2021**, *325*, 115254. [CrossRef]
13. Dai, C.; Zhang, G.; Zhao, F. Study on the factors affecting the freezing of aldehyde gel. *Oilfield Chem.* **2001**, *24–26*, 47. [CrossRef]
14. Hubbard, S.; Roberts, L.J.; Sorbie, K.S. Experimental and Theoretical Investigation of Time-Setting Polymer Gels in Porous Media. *SPE Reserv. Eng.* **1988**, *3*, 1257–1267. [CrossRef]
15. Seno, N.N.; Zheng, H. Chemical glue water plugging design and field application in oil wells of natural fractured carbonate reservoirs. *Oil Recovery Process Inf.* **1990**, *3*, 11–25.
16. Southard, M.Z.; Green, D.W.; Willhite, G.P. *Kinetics of the Chromium(VI) Thiourea Reaction in the Presence of Polyacrylamide*; SPE: Tulsa, OK, USA, 1984; p. SPE-12715-MS.

17. Sydansk, R.D. *A New Conformance-Improvement-Treatment Chromium(III) Gel Technology*; Society of Petroleum Engineers: Tulsa, OK, USA, 1988.
18. Weisser, D.F.; Shakeel, A.; Mayer, D.; Schmid, J.; Heienbrock, S.J.; Deckert, M.H.; Rapp, B.E. Gel Point Investigation of Liquid Silicone Rubber Using Rheological Approaches. *Polymer* **2023**, *283*, 126286. [CrossRef]
19. Zhou, H.; Jia, H.; Cao, J.; Liu, X.; Zhu, Y.; Jiang, K. Quick Measurement of Gel Strength by Breakthrough Vacuum Value Method through Viscosity Quantification. *Oil Drill. Prod. Technol.* **2015**, *37*, 116–119. [CrossRef]
20. Zhu, H.; Yu, C.; Zhao, F. Zirconium Gelling System for Water Shutoff. *Oilfield Chem.* **1989**, *1*, 27–31. [CrossRef]
21. Zhao, Q. Studies on Behavior Evaluation Methods and Suitable Conditions of Polymer Gels. Master's Thesis, Daqing Petroluem Institute, Daqing, China, 2004.
22. Ai, W. Laboratory Study on Low Temperature Chrome Gel. Master's Thesis, Daqing Petroluem Institute, Daqing, China, 2001.
23. Purkaple, J.D.; Summers, L.E. Evaluation of Commercial Crosslinked Polyacrylamide Gel Systems for Injection Profile Modification. In *SPE Enhanced Oil Recovery Symposium*; SPE: Tulsa, OK, USA, 1988.
24. Aldhaheri, M.; Wei, M.; Zhang, N.; Bai, B. A Review of Field Oil-Production Response of Injection-Well Gel Treatments. *SPE Reserv. Eval. Eng.* **2019**, *22*, 597–611. [CrossRef]
25. Kang, C.; Guo, J.; Sun, X.; Fei, D. Progress in Evaluation Methods of Gel Plugging Agent. *Appl. Chem. Ind.* **2022**, *10*, 2952–2955. [CrossRef]
26. Tang, X.; Liu, Y.; Xiang, W.; Liu, G.; Lin, Y.; Hu, P. Performance Evaluation on Indexes and Measures of Cross—Linked Polymer Gel for Water Shutoff and Profile Control. *Oil Drill. Prod. Technol.* **2004**, *49–53*, 84. [CrossRef]
27. Seright, R.; Brattekas, B. Water Shutoff and Conformance Improvement: An Introduction. *Pet. Sci.* **2021**, *18*, 450–478. [CrossRef]
28. Lai, N.; Yuan, L.; Li, W.; Xiong, W.; Yu, G.; Du, Z.; Song, Z.; Wen, Y. Research on the Properties of Natural Gas Foamed Gel as a Profile Control and an Oil Displacement Agent in Strong Heterogeneous Reservoirs. *Energy Fuels* **2021**, *35*, 7738–7755. [CrossRef]
29. Konoplin, A.Y.; Pushkarev, A.V.; Shakurov, A.V.; Baurova, N.I. Evaluation of Frost Resistance of Organosilicon Sealant at Ultralow Temperatures. *Polym. Sci. Ser. D* **2021**, *14*, 335–339. [CrossRef]
30. Li, X.; Fu, M.; Hu, J. Preparation and Performance Evaluation of Temperature-Resistant and Salt-Resistant Gels. *Gels* **2024**, *10*, 337. [CrossRef] [PubMed]
31. Singh, W.P.; Singh, R.S. Gelation-Based Visual Detection of Analytes. *Soft Mater.* **2019**, *17*, 93–118. [CrossRef]
32. Wang, Z.; Zhao, X.; Bai, Y.; Gao, Y. Study of a Double Cross-Linked HPAM Gel for in-Depth Profile Control. *J. Dispers. Sci. Technol.* **2016**, *37*, 1010–1018. [CrossRef]
33. Liu, Y.; Bai, B.; Wang, Y. Applied Technologies and Prospects of Conformance Control Treatments in China. *Oil Gas Sci. Technol.—Rev. IFP Energ. Nouv.* **2010**, *65*, 859–878. [CrossRef]
34. Liu, P.; Li, W.; Wei, F.; Hu, F.; Zhu, X.; Jia, Z. Preparation of a Fluid Diversion Agent for Profile Control in Elevated Temperature and High Salinity Reservoirs. *J. Appl. Polym. Sci.* **2021**, *138*, 50875. [CrossRef]
35. Ge, J.; Guo, H.; Zhang, T.; Zhou, D.; Xu, Y.; Lü, Q. Development of Temperature and Salinity Resistant Phenolic Gel and Its Performance Regulation Mechanism. *Acta Pet. Sin.* **2022**, *43*, 1145–1157. [CrossRef]
36. Farasat, A.; Younesian-Farid, H.; Sadeghnejad, S. Conformance Control Study of Preformed Particle Gels (PPGs) in Mature Waterflooded Reservoirs: Numerical and Experimental Investigations. *J. Pet. Sci. Eng.* **2021**, *203*, 108575. [CrossRef]
37. Wang, Z.; Bai, B.; Sun, X.; Wang, J. Effect of Multiple Factors on Preformed Particle Gel Placement, Dehydration, and Plugging Performance in Partially Open Fractures. *Fuel* **2019**, *251*, 73–81. [CrossRef]
38. Leng, J.; Wei, M.; Bai, B. Review of Transport Mechanisms and Numerical Simulation Studies of Preformed Particle Gel for Conformance Control. *J. Pet. Sci. Eng.* **2021**, *206*, 109051. [CrossRef]
39. Xie, K.; Su, C.; Liu, C.; Cao, W.; He, X.; Ding, H.; Mei, J.; Yan, K.; Cheng, Q.; Lu, X. Synthesis and Performance Evaluation of an Organic/Inorganic Composite Gel Plugging System for Offshore Oilfields. *ACS Omega* **2022**, *7*, 12870–12878. [CrossRef] [PubMed]
40. Yang, S.; Shu, Z.; Zhang, Q.; Zhao, Y.; Chen, Z.; Zhang, L. Experimental Study on Compatibility of Reservoir and Polymer Gel Profile Control Agent. *Front. Earth Sci.* **2022**, *10*, 1029309. [CrossRef]
41. Zhang, Y.; Yang, L.; Xi, R.; Geng, Z.; Zhang, Q.; Ma, X. CO_2-H_2O Two-Phase Displacement Characteristics of Low Permeability Core Using Nuclear Magnetic Resonance and Magnetic Resonance Imaging Techniques. *RockandSoilMechanics* **2023**, *44*, 1636–1644. [CrossRef]
42. Liu, Y.; Wang, Z. *Chemical Profile Control Water Plugging Technology*; Petroleum Industry press: Beijing, China, 1999; ISBN 978-7-5021-2740-4.
43. Wu, W.; Hou, J.; Li, G.; Chen, L. Effect of Temperature and Inorganic Salts Concentration on Syneresis Rate of AM/DAC Hydrogel. *Colloids Surf. A Physicochem. Eng. Asp.* **2022**, *641*, 128561. [CrossRef]

gels

MDPI

Article

Preparation and Effect of CO_2 Response Gel for Plugging Low-Permeability Reservoirs

Huipeng Wang, Changhua Yang *, Yongwei Zhang and Chen Wang

School of Petroleum Engineering, Xi'an Shiyou University, Xi'an 710065, China; 15965803805@163.com (H.W.); 13571453944@163.com (Y.Z.); 15238733188@163.com (C.W.)
* Correspondence: ych569@126.com

Abstract: In order to solve the problem of gas channeling during CO_2 flooding in low-permeability reservoirs, a novel CO_2 responsive gel channeling system was prepared by using carrageenan, branched polyethylene imide and ethylenediamine under laboratory conditions. Based on the Box–Behnken response surface design method, the optimal synthesis concentration of the system was 0.5 wt% carrageenan, 2.5 wt% branchized polyethylenimide and 6.5 wt% ethylenediamine. The micromorphology of the system before and after response was characterized by scanning electron microscopy. The rheology and dehydration rate were tested under different conditions. The channeling performance and enhanced oil recovery effect of the gel system were simulated by a core displacement experiment. The experimental results show that the internal structure of the system changes from a disordered, smooth and loosely separated lamellae structure to a more uniform, complete and orderly three-dimensional network structure after exposure to CO_2. The viscosity of the system was similar to aqueous solution before contact with CO_2 and showed viscoelastic solid properties after contact with CO_2. The experiment employing dehydration rates at different temperatures showed that the internal structure of the gel would change at a high temperature, but the gel system had a certain self-healing ability. The results of the displacement experiment show that the plugging rate of the gel system is stable at 85.32% after CO_2 contact, and the recovery rate is increased by 17.06%, which provides an important guide for the development of low-permeability reservoirs.

Keywords: gas channeling; CO_2 response gel; response surface; three-dimensional network structure

1. Introduction

Of the known reservoir reserves, low-permeability reservoirs account for about 38% of the world's total, and China accounts for about two-thirds of them [1,2]. The characteristics of small pore throat size, complex pore structure, natural micro-fracture development, and the wide spatial scale distribution of fractures and pore throats in low-permeability reservoirs make them more difficult to exploit [3,4]. The practice at home and abroad shows that CO_2 flooding is one of the key technologies used for the effective exploitation of low-permeability reservoirs [5]. The high fluidity, viscosity reduction, volume expansion and interfacial tension reduction of CO_2 gas make it more effective than water flooding in the process of oil flooding to reduce the residual oil saturation in the pores [6,7]. However, the CO_2 flooding method has some disadvantages, such as low sweep efficiency, mainly due to the heterogeneity of the reservoir, and it is easy for gas to form a channel in the hyperpermeability area, and the viscous fingering and gravity overlap are caused by the viscosity difference and density difference between crude oil and CO_2. The main reason for CO_2 displacement channeling is that the viscosity of CO_2 is much lower than that of formation water and crude oil, resulting in the difference in fluidity between them [8]. This difference causes the injected CO_2 to bypass the crude oil and cross-flow occurs, reducing the sweep efficiency. Secondly, the heterogeneity of the reservoir means the injected CO_2 tends to form a channeling channel along the high-permeability zone, which further affects the displacement effect. In addition, there are also problems such as gravity overlap, oil and gas viscosity differences, and non-homogeneous layer capillary forces, which not only limit the gas displacement efficiency but also increase the difficulty of plugging gas channeling [9].

To solve the above problems, domestic and foreign scholars have accumulated numerous anti-gas channeling methods and applied them to low-permeability reservoirs at home and abroad on a large scale [10]. At present, the main methods to prevent CO_2 gas channeling include alternating injections of water and gas (WAG), polymer gel channeling [11,12], chemical precipitation, CO_2 foam [13] and CO_2 thickening [14]. These methods have played a positive role in controlling gas channeling, but there are also great limitations. Alternate injections of water and air are not suitable under severe heterogeneous conditions because the fluidity instability during gas injection can lead to cavitation and device corrosion problems [14]. Gel injection is limited by its performance, and it is difficult to effectively plug deep reservoirs. The controllability of chemical precipitation is poor; it is difficult to effectively control the improvement effect, and this may cause damage to the reservoir [15]. The strength of CO_2 foam is low; it is difficult to control foam formation under reservoir conditions, and it is not suitable for severely heterogeneous conditions. CO_2 thickening methods are costly and complex to configure. In recent years, a new class of smart polymers, namely CO_2 stimulus responsive polymers, has emerged [16]. These polymers have a special structure and contain groups that can respond to CO_2, such as amidinoyl, guanidine and amine groups [17,18]. Li's team synthesized a CO_2-responsive long-chain polyamine (ODPTA) and applied this long-chain polyamine to foam generation, achieving excellent plugging performance [19]. Although CO_2-responsive foams have shown remarkable results in enhancing WAG projects, the resulting foams are often thermodynamically unstable and have a weak plugging strength compared to gels, limiting their application in the field. For the first time, Wan-fen Pu et al. [20] prepared polyacrylamide colloid (PAASP) by using methylene bisacrylamide as a crosslinking agent. Subsequently, they polymerized and cross-linked CO_2 response monomers, acrylamide, etc., with the first network in situ to form an interpenetrating network polymeric gel (IPN-ASAP), and they revealed its response mechanism. The study showed that the protonation reaction occurred after the contact of IPN-ASAP with CO_2, which made the polymer coil expand. At the same time, due to the hydrophilic change and electrostatic repulsion, more water molecules entered the gel interior and increased its particle size. In laboratory EOR experiments, the injection pressure of IPN-ASAP increased from 5.0 KPa before exposure to CO_2 to 342.0 KPa. The injected CO_2 successfully penetrated the matrix and displaced the oil, resulting in an 8.2% increase in recovery in the tight core (permeability 0.85×10^{-3} μm^2). Aiming at the problem of gas channeling in the late stage of a low-permeability reservoir, Demingley et al. [21] synthesized a CO_2-responsive gel in the laboratory. The results of environmental scanning electron microscopy showed the formation of three-dimensional network aggregates. After protonation, it was confirmed that bola-type amines bridge two anionic surfactants through non-covalent force to form pseudo-gemini surfactants. This causes wormlike micelles to aggregate and tangle with each other, eventually forming a gel system. The results show that the sealing efficiency of the gel system is more than 90%. Yong-Guang Jia et al. [22] synthesized hydrogels with self-healing properties using cholic acid dimers as guest crosslinkers and a β-cyclodextrin polymer as the main body. To provide the system responsiveness, they introduced protonable benzimidazole. Under alternating CO_2/N_2 bubbling, the reversible transition between solution and gel was realized due to the dynamic complexation between cholic acid, benzimidazole and β-cyclodextrin. These laboratory experiments demonstrate the potential of CO_2-responsive polymers in improving reservoir heterogeneity and improving oil displacement efficiency. However, given the application in oilfield construction sites, the monomer cost of CO_2-responsive polymers is high, and the synthesis of polymers with complex supramolecular structures is cumbersome and rigorous for industrial production conditions. Therefore, much research is still needed to verify the applicability of CO_2-responsive polymers in the oilfield field.

As one of the important components of CO_2 geological utilization in CCUS technology, CO_2 flooding is also one of the key technologies for enhanced oil recovery, so it is particularly important to solve the gas channeling problem [23]. CO_2 flooding can effectively utilize and store a large amount of CO_2, thus reducing emissions to the atmosphere, which is of great significance for mitigating climate change. A novel CO_2 response gel channeling system was prepared by using carreenan, branched polyethylenimide and ethylenediamine. The protonation process of amine functional groups was understood by means of microscopic morphology characterization; the reason for the sharp increase in the viscosity of the system after contact with CO_2 was revealed from a microscopic point of view; and the rheological properties of the CO_2 response gel system and the performance of plugging to enhance oil recovery were clarified [24]. This provides a new way to solve the problem of gas channeling in CO_2 flooding technology in low-permeability reservoirs and also brings important development opportunities for CO_2 capture, utilization and storage (CCUS) [25].

2. Results and Discussion

2.1. Preparation of the Gel System

The physical drawings before and after the response of the three aqueous systems to CO_2 exposure are shown in Figure 1. Before contact with CO_2, we have a transparent clear solution with a low viscosity, close to water; however, once fully in contact with CO_2, the viscosity of the system rises sharply, forming a CO_2 responsive gel sealing system. Carrageenan is a negatively charged biopolymer that can interact with positively charged polymers, and carrageenan itself is a thermally reversible physical gel. Branched polyethyleneimine is a polycation, and the polymer chain contains a large number of amine groups, such as primary amines, secondary amines, tertiary amines, etc. [26,27], in the CO_2 due to the acidic conditions of protonation, thereby interacting with the negatively charged polyanion, resulting in the ordered cross-linking of the polymer chain. At the same time, ethylenediamine will chemically react with CO_2 to form organic amine salt (ammonium carbamate) with high viscosity, which further improves the strength of the gel system. The reaction principle is shown in Figure 2 [28].

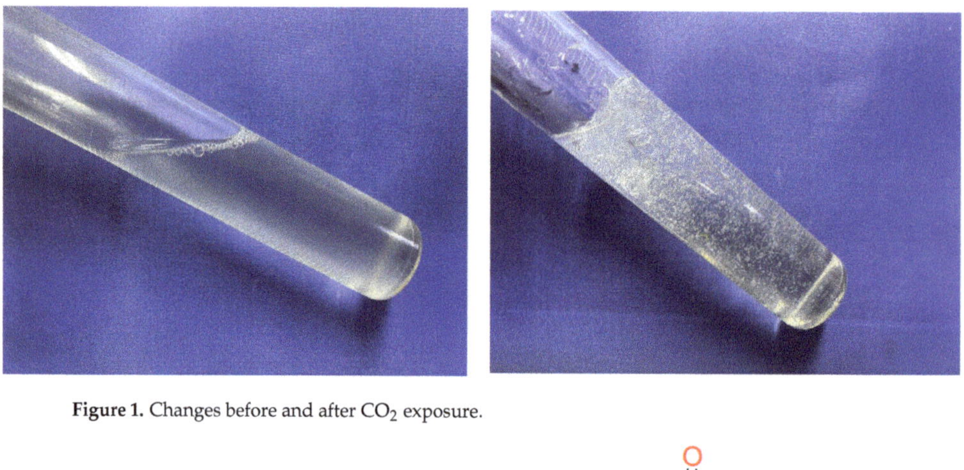

Figure 1. Changes before and after CO_2 exposure.

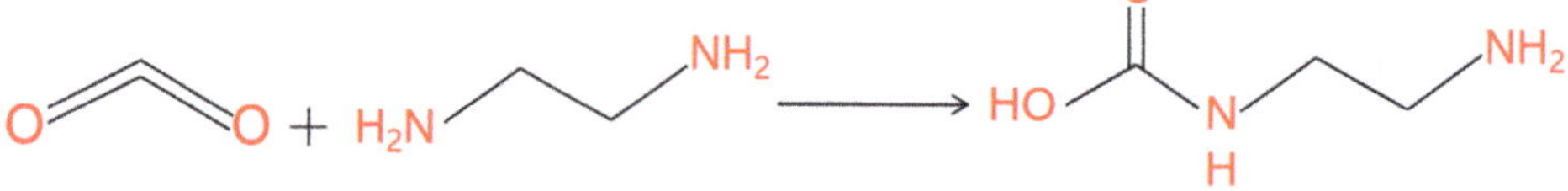

Figure 2. Ethylenediamine principle diagram of the reaction with CO_2.

2.2. Concentration Optimization of Gel System

2.2.1. Results of Response Surface Experiment Design

According to the results of the single-factor experiment, three single-factor variables with the greatest influence on the gel strength were selected. Based on the Box–Behnken response surface design method, an experiment with three factors, three levels and seventeen experimental points was designed. The design and results of the response surface experiment are shown in Table 1.

Table 1. Design and results of response surface experiments.

The No.	A	B	C	Y Intensity Score/(Points)
1	−1	−1	0	6.5
2	1	−1	0	6.6
3	−1	1	0	6.3
4	1	1	0	6.8
5	−1	0	−1	7.2
6	1	0	−1	7.4
7	−1	0	1	7.7
8	1	0	1	8
9	0	−1	1	8.5
10	0	1	−1	7
11	0	−1	1	7.5
12	0	1	1	8.9
13	0	0	0	9.6
14	0	0	0	9.3
15	0	0	0	9.4
16	0	0	0	9.2
17	0	0	0	9.5

2.2.2. Model Establishment and Significance Analysis

Design Expert version 13 software was used to analyze the variance of the experimental results of the response gel, and the quadratic polynomial regression equation obtained by multiple regression fitting of the data in Table 2 is as follows:

$$Y = 9.4 + 0.1375A - 0.0125B + 0.25C + 0.1AB + 0.025AC + 0.725BC - 1.63A^2 - 1.23B^2 - 0.2C^2. \quad (1)$$

In Formula (1), Y—Intensity Score, points; A—Carrageenan Addition, wt%; B—Branched Polyvinylimide Addition, wt%; and C—Ethylenediamine Addition, wt%.

The results of the regression model ANOVA are shown in Table 2 [29].

Table 2 shows that significant items for A and C^2. The most significant terms were C, BC, A^2 and B^2. The insignificant terms were B, AB, and AC. The *p*-value of the model was less than 0.0001, indicating that the model was extremely significant. The loss of fit term was not significant ($p = 0.9559 > 0.05$), indicating that the degree of nonlinear fitting was relatively high. The coefficient of determination ($R^2 = 0.9951$) indicated that the actual experimental values were well correlated with the predicted values. Therefore, this model can better reflect the relationship between monomer addition and gel strength in the system and predict the best reaction conditions. According to the *F* value, the order of influence of the three factors on the gel strength score is C > A > B; that is, the amount of ethylenediamine added > the amount of carrageenan added > the amount of branched polyethylenimine added.

Figure 3 shows the comparison between the predicted values of the model and the actual values. Figure 3 shows that the actual and predicted values of the established model are almost on the line Y = X, indicating that this quadratic regression model can be used for response surface analysis and the optimization of CO_2 responsive gels.

Table 2. Regression model analysis of variance.

Source of Variation	Sum of Squares	Degree of Freedom	Mean Square	*F* Number	*p* Number	Significance
Model	21.71	9	2.41	157.08	<0.0001	**
A	0.1513	1	0.1513	9.85	0.0164	*
B	0.0012	1	0.0012	0.0814	0.7837	
C	0.5000	1	0.5000	32.56	0.0007	**
AB	0.0400	1	0.0400	2.60	0.1506	
AC	0.0025	1	0.0025	0.1628	0.6986	
BC	2.10	1	2.10	136.91	<0.0001	**
A^2	11.12	1	11.12	723.99	<0.0001	**
B^2	6.32	1	6.32	411.43	<0.0001	**
C^2	0.1684	1	0.1684	10.97	0.0129	*
Residual	0.1075	7	0.0154			
Lack of Fit	0.0075	3	0.0025	0.1000	0.9559	
Pure Error	0.1000	4	0.0250			
Cor Total	21.82	16				
R^2	0.9951					
R^2_{Adj}	0.9887					

Note: * indicates significant ($p < 0.05$), ** indicates extremely significant ($p < 0.01$).

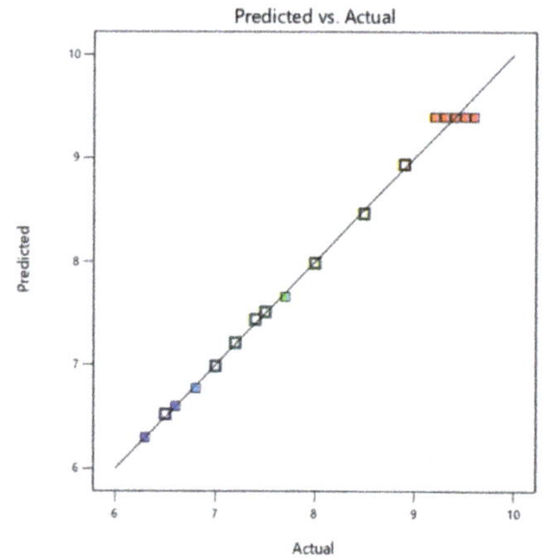

Figure 3. Comparison between the predicted values of the model and the actual values.

2.2.3. Response Surface Analysis

The use of the Design Expert software was used to show interactions between various factors, focusing on their impact on the CO_2 response gel strength of the contour plots and response surface figures, as shown in Figure 4 [30].

Figure 4 shows that response surface figure of the degree of curvature of curve can reflect the study of the strength of the factors that affect the results. The more curved the curve, the greater the influence of this factor on the results. If the contour shows an oval shape, it means that there is a significant interaction between the study factors. If it is circular, the interaction is less significant. Therefore, based on the above analysis of contour lines and response surface plots, it can be concluded that the interaction between the dosage of ramped polyethylenimine and ethylenediamine was extremely significant ($p < 0.01$), and the interaction between other factors was not significant. The single-factor addition of ethylenediamine had the most obvious effect on the strength of the response gel, followed by the addition of carrageenan, and the least effect was produced by the addition of branched polyethylenimine.

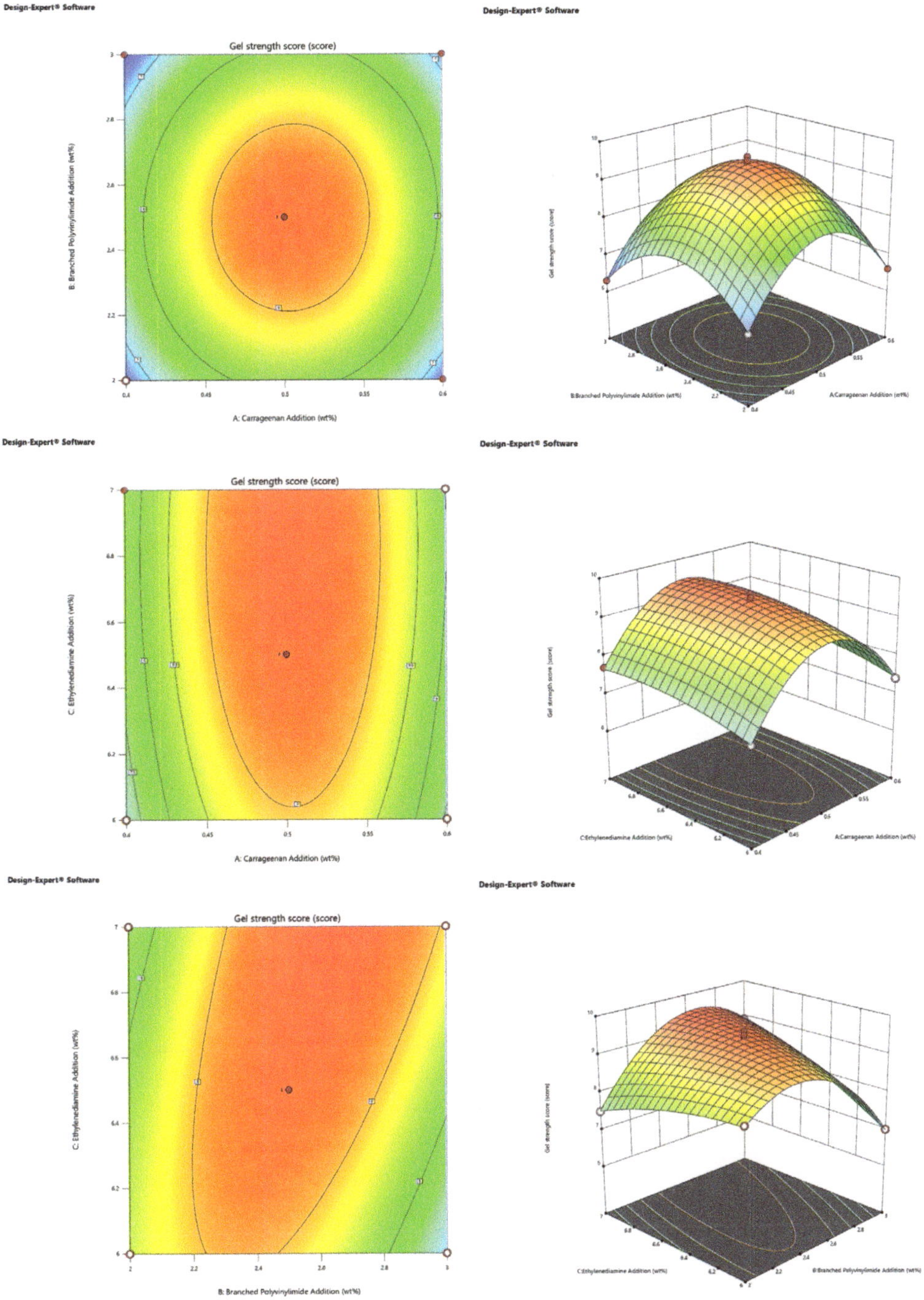

Figure 4. Contour plot and response surface plot of the interaction between the three factors.

2.2.4. Prediction and Verification of Optimal Process Conditions

The experimental data were optimized and predicted by Design Expert software. The best formula of CO_2 response gel was 0.506 wt% for carrageenan, 2.548 wt% for polyethylenimine, and 6.602 wt% for ethylenediamine, and the predicted value of CO_2 response gel strength score was 9.559. Combined with the actual operation, the selection of a carrageenan added amount of 0.5 wt%, a branched polyethylene imine added amount of 2.5 wt %, and an ethylenediamine added amount of 6.5 wt % is close to the model prediction and shows that the model can be used for the CO_2 response to the strength of gel process optimization.

2.3. *Microstructure of the CO_2 Responsive Gel*

Using scanning electron microscopy (SEM) with a carrageenan added amount of 0.5 wt%, a branched polyethylene imine added amount of 2.5 wt%, and an ethylenediamine added amount of 6.5 wt%, as well as a volume ratio of 2:1:1 for gel sealing channeling system components before and after contact with CO_2, we can observe the response of the internal structure through electron microscopy (SEM) images, as shown in Figure 5.

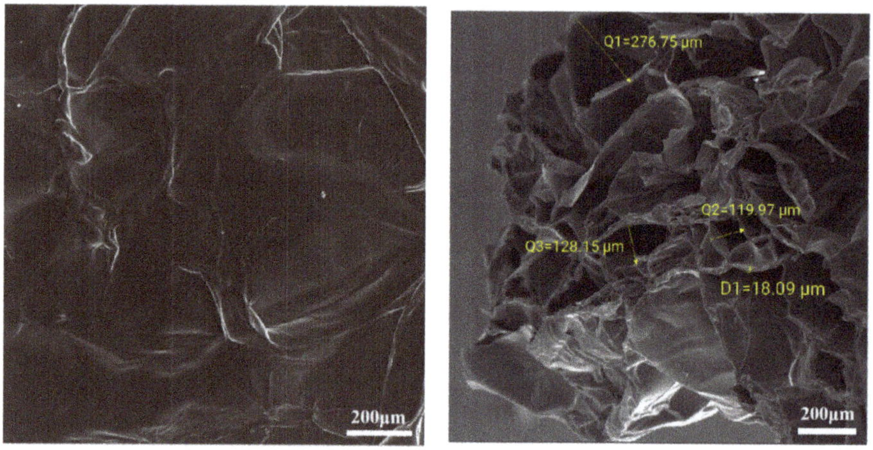

Figure 5. Electron microscopic images of the CO_2 responsive gel system before and after response.

As shown in Figure 5, in the electron microscope picture of the system before CO_2 response, it can be seen that the lamellar structures are disorderly and loosely separated from each other, and the surface is relatively flat and smooth [31]. Before CO_2 gas, carrageenan and branched polyethylenimine, two long-chain polymers, are physically wrapped, the connection points are not evenly distributed, and CO_2 cannot chemically react with ethylene diamine. Due to the low degree of cross-linking, the internal support is insufficient, and the structure is easily damaged by the influence of the external environment. With the penetration of CO_2 gas, the polycation undergoes protonation and interacts with the negative group on the polyanion. This electrostatic interaction causes the whole system to spread out evenly, forming a more uniform, complete and neatly arranged three-dimensional network structure, and the parts are closely connected to each other. In addition, the CO_2 and ethylene diamine reaction increases the system viscosity. Therefore, the overall structure of the gel system is more stable. By measuring the distance between the two reticular structures, it was found that the distance between each two reticular structures was between 15 and 130 μm [32], indicating that the reticular structures were closely arranged and the strength of the gel was stronger, and it was not easy to break.

2.4. Rheological Test

Under 30 °C, a carrageenan added amount of 0.5 wt%, a branched polyethylene imine added amount of 2.5 wt%, and an ethylenediamine added amount of 6.5 wt%, as well as a volume ratio of 2:1:1 for system components before and after contact with CO_2, we can observe the response of viscosity change with the shear rate, as shown in Figure 6. The response gel was scanned at different frequency ranges under a shear stress of 0.2 Pa [33], and the viscoelastic curves are shown in Figure 7.

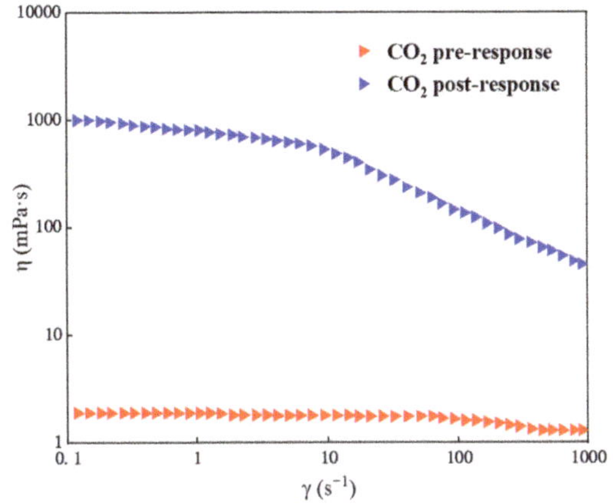

Figure 6. Flow curves of the system before and after CO_2 response.

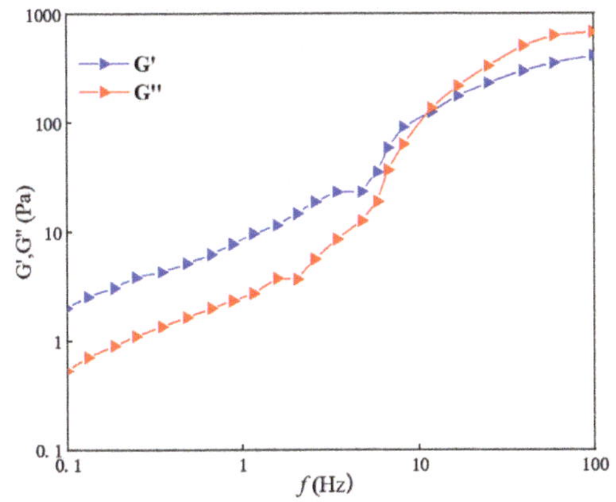

Figure 7. Response gel viscoelastic curves.

Figure 6 shows that before the gel system is exposed to CO_2, the viscosity of the gel is similar to that of water (about 1.7 mPa·s), and the viscosity is almost not affected by the shear rate, which belongs to a typical Newtonian fluid. However, once in contact with CO_2, the viscosity of the gel began to decrease with increasing shear rate, showing the characteristic of shear thinning. When the shear rate is low, within 10 s^{-1}, the change in viscosity is relatively small and the reaction of the gel is weak. But the gel high-speed shear viscosity will be significantly reduced, eventually becoming steady at about 45 mPa·s.

Figure 7 shows that both the storage modulus G' and the loss modulus G'' of the CO_2 response gel increase with the increase in the scanning frequency. In the low-frequency region, the CO_2 response gel energy storage modulus G' is greater than the loss modulus G'', which indicates that system gives priority to a viscoelastic solid and has obvious elasticity, and shear oscillation will not destroy the response to the internal structure of the gel. In the high-frequency region, the two curves intersect, which indicates that a local collapse of the internal structure of the response gel begins, gradually changing from a viscoelastic solid to a viscoelastic liquid. At the intersection point, after CO_2 response gel energy storage modulus G' is less than the loss modulus G'', the system gives priority to a viscoelastic fluid and the observed system becomes obviously thinner; at the same time, it also has a certain viscosity.

2.5. CO_2 Responds to the Dehydration Rate of the Gel

The prepared CO_2 response gel sealing channeling system was placed in an oven at 50 °C, 60 °C, and 70 °C, and the water loss mass of the system was recorded every five days during the aging process, with a total aging time of 30 days. The experimental results are shown in Figure 8.

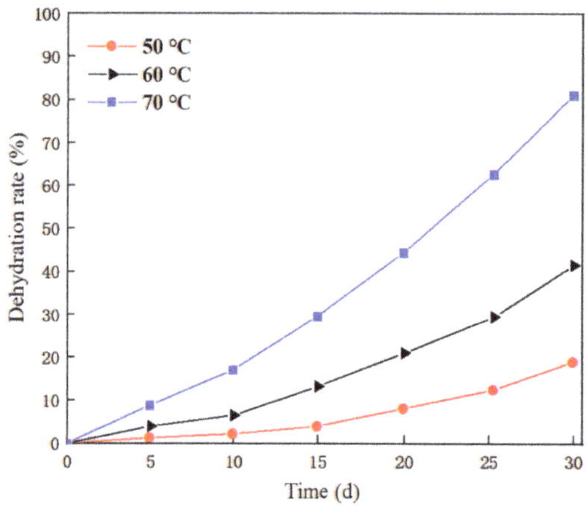

Figure 8. Gel dehydration rates at different temperatures.

Figure 8 shows that with the increase in aging days, the dehydration rate at all three temperatures becomes larger. At 50 °C, the dehydration rate remained below 10% for the initial 15 days, while after 30 days, the dehydration rate of the system eventually stabilized at about 18%. At 60 °C, the dehydration rate of the system did not change much during the first 10 days but gradually increased thereafter, eventually reaching around 40% after 30 days. However, at 70 °C, the dehydration rate of the system began to increase significantly on day 5, showing poor water retention, and the final dehydration rate was around 80% after 30 days. This suggests that high temperature may result in changes in the molecular structure of the polymer, such as the cross-linking of the polymer chain or fracture, thus affecting the material properties. It is worth noting that when the temperature returned to room temperature, the gel system returned to the previous state, indicating that the response gel had some self-repair ability [34,35].

2.6. Response Gel Gas Channeling Plugging Ability

The injectivity and pluggability of the CO_2-responsive gel were evaluated using a core displacement device, and the experimental results are shown in Figures 9 and 10 [36].

Figures 9 and 10 show that during the primary CO_2 displacement stage, gas injection was 1.7 times the core pore volume, and subsequent recovery was stable at 47.6%. When a gel system of 0.4 times the pore volume was injected into the system, the pressure at the inlet of the core gripper rose rapidly and finally stabilized at about 12 MPa [37]. This indicates that during the injection process, the gel-based liquid reacts with the CO_2 gas in the core pores to form a gel system, which effectively seals the large pores in the core, resulting in a recovery rate of 18.13% during the injection phase [38]. Subsequent gas drives the remaining oil in the pore out, and the subsequent stage of gas drives the recovery efficiency up to 17.06%, and then it remained stable. The overall recovery efficiency reached 82.76%. After the CO_2 contact system, the blocking rate of the laboratory reached 98.73% and then decreased during the gas flooding process, and the final blocking rate stabilized at 85.32%, which indicated that the response gel system could effectively block the gas channelization channel.

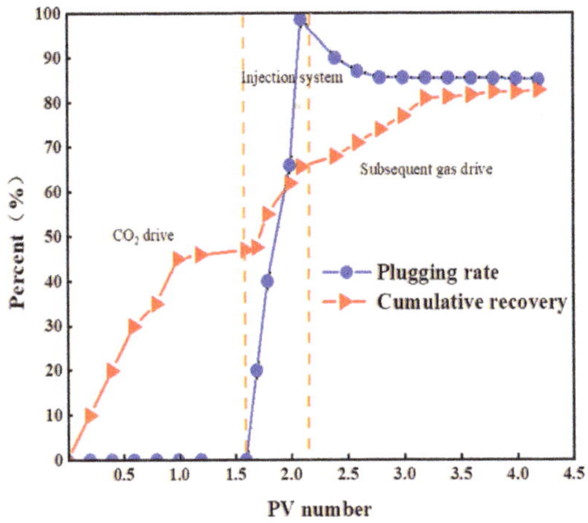

Figure 9. Cumulative recovery and laboratory sealing rate.

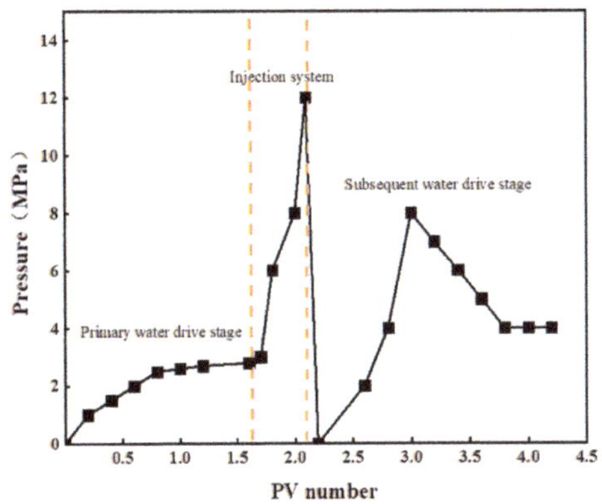

Figure 10. Core displacement pressure changes.

3. Conclusions

(1) A new CO_2 response gel sealing and channelizing system was prepared by using carrageenan, polyethylenimine and ethylenediamine. The Box–Behnken response surface design method was used to establish the regression equation of the relationship between the gel strength score and the addition of the three agents, and the optimal formula of the gel system was obtained: carrageenan added 0.5 wt%, branched polyethylenimine added 2.5 wt%, and ethylenediamine added 6.5 wt%.

(2) Scanning electron microscopy was used to characterize the internal structure of the gel sealing and channeling system before and after contact with CO_2. The surface of the system was smooth before contact with CO_2. As the CO_2 gas bubbled into

the more uniform, complete and neat rows of the 3D mesh structure, each part was closely connected.

(3) Rheological testing was carried out through the gel system. Before CO_2 contact, the gel system behaves as a Newtonian fluid with a viscosity of about 1.7 mPa·s, which is almost unaffected by the shear rate. In the low-frequency region, the gel is mainly a viscoelastic solid, and the storage modulus G' is greater than the loss modulus G'', showing obvious elasticity. In the high-frequency region, with the increase in scanning frequency, the internal structure of the gel changes from a viscoelastic solid to a viscoelastic liquid. The storage modulus G' is smaller than the loss modulus G'', and the system shows viscous characteristics.

(4) The dehydration rate of the response gel at different temperatures was tested. With the increase in aging time, the dehydration rate of the gel system at different temperatures gradually increased, indicating that a high temperature may cause structural changes in the polymer. However, when the polymer was restored to normal temperature, it returned to the original state, and the gel system showed strong self-repair ability.

(5) In the CO_2 displacement process, the gel system successfully blocked the macropores in the core, resulting in a recovery rate of 18.13%. A later stage of gas drives the effective expulsion of the remaining oil in the small pore, producing a recovery efficiency up to 17.06%, and the overall recovery rate reached 82.76%. Laboratory results showed that the plugging rate of the response gel system was as high as 98.73% after CO_2 contact, decreased slightly during gas flooding, and finally stabilized at 85.32%.

4. Materials and Methods

4.1. Experimental Instruments and Raw Materials

A HAAKE RS600 rheometer was provided by Thermo Fisher Technology (Shanghai, China) Co., LTD., with a magnetic stirring device from Shanghai Jiecheng Experimental Instrument (Shanghai, China) Co., LTD., a blast drying oven from Gongyi Yuhua Instrument (Zhengzhou, China) Co., LTD., and a freeze dryer from Sihuan Scientific Instrument (Tianjin, China) Technology Co., LTD, as well as an electronic analytical balance from Nanjing Donglong Machinery Technology (Nanjing, China) Co., LTD and a field emission scanning electron microscope from FEI Company (Houston, TX, USA).

Ethylenediamine and carrageenan (purity 99.99%, analytical purity) were supplied by Aladdin Reagent Shanghai (Shanghai, China) Co., LTD. Branched polyethylenimine (molecular weight 700,000, 50 wt% aqueous solution) was from Beijing Huawei Ruike Chemical (Beijing, China) Co., LTD. and high-purity CO_2 gas (purity 99.99%) was provided by Shaanxi Kuaite Gas (Xi'an, China) Co., LTD. Deionized water was homemade in the laboratory [39].

4.2. Preparation of CO_2-Responsive Gel System

A certain concentration of carrageenan solution was prepared by taking a certain amount of carrageenan powder and dissolving it in a certain amount of deionized water under the condition of 50 °C, using a digital-display constant-temperature mixer for about 30 min. Then, a certain amount of branched polyethylenimide aqueous solution was added to an appropriate amount of deionized water and diluted to a specific concentration. The mixture was placed in a digital-display constant-temperature stirring device for about 45 min to prepare a certain concentration of branched polyethylenimide aqueous solution [40]. Finally, a certain amount of ethylenediamine solution was added to an appropriate amount of deionized water and diluted to a specific concentration. The mixture was stirred in a digital-display constant-temperature stirring device for about 20 min to prepare a certain concentration of ethylenediamine aqueous solution. The ratio of the three aqueous solutions was V (Carrageenan solution):V (branched polythylenimine aqueous solution):V (ethylenediamine aqueous solution) = 2:1:1. The solution system was fully stirred in a digital-display constant-temperature stirring device for about 30 min, and the prepared solution system was transferred to a high-temperature test tube. Then, CO_2 gas was injected at a flow rate of 5 mL/min for 3 min to form a CO_2 responsive gel system.

4.3. System Optimization

The gel system was optimized based on the Box–Behnken response surface design method [41,42]. According to a large number of single-factor experimental results, the optimal concentration of Carrageenan solution was 0.4–0.6 wt% under the condition that three aqueous solutions V (carrageenan solution), V (branched-polyethyleneimine aqueous solution) and V (ethylenediamine aqueous solution) = 2:1:1. The optimal concentration of branched polyethylenimide solution was 2.0–3.0 wt%. The optimal concentration of ethylenediamine aqueous solution is 6.0–7.0 wt%. Therefore, a three-factor, three-level experiment was designed to observe the gel strength based on the GSC visual method proposed by Sydansk, and the gel strength grade was divided into A–J. For convenience, the gel strength score of 1–10 was taken as the response value, and the higher the score, the greater the gel strength. The dosage of carrageenan, branched polyethylenimide and ethylenediamine were selected as variable factors. Test factors and levels are shown in Table 3.

Table 3. Experimental factors and levels of response surface.

Level	Factor		
	A Carrageenan Addition (wt%)	B Branched Polyvinylimide Addition (wt%)	C Ethylenediamine Addition (wt%)
−1	0.4	2.0	6.0
0	0.5	2.5	6.5
1	0.6	3.0	7.0

4.4. Testing and Characterization

(1) The scanning electron microscope. The experimental pre-and post-response system samples were rapidly quenched in liquid nitrogen to maintain their morphology. The samples were subsequently transferred to a freeze-drying oven and freeze-dried at −75 °C for approximately 36 h to obtain freeze-dried samples of the different systems. Next, an appropriate amount of dry sample was prepared as thin slices, and the samples were sprayed with gold. Finally, conductive glue was used to paste the sample on the sample stage for further experimental analysis.

(2) Rheological test. At 30 °C, a HAAKE RS600 rheometer was used to test the system before and after CO_2 response at different shear rates ranging from 0.01 to 1000 s^{-1}, and the viscosity change was observed. We observed the response to CO_2 gels in different frequency ranges, measuring the response of the gel energy storage modulus (G') and loss modulus (G''), the test of a 0.1~100 Hz frequency range, and 0.2 Pa shear stress.

(3) Dehydration rate test. Temperature has a certain effect on the rheological behavior of the gel system. The CO_2 response gel system was selected and placed in an oven at temperatures of 50 °C, 60 °C, and 70 °C, respectively. The dehydration mass was recorded at intervals for a total of 30 days and the dehydration rate was calculated.

4.5. Gas Channeling Plugging Capacity

A sandstone core 10 cm long and 2.5 cm in diameter was used with a permeability of 50×10^{-3} μm^2. The sealing performance of the CO_2 response gel plugging system and improvements in the effect of recovery were studied [43]. First, we determined the core dry weight, vacuum-extracted the formation water, determined the saturation to simulate, and calculated the core pore volume. A saturated oil experiment was then performed to record the saturated volume of oil. A CO_2 displacement experiment was carried out at 70 °C, at a rate of 0.5 mL/min for CO_2 flooding, with gas channeling out until exports. Then, we examined the response at a speed of 0.2 mL/min 0.5 PV of the gel system and waited for the gel formation. Finally, we released the pressure of the injection side, continued to use the 0.5 mL/min rate of CO_2 gas displacement, recorded the change in the pressure, and calculated the total recovery factor and plugging rate [44].

Author Contributions: Conceptualization, H.W. and C.Y.; methodology, H.W.; software, C.Y.; validation, C.Y.; formal analysis, H.W.; investigation, C.W.; resources, C.Y.; data curation, Y.Z.; writing—original draft preparation, H.W.; writing—review and editing, H.W.; visualization, C.W.; supervision, C.W.; project administration, Y.Z.; funding acquisition, C.Y. All authors have read and agreed to the published version of the manuscript.

Funding: Graduate Student Innovation Fund Project of Xi'an Shiyou University (No.: YCX2413057); The Natural Science Basic Research Program of Shaanxi Province (2024JC-YBQN-0426); China Postdoctoral Science Foundation (2023MD744257); State-sponsored Postdoctora Researcher program (GZC20232142); Shaanxi Province Postdoctoral Research Project (2023BSHTBZZ45).

Institutional Review Board Statement: Not applicable.

Informed Consent Statement: Not applicable.

Data Availability Statement: The data presented in this study are openly available in the article.

Conflicts of Interest: The authors declare no conflicts of interest.

References

1. Wang, Z.S.; Zeng, L.B.; Yu, J.T.; Zhang, Z.G.; Yang, S.Y.; Liu, Y.C. Key geological factors controlling oil displacement efficiency of CO_2 injection in low-permeability reservoirs. *Energy Explor. Exploit.* **2021**, *39*, 993–1009. [CrossRef]
2. Yin, D.Y.; Wang, D.Q.; Zhou, Y.Z.; Zhang, C.L. Pore Structure Characteristics of Ultra-Low Permeability Reservoirs. *Nat. Resour. Res.* **2021**, *30*, 451–462. [CrossRef]
3. Liao, H.Y. Application of a Multistage Plugging System in a Fractured Reservoir of CO_2 Flooding. *ACS Omega* **2024**, *9*, 4673–4683. [CrossRef] [PubMed]
4. Ma, F.; He, S.; Zhu, H.; Xie, Q.; Jiao, C. The Effect of Stress and Pore Pressure on Formation Permeability of Ultra-Low-Permeability Reservoir. *Pet. Sci. Technol.* **2012**, *30*, 1221–1231. [CrossRef]
5. Liu, B.L.; Yang, L.; Liu, Y.; Lei, Z.D.; Chen, X.B.; Wang, K.; Yu, Q. Main Factors for the Development of Ultra-Low Permeability Reservoirs. *J. Coast. Res.* **2020**, *104*, 435–439. [CrossRef]
6. Wu, Z.Y.; Feng, Q.H.; Tang, Y.L.; Zhou, D.Y.; Lian, L.M. Experimental Study on Carbon Dioxide Flooding Technology in the Lunnan Oilfield, Tarim Basin. *Energies* **2024**, *17*, 386. [CrossRef]
7. Xu, B. CO_2 miscible flooding in low permeability sandstone reservoirs and its influence on crude oil properties. *Pet. Sci. Technol.* **2017**, *35*, 2024–2029. [CrossRef]
8. Su, B.Y.; Yu, J.C.; Liu, S.C.; Zhang, J.B. Pseudo-Percolation Threshold Theory in evaluation of low-porosity and low-permeability oil reservoir. *Energy Explor. Exploit.* **2014**, *32*, 621–633. [CrossRef]
9. Neuma de Castro Dantas, T.; Viana, F.F.; Thaise Costa de Souza, T.; Dantas Neto, A.A.; Aum, P.T.P. Study of single-phase polymer-alkaline-microemulsion flooding for enhancing oil recovery in sandstone reservoirs. *Fuel* **2021**, *302*, 121176. [CrossRef]
10. Chen, L.L.; Liao, X.W.; Tang, S.F.; Feng, S.Y.; Tang, R.J.; Jiang, S.J.; Dong, Y.W. CO_2-Low Interfacial Tension Viscoelastic Fluid Synergistic Flooding in Tight Reservoirs. *ACS Omega* **2022**, *7*, 6271–6279. [CrossRef]
11. Bai, L.; Shi, C.Y.; Tang, K.; Xie, H.; Yang, S.L.; Zhu, D.Y. Study on migration and plugging performance of polymer gel in fractured cores using nuclear magnetic resonance technology. *Geoenergy Sci. Eng.* **2023**, *227*, 211891. [CrossRef]
12. Bai, Y.Y.; Pu, W.F.; Jin, X.; Shen, C.; Ren, H.L. Review of the micro and Macro mechanisms of gel-based plugging agents for enhancing oil recovery of unconventional water flooding oil reservoirs. *J. Mol. Liq.* **2024**, *399*, 124318. [CrossRef]
13. Gao, Q.; Xu, X.G.; Liu, S.; Mmbuji, A.O.; Wu, Y.K. CO2-Responsive Foam (CRF) for Mitigating CO_2 Channeling in Heterogenous Reservoirs. *Energy Fuels* **2024**, *38*, 3755–3768. [CrossRef]
14. Dong, L.F.; Yue, X.a.; Su, Q.; Qin, W.; Song, W.X.; Zhang, D.X.; Zhang, Y. Study on the Plugging Ability of Polymer Gel Particle for the Profile Control in Reservoir. *J. Dispers. Sci. Technol.* **2016**, *37*, 34–40. [CrossRef]
15. Li, X.D.; Fu, M.L.; Hu, J. Preparation and Performance Evaluation of Temperature-Resistant and Salt-Resistant Gels. *Gels* **2024**, *10*, 337. [CrossRef] [PubMed]
16. Lin, S.; Theato, P. CO_2-Responsive Polymers. *Macromol. Rapid Commun.* **2013**, *34*, 1118–1133. [CrossRef] [PubMed]
17. Darabi, A.; Jessop, P.G.; Cunningham, M.F. CO_2-responsive polymeric materials: Synthesis, self-assembly, and functional applications. *Chem. Soc. Rev.* **2016**, *45*, 4391–4436. [CrossRef] [PubMed]
18. Liu, F.; Urban, M.W. Recent advances and challenges in designing stimuli-responsive polymers. *Prog. Polym. Sci.* **2010**, *35*, 3–23. [CrossRef]
19. Fan, X.Y.; Zhao, P.F.; Zhang, Q.G.; Zhang, T.; Zhu, K.; Zhou, C.H. A Polymer Plugging Gel for the Fractured Strata and Its Application. *Materials* **2018**, *11*, 856. [CrossRef]

20. Han, X.D.; Zhong, L.G.; Liu, Y.G.; Wang, Q.X.; Wang, C. Experimental study of polyacrylamide polymer gel for profile controlling in high-temperature offshore reservoirs. *Energy Sources Part A Recovery Util. Environ. Eff.* **2020**. [CrossRef]
21. Liu, H.B.; Lin, S.J.; Feng, Y.J.; Theato, P. CO_2-Responsive polymer materials. *Polym. Chem.* **2017**, *8*, 12–23. [CrossRef]
22. Leng, G.Y.; Yan, W.; Li, F.; Wang, B.; Yuan, G.Y.; Chu, C.C.; Gao, C.H. Improved oil recovery in sandstone reservoirs with deep profile control technology: A comparative study between modified starch gel and polymer gel. *Geoenergy Sci. Eng.* **2023**, *226*, 211697. [CrossRef]
23. Cui, L.; Hou, J.R.; Yin, X.W. Feasibility of gas drive in Fang-48 fault block oil reservoir. *Pet. Sci.* **2007**, *4*, 51–56. [CrossRef]
24. Dong, S.Y.; Li, L.; Wu, Y.J.; Huang, X.L.; Wang, X.F. Preparation and Study of Polyvinyl Alcohol Gel Structures with Acrylamide and 2-Acrylamido-2-methyl-1-propanesulfonic Acid for Application in Saline Oil Reservoirs for Profile Modification. *ACS Appl. Mater. Interfaces* **2023**, *15*, 14788–14799. [CrossRef] [PubMed]
25. Kang, W.L.; Kang, X.; Lashari, Z.A.; Li, Z.; Zhou, B.B.; Yang, H.B.; Aidarova, S. Progress of polymer gels for conformance control in oilfield. *Adv. Colloid Interface Sci.* **2021**, *289*, 102363. [CrossRef] [PubMed]
26. Zhang, S.; Wei, F.L.; Liu, P.D.; Shao, L.M.; Li, W.T. Plugging mechanisms of polymer gel used for hydraulic fracture water shutoff. *e-Polymers* **2020**, *20*, 411–422. [CrossRef]
27. Zhu, T.; Yang, S.; Choi, D.K.; Row, K.H. Adsorption of carbon dioxide using polyethyleneimine modified silica gel. *Korean J. Chem. Eng.* **2010**, *27*, 1910–1915. [CrossRef]
28. Pan, Y.; Cui, X.L.; Wang, H.; Lou, X.; Yang, S.C.; Oluwabusuyi, F.F. Research Progress of Intelligent Polymer Plugging Materials. *Molecules* **2023**, *28*, 2975. [CrossRef]
29. Pan, C.F.; Chen, Z.; Tang, Q.M.; He, Z.G.; Wang, L.M.; Li, H.H.; Zhou, W.Q. Heat Dissipation Improvement of Lithium Battery Pack with Liquid Cooling System Based on Response-Surface Optimization. *J. Energy Eng.* **2022**, *148*, 04022022. [CrossRef]
30. Shi, L.; Yang, R.-J.; Zhu, P. An adaptive response surface method for crashworthiness optimization. *Eng. Optim.* **2013**, *45*, 1365–1377. [CrossRef]
31. Song, H.K.; Cho, K.W.; Lee, K.H. Adsorption of carbon dioxide on the chemically modified silica adsorbents. *J. Non-Cryst. Solids* **1998**, *242*, 69–80. [CrossRef]
32. Shi, B.; Zhang, G.M.; Zhang, L.; Wang, C.J.; Li, Z.H.; Chen, F.P. Study on a Strong Polymer Gel by the Addition of Micron Graphite Oxide Powder and Its Plugging of Fracture. *Gels* **2024**, *10*, 304. [CrossRef]
33. Yang, H.B.; Iqbal, M.W.; Lashari, Z.A.; Cao, C.X.; Tang, X.C.; Kang, W.L. Experimental research on amphiphilic polymer/organic chromium gel for high salinity reservoirs. *Colloids Surf. A Physicochem. Eng. Asp.* **2019**, *582*, 123900. [CrossRef]
34. Tang, L.H.; Wang, T.Y.; Song, H.F.; Xu, M.M.; Xu, Y.B.; Wang, Y.Y. A temperature-sensitive plugging material composed of shape memory polymer and self-made gel. *Mater. Res. Express* **2023**, *10*, 065701. [CrossRef]
35. Wang, Q.Y.; Cai, J.; Wang, J.N.; Zhou, C.H.; Wen, X.X.; Zhang, J.; Mao, H. Development and Application of the Anti-High-Temperature Delayed Crosslinking Polymer as a Gel Plugging Additive for Drilling Fluid. *Gels* **2024**, *10*, 73. [CrossRef] [PubMed]
36. Zhang, D.; Wei, J.G.; Fang, X.R. Study on the variation of rock pore structure after polymer get flooding. *e-Polymers* **2020**, *20*, 32–38. [CrossRef]
37. Tong, L.; Fan, R.G.; Yang, S.C.; Zhang, Q.S.; Pan, Y. Experimental Study of Oilfield Water Plugging with Organic Chromium Gel. *Chem. Technol. Fuels Oils* **2022**, *57*, 991–999. [CrossRef]
38. Wang, K.L.; Guo, Y.Q.; Wen, J.; Yang, H.; Zhang, H. Magnetic smart polymer gel with directional plugging for conformance control in oil reservoirs. *J. Mol. Liq.* **2024**, *405*, 125046. [CrossRef]
39. Zhang, Y.; Cai, H.Y.; Li, J.G.; Cheng, R.; Wang, M.J.; Bai, X.; Dai, C.L. Experimental study of acrylamide monomer polymer gel for water plugging in low temperature and high salinity reservoir. *Energy Sources Part A Recovery Util. Environ. Eff.* **2018**, *40*, 2948–2959. [CrossRef]
40. Huang, B.; Zhang, W.S.; Zhou, Q.; Fu, C.; He, S.B. Preparation and Experimental Study of a Low-Initial-Viscosity Gel Plugging Agent. *ACS Omega* **2020**, *5*, 15715–15727. [CrossRef]
41. Costa, M.C.; Pereira, M.L.; Maréchal, Y.; Coulomb, J.L.; Cardoso, J.R. Optimization of grounding grids by response surfaces and genetic algorithms. *IEEE Trans. Magn.* **2003**, *39*, 1301–1304. [CrossRef]
42. Lee, D.-H.; Jeong, I.J.; Kim, K.-J. Methods and Applications of Dual Response Surface Optimization: A Literature Review. *J. Korean Inst. Ind. Eng.* **2013**, *39*, 342–350.
43. Lyu, C.H.; Zhong, L.G.; Wang, Q.; Zhang, W.; Han, X.D.; Chen, M.Q.; Yang, J.W. Core scale analysis of low viscosity oil injection in enhancing oil recovery of heavy oil reservoirs. *Energy* **2023**, *275*, 127432. [CrossRef]
44. Beeson, D.M.; Ortloff, G.D. Laboratory investigation of the water-driven carbon dioxide process for oil recovery. *Trans. Am. Inst. Min. Metall. Eng.* **1959**, *216*, 388–391. [CrossRef]

gels

MDPI

Article

High-Temperature-Resistant Profile Control System Formed by Hydrolyzed Polyacrylamide and Water-Soluble Phenol-Formaldehyde Resin

Xuanran Li [1], Shanglin Liu [1], Juan Zhang [2,3,*], Shujun Han [1], Lun Zhao [1], Anzhu Xu [1], Jincai Wang [1], Fujian Zhou [2,3] and Minghui Li [1,2,3]

1 Research Institute of Petroleum Exploration & Development, Beijing 100083, China
2 National Key Laboratory of Petroleum Resources and Engineering, China University of Petroleum (BeiJing), Beijing 102249, China
3 Unconventional Petroleum Research Institute, China University of Petroleum (Beijing), Beijing 102249, China
* Correspondence: juanzhang@cup.edu.cn

Abstract: To realize the effective profile control of a heavy oil reservoir, hydrolyzed polyacrylamide (HPAM) and water-soluble phenol-formaldehyde resin (PR) were chosen to prepare the profile control system, which gelled at medium or low temperatures and existed stably at high temperatures in the meantime. The effects of phenolic ratios, PR concentration, and HPAM concentration on the formation and strength of the gels were systematically studied by the gel-strength code method and rheological measurements. And the microstructure of the gels was investigated by scanning electron microscope measurements. The results showed that the gelling time of the HPAM-PR system was 13 h at 70 °C. The formed gel could stay stable for 90 days at 140 °C. In addition, the gels showed viscoelastic properties, and the viscosity reached 18,000 mPa·s under a 1.5 s^{-1} shearing rate due to their three-dimensional cellular network structure. The formation of the gels was attributable to the hydroxyl groups of the PR crosslinking agent, which could undergo the dehydration condensation reaction with amide groups under non-acidic conditions and form intermolecular crosslinking with HPAM molecules. And the organic crosslinker gel system could maintain stability at higher temperatures because covalent bonds formed between molecules.

Keywords: heavy oil reservoir; profile control; hydrolyzed polyacrylamide; phenol-formaldehyde resin; gel

1. Introduction

In recent years, the steam-driven recovery of thick oil reservoirs has received increasing attention. In order to improve the non-homogeneous formation environment, some technologies have been performed, such as acidizing, profile control technology et al. [1,2]. Deep profile control technology is achieved by injecting a conditioning agent to seal the hypertonic layer. There are many types of plugging agents, among which gel-type plugging agents are most widely used in practical applications [3]. The gel-plugging agent is usually made of the polymer as the main agent, a crosslinking agent, stabilizer, additives, etc., which is relatively inexpensive, stratigraphically selective, and has a good plugging effect. During the injection process, the crosslinking solution is prioritized to enter the high-permeability layer to form gel blocking. During the migration of the gel in the pore throat, the gel is stretched due to the space limitation of the pore throat. Meanwhile, the gel can change shape to pass through the pore throat; thereby, it has a strong flow capacity. And the strength of the gel can prevent the occurrence of the scuttling phenomenon so that the steam enters the medium and low permeability reservoirs and improves the recovery rate of the reservoir [4]. High-temperature steam experiences heat loss during the injection process and becomes hot water or steam at 140 °C~190 °C after reaching the target formation. In order to adapt to the high-temperature environment after steam throughput, the gel-plugging agent should meet the following two requirements: firstly, a good gel-forming performance in low and medium-temperature formations to meet the plugging of high-permeability formations; secondly, the performance of the gel should remain stable in the high-temperature steam environment [5,6].

At present, the gel-plugging technology applicable to medium- and low-temperature reservoirs with formation temperature no higher than 90 °C is relatively mature, and the most widely used gel-plugging agents mainly include inorganic plugging gels, PAM/metal ion crosslinking gels, PAM/organic crosslinking agent gels, etc. However, these gel-type plugging systems show clear shortcomings in their comprehensive performance. On the one hand, the gel is good, but the maximum temperature resistance does not exceed 130 °C; on the other hand, the temperature performance is good, but the gel is too fast (<30 min) and cannot realize the effective injection of the scene [7]. For high-temperature reservoirs with temperatures higher than 90 °C, in order to improve the non-homogeneous environment of the formation and enhance the oil recovery effect, the research and development of the high-temperature profiling agent has become an important topic. In order to ensure the sealing effect on the high permeability layer, such regulators are required to have good temperature resistance [8]. However, the currently reported high-temperature-resistant regulator system has too high gel-forming temperatures due to the organic crosslinking agent and cannot be gel-forming at low temperatures [9]. Therefore, there is an urgent need to develop a gel-plugging agent with high gel formation strength at low and medium temperatures and good temperature resistance at high temperatures in order to satisfy the demand for the steam throughput of thick oil reservoirs and to improve crude oil recovery [10].

The commonly used gel crosslinking systems mainly include the HPAM/inorganic metal ion gel system and phenolic resin organic crosslinking system [11]. The HPAM/inorganic metal ion gel system is usually used as an anatomical water-plugging agent or fracturing fluid in oilfield production, and the reaction mechanism is that carboxylate and high-valent metal ions in the polyacrylamide undergo a crosslinking reaction through ligand bonds [12–15], among which the most reported related studies use a chromium acetate crosslinker. DiGiacomo [16] characterized polyacrylamide and trivalent chromium ion gels by a ^{13}C nuclear magnetic resonance based on the paramagnetism of Cr^{3+} and found that chromium ions in the complexed or free state crosslinked with carboxylates on HPAM molecules through ligand bonding to form a three-dimensional reticulated gel skeleton, yielding a strong gel with good viscoelasticity. The chromium ion crosslinking agent is usually used in reservoirs with formation temperatures below 70 °C. Too-high gel formation temperatures will lead to a violent reaction, and it is difficult to control its gel formation time and strength. Due to the fast reaction rate of Cr^{3+} and the polymer, the crosslinking time can be prolonged by introducing divalent chromium ions together with the use of reducing agents to reduce hexavalent chromium ions to trivalent chromium ions in the formation or injection process in the related research [17]. However, due to the carcinogenic properties of divalent chromium ions, the complexes of Cr^{3+} (chromium acetate [18,19], chromium malonate, etc.) are usually used as a substitute, which can also effectively delay the gel formation time. The ligand bonds formed by COO- and Cr^{3+} in chromium gels are easy to break at high temperatures above 100 °C. Compared with metal ion crosslinker gel systems, organic crosslinker gel systems are able to remain stable at higher temperatures through the formation of covalent bonds between molecules [20]. One of the most widely used crosslinkers is phenolic resin crosslinkers and commonly used phenols include phenol, hydroquinone, and resorcinol, etc., and commonly used aldehydes include formaldehyde and hexamethylene tetramine (urotropin). The phenolic resin gel conditioning system has good temperature resistance regarding the gel formed at high temperatures, but its gel formation is slow and weak at low temperatures, which is due to the high activation energy of the organic reactive groups and the lower gel formation temperature makes the crosslinking reaction proceed slowly [21,22].

Therefore, in this paper, starting from the synthesis method of phenolic resins, a resin crosslinker with higher activity was synthesized by adjusting the phenolic-to-formaldehyde ratio, which, in turn, improved its gel formation rate at low temperatures to prepare a high-strength temperature-resistant gel system. A temperature-resistant HPAM was selected, and the properties of HPAM and different phenolic resin crosslinker systems in terms of gel-forming strength, gel-forming time, dehydration rate, and stability at different temperatures were investigated to prepare a PR crosslinking system that was low-cost, highly stable, and suitable for steam drive in medium-to low-temperature reservoirs.

2. Results and Discussion

2.1. Effect of Phenolic Ratio of the PR Crosslinkers on the Gel Formation and Strength

The gelation of the 0.8 wt% PR crosslinker and HPAM with different concentrations at 50 °C and the stability of the obtained gels at 140 °C are shown in Table 1. It was seen that the bigger the $r_{F/P}$ value, the stronger the gel strength. When $r_{F/P}$ was one, the maximum gel strength was E, while when $r_{F/P}$ was three, the maximum gel strength was G at 50 °C. This was because, with the increase in the formaldehyde amount, the same volume of resin contained more hydroxymethyl groups, which had high chemical activity. Under alkaline conditions, these hydroxymethyl groups could react with the amide groups on the polyacrylamide molecular chain, resulting in crosslinking between molecules, thereby improving the reaction rate and enhancing the strength of gel formation. The gels with high strength (≥F) were stored at 140 °C to evaluate their high-temperature-resistance performance. Dehydration occurred for the gels with 0.6 wt% HPAM. The strength of the gels with 0.8 wt% HPAM and $r_{F/P} \leq 3$ was D grade after 30 days. It was worth noting that the strength of the gels with $r_{F/P} = 4$ reached E grade even after 90 days. Moreover, the $r_{F/P}$ had a great effect on the gelation time. In particular, when the HPAM concentration was 0.6 wt%, with the increase in $r_{F/P}$, the gelation time decreased from 160 h to 31 h. It could be seen that the overall reaction time of the PR-HPAM gel was still slow at the reaction temperature of 50 °C. Increasing the $r_{F/P}$ to improve the reactivity of oligomers of resin crosslinkers was an effective method to accelerate the reaction rate and increase the gel strength at medium and low temperatures.

Table 1. Optimizing phenolic ratios for the PR gel system (T = 50 °C).

HPAM Conc.	$r_{F/P}$	Gelation Time/h	0 Day	140 °C 30 Days	60 Days	90 Days
0.4%	1:1	186	C	-	-	-
	1:2	132	C	-	-	-
	1:3	75	D	-	-	-
	1:4	38	D	-	-	-
0.6%	1:1	160	E	-	-	-
	1:2	118	E	-	-	-
	1:3	63	F	Dehydration	Dehydration	Dehydration
	1:4	31	F	Dehydration	Dehydration	Dehydration
0.8%	1:1	124	E	-	-	-
	1:2	90	F	F	C	-
	1:3	47	G	F	D	D
	1:4	24	G	F	F	E
1.0%	1:1	102	E	-	-	-
	1:2	72	F	F	C	-
	1:3	36	G	F	E	D
	1:4	16	G	F	F	E

The gelation of the 0.8 wt% PR crosslinker and HPAM with different concentrations at 90 °C and the stability of the obtained gels at 140 °C are shown in Table 2. The gelation temperature also had clear effects on the gelation time, gel strength, and high-temperature resistance performance. The influence of the $r_{F/P}$ on the profile control system at 90 °C was clearer than that at 50 °C, which reflected the higher gel strength, faster gelation time, and longer temperature resistance at 140 °C. This was because in the high-temperature environment, the activation energy of the amide group of polyacrylamides and the hydroxymethyl in the resin was significantly increased, and the intermolecular movement was more intense, which promoted the crosslinking and entanglement of the thermoplastic skeleton of the water-soluble resin and the polyacrylamide skeleton, and thus improved the overall stability of the gel. It could be seen that when the HPAM concentration was 0.8 wt%, with $r_{F/P}$ increasing, the gel strength increased from E to G grade, and the shear viscosity increased from 10,000 mPa·s to 19,000 mPa·s, as shown in Figure 1. In addition, when the HPAM concentration was higher than 0.8 wt%, the gel with $r_{F/P} < 2$ showed poor temperature resistance, while the gel strength was F when $r_{F/P} = 4$. Therefore, the ideal $r_{F/P}$ was four in order to obtain the profile gel system with high gel strength and good temperature resistance performance.

Table 2. Optimizing phenolic ratios for the PR gel system (T = 90 °C).

HPAM Conc.	$r_{F/P}$	Gelation Time/h	140 °C 0 Day	30 Days	60 Days	90 Days
0.4%	1:1	15	D	-	-	-
	1:2	13	D	-	-	-
	1:3	10	E	-	-	-
	1:4	8	E	-	-	-
0.6%	1:1	12	E	-	-	-
	1:2	10	F	D	B	-
	1:3	8	F	E	E	C
	1:4	5	G	Dehydration	Dehydration	Dehydration
0.8%	1:1	10	E	-	-	-
	1:2	8	F	F	C	-
	1:3	6	G	F	E	C
	1:4	4	G	G	F	F
1.0%	1:1	8	F	E	C	-
	1:2	6	G	F	C	-
	1:3	5	G	F	F	C
	1:4	3	H	G	F	F

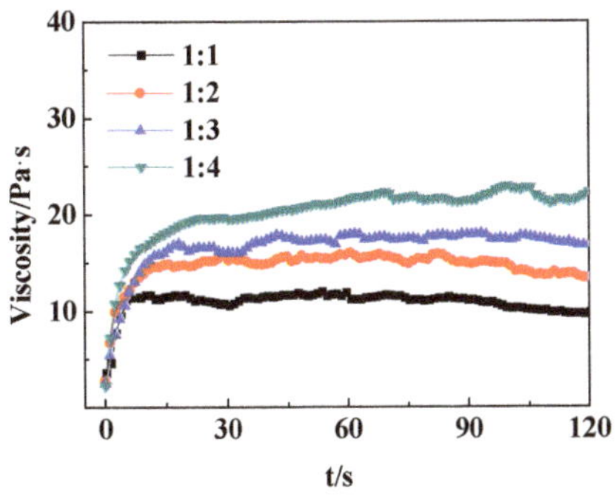

Figure 1. Effect of different phenolic ratios on gel viscosity (0.8 wt% HPAM + 0.8 wt% PR, T = 90 °C).

2.2. Effect of HPAM Concentration on the Gel Formation and Strength

The gelation of 0.8 wt% PR with $r_{F/P} = 4$ and HPAM with different concentrations at 90 °C and the stability of the obtained gels at 140 °C are shown in Table 3. The results show that the gel strength increased with the increase in the HPAM concentration. When the

HPAM concentration was ≥0.6 wt%, the gel strength was G or H grad. However, the gel with 0.6 wt% HPAM became severely dehydrated at 140 °C. The gels with 0.6 and 0.8 wt% HPAM showed good temperature resistance and remained stable after 90 days. Figure 2 presents the shearing viscosities of the gels with 0.8 and 1.0 wt% HPAM at about 20 Pa·s and 30 Pa·s, respectively, which meets the viscosity and strength requirements of the profile control system.

Table 3. Optimizing polymer concentration for PR gel system (T = 90 °C).

$r_{F/P}$	HPAM Conc.	Gelation Time/h	0 Day	140 °C 30 Days	60 Days	90 Days
1:4	0.4%	8	E	-	-	-
	0.6%	5	G	Dehydration	Dehydration	Dehydration
	0.8%	4	G	G	F	F
	1.0%	3	H	G	F	F

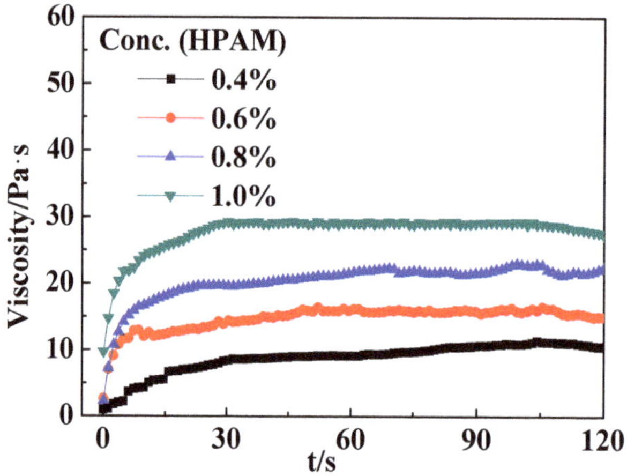

Figure 2. Effect of different polymer concentrations on gel viscosity (T = 90 °C; $\gamma = 1\ s^{-1}$).

2.3. Effect of PR Crosslinker Concentration on the Gel Formation and Strength

In total, 0.8 wt% of the HPAM solution and the 0.3 wt%–1.0 wt% PR crosslinker ($r_{F/P} = 4$) were mixed together to gel at 50 °C, 70 °C, and 90 °C, and the gelation state was observed. Gels with a strength greater than F were placed in a constant temperature oven at 140 °C to assess their thermal stability. The results are shown in Table 4. It was observed that the greater the content of the PR crosslinker, the higher the gel strength. When the PR resin mass fraction was 0.3 wt%, the gelation rate of the mixture at 50 °C and 70 °C was slow, indicating that the low concentration of the crosslinker and the low crosslink density with the polymer backbone resulted in weak gel strength. When the PR resin mass fraction was ≥0.8 wt%, i.e., when the mass ratio of PR resin to polymer was one, the gel strength reached the G level. Therefore, the PR crosslinker mass fraction needed to be at least 0.8 wt%. The thermal stability test results show that when the PR crosslinker mass fraction was 0.8 wt%, the gel exhibited good thermal stability and retained E-level strength after 90 days. When the PR crosslinker mass fraction was 1.0 wt%, the gel was dehydrated at high temperatures, and the gel volume shrank significantly. Considering all factors, the optimal concentration of the PR crosslinker was determined to be 0.8 wt%, at which point the mass ratio of HPAM to crosslinker was exactly one.

Table 4. Optimizing crosslinker PR concentration for the PR gel system.

Temperature	PR Conc.	Gelation Time/h	140 °C			
			0 Day	30 Days	60 Days	90 Days
50 °C	0.3%	-	C	-	-	-
	0.5%	39	E	-	-	-
	0.8%	24	G	F	F	E
	1.0%	20	G	Dehydration	F	E
70 °C	0.3%	-	D	-	-	-
	0.5%	20	E	-	-	-
	0.8%	13	G	G	F	E
	1.0%	10	G	Dehydration	Dehydration	Dehydration
90 °C	0.3%	7	E	-	-	-
	0.5%	6	F	E	B	B
	0.8%	4	G	G	F	E
	1.0%	3	H	Dehydration	Dehydration	Dehydration

According to the above experimental results, when the resin crosslinking agent content was low, the crosslinking with the active sites on the polymer was incomplete, resulting in an unstable gel structure. Conversely, when the crosslinking agent content was too high, the actual number of crosslinking points involved in the reaction exceeded the number required for forming a stable three-dimensional structure, leading to a deterioration in the stability of the gel structure.

2.4. Rheological Properties of the Gels

A crosslinking system was prepared with 0.4 wt%–1.0 wt% HPAM and 0.08 wt% PR resin ($r_{F/P}$ = 4:1). The system was placed in an oven at 70 °C for gelation over 3 days. The viscoelasticity of the formed gel was measured using a HAAKE rheometer, and the test results are shown in Figure 3. From Figure 3, it can be seen that in the frequency sweep range of 0–10 Hz, the elastic modulus G' and the viscous modulus G'' values of the PR gels showed typical properties of gels. Thhe elastic modulus was higher than the viscous modulus within the measured range. As the HPAM concentration increased from 0.4 wt% to 1.0 wt%, the G' values of the gels initially increased uniformly and then increased slowly, indicating that the gel strength was significantly enhanced by a moderate increase in polymer concentration. When the HPAM concentration reached 0.8 wt%, the gel strength began to increase slowly, suggesting that the gel became well crosslinked at this concentration. The viscous modulus G'' of the gel showed a trend of first increasing and then decreasing with the increase in shear frequency, likely due to the shear thinning of the gel under high-frequency shear [23]. As the polymer concentration increased, the G'' values of the gel increased uniformly, indicating that the increase in polymer concentration promoted higher viscosity. This was consistent with the results observed using the visual inspection method.

The effects of PR crosslinker concentration on the rheological properties of the gels are shown in Figure 4. With the increase in the PR crosslinker concentration, both G' and G'' increased. At the same time, both the elastic modulus and the viscous modulus increased with the increase in frequency, and the amplitude of the increase remained the same, which is a typical rheological property of the hydrogel phase.

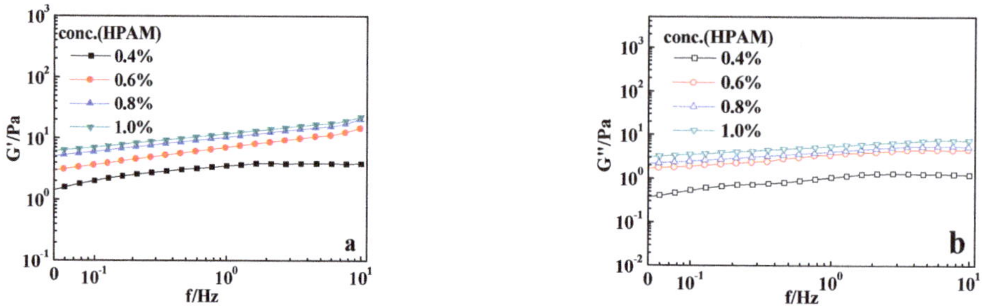

Figure 3. Oscillatory rheological results of the typical samples with different HPAM concentrations: (**a**) the elastic modulus G′; (**b**) the viscous modulus G″. The PR resin ($r_{F/P}$ = 4:1) concentration was 0.08 wt%. T = 90 °C.

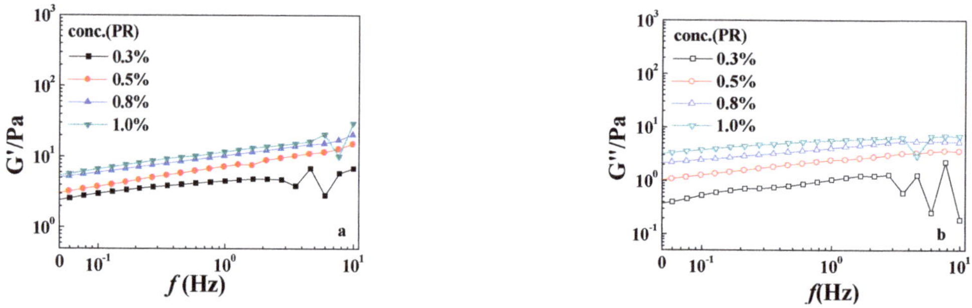

Figure 4. Oscillatory rheological results of the typical samples with different PR concentrations ($r_{F/P}$ = 4): (**a**) the elastic modulus G′; (**b**) the viscous modulus G″. The HPAM concentration was 0.8 wt%.

2.5. Microstructure of the Gels

The microstructures of the sample with 0.8 wt% HPAM and 0.8 wt% PR ($r_{F/P}$ = 4) before and after gelation at 90 °C are shown in Figure 5. Figure 5a shows that the solution before gelation displayed a chaotic branch-shaped mesh structure with sparse skeletons and large pores. However, after gelation, as shown in Figure 5b, the gel represented a high-density three-dimensional mesh structure similar to a honeycomb, which was attributed to its high viscoelasticity and high-temperature stability. Furthermore, it can be seen from Figure 5c,d that the gel had tight connections between the pores, a higher density of three-dimensional pores, and pore sizes ranging from 20 to 40 μm, which also made the gel more stable at high temperatures.

Based on the above results, it was speculated that the formation of the gels was attributable to the hydroxyl groups of the PR crosslinking agent, which could undergo a dehydration condensation reaction with amide groups under non-acidic conditions and undergo intermolecular crosslinking with HPAM molecules [20]. And the organic crosslinker gel system could maintain stability at higher temperatures because covalent bonds formed between the molecules.

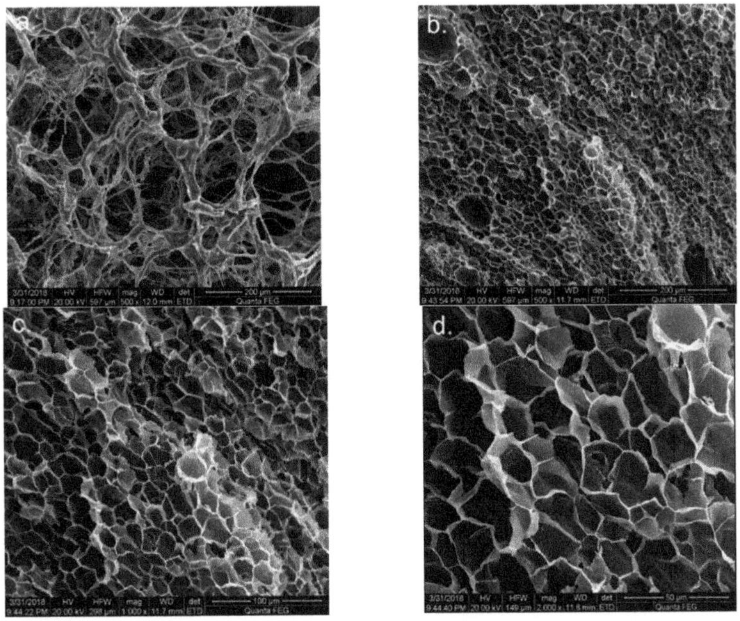

Figure 5. The microstructures of the gel system with 0.8 wt% HPAM and 0.8 wt% PR ($r_{F/P}$ = 4): (**a**) the SEM image of the solution before gelation; (**b–d**) the SEM images of the solution after gelation with different magnification.

3. Conclusions

The phenolic ratios, PR concentration, and HPAM concentration had great effects on the formation and strength of the gel profile control system. Specifically, with the increase in the phenolic ratio, PR concentration, and HPAM concentration, the gel strength was significantly enhanced, and a longer temperature resistance time was obtained. The preferred formulation of the gel system was 0.8 wt% HPAM and 0.8 wt% of the PR crosslinker, which could stay stable for 90 days at 140 °C. In addition, the viscosity of the gel system reached 18,000 mPa·s under a 1.5 s^{-1} shearing rate due to its three-dimensional cellular network structure with a pore diameter of 20 μm.

4. Materials and Methods

4.1. Materials

Phenol (purity > 99%) was purchased from Beijing Fuchen Chemicals Co., Ltd. (Beijing, China). Formaldehyde (37% solution in water) and calcium silicate (purity > 99%) were obtained from Beijing Yili Fine Chemicals Co., Ltd. (Beijing, China) Sodium hydroxide (purity > 99%) was bought from Beijing Modern Oriental Fine Chemicals Co., Ltd. (Beijing, China) HPAM with a molecular weight of 5–7 million and hydrolysis degree of 25% was obtained from Aisen Flocculant (Taizhou, China) Co., Ltd. Deionized water was made in a laboratory.

Different types of water-soluble PR crosslinkers were synthesized by adjusting the ratio of formaldehyde and phenol ($r_{F/P}$) to 1:1, 1:2, 1:3, and 1:4, respectively. Firstly, the melted phenol was placed in a three-necked, round-bottomed flask with the water running through the condensing pipe. Then, the reaction catalyst NaOH solution was dropped into the flask and reacted with phenol for 20 min under stirring conditions. Nitrogen was continuously injected into the reaction system at a certain rate during the reaction process. After the NaOH drops were added, the temperature of the water bath was raised to 75.5 °C. Next, a certain amount of formaldehyde solution was dropped into the flask with a constant pressure drip funnel, and the reaction was carried out for 3 h to prepare the base-catalyzed phenolic resin. The obtained product was stored at 5 °C.

4.2. Preparation of the Gels

A certain mass of HPAM was added into deionized water, which was then stirred continuously for 4 h at the stirring rate of 400 r/min until the solution was homogeneous and transparent. Then, the solution was stored for 12 h to make it fully hydrolyze. Next, a certain amount of PR crosslinkers and heat stabilizer was added to the polymer solution, which was stirred at 400 r/min. The mixed solution was stirred for 2 h and then injected into ampoules, which were vacuumed and sealed. All the ampoules were then stored under a set temperature to observe the gelation of the solutions.

4.3. The Gel-Strength Code

The strength of the gels was evaluated by the gel-strength code reported by Sydansk [24], as shown in Table 5. When using the gel-strength code method, the same volume of the samples should be placed in the same container for observation. Specifically, the gel strength was evaluated visually by reversing the ampoule containing the gel at 180° and observing the flow state of the gel inside the bottle.

Table 5. The gel-strength code [24].

Code	Gel Types	Gel Strength Description
A	No detectable gel formed	The gel appears to have the same viscosity (fluidity) as the original polymer solution, and no gel is visually detectable.
B	Highly flowing gel	The gel appears to be only slightly more viscous (less fluid) than the initial polymer solution.
C	Flowing gel	Most of the clearly detectable gel flows to the bottle cap upon inversion.
D	Moderately flowing gel	Only a small portion (about 5 to 15%) of the gel does not readily flow to the bottle cap upon inversion, which is usually characterized as a "tonguing" gel.
E	Barely flowing gel	The gel can barely flow to the bottle cap and/or a significant portion (>15%) of the gel does not flow upon inversion.
F	Highly deformable nonflowing gel	The gel does not flow to the bottle cap upon inversion.
G	Moderately deformable nonflowing gel	The gel flows about halfway down the bottle upon inversion.
H	Slightly deformable nonflowing gel	The gel surface only slightly deforms upon inversion.
I	Rigid gel	There is no gel-surface deformation upon inversion.
J	Ringing rigid gel	A tuning-fork-like mechanical vibration can be felt after tapping the bottle.

4.4. Evaluation Method of Gel Formation Time

The ampoule containing the solution was put in a constant temperature oven to observe the gel formation process every hour. The gel formation time was identified when the ampoule was tilted at 45°, and the liquid surface appeared uneven. The final gelation time was identified when the ampoule was placed horizontally for 30 s, and the gel did not flow.

4.5. Evaluation Method of High-Temperature Stability of Gels

The high-temperature stability of the gel was evaluated by the visual inspection method. The ampoule that had been gelatinized at a low temperature was placed in a high-temperature oven at 140 °C. The ampoule was turned over at certain intervals to record the gel flow, degradation and dehydration, and the change in gel strength was determined by the strength code method; the high-temperature stability of the gel was also evaluated by observing the gel strength, and dehydration rate.

4.6. Rheological Properties of Gels Measurements

The experiment was carried out by the RS600 rheometer (Thermo Fisher Scientific, Waltham, MA, USA), the cone-plate-measuring rotor system C60/1°Ti was selected, and the experimental measurements were all carried out at 25 °C. The test sampling process should avoid the use of syringes and other apparatus that would produce pre-shear interference with the sample. Therefore, a small spoon was used to take a spoonful of the sample placed in the center of the sample stage. Conditions for the determination of the viscosity of the gel conditioner are as follows: temperature T = 25 °C and shear rate γ = 1.5 s^{-1} under the continuous measurement of 120 s to obtain the viscosity curve of the gel, which tends to stabilize the viscosity of the gel. The conditions for the determination of the viscoelasticity of the gel conditioner are as follows: the viscoelastic curve of the gel is obtained after measuring the frequency f = 0.1–10 Hz under the temperature T = 25 °C and stress τ = 1 Pa, which is the viscoelastic modulus curve of the gel.

4.7. Scanning Electron Microscope (SEM) Measurements

A Quanta 200F (FEI Company, Hillsboro, OR, USA) scanning electron microscope was used for observation, with the accelerating voltage set to 20 kV and the scanning mode in a high vacuum mode. The sample was first freeze-dried, and then a sample sheet of approximately 4 mm^2 was taken and affixed to the sample plate using conductive tape in sequence. The sample was frozen with liquid nitrogen and then quickly freeze-dried in a freeze-dryer, sublimating the water to create a dry sample. Gold spray coating was applied to render the sample conductive and prevent charge accumulation, as well as to mitigate thermal damage to the sample. The sample plate was placed within the SEM sample chamber to obtain the photographs.

Author Contributions: Conceptualization, S.H.; Methodology, X.L.; Validation, S.L.; Investigation, L.Z.; Resources, J.W. and F.Z.; Data curation, J.Z. and A.X.; Writing—review & editing, M.L. All authors have read and agreed to the published version of the manuscript.

Funding: This research received no external funding.

Data Availability Statement: The original contributions presented in the study are included in the article, further inquiries can be directed to the corresponding authors.

Conflicts of Interest: The authors declare no conflict of interest.

References

1. Khormali, A.; Ahmadi, S. Experimental and modeling analysis on the performance of 2-mercaptobenzimidazole corrosion inhibitor in hydrochloric acid solution during acidizing in the petroleum industry. *J. Pet. Explor. Prod. Technol.* **2023**, *13*, 2217. [CrossRef]
2. Dai, C.L.; Liu, Y.F.; Zou, C.W.; You, Q.; Yang, S.; Zhao, M.W.; Zhao, G.; Wu, Y.N.; Sun, Y.P. Investigation on matching relationship between dispersed particle gel (DPG) and reservoir pore-throats for in-depth profile control. *Fuel* **2017**, *207*, 109–120. [CrossRef]
3. Ma, B. Research and Performance Evaluation of Inorganic Precipitation Profile Control System Injected in Single Slug. *Oilfield Chem.* **2019**, *36*, 90.
4. Chen, L.F.; Zhu, X.M.; Fu, M.L.; Zhao, H.; Li, G.; Zuo, J.Q. Experimental study of calcium-enhancing terpolymer hydrogel for improved oil recovery in ultradeep carbonate reservoir. *Colloids Surf. A Physicochem. Eng. Asp.* **2019**, *570*, 251–259. [CrossRef]
5. Zhang, X.Y.; Zhang, S.S.; Li, L.; Wu, R.N.; Liu, D.M.; Wu, J.H.; Wu, W. High-temperature-resistant polymer gel system with metal–organic mixed cross-linking agents. *J. Appl. Polym. Sci.* **2015**, *132*, 42261. [CrossRef]
6. Hou, S.W.; Wang, Y.; Rao, Z.; Zhang, Z.G. Experimental study on deep prolile control and oil displacement mechanism of high water cut reservoir. *Sci. Technol. Eng.* **2020**, *20*, 13163.
7. Cai, W.S.; Huang, R.H. Slow Gelation of Titanium (IV) with Partially Hydrolyzed Polyacrylamide. Crosslinking Reaction and Gel Properties. *Polym. J.* **2001**, *33*, 330–335. [CrossRef]
8. Xu, D. Development Application of FP-fly Ash Profile Control Agent. *J. Jiang Han Pet. Inst.* **2004**, *26*, 108.
9. Fox, K.B.; Evans, A.J. Acceleration of Gelation of Water Soluble Polymers. U.S. Patent No. 5,447,986, 2 January 1996.

10. Luo, J. Research and Application of Adaptive Multifunctional Conditioning Technology for Tight Oil Reservoirs in Ku Dong Oilfield. *Inn. Mong. Petrochem. Ind.* **2010**, *18*, 151.
11. Li, C.X.; Chen, H.J. Research progress of high temperature resistant and high salinitytolerant gel profile control technology. *Appl. Chem. Ind.* **2013**, *6*, 114.
12. Prud, A.; Homme, R.K.; Uhl, J.T.; Halverson, F. Rheological Monitoring of the Formation of Polyacrylamide/Cr^{+3} Gels. *Soc. Pet. Eng. J.* **1983**, *5*, 804–808. [CrossRef]
13. Sydansk, R.D. A Newly developed Chromium (III) Gel Technology. *SPE Reserv. Eng.* **1990**, *5*, 346–352. [CrossRef]
14. Lockhart, T.P. Chemical Properties of Chromium/Polyacrylamide Gels. *SPE Adv. Technol. Ser.* **1994**, *2*, 199–205. [CrossRef]
15. Sydansk, R.D.; Southwell, G.P. *More Than 12 Years of Experience with a Successful Conformance-Control Polymer Gel Technology: SPE/AAPG Western Regional Meeting*; Society of Petroleum Engineers: Long Beach, CA, USA, 2000.
16. DiGiacomo, P.M.; Schramm, C.M. *Mechanism of Polyacrylamide Gel Syneresis Determined by C-13 NMR: SPE Oilfield and Geothermal Chemistry Symposium*; Society of Petroleum Engineers: Denver, CO, USA, 1983.
17. Lei, Z.; Wu, F.P.; Pu, C.S. Mechanism of formation of chromium acetate (Cr^{3+})/phenol-formaldehyde resin prepolymer (PRP) complex and its compound cross-linking reaction with polymer for conformance control. *J. Pet. Sci. Eng.* **2019**, *179*, 675–683.
18. Fulleylove, R.J.; Morgan, J.C.; Stevens, D.G.; Thrasher, D.R. *Water Shut-Off in Oil Production Wells—Lessons from 12 Treatments: Abu Dhabi International Petroleum Exhibition and Conference*; Society of Petroleum Engineers: Abu Dhabi, United Arab Emirates, 1996.
19. Romero-Zeron, L.B.; Hum, F.M.; Kantzas, A. Characterization of Crosslinked Gel Kinetics and Gel Strength by Use of NMR. *SPE Reserv. Eval. Eng.* **2008**, *11*, 439–453. [CrossRef]
20. Hutchins, R.D.; Dovan, H.T.; Sandiford, B.B. *Field Applications of High Temperature Organic Gels for Water Control: SPE/dOE Improved Oil Recovery Symposium*; Society of Petroleum Engineers: Tulsa, OK, USA, 1996; Volume 4, p. 21.
21. Zhang, S.L.; Guo, J.X.; Gu, Y.; Zhao, Q.; Yang, R.; Yang, Y. Polyacrylamide gel formed by Cr (III) and phenolic resin for water control in high-temperature reservoirs. *J. Pet. Sci. Eng.* **2020**, *194*, 107423. [CrossRef]
22. Yu, H.; Wang, Y.; Ji, W.; Zhang, J.; Zhang, P.; Chen, W.; Qi, Z. Study of a profile control agent applied in an offshore oilfield. *Pet. Sci. Technol.* **2011**, *29*, 1285–1297. [CrossRef]
23. Hamza, A.; Shamlooh, M.; Hussein, I.A.; Nasser, M.S.; Salehi, S. Rheology of triamine functionalized silica reinforced polymeric gels developed for conformance control applications. *Energy Fuels* **2020**, *34*, 1093–1098. [CrossRef]
24. Sydansk, R.D. *A New Conformance-Improvement-Treatment Chromium(III) Gel Technology*; SPE/DOE: Tulsa, OK, USA, 1988.

gels

Review

Progress of Research into Preformed Particle Gels for Profile Control and Water Shutoff Techniques

Wei Ma [1,2,3], Yikun Li [1,*], Pingde Liu [1], Zhichang Liu [2] and Tao Song [1]

1 Key Laboratory of Oilfield Chemistry of China National Petroleum Corporation, Research Institute of Petroleum Exploration and Development, PetroChina, Beijing 100083, China; mw17860271013@163.com (W.M.)
2 College of Chemical Engineering and Environment, China University of Petroleum (Beijing), Beijing 102249, China
3 National Elite Institute of Engineering, Beijing 100096, China
* Correspondence: liyikun@petrochina.com.cn

Abstract: Gel treatment is an economical and efficient method of controlling excessive water production. The gelation of in situ gels is prone to being affected by the dilution of formation water, chromatographic during the transportation process, and thus controlling the gelation time and penetration depth is a challenging task. Therefore, a novel gel system termed preformed particle gels (PPGs) has been developed to overcome the drawbacks of in situ gels. PPGs are superabsorbent polymer gels which can swell but not dissolve in brines. Typically, PPGs are a granular gels formed based on the crosslinking of polyacrylamide, characterized by controllable particle size and strength. This work summarizes the application scenarios of PPGs and elucidates their plugging mechanisms. Additionally, several newly developed PPG systems such as high-temperature-resistant PPGs, re-crosslinkable PPGs, and delayed-swelling PPGs are also covered. This research indicates that PPGs can selectively block the formation of fractures or high-permeability channels. The performance of the novel modified PPGs was superior to in situ gels in harsh environments. Lastly, we outlined recommended improvements for the novel PPGs and suggested future research directions.

Keywords: preformed particle gels (PPGs); profile control and displacement; water shutoff; fracture

1. Introduction

As oil field development progresses, the production rates of oil wells gradually decrease. Existing techniques to enhance oil production rate mainly include fracturing [1], acidizing [2], profile control, and water shutoff. Excessive water production can lead to reduced oil production or even well abandonment, primarily due to reservoir heterogeneity and especially due to the presence of high-permeability fractures. Profile control and water shutoff is a crucial technology in oil field development for enhancing sweep efficiency and reducing water invasion in formations. Gel treatment is an economical and efficient method for controlling excessive water production and has been widely applied in the field. Several polymer gel systems have been developed, including in situ gels and PPGs. In situ gel treatment involves the simultaneous or sequential injection of polymer and crosslinker solutions, termed gelants, into the formation. Gelants can change to gels under the stimuli of temperature or salinity to block high-permeability channels. This method has several inherent drawbacks; to be more precise, the gelation time, gel quality, and injection depth are difficult to precisely control. Moreover, it may potentially damage the reservoir, resulting in the simultaneous plugging of oil and water [3,4]. To overcome the limitations of gel treatment, researchers have developed preformed particle gels (PPGs), which hold significant importance in the area of oil field profile control and water shutoff.

PPGs possess strong hydrophilic groups (such as amide and carboxyl groups) and a three-dimensional network structure. When in contact with water, hydrogen bonds form between water molecules and the hydrophilic groups, causing the molecular chains between crosslinking points to unfold and generate cohesion. At equilibrium, PPGs can swell from several times to several hundred times their dry diameter. As shown in Figure 1. PPGs can be synthesized in several ways.

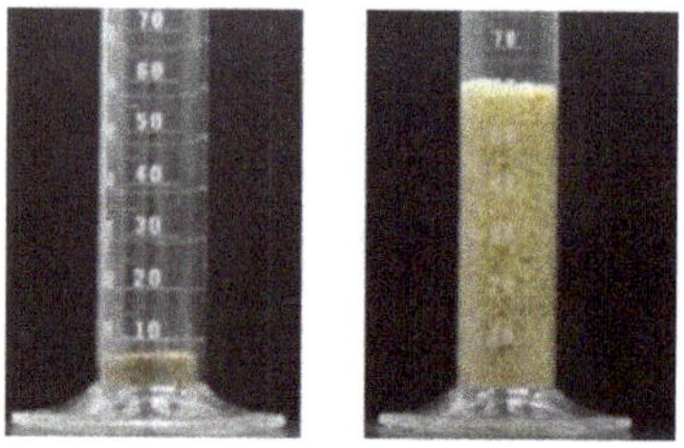

(a) (b)

Figure 1. (**a**) Dried particles; (**b**) swelling particles [5].

The first method involves the formation of high-strength gel clusters through aqueous radical polymerization, followed by granulation, drying, crushing, and sieving to obtain solid particles. This process is simple and cost-effective. Another method involves obtaining PPG emulsions through inverse emulsion polymerization. The PPGs obtained using this method have small particle size, uniform distribution, and high roundness, but a higher cost [6]. The third method involves preparing bulk gels followed by shear cutting to obtain multiscale dispersed particle gels, characterized by low viscosity and negatively charged surfaces [7]. Since PPGs are prepared on the ground, their particle size and gel strength are controllable.

During the process of PPGs absorbing water and swelling, when the water absorption reaches its maximum, the energy of the PPGs is at its lowest, indicating a stable state. The polymer chains constituting PPGs contain numerous detachable functional groups. After water absorption and swelling, they generate polymer anions and many positively charged cations. These cations disperse randomly around the anions, forming a stable electric field. When external cations are introduced, they shield the negative charges, reducing the intermolecular forces between polymer molecules, making stabilization easier and decreasing water absorption capacity. The influence of divalent cations on PPG performance is particularly significant. Compared to gels, they have better compatibility with formation water and shear resistance [8,9]. Unlike traditional polymer gels, PPGs exhibit a lower swelling ratio in high-salinity water, leading to an increase in crosslinking density and enhanced thermal stability. The introduction of inorganic fillers and molecular structure design can improve their temperature resistance by up to 120 °C [5,10–12]. The properties of PPGs are influenced by their composition, salinity, pH, and temperature. As the dosage of the main agent and crosslinking agent increases and the dosage of the initiator decreases, the swelling ratio of the PPGs decreases [13]. When the temperature increases, the swelling capacity of PPGs increases but their strength decreases. With an increase in the salinity of formation water, the swelling capacity of PPGs decreases while their strength increases. The influence of salinity on the plugging performance of PPGs is greater than that of particle size [14]. PPG suspensions contain only one effective component, simplifying injection equipment and reducing operational costs.

Mature oil fields with high water content commonly have high-permeability zones or fractures. PPGs of appropriate size and strength can preferentially enter high-permeability channels, causing minimal damage to low-permeability oil-producing layers. The main difference between the various types of PPGs lies in their particle size, ranging from nano- to millimeter-scale PPGs and microgels. Nanoscale PPGs and microgels can be used to plug pores with permeabilities less than 1 Darcy. Millimeter-scale PPGs can effectively block various fractures or high-permeability channels in reservoirs [5,15]. Key factors in selecting PPGs for application include compatibility with produced water, swelling ratio in injected water, strength after swelling and initial particle size. Additionally, combining with water flooding, gas displacement, and polymer displacement can enhance both the plugging performance of PPGs and the displacement efficiency of oil-displacing agents [16–21].

This paper first explains that PPGs, with their special properties, can be used for profile control and water shutoff in oilfields. It then introduces the mechanisms and application examples of PPGs for in selectively plugging high-permeability water flow channels in formations. The main focus of this study is to systematically summarize recent research into novel PPGs, aiming to identify the research directions of PPGs and the methods and strategies used for optimizing their performance, guiding their practical applications.

2. The Plugging Mechanism of PPGs

After being injected into a formation, PPGs preferentially enter high-permeability channels under injection pressure, accompanied by swelling, fragmentation, and deformation. As the distance into the pores increases, the pressure on PPGs gradually decreases. When the resistance experienced by PPGs exceeds the driving force, they are retained, plugging the preferential water flow channels. As shown in Figure 2.

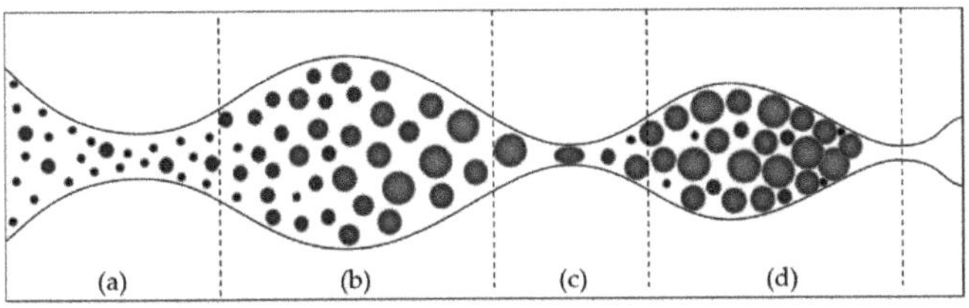

Figure 2. Migration of PPGs in formation channels [4]. (**a**) Passage of PPGs before swelling; (**b**) passage of PPG after swelling; (**c**) passage through pore throats after PPGs deformation or fragmentation; (**d**) lodging or entrapment of PPGs resulting in blockages.

On a macroscopic scale, PPG migration in porous media has three modes: direct passage, passage after fragmentation, and blockage [22]. The greatest resistance that PPGs face during migration occurs at the pore throats. Expanded PPGs are elastic and can pass through pore throats smaller than their own particle size. During PPG migration, particle strength has a greater impact than particle size. Fully expanded PPGs have better injectability than partially expanded PPGs with a larger diameter [23–25]. In larger channels, PPGs can be retained by adsorption, and the smaller the PPG particle size, the higher the retention rate. When PPGs encounter pore throats, if the ratio of particle size to pore size becomes too large, interception occurs. Retention and interception lead to blockage, and the blockage rate under both modes decreases with increases in injection rate, formation permeability, and temperature [26].

Zhao et al. [27,28] indicated that the threshold penetration pressure (ΔPth) is a key factor influencing the selective injection of PPGs. ΔPth is the minimum driving pressure difference required for PPGs to enter channels or matrices, depending on the strength of the expanded PPGs and the ratio of particle diameter to pore size. For micrometer-scale PPGs in targeted formation channels, the ratio of particle diameter to pore size should be less than 2, while for non-targeted formations, the ratio should exceed 5 to prevent damage to the matrices. Dai et al. [29] revealed the matching pattern between the average diameter of PPGs and the size of pore throats by introducing a matching coefficient and using injection performance and regulation ability as indicators. In simulation core experiments conducted by Seright [30,31], tracer studies and permeability measurements indicated that PPGs are more efficient than gels in crack injection, are more conducive to plugging water in cracks, and cause less damage to low-permeability reservoirs. Coste et al. [32] discovered that after PPGs expand within channels, they can displace some of the residual oil within pores near the well zone. In distant well areas, where the pressure decreases, PPGs remain and to block the channels. Elsharafi et al. [33,34] discovered that high-strength PPGs cause less damage to low-permeability reservoirs compared to low-strength PPGs. Li et al. [35], using the Pressure Transfer Test (PTT), discovered that, when used to treat naturally fractured tight sandstone reservoirs, PPGs had a 12% higher plugging effect on vertical and oblique fractures compared to horizontal fractures. Moreover, the use of high injection pressure and low-salinity injection water could enhance the plugging effect.

3. Application Example

In 1999, Li et al. [36] demonstrated the feasibility of applying PPGs in the field of water shutoff and profile control. In the same year, the Zhongyuan Oilfield successfully used PPGs for the first time to adjust injection wells. The reservoir rock type was sandstone, with a reservoir temperature of 107 °C, total formation water salinity of 15 × 10^4 mg/L, and an average permeability of 121 mD. The reservoir had been developed through water injection without hydraulic fracturing or the presence of natural fractures, and the injection well profiles showed strong reservoir heterogeneity. The candidate wells were two adjacent injection wells in Pucheng Oilfield: W51-75 and P-72. The average water cut of the corresponding production well prior to profile control was higher than 85%. After profile control with PPGs, the injection pressure of the injection wells increased, improving the injection profile. The daily oil production increased from 40 t/d to 60 t/d, with a total increased oil production of 3239 tons. Each ton of injected PPGs increased oil production by 158 tons, demonstrating favorable economic benefits.

In 2000, Daqing Oilfield first utilized PPGs for profile control. The selected injection well was Xingbei Oilfield Well Xing-7-24. The reservoir temperature was 45 °C and the formation water salinity was 4500 mg/L. The corresponding four production wells had a water output exceeding 700 m^3/d, with an average water cut exceeding 90%. Approximately 85% of the injected water directly exited through high-permeability zones that accounted for less than one-fifth of the reservoir thickness. This resulted in severely ineffective cycling. After trials, a total of 15.5 tons of 1.5–5 mm PPGs were injected into the well and the injection pressure increased from 5.0 MPa to 11.6 MPa. The water cut of the corresponding production well decreased by 8%, and the heterogeneity of the reservoir improved. The effectiveness lasted for over 6 months, with a total increase in oil production of approximately 2400 tons [37].

In a certain block of the Lamadian Reservoir in Daqing Oilfield, the reservoir temperature was approximately 40 °C, the formation water salinity was about 4000 mg/L, and the reservoir was highly heterogeneous. The average water cut in the production wells reached 95.4% and showed a rising trend. Between 2003 and 2004, WT (millimeter-scale PPGs) were used for conformance control in the four injection wells 7-1827, 7-1927, 8-1827, and 9-1927. These four water injection wells adopted the method of carrying PPGs with a polymer solution and cumulatively injected 132 tons of PPG suspension over a period of 5 months. After the PPG conformance control, the water injection profiles were improved. The oil production increased by 34.8 tons per day and the water cut decreased by 0.94%. By March 2005, the cumulative increase in oil production was about 15,000 tons, equivalent to 113 tons of oil per ton of PPGs used [38].

The Luo-1 Chang-8 block in Changqing Oilfield is a low-permeability reservoir which has been developed since 2007. Fractures are present in the central and northcentral areas of the block. By the end of 2011, the number of water-producing wells amounted to 38, accounting for 20.5% of the total wells, and indicating severe production capacity losses. In 2012, 17 injection wells in this block underwent conformance control. The injection process used "weak gel + PPGs + inorganic system" as the primary plugging components. After treatment, the average injection pressure of the 17 wells increased by 1.6 MPa, and 36 of the corresponding 93 production wells showed effective results. Cumulatively, these wells produced an additional 2733 tons of oil and showed a decrease of 1894 m^3 in water production [39].

Alaska's West Sak oil field had Void Space Conduits, which presented "short-circuiting" issues during water flooding. Traditional PPGs could plug pores, but had a short effective duration and lacked sufficient strength to control water flow. To enhance the plugging performance of PPGs, researchers developed a new type of re-crosslinkable preformed particle gel, RPPG [40]. Simulation core experiments showed that the breakthrough pressure in 2 mm fractures was more than five times that of conventional PPGs, with stability lasting up to 300 days. The target formation temperature in West Sak was approximately 24 °C, with permeability ranging from 20 to 3000 mD. Between 2017 and 2019, RPPG profile control treatments were conducted in the target formation. During the production process, no RPPGs were detected. The economic benefits of RPPGs were found to be 23% higher compared to traditional PPGs [41].

4. The Current Research Status

4.1. High-Temperature-Resistant PPGs

Traditional PPG, based on acrylamide homopolymer networks, dehydrates and degrades rapidly under high-temperature and high-salinity conditions [42]. There are typically three ways to enhance the thermal resistance of PPGs: copolymerizing with functional monomers, improving crosslinkers, and introducing inorganic materials. 2-Acrylamido-2-methylpropane sulfonic acid (AMPS) is a commonly used high-temperature-resistant and salt-resistant monomer; options also include N-vinyl pyrrolidone (NVP), sodium styrene sulfonate, 4-vinylimidazole, and vinylimidazole, which contain cyclic rigid groups. Using high-temperature-resistant crosslinkers can also enhance the thermal stability of PPGs.

Recent studies have found that chitosan can act as a crosslinker for high-temperature-resistant acrylamide-based gel [43,44], and Elaf et al. [45] developed biodegradable PPGs composed of polyacrylamide and chitosan cross-linkers, named PAM/Cs. Experimental verification confirmed that PAM/Cs exhibited excellent swelling performance and mechanical strength at specific compositional ratios and concentrations. However, their performance declined, and they were prone to degradation in high-temperature, high-salinity environments. Elaf et al. [46] subsequently incorporated N,N-methylenebisacrylamide (MBA) into the system and employed a microwave-assisted synthesis process to graft acrylamide (AM)/MBA onto the chitosan framework, thereby preparing environmentally friendly PPGs with temperature and salt resistance (Cs/PAMBAs). Its particle size range could be controlled between 10 μm and 1 mm. As a near-wellbore plugging agent, the swelling ratio in saline water with a salinity of 6×10^4 mg/L was 11 within one hour. Cs/PAMBAs' swelling performance and mechanical strength remained stable or slightly decreased under conditions of 200,000 mg/L salinity and temperatures $\leq$ 130 °C, indicating long-term stability.

The research conducted by Zhu et al. [47] indicated that the gel prepared from an AM/NVP/AMPS terpolymer using a high-temperature-resistant crosslinker exhibited excellent heat resistance. Ahdaya et al. [48] found that adding additives such as mica, walnut shells, and bentonite to PPGs significantly enhanced their strength. Durán-Valencia et al. [49] incorporated modified bentonite (MB) dispersion into the AM/NVP/AMPS system and synthesized modified PPGs through free radical polymerization at room temperature. This approach enhanced the mechanical strength of PPGs. SEM images before and after swelling are shown in Figure 3. In this system, the molar ratio of AM, NVP, and AMPS monomers was 1:1:1, with a 0.5% mass concentration of the crosslinker MBA, and a 2% concentration of modified bentonite (MB). When used at a concentration of 30%, it could remain stable for three months at 130 °C in brine with a salinity of approximately 25.5×10^4 mg/L. Saghafi et al. [11] introduced N,N-dimethylacrylamide as a new monomer into the AM/NVP/AMPS system. The monomer ratio of AM/DMA/NVP/AMPS was set at 2:1:1:2, with a free radical polymerization mass concentration of 30%. Additionally, 0.45% MBA and 2.5% nano-clay was added to synthesize high-temperature-resistant PPGs. In formation water with a salinity of 22.5×10^4 mg/L, the swelling ratio was 14, and the suspension maintained 75% of its original volume after aging for 120 days at 145 °C, indicating good temperature resistance.

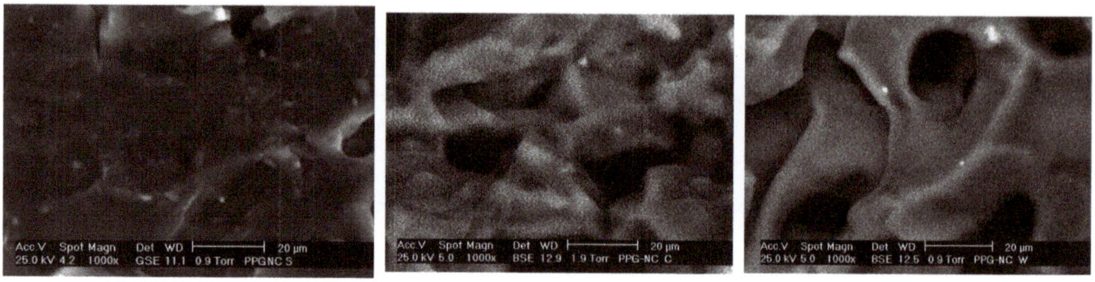

(a) (b) (c)

Figure 3. SEM image of modified PPG sample. (**a**) Dried sample; (**b**) sample swollen in production water; (**c**) sample swollen in distilled water [49].

Salunkhe et al. [50] used N,N-dimethylacrylamide and sodium styrenesulfonate (NaSS) as monomers, with divinylbenzene (DVB) as the crosslinker, and synthesized a high-temperature-resistant PPG (HT-PPG) after ratio optimization. Polydimethylacrylamide offered better thermal and hydrolytic stability than polyacrylamide, and DVB provided stable covalent crosslinks within the gel particles. After 18 months of aging at 150 °C, as shown in Figure 4, no degradation or dehydration was observeda nd the color change was due to residual oxygen. NMR spectra and SEM images indicated that the original chemical composition and crosslinking network structure were maintained even after aging; Schuman et al. [51] investigated the performance of HT-PPGs under North Sea reservoir conditions, where the formation water salinity was approximately 7.6×10^4 mg/L, and the reservoir temperature ranged from 130 to 150 °C. The swelling ratio of HT-PPGs could reach over 30, and its stability lasted for more than 18 months, demonstrating excellent thermal stability and chemical stability. In simulation core experiments, HT-PPGs were able to reduce the permeability of 2 mm fractures to a millidarcy level.

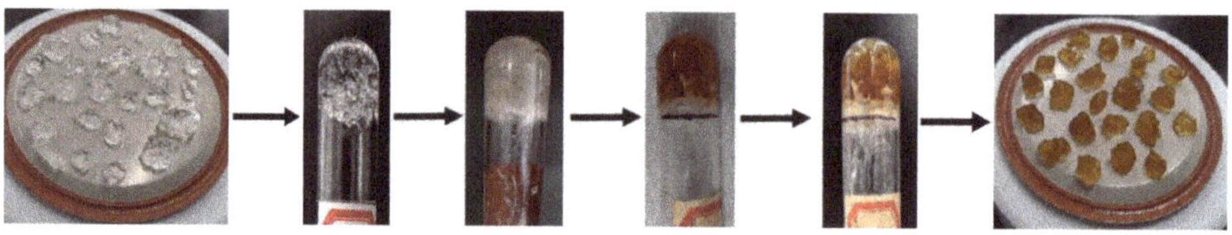

(a) (b) (c) (d)

Figure 4. HT-PPG remains intact after aging at 150 °C for 18 months. (**a**) Before aging; (**b**) aging for 6 months; (**c**) aging for 12 months; (**d**) aging for 18 months [50].

4.2. Re-Crosslinkable PPG

Conventional PPGs, due to their lack of inter-particle cohesion and particle-rock adhesion, are prone to being flushed out by water flow in large fractures and high-permeability channels, thus limiting their plugging efficiency. RPPGs can undergo secondary crosslinking to form bulk gels, as shown in Figure 5, significantly increasing their plugging efficiency. Through molecular structure design, the addition of additives, or the use of multiple crosslinkers, PPGs can undergo secondary crosslinking after being injected into formations.

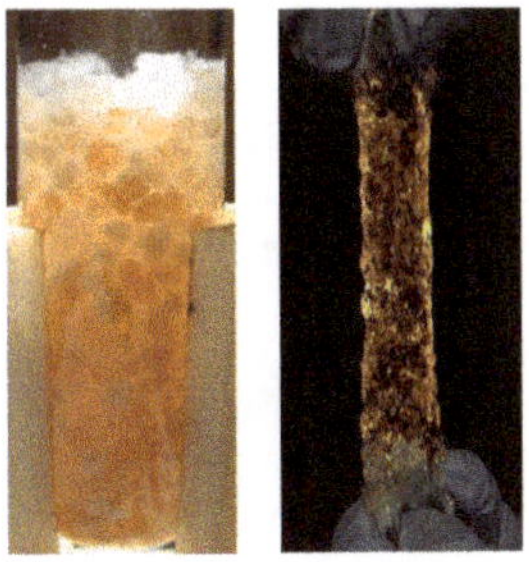

(a) (b)

Figure 5. Comparison of morphology between PPGs (**a**) and RPPGs (**b**) [49,52].

Branched PPG (B-PPG) are a type of PPG with branching structures between particles, which has excellent conformance control functionality in the "PPG–polymer–surfactant" ternary composite flooding system and has been successfully applied in the Shengli Oilfield [53–56]. Gong et al. [57] found in a sand pack plugging test that the B-PPG/polyacrylamide/surfactant system could further enhance oil recovery after polymer flooding.

Xu Long et al. [58] introduced microbial polysaccharides (Diutan gum or xanthan gum) into B-PPG suspensions, enhancing the viscosity and stability of the system. A gel network based on double helix structures and gel particles could be formed between Diutan gum and B-PPG through van der Waals forces and hydrogen bonds. At low concentrations, Diutan gum could form a solid mesh layer on the surface of particles, enhancing the viscoelasticity, temperature resistance, and salt tolerance of the system. In heterogeneous sand pack heavy oil displacement experiments under conditions of 90 °C and a salinity of approximately 24,400 mg/L, the single B-PPG system was affected by high temperature and high salinity. The ultimate recovery ratio was 44.9%, only slightly higher than that of water flooding. However, the Diutan gum/B-PPG system showed a significant increase in injection pressure, leading to a final recovery ratio of 57.1%. Polysaccharide/B-PPG composite materials could form high-strength gels in high-permeability channels, demonstrating excellent conformance control performance.

Zhai et al. [59] added sodium d-gluconate (GA) and polyvinyl alcohol (PVA) to the acrylamide homopolymer system and developed temperature- and salt-resistant RPPGs (GA-RPPGs). The crosslinker system was composed of MBA, polyethylene glycol (400) diacrylate (PEG400DA), GA, and PVA. Under ultraviolet irradiation, MBA and PEG400DA formed a three-dimensional network structure of polymer chains, while the hydroxyl groups of GA molecules and the semi-interpenetrating structure of PVA also initiated crosslinking. At a certain temperature, GA molecules chelated with metal ions in saline solution, causing the boundaries of GA-RPPGs to disappear, and weak gel formation occurred through mutual stacking and contact. At 130 °C and in a 10% salt solution, GA-RPPGs exhibited good mechanical properties and re-crosslinking capabilities, retaining elasticity even after aging for 45 days.

Song et al. systematically evaluated two types of RPPGs: SR-RPPGs and HT-BRPPGs. When the concentration of divalent cations such as Ca^{2+} and Mg^{2+} in formation water was too high, the carboxyl groups in the hydrolyzed gel chelated with the divalent cations, causing severe synaeresis. The performance of conventional PPGs is also affected by excessive divalent cations. SR-RPPGs are suitable for plugging high-salinity fractured reservoirs [60]. SR-RPPGs were synthesized using AM and AMPS monomers, crosslinked with MBA. AMPS inhibited the hydrolysis of amide groups and enhanced thermal stability and salt tolerance [61]. During the production process, the addition of xanthan gum enhanced the system's salt resistance and strength [62], while the addition of Cr^{3+} during injection facilitated re-crosslinking in fractures. Laboratory simulations of conditions in Middle Eastern fractured reservoirs (total salinity of formation water was approximately 21×10^4 mg/L, with Ca^{2+} content of 19,000 mg/L and Mg^{2+} content of 2411 mg/L).

SR-RPPGs with particle sizes of 1–2 mm were used, and the bulk gel formed after re-crosslinking could expand to over 30 times its original volume in formation water. After aging at 100 °C for 220 days, the strength grade was I, demonstrating good thermal stability and phase stability in Middle Eastern formation water and 5% $CaCl_2$ solution. In cores with a fracture width of 1.5 mm, the breakthrough pressure gradient could reach 20.98 MPa/m, indicating excellent plugging performance.

HT-BRPPG were suitable for plugging high-temperature fractured reservoirs [52]. Laboratory simulations were conducted under the high-temperature sandstone reservoir conditions of the North Sea Ekofisk field, with a simulated formation water salinity of approximately 70,000 ppm. Within 2 h at 130 °C, HT-BRPPG exhibited a swelling ratio of up to 41, with minimal influence from formation water salinity and pH shown on its swelling performance. HT-BRPPG underwent transamidation reactions in the presence of branched polyethyleneimine for re-crosslinking, controlled by temperature. The bulk gel formed after re-crosslinking at 130 °C and aged for 450 days achieved a strength code of H. SEM images showed that it still retained a three-dimensional network structure, demonstrating excellent long-term thermal stability. In 3 mm fractured cement cores, the breakthrough pressure gradient was 8.77 MPa/m.

4.3. Delayed-Swelling PPGs

Conventional PPGs have a rapid water absorption rate, causing them to expand before entering formation pores, which affects the transportability of PPGs. This makes it difficult for PPGs to reach deep formations, reducing the efficiency of plugging. Using a dual crosslinker system of stable and degradable crosslinkers during PPG production could effectively delay the swelling time of PPGs. Additionally, incorporating nanoparticles also could enhance the delayed swelling property of PPGs [63].

Wu et al. [64] used degradable crosslinkers and hydrophobic monomers in the synthesis of PPGs, effectively delaying the swelling time of PPGs. Monomers included AM, AMPS, tert-butyl acrylate (tBA), and dimethylaminoethyl methacrylate (DMAEMA). The component ratio was AM:AMPS:DMAEMA:tBA = 4:1:1:1. MBA and polyethylene glycol (600) diacrylate (PEG600DA) were selected as crosslinkers, and salt-resistant delayed swelling PPGs were prepared through inverse emulsion polymerization. The inclusion of tBA monomers enhanced the hydrophobicity of PPGs, restricting their water absorption during transportation. Interaction between DMAEMA and AMPS increased the crosslinking network density and salt resistance. The use of dual crosslinker also increased the crosslinking network density of PPG. Upon entering the formation, the PEG600DA and tBA monomers hydrolyzed, restoring and enhancing the water absorption and swelling capacity of PPGs. At 90 °C, PPGs using only MBA as the crosslinker and PPGs using a dual crosslinker were dispersed in a sodium chloride solution with a salinity of 15×10^4 mg/L. Two days later, the volume-swelling ratio of the dual-crosslinked PPGs remained at 2.36, while that of the single-crosslinked PPGs was close to 9. After twenty days, the volume-swelling ratio of the dual-crosslinked PPGs reached 9.49. The dual-crosslinked PPGs demonstrated excellent transportability and good plugging performance in formation pores.

In low-permeability reservoirs undergoing CO_2 flooding, the presence of fractures can lead to gas channeling issues. Bai et al. [65] developed and evaluated a series of PPGs with controllable swelling rates, which could enhance the efficiency of CO_2 flooding in reservoirs. Zhou et al. [66] utilized AM, sodium styrene sulfonate, and dimethyl diallyl ammonium chloride as monomers, while MBA and polyethylene glycol (200) diacrylate (UCA) were chosen as crosslinkers. Dual-crosslinked nano-particle microgels (DCNPM-As) were prepared by inverse microemulsion polymerization, exhibiting delayed swelling and anti-gas-channeling properties. The dual-crosslinking system could maintain smaller swelling ratios of DCNPM-As at low temperatures. UCA, as an unstable crosslinker, resulted in successive ester bond cleavage both externally and internally as the temperature increased (>80 °C). It exhibited secondary swelling characteristics, ultimately achieving higher swelling ratios compared to the microgels prepared using a single crosslinker. In an experiment involving supercritical CO_2 flooding gas channeling resistance, DCNPM-As demonstrated a plugging efficiency of 95.48% and exhibited delayed swelling characteristics.

4.4. Augmented PPGs

Conventional PPGs are based on covalent bonds to form a crosslinked network structure, which may be susceptible to shear damage during transportation, affecting the plugging performance. Adding inorganic materials can improve the strength of PPGs (As shown in Table 1), but the enhancement is limited, and compatibility issues between the inorganic materials and polymers may arise during synthesis and use. Improving the crosslinking agent system of PPGs and designing their molecular structure can avoid compatibility issues while significantly enhancing their strength and toughness.

Table 1. Enhancing the performance of PPGs by introducing inorganic materials.

Researcher	Study on PPGs	Improvement
Kaio A. B. Pereira [67]	The Influence of Fly Ash (CFA) on Polyacrylamide Systems.	Viscoelasticity and temperature resistance are improved.
Peyman Abbasi Khoshkar [68]	The effect of nano-clay and nano-silica on AM-AMPS system.	The equilibrium swelling ratio and temperature resistance were improved, the pH and CO_2 tolerance were improved, and the synaeresis of PPG was inhibited.
Abhinav Kumar [69]	The Reinforcing Effect of Nanomontmorillonite on Polyacrylamide Systems.	The swelling performance, thermal stability and strength are improved, and the water shutoff performance in sandstone is enhanced.
Aminsadegh Paprouschi [70]	The effect of Sodium silicate and graphene nanosheets on AM-AMPS system.	The storage modulus is greatly improved, and the strength, thermal stability, and dehydration tolerance are improved.
Sthéfany Z.S. do Amparo [71]	Graphene oxide- and carbon nanotube-reinforced polyacrylamide PPGs.	The shear deformation resistance and swelling performance are improved.
Li [72]	The effect of laponite on AM-AMPS system.	The swelling rate and mechanical properties are enhanced, and the thermal stability is also slightly improved.
Kang [73]	Effect of sawdust on acrylamide/acrylic acid system.	Viscoelasticity, shear resistance, and yield stress are improved, and the plugging rate in cracks is improved.
Abhinav Kumar [74]	Effect of halloysite nanotubes on polyacrylamide PPGs.	The swelling capacity, elastic modulus and temperature resistance are improved, and the plugging effect of permeability in the 2-6 D sand pack plugging test is enhanced.
Yugal Kishor Pandit [75]	Effect of bentonite and silica on AM-NVP-AMPS system.	Long-term thermal stability, elastic modulus and plugging performance are enhanced.

Ai et al. [76] synthesized a polyrotaxane crosslinker (PRc) and introduced a dynamic crosslinking structure into the acrylamide network, enhancing the deformability of acrylamide PPGs. In tensile experiments, the elongation at break of bulk gel prepared using MBA as the crosslinker was 332%, while that of bulk gel crosslinked by PRc (PR-G) reached 937%. After water absorption, the elongation at break of PR-G was 346%, demonstrating good tensile recovery. The crosslinking points of PR-G were on the cyclodextrin within the polyrotaxane crosslinker. When subjected to external forces, cyclodextrins could balance the tension on polymer chains like pulleys. At 90 °C, the swelling ratio in distilled water was 39.89, while in a 12% $CaCl_2$ solution, it was 7.72. In a heterogeneous sand pack plugging test, PR-G exhibited good plugging capability for high-permeability layers.

Hao et al. [77] synthesized a host-guest inclusion supramolecular gel particle (S-PPG) using allyl-β-cyclodextrin (allyl-β-CD), cetyltrimethylammonium chloride (C16DMAAC), MBA, and AM as the main materials. The allyl-β-CD, as the host molecule, had an amphiphilic structure capable of encapsulating guest molecules such as C16DMAAC to form supramolecular gels. S-PPG possessed a synergistic crosslinking network structure of covalent and non-covalent bonds, which prevented it from breaking easily or self-bonding after fracture when passing through pores, ensuring its plugging performance. For the S-PPG system with a mass fraction of 0.6%, the ratios of host and guest monomers in the polymer were both 2.5%, with an average particle size of 34.56 μm. Under a strain amplitude of 1% and scanning frequencies ranging from 0.5 to 10.0 Hz, the elastic modulus was 73 Pa, which was 3.04 times that of conventional acrylamide PPGs. Its shear recovery performance and yield performance were four times that of conventional PPGs. After 24 h of swelling in formation water with a salinity of 6726 mg/L at 50 °C, the swelling ratio reached 23.

Deng et al. [78] developed a highly polymer-compatible crosslinker and enhancer: vinyl-functionalized silica nanoparticles (VSNPs). The polymer synthesis monomers included AM, N,N-dimethylacrylamide, and N-vinylpyrrolidone, with the addition of sodium alginate (SA). Using MBA and VSNPs as crosslinkers, a high-strength, high-temperature-resistant, and salt-resistant type of PPG (MC-PPG) was developed. MC-PPGs possessed a multi-crosslinking structure, including the nano-crosslinking of VSNPs, chemical crosslinking of MBA, self-crosslinking of DMA, and physical crosslinking of SA, which significantly improved their strength. DMA and NVP, as functional monomers, enhanced the temperature resistance and salt tolerance of MC-PPGs. MC-PPGs exhibited a honeycomb-like porous structure, with an equilibrium swelling ratio of up to 760% in deionized water. Compared to other PPGs, it had a higher elastic modulus (approximately 110 kPa) and stronger shear resistance and deformability. Additionally, the salinity and pH of the formation water had minimal effects on the swelling performance and strength of MC-PPGs. It remained stable for over 13 months at 120 °C and over 1 month at 160 °C. The high-temperature degradation products of MC-PPGs were liquid, causing minimal damage to the matrix. A core-flooding experiment demonstrated a breakthrough pressure gradient of 15.89 MPa/m at 20 °C and still exceeded 14 MPa/m at 120 °C, effectively plugging fractures and high-permeability channels.

4.5. Degradable PPG

During the oil recovery process, non-degradable substances can damage the reservoir [79].Conventional PPGs are difficult to degrade after plugging, and the residue can cause permanent damage to reservoirs. Degradable preformed particle gels (DPPGs) can degrade into low-viscosity liquids after completing plugging tasks, with minimal or no solid residues or by-products as temporary plugging agents in acidizing and fracturing operations.

Zhao et al. [80] introduced acid-resistant functional monomers and self-degradable crosslinking structures into PPGs, which could degrade into liquid after acidification. The DPPGs had an equilibrium swelling ratio of up to 70 in water and exhibited good plugging and self-degradation capabilities under acidic conditions in core experiments. The self-degradation time and strength could be controlled by adjusting the concentrations of monomers, initiators, and crosslinkers [81]. Zhang et al. [82] introduced sodium alginate (SA) into the polyacrylamide network, with MBA and $Fe(NO_3)_3 \cdot 9H_2O$ as crosslinkers, to synthesize a DPPG (d-PPG). With the strain amplitude fixed at 0.5%, the elastic modulus increased proportionally within the scanning frequency range of 0.1–100 Hz, reaching up to 86,445 Pa. The equilibrium swelling ratio and swelling rate increased when the temperature rose and salt concentration decreased. Dried d-PPGs exhibited an equilibrium swelling ratio of approximately 375% after 5 h in a 50000 ppm NaCl solution at 80 °C, with a degradation time of 501 h. When the temperature rose to 100 °C, the degradation time decreased to 107 h. In core-flooding experiments, d-PPGs demonstrated a high breakthrough pressure, with a plugging efficiency of up to 99.83%. After degradation, the liquid viscosity was less than 5 mPa·s, causing minimal damage to the formation.

5. Discussion

PPGs' mechanism of selective plugging of water channels differs from traditional polymer gels. Traditional polymer gel plugging agents form a gel in the formation pores, adsorbing onto the pore walls, and selectively plugging water relying on their property of "swelling in water and remaining unchanged in oil". PPGs selectively enter high-permeability water flow channels, swell upon water absorption, and rely on adsorption, retention, and interception to achieve plugging. In the production of PPGs, the polymer crosslinking process occurs in surface equipment, with controllable particle size and strength, overcoming the challenges of precise control over gelation time, gel quality, and injection depth faced by traditional polymer gels. Moreover, PPG suspension has only one effective component, making injection equipment simple and production operation costs low.

The improvement of PPG performance can be achieved through several pathways. Firstly, the molecular structure design of polymers can enhance the strength and high-temperature resistance of PPGs. Molecular structure design involves copolymerization with monomers containing functional groups into the original polymer system. For example, the sulfonic acid group on the AMPS monomer provides large steric hindrance and electrostatic repulsion, enhancing the flexibility of polymer molecular chains and inhibiting polymer hydrolysis at high temperatures; cyclic rigid groups not only provide significant steric hindrance but also absorb heat from other parts of the molecule through resonance absorption.

Furthermore, PPGs can be endowed with special functionalities by improving the crosslinking agent system. The use of various specific crosslinking agents (such as the crosslinking agent system composed of MBA and Cr^{3+}) can induce secondary crosslinking of RPPGs after injection into formation pores, forming bulk gels under environmental influence and greatly enhancing their plugging performance. The use of conventional crosslinking agents and degradable crosslinking agents can delay the expansion time of PPGs. The use of dual crosslinking agents allows PPGs to achieve a higher crosslinking network density before reaching the target formation. Upon reaching the target formation, one of the crosslinking agents undergoes degradation under environmental influence, enhancing the swelling capability of PPGs. Additionally, the use of multiple crosslinking agents or the introduction of special crosslinking structures can significantly enhance the strength and toughness of PPGs.

Lastly, adding a certain amount of inorganic material (such as nanoscale bentonite, etc.) can also enhance the thermal resistance and strength of PPGs. However, inappropriate amounts can lead to compatibility issues, adversely affecting the performance of PPGs.

In conclusion, the main approaches to improving the high-temperature resistance of PPGs are molecular structure design and improvements in the crosslinking agent system. It is worth noting that while functional monomers endow PPGs with excellent performance, the presence of large functional groups can have adverse effects on the polymerization process. Therefore, it is necessary to carefully select the optimal synthesis conditions when preparing PPGs. Re-crosslinkable PPGs, delayed swelling PPGs, augmented PPGs, and degradable PPGs are primarily prepared through the design of crosslinking agent systems. The addition of inorganic materials typically serves as auxiliary reinforcement. Finally, the performance evaluation of polymer gel and various PPGs is shown in Table 2.

Table 2. Performance evaluation of polymer gels and various PPGs.

	Conventional Polymer Gels	Conventional PPGs	High-Temperature Resistant PPGs	Re-Crosslinkable PPGs	Delayed Swelling PPGs	Augmented PPGs	Degradable PPGs
Temperature tolerance	×	√	√√	√	√	√	√
Salt tolerance	×	√	√	√	√	√	√
Selective water plugging	√	√	√	√	√	√	√
Long-term stability	×	√	√	√	√	√√	×
Migration capability	×	√	√	√	√√	√	√
Convenient construction	×	√	√	√	√	√	√
Applicability of large fractures	×	×	×	√	×	×	×

'×' represents suboptimal; '√' represents good; '√√' represents excellent.

6. Conclusions and Foresight

PPGs, relying on their controllable properties of swelling, size, and strength, can selectively enter and plug high-permeability zones in reservoirs. They are an important technology for profile control and displacement and water shutoff in oil fields. PPGs, with thier excellent performance and functionality, have been widely applied in oil fields with positive results.

Applicative prospects or research directions for PPGs:

(1) Exploring functional monomers and crosslinking agents resistant to high temperatures, targeting high-temperature reservoirs, and developing new types of high-temperature-resistant PPGs while ensuring plugging performance and long-term stability.

(2) PPGs are prone to being flushed out when dealing with large fractures, resulting in poor plugging effectiveness. High-strength RPPGs can effectively plug large fractures or excessively high-permeability channels.

(3) Conventional PPGs expand too rapidly in formation pores, lacking deep profile control. Delayed-swelling PPGs are easier to inject into deeper reservoirs, significantly enhancing plugging efficiency.

(4) PPGs have potential applications in gas reservoir water shutoff.

Author Contributions: Writing—original draft, W.M.; writing—review and editing, Y.L., P.L., Z.L. and T.S. All authors have read and agreed to the published version of the manuscript.

Funding: This research received no external funding.

Data Availability Statement: No new data were created or analyzed in this study. Data sharing is not applicable to this article.

Conflicts of Interest: Author W.M., Y.L., P.L., T.S. was employed by China National Petroleum Corporation. The remaining authors declare that the research was conducted in the absence of any commercial or financial relationships that could be construed as a potential conflict of interest.

References

1. He, J.; Wang, H. Design and Effect of Fracture-flooding in Class III Oil Reservoirs. *J. Southwest Pet. Univ. (Sci. Technol. Ed.)* **2018**, *40*, 95–104. [CrossRef]
2. Khormali, A.; Ahmadi, S. Experimental and modeling analysis on the performance of 2-mercaptobenzimidazole corrosion inhibitor in hydrochloric acid solution during acidizing in the petroleum industry. *J. Pet. Explor. Prod. Technol.* **2023**, *13*, 2217–2235. [CrossRef]

3. Cottin, C.; Al-Amrie, O.; Barrois, E.; Templier, A.; Salehi, N.; Zaitoun, A. Chemical water shutoff pilot in a mature offshore carbonate field. In Proceedings of the Abu Dhabi International Petroleum Exhibition and Conference, Abu Dhabi, United Arab Emirates, 13 November 2017. [CrossRef]
4. Guo, P.; Tian, Z.; Zhou, R.; Chen, F.; Du, J.; Wang, Z.; Hu, S. Chemical water shutoff agents and their plugging mechanism for gas reservoirs: A review and prospects. *J. Nat. Gas Sci. Eng.* **2022**, *104*, 104658. [CrossRef]
5. Bai, B.; Wei, M.; Liu, Y. Field and Lab Experience with a Successful Preformed Particle Gel Conformance Control Technology. In Proceedings of the SPE Production and Operations Symposium, Oklahoma City, OK, USA, 23–26 March 2013. [CrossRef]
6. Bai, B.; Liu, W.; Li, L.; Liu, G.; Tang, X. An analysis on intrinsie factors influencing the propertiesof pre-crosslinking gelled particles. *Pet. Explor. Dev.* **2002**, *29*, 3. [CrossRef]
7. Zhao, G.; You, Q.; Gu, C.; Lv, Y.; Dai, C. Preparation mechanism of multiscale dispersed particle gel. *Acta Pet. Sin.* **2017**, *38*, 821–829. [CrossRef]
8. Feng, Y.; Tabary, R.; Renard, M.; Le Bon, C.; Omari, A.; Chauveteau, G. Characteristics of Microgels Designed for Water Shutoff and Profile Control. In Proceedings of the International Symposium on Oilfield Chemistry, Houston, TX, USA, 5 February 2003. [CrossRef]
9. Seright, R.S.; Lane, R.H.; Sydansk, R.D. A Strategy for Attacking Excess Water Production. *SPE Prod. Facil.* **2003**, *18*, 158–169. [CrossRef]
10. Goudarzi, A.; Zhang, H.; Varavei, A.; Hu, Y.; Delshad, M.; Bai, B.; Sepehrnoori, K. Water Management in Mature Oil Fields using Preformed Particle Gels. In Proceedings of the SPE Western Regional & AAPG Pacific Section Meeting 2013 Joint Technical Conference, Monterey, CA, USA, 19 April 2013. [CrossRef]
11. Saghafi, H.R.; Naderifar, A.; Gerami, S.; Emadi, M.A. Improvement in thermo-chemical stability of nanocomposite preformed particle gels for conformance control in harsh oil reservoir conditions. *Can. J. Chem. Eng.* **2016**, *94*, 1880–1890. [CrossRef]
12. Tongwa, P.; Nygaard, R.; Bai, B. Evaluation of a Nanocomposite Hydrogel for Water Shut-Off in Enhanced Oil Recovery Applications: Design, Synthesis, and Characterization. *J. Appl. Polym. Sci.* **2012**, *128*, 787–794. [CrossRef]
13. Kang, W.; Xu, H.; Yang, H.; Zhao, W.; Du, S.; Gong, X.; Liu, X. Study on expansion property of pre-crosslinked particle gel with low elasticity and high viscosity. *Chem. Eng. Oil Gas* **2014**, *43*, 662–665+674. [CrossRef]
14. Imqam, A.; Bai, B. Optimizing the strength and size of preformed particle gels for better conformance control treatment. *Fuel* **2015**, *148*, 178–185. [CrossRef]
15. Qiu, Y.; Wu, F.; Wei, M.; Kang, W.; Li, B. Lessons Learned from Applying Particle Gels in Mature Oilfields. In Proceedings of the SPE Improved Oil Recovery Symposium, Tulsa, OK, USA, 12 April 2014. [CrossRef]
16. Al Brahim, A.; Eriyagama, Y.; Bai, B.; Schuman, T. Feasibility Study of Recrosslinkable Preformed Particle Gels for Natural Gas Injection Profile Control. *SPE J.* **2023**, *28*, 1829–1841. [CrossRef]
17. Alhuraishawy, A.K.; Bai, B.; Imqam, A.; Wei, M. Experimental study of combining low salinity water flooding and preformed particle gel to enhance oil recovery for fractured carbonate reservoirs. *Fuel* **2018**, *214*, 342–350. [CrossRef]
18. Alhuraishawy, A.K.; Sun, X.; Bai, B.; Wei, M.; Imqam, A. Areal sweep efficiency improvement by integrating preformed particle gel and low salinity water flooding in fractured reservoirs. *Fuel* **2018**, *221*, 380–392. [CrossRef]
19. Kang, W.; Shao, S.; Yang, H.; Chen, C.; Gabdullin, M. The effect of stepwise increasing of water injection rates on enhanced oil recovery after preformed particle gel treatment. *J. Pet. Sci. Eng.* **2019**, *182*, 106239. [CrossRef]
20. Sun, X.; Alhuraishawy, A.K.; Bai, B.; Wei, M. Combining preformed particle gel and low salinity waterflooding to improve conformance control in fractured reservoirs. *Fuel* **2018**, *221*, 501–512. [CrossRef]
21. Zhao, Y.; Leng, J.; Lin, B.; Wei, M.; Bai, B. Experimental Study of Microgel Conformance-Control Treatment for a Polymer-Flooding Reservoir Containing Superpermeable Channels. *SPE J.* **2021**, *26*, 2305–2317. [CrossRef]
22. Bai, B.; Liu, Y.; Coste, J.-P.; Li, L. Preformed Particle Gel for Conformance Control: Transport Mechanism Through Porous Media. *SPE Reserv. Eval. Eng.* **2007**, *10*, 176–184. [CrossRef]
23. Yao, C.; Lei, G.; Cathles, L.M.; Steenhuis, T.S. Pore-Scale Investigation of Micron-Size Polyacrylamide Elastic Microspheres (MPEMs) Transport and Retention in Saturated Porous Media. *Environ. Sci. Technol.* **2014**, *48*, 5329–5335. [CrossRef]
24. Imqam, A.; Bai, B.; Delshad, M. Micro-particle gel transport performance through unconsolidated sandstone and its blocking to water flow during conformance control treatments. *Fuel* **2018**, *231*, 479–488. [CrossRef]
25. Li, Y.; Sarıyer, O.S.; Ramachandran, A.; Panyukov, S.; Rubinstein, M.; Kumacheva, E. Universal behavior of hydrogels confined to narrow capillaries. *Sci. Rep.* **2015**, *5*, 17017. [CrossRef]
26. Saghafi; Reza, H. Retention characteristics of enhanced preformed particle gels (PPGs) in porous media: Conformance control implications. *J. Pet. Sci. Eng.* **2018**, *166*, 962–968. [CrossRef]
27. Zhao, Y.; Bai, B. Selective penetration behavior of microgels in superpermeable channels and reservoir matrices. *J. Pet. Sci. Eng.* **2022**, *210*, 109897. [CrossRef]
28. Zhao, Y.; Bai, B. Laboratory Evaluation of Placement Behavior of Microgels for Conformance Control in Reservoirs Containing Superpermeable Channels. *Energy Fuels* **2022**, *36*, 1374–1387. [CrossRef]
29. Dai, C.; Zou, C.; Liu, Y.; You, Q.; Tong, Y.; Wu, C.; Shan, C. Matching principle and in-depth profile control mechanism between elastic dispersed particle gel and pore throat. *Acta Pet. Sin.* **2018**, *39*, 427–434. [CrossRef]

30. Seright, R.S. Gel Placement in Fractured Systems. *SPE Prod. Facil.* **1995**, *10*, 241–248. [CrossRef]
31. Seright, R.S. Gel Propagation Through Fractures. *SPE Prod. Facil.* **2001**, *16*, 225–231. [CrossRef]
32. Coste, J.-P.; Liu, Y.; Bai, B.; LI, Y.; Shen, P.; Wang, Z.; Zhu, G. In-Depth Fluid Diversion by Pre-Gelled Particles. Laboratory Study and Pilot Testing. In Proceedings of the SPE/DOE Improved Oil Recovery Symposium, Tulsa, OK, USA, 3 April 2000. [CrossRef]
33. Elsharafi, M.O.; Bai, B. Effect of weak preformed particle gel on unswept oil zones/areas during conformance control treatments. *Ind. Eng. Chem. Res.* **2012**, *51*, 11547–11554. [CrossRef]
34. Elsharafi, M.O.; Bai, B. Effect of Strong Preformed Particle Gel on Unswept Oil Zones/Areas During Conformance Control Treatments. In Proceedings of the EAGE Annual Conference & Exhibition incorporating SPE Europec, London, UK, 10 June 2013. [CrossRef]
35. Li, M.; Guo, J.; Zhou, F.; Cao, L.; Li, B. Investigating the effect of preformed particle gel on conformance control of naturally fractured tight sandstone reservoirs: An experimental study using pressure transfer test method. *Fuel* **2023**, *347*, 128479. [CrossRef]
36. Li, Y.; Liu, Y.; Bai, B.; Liu, G. Research of expansion grain chemical for water shutoff and profile control. *Oil Drill. Prod. Technol.* **1999**, *3*, 65–68+114. [CrossRef]
37. Bai, B.; Li, L.; Liu, Y.; Liu, H.; Wang, Z.; You, C. Preformed Particle Gel for Conformance Control: Factors Affecting Its Properties and Applications. *SPE Reserv. Eval. Eng.* **2007**, *10*, 415–422. [CrossRef]
38. Bai, B.; Huang, F.; Liu, Y.; Seright, R.S.; Wang, Y. Case study on preformed particle gel for in-depth fluid diversion. In Proceedings of the SPE Improved Oil Recovery Conference, Tulsa, OK, USA, 20 April 2008. [CrossRef]
39. Ma, H.; Xu, Y.; Wei, J.; Liu, M.; Yao, C. Profile control technology in Luo1 Chang8 oilfield and its effect. *Petrochem. Ind. Appl.* **2013**, *32*, 57–59+71. [CrossRef]
40. Pu, J.; Bai, B.; Alhuraishawy, A.; Schuman, T.; Chen, Y.; Sun, X. A Recrosslinkable Preformed Particle Gel for Conformance Control in Heterogeneous Reservoirs Containing Linear-Flow Features. *SPE J.* **2019**, *24*, 1714–1725. [CrossRef]
41. Targac, G.; Gallo, C.; Smith, D.; Huang, C.-K.; Autry, S.; Peirce, J.; Baohong, L. Case History of Conformance Solutions for West Sak Wormhole/Void Space Conduit with a New Reassembling Pre-formed Particle Gel RPPG. In Proceedings of the SPE Annual Technical Conference and Exhibition, Virtual, 19 October 2020. [CrossRef]
42. Xiong, B.; Loss, R.D.; Shields, D.; Pawlik, T.; Hochreiter, R.; Zydney, A.L.; Kumar, M. Polyacrylamide degradation and its implications in environmental systems. *NPJ Clean Water* **2018**, *1*, 17. [CrossRef]
43. Magzoub, M.I.; Salehi, S.; Hussein, I.A.; Nasser, M.S. Loss circulation in drilling and well construction: The significance of applications of crosslinked polymers in wellbore strengthening: A review. *J. Pet. Sci. Eng.* **2020**, *185*, 106653. [CrossRef]
44. Negi, H.; Verma, P.; Singh, R.K. A comprehensive review on the applications of functionalized chitosan in petroleum industry. *Carbohydr. Polym.* **2021**, *266*, 118125. [CrossRef] [PubMed]
45. Elaf, R.; Ben Ali, A.; Saad, M.; Hussein, I.A.; Nimir, H.; Bai, B.J. Biodegradable Preformed Particle Gel (PPG) Made of Natural Chitosan Material for Water Shut-Off Application. *Polymers* **2023**, *15*, 1961. [CrossRef] [PubMed]
46. Elaf, R.; Ben Ali, A.; Saad, M.; Hussein, I.A.; Bai, B.J. Development of eco-friendly chitosan-g-polyacrylamide preformed particle gel for conformance control in high-temperature and high-salinity reservoirs. *Geoenergy Sci. Eng.* **2023**, *230*, 212136. [CrossRef]
47. Zhu, D.Y.; Hou, J.R.; Wei, Q.; Wu, X.; Bai, B.J. Terpolymer Gel System Formed by Resorcinol Hexamethylenetetramine for Water Management in Extremely High Temperature Reservoirs. *Energy Fuels* **2017**, *31*, 1519–1528. [CrossRef]
48. Ahdaya, M.; Al Brahim, A.; Song, T.; Bai, B.J.; Eriyagama, Y.; Schuman, T.; Zhang, Y. Impact of different additives on low-temperature recrosslinkable preformed particle gel to mitigate fluid losses in fractured formations. *Geoenergy Sci. Eng.* **2023**, *224*, 211597. [CrossRef]
49. Durán-Valencia, C.; Bai, B.; Reyes, H.; Fajardo-López, R.; Barragán-Aroche, F.; López-Ramírez, S. Development of enhanced nanocomposite preformed particle gels for conformance control in high-temperature and high-salinity oil reservoirs. *Polym. J.* **2014**, *46*, 277–284. [CrossRef]
50. Salunkhe, B.; Schuman, T.; Al Brahim, A.; Bai, B.J. Ultra-high temperature resistant preformed particle gels for enhanced oil recovery. *Chem. Eng. J.* **2021**, *426*, 130712. [CrossRef]
51. Schuman, T.; Salunkhe, B.; Al Brahim, A.; Bai, B. Evaluation of Ultrahigh-Temperature-Resistant Preformed Particle Gels for Conformance Control in North Sea Reservoirs. *SPE J.* **2022**, *27*, 3660–3673. [CrossRef]
52. Song, T.; Ahdaya, M.; Zhao, S.; Zhao, Y.; Schuman, T.; Bai, B. Evaluation of a Novel Recrosslinkable Hyperbranched Preformed Particle Gel for the Conformance Control of High-Temperature Reservoirs with Fractures. *SPE J.* **2022**, *27*, 3598–3610. [CrossRef]
53. An, Z.B.; Zhou, K.; Wu, D.J.; Hou, J. Production characteristics and displacement mechanisms of infilling polymer-surfactant-preformed particle gel flooding in post-polymer flooding reservoirs: A review of practice in Ng3 block of Gudao Oilfield. *Pet. Sci.* **2023**, *20*, 2354–2371. [CrossRef]
54. Cui, X.-h.; Li, Z.-q.; Cao, X.-I.; Song, X.-w.; Zhang, X. A Novel PPG Enhanced Surfactant-Polymer System for EOR. In Proceedings of the SPE Enhanced Oil Recovery Conference, Kuala Lumpur, Malaysia, 19 July 2011. [CrossRef]
55. Sang, Q.; Li, Y.J.; Yu, L.; Li, Z.Q.; Dong, M.Z. Enhanced oil recovery by branched-preformed particle gel injection in parallel-sandpack models. *Fuel* **2014**, *136*, 295–306. [CrossRef]
56. Chen, X. Method study for application of heterogeneous oil displacement agent. *Oil Drill. Prod. Technol.* **2009**, *31*, 85–88. [CrossRef]

57. Gong, H.J.; Zhang, H.; Xu, L.; Li, K.N.; Yu, L.; Li, Y.J.; Dong, M.Z. Further enhanced oil recovery by branched-preformed particle gel/HPAM/surfactant mixed solutions after polymer flooding in parallel-sandpack models. *Rsc Adv.* **2017**, *7*, 39564–39575. [CrossRef]
58. Xu, L.; Qiu, Z.; Gong, H.J.; Zhu, C.F.; Sang, Q.; Li, Y.J.; Dong, M.Z. Synergy of microbial polysaccharides and branched-preformed particle gel on thickening and enhanced oil recovery. *Chem. Eng. Sci.* **2019**, *208*, 115138. [CrossRef]
59. Zhai, Y.; Fang, Z.; Feng, J.; Sun, C.; Deng, W.; Wen, Y. Development and performance evaluation of re-crosslinkable preformed particle gel under high temperature and high salt conditions. *Polymer* **2024**, *296*, 126824. [CrossRef]
60. Song, T.; Ahdaya, M.; Zhai, Z.M.; Schuman, T.; Bai, B.J. Comprehensive evaluation of a novel re-crosslinkable preformed particle gel for the water management of reservoir with concentrated divalent ions. *Fuel* **2023**, *331*, 125974. [CrossRef]
61. Guo, H.; Ge, J.; Xu, Y.; Lv, Q.; Li, Z.; Zhou, D.; Tao, Z. Preparation and Mechanism of Stability for High-Temperature and High-Salinity Gels. *SPE J.* **2022**, *27*, 3565–3578. [CrossRef]
62. Jia, H.; Kang, Z.; Li, Z.; Zhao, J. The Potential of Ultrahigh Strength Gel through Novel Multistage Reinforcement Method for Sealing Operations in Medium to Ultralow Temperature Reservoirs. *SPE J.* **2022**, *27*, 2145–2160. [CrossRef]
63. Qin, Y.; Zhang, D.; Liu, Y.; Cheng, L. Synthesis and performance evaluation of nanocomposite gel particle with salt tolerance andretarding swelling. *Mod. Chem. Ind.* **2018**, *38*, 171–174. [CrossRef]
64. Wu, Y.; Zhang, H.; Zhang, L.; Huang, Y.; Zhao, M.; Dai, C. Development of Novel Delayed Swelling Polymer Gel Particles with Salt Resistance for Enhanced In-Depth Permeability Control. *SPE J.* **2024**, *29*, 2060–2075. [CrossRef]
65. Bai, B.; Sun, X. Development of Swelling-Rate Controllable Particle Gels to Control the Conformance of CO_2 Flooding. In Proceedings of the SPE Improved Oil Recovery Conference, Virtual, 30 August 2020. [CrossRef]
66. Zhou, B.B.; Kang, W.L.; Jiang, H.Z.; Yang, H.B.; Li, Z.; Lv, Z.Q.; Xu, Z.; Ning, C.Y.; Wang, H.Z.; Xie, S.T. Preparation and crosslinking mechanism of delayed swelling double-crosslinking nano polymer gel microsphere for anti-CO_2 gas channeling. *J. Pet. Sci. Eng.* **2022**, *219*, 111122. [CrossRef]
67. Pereira, K.A.B.; Pereira, K.A.B.; Oliveira, P.F.; Mansur, C.R.E. Behavior of partially hydrolyzed polyacrylamide/polyethyleneimine reinforced with coal fly ash for preformed particle hydrogels. *J. Appl. Polym. Sci.* **2020**, *137*, 49423. [CrossRef]
68. Khoshkar, P.A.; Fatemi, M.; Ghazanfari, M.H. Static and dynamic evaluation of the effect of nanomaterials on the performance of a novel synthesized PPG for water shut-off and improved oil recovery in fractured reservoirs. *J. Pet. Sci. Eng.* **2020**, *189*, 107019. [CrossRef]
69. Kumar, A.; Mahto, V.; Sharma, V.P. Reinforced preformed particle gel: Synthesis, characterization and performance evaluation for water shut-off jobs in heterogeneous reservoir. *J. Pet. Sci. Eng.* **2020**, *193*, 107408. [CrossRef]
70. Paprouschi, A.; Fatemi, M.; Ghazanfari, M.H. Effect of silicate sodium and graphene nanoplatelets on morphology and rheology characteristics of new synthesized preformed particle gel (PPG) for water shut-off treatment. *J. Pet. Sci. Eng.* **2021**, *204*, 108736. [CrossRef]
71. Do Amparo, S.Z.S.; de Vasconcelos, C.K.B.; Almeida, A.; Sena, L.E.B.; Lima, M.; Medeiros, F.S.; Caliman, V.; Silva, G.G.; Viana, M.M. Microwave-assisted synthesis of PAM preformed particle gels reinforced with carbon nanomaterials for conformance control in oil recovery. *Fuel* **2022**, *330*, 125650. [CrossRef]
72. Cheng, L.; Qin, Y.; Gao, K.J.; Zhang, L.Y.; Zhou, J.S.; Zhang, D.X.; Liao, R.Q.; Li, Z. Experimental Investigation of a Novel Nanocomposite Particle Gel for Water Shutoff Treatment in Mature Oilfields. *Acs Omega* **2022**, *7*, 8887–8895. [CrossRef]
73. Kang, W.L.; Wang, J.Q.; Ye, Z.Q.; Gu, G.J.; Li, W.M.; Yang, H.B.; Li, Z.; Xu, H.X.; Lv, Z.Q.; Sarsenbekuly, B. Study on preparation and plugging effect of sawdust gel particle in fractured reservoir. *J. Pet. Sci. Eng.* **2022**, *212*, 110358. [CrossRef]
74. Kumar, A.; Mahto, V.; Sharma, V.P. A systematic multi-scale experimental investigation of a newly developed halloysite nanotubes reinforced preformed particle gel for improving oil recovery from heterogeneous reservoir: Synthesis, characterization, optimization and performance assessment. *Geoenergy Sci. Eng.* **2023**, *226*, 211786. [CrossRef]
75. Pandit, Y.K.; Mahto, V.; Udayabhanu, G.; Ravishankar, B.; Dhandi, M. Preparation, characterization, and evaluation of a nanomaterial reinforced particle gel system for the control of excessive water production in the mature oil fields. *Geoenergy Sci. Eng.* **2024**, *235*, 212707. [CrossRef]
76. Ai, L.Q.; Yang, H.; Wang, S.W.; Zheng, L.J.; Yuan, H.Y.; Yu, X.R.; Su, G.S. Acrylamide preformed particle gels with movable-crosslinking for conformance control. *J. Mol. Liq.* **2023**, *384*, 122217. [CrossRef]
77. Hao, J.; Fan, H.; Kang, W.; Li, Z.; Yang, H.; Sarsenbekuly, B. Preparation and Property Evaluation of Host-guest Inclusion Supramolecular Gel Particles Profile Pontrol Agent. *Oilfield Chem.* **2023**, *40*, 650–655. [CrossRef]
78. Deng, J.A.; Lian, H.Y.; Zhuang, Y.; Zhao, H.H.; Wang, Z.Y.; Tian, Y.J.; Lin, C.; Yuan, H.Z.; Han, M.; Lu, G.W.; et al. Synthesis and performance evaluation of multi-crosslinked preformed particle gels with ultra-high strength and high-temperature and high-salinity resistance for conformance control. *Fuel* **2024**, *357*, 130027. [CrossRef]
79. Khormali, A.; Ahmadi, S.; Kazemzadeh, Y. Inhibition of Barium Sulfate Precipitation During Water Injection into Oil Reservoirs Using Various Scale Inhibitors. *Arab. J. Sci. Eng.* **2023**, *48*, 9383–9399. [CrossRef]
80. Zhao, S.D.; Zhu, D.Y.; Bai, B.J. Experimental study of degradable preformed particle gel (DPPG) as temporary plugging agent for carbonate reservoir matrix acidizing to improve oil recovery. *J. Pet. Sci. Eng.* **2021**, *205*, 108760. [CrossRef]

81. Zhu, D.Y.; Xu, Z.H.; Sun, R.X.; Fang, X.Y.; Gao, D.W.; Jia, X.B.; Hu, J.R.; Weng, J.T. Laboratory evaluation on temporary plugging performance of degradable preformed particle gels (DPPGs). *Fuel* **2021**, *289*, 119743. [CrossRef]
82. Zhang, X.; Deng, J.N.; Yang, K.; Li, Q.; Meng, S.Y.; Sun, X.X.; Song, Z.Z.; Tian, Y.D.; Zhang, S.A.; Liu, X.J.; et al. High-strength and self-degradable sodium alginate/polyacrylamide preformed particle gels for conformance control to enhance oil recovery. *Pet. Sci.* **2022**, *19*, 3149–3158. [CrossRef]

gels

MDPI

Article

Analysis of the Distribution Pattern of Remaining Oil and Development Potential after Weak Gel Flooding in the Offshore LD Oilfield

Lizhen Ge [1], Xiaoming Chen [1], Gang Wang [1], Guohao Zhang [1], Jinyi Li [1], Yang Liu [2], Lixiao Xiao [2], Yuchen Wen [2], Weifeng Yuan [2,*], Ming Qu [3] and Mingxing Bai [4]

1. Tianjin Branch of CNOOC Ltd., Tianjin 300459, China
2. Research Institute of Unconventional Petroleum Science and Technology, China University of Petroleum (Beijing), Beijing 102249, China
3. SANYA Offshore Oil and Gas Research Institute, Northeast Petroleum University, Sanya 572025, China; m.qu@foxmail.com
4. College of Petroleum Engineering, Northeast Petroleum University, Daqing 163318, China

* Correspondence: yuanweifeng_0121@163.com

Abstract: The LD oilfield is one of the representative offshore oilfields. After weak gel flooding, the recovery rate is significantly improved. However, the oilfield is then in a medium- to high-water content stage, presenting a complex distribution of the remaining oil. The measures for further enhanced oil recovery (EOR) are uncertain. As a result, it is necessary to clarify the distribution pattern and development potential of the remaining oil during the high-water content period after weak gel flooding. In this study, an online nuclear magnetic resonance (NMR) oil displacement experiment and microscopic oil displacement experiment were conducted, and the mechanisms of weak gel flooding and the distribution pattern of the remaining oil were clarified in the LD oilfield. Additionally, high-multiple water flooding and numerical simulation experiments were conducted to analyze the development potential after weak gel flooding. The results show that the effect of weak gel flooding was more significant in the core of 1500 mD, with an increase in oil recovery of 9% compared to 500 mD. At a permeability of 500 mD, the degree of crude oil mobilization in micropores and small pores caused by weak gel flooding was improved by 29.64% and 23.48%, respectively, compared with water flooding. At 1500 mD, the degree of crude oil mobilization in small pores caused by weak gel flooding was increased by 37.79% compared to water flooding. After weak gel flooding, the remaining oil was primarily distributed in medium and large pores. Microscopically, the remaining oil was dominated by cluster residual oil, accounting for 16.49%, followed by columnar, membranous, and blind-end residual oil. High multiple water flooding experiments demonstrated that weak gel flooding could significantly reduce development time. The ultimate oil recovery efficiency of 500 mD and 1500 mD reached 71.85% and 80.69%, respectively. Numerical simulation results show that the ultimate oil recovery efficiency increased from 62.04% to 71.3% after weak gel flooding. This indicated that the LD oilfield still had certain development potential after weak gel flooding. The subsequent direction for enhanced oil recovery focuses mainly on mobilizing oil in medium pores or clustered remaining oil. This will play a crucial role in further exploring methods for utilizing the remaining oil and increasing the recovery rate.

Keywords: weak gel flooding; online nuclear magnetic resonance; microscopic oil displacement; remaining oil; development potential

1. Introduction

Prolonged water injection in reservoirs often results in the formation of dominant channels, exacerbating reservoir heterogeneity and leading to fluid channeling. This phenomenon reduces the efficiency of waterflooding within the reservoir [1–3]. Consequently, deep profile control and displacement in the reservoir are essential to mitigate water channeling and improve reservoir heterogeneity [4–7].

In recent years, research on reservoir profile control and displacement technologies has primarily focused on polyacrylamide-based plugging and surfactant-based washing [8]. The introduction of polymers or cross-linked polymer gel plugs effectively seals off water channeling pathways and facilitates subsequent water injection to bypass these channels. The addition of surfactants to the injected water reduces the interfacial tension (IFT) or forms emulsions, thereby leading to enhanced oil recovery (EOR). Microemulsions exhibit interesting physicochemical properties and have a wide range of applications [9–11]. Moreover, surfactant-based microemulsion flooding, a miscible displacement process, significantly reduces IFT and decreases the capillary forces acting on oil droplets, thereby becoming a potential candidate for EOR [12]. The injection of polymer solutions, despite improving the mobility ratio and expanding the sweep efficiency, faces significant shear forces within porous media, leading to considerable viscosity reduction [8,13]. Core flooding experiments have shown that polymer solutions have a limited ability to alter fluid flow due to shear effects. In summary, the direct application of polymer flooding, surfactant flooding, or microemulsion flooding presents certain limitations in strongly heterogeneous reservoirs, including the risk of breakthrough.

To effectively plug water flow channels, it is common to inject plugging agents with greater sealing strength into water injection wells, thereby improving water injection efficiency and increasing oil recovery. Liu et al. [14] proposed a sodium silicate water shut-off agent that can precipitate in the formation, partially or completely blocking the pores in the formation. Bai et al. [15] proposed a polymer gel plugging agent for in situ cross-linking of HPAM with chromium (Cr^{3+}). Zhu et al. [16] recently developed preformed particle gels (PPGs) in which the conformance control mechanism mainly involves physical plugging, including trapping and flocculation. Gallo et al. [17] proposed a resin-type water shut-off agent that can enter the formation after dilution. Under formation temperature, it exhibits extremely high strength after aging and is capable of plugging pores and throats in the formation with the assistance of hardening agents. Recent foreign applications of gel systems, such as in-depth fluid diversion and colloidal dispersion gel techniques, have revolutionized conventional polymer flooding. It can take the advantages of cross-linked polymers to modulate internal flow dynamics in reservoirs [18–21].

Weak gel-based displacement agents are developed from colloidal dispersion gel and bulk gel systems. This is a three-dimensional network structure consisting of low-concentration polymers, cross-linkers, and stabilizers, which are formed mainly via intermolecular cross-linking, supplemented by intramolecular cross-linking [22,23]. The weak gel systems exhibit excellent injectability, transportability, and deep profile control capabilities, allowing them to plug deep reservoir layers and alter the flow channels of injected water [24,25]. When polymers and cross-linkers are injected into heterogeneous porous media, they selectively flow towards higher permeability layers, forming in situ weak gels, increasing flow resistance, and even blocking high-permeability channels. As a result, the injection pressure gradient is increased, forcing subsequently injected water to overcome capillary forces and enter smaller pore channels, mobilizing the remaining oil [26]. Research by Seright and others [27] has extensively investigated the sealing effect of weak gels on fractures, indicating that partially cross-linked weak gels not only reduce damage to non-target layers but also enhance the sealing effect on fractures. Cheng et al. [28] carried on microscopic visualization experiments and numerical simulations and indicated that weak gels could effectively reduce the water-to-oil mobility ratio. High-permeability layers and large pore channels are selectively blocked, subsequent water flow is forced to redirect, and viscoelastic movement is used to generate negative pressure for oil suction. Cao et al. [29] developed a formula for weak gel systems based on the characteristics of high temperature, high permeability, and large pores in the Y1 block of the Daqing Yushulin Oilfield. These systems undergo cross-linking reactions upon entering deep strata, achieving deep control over oil displacement. Core plugging and oil recovery tests indicated that the gel system could be easily injected into porous media and then formed into gel in situ. Afterwards, the high-permeability layers were selectively plugged, and subsequent water flowed into low-permeability layers, thus improving sweep efficiency and ultimate oil recovery [30]. Wang et al. [31] evaluated weak gel under different conditions, demonstrating its superior profile control and oil displacement performance compared to polymers. Successful field tests of weak gel flooding were conducted in the LD oilfield, with most production wells around the weak gel injection wells responding post-injection. The injection pressure significantly increased, and the average daily oil production of the responding wells increased by 33.2 m^3/d, indicating a notable increment in oil recovery.

Laboratory experiments and field pilot tests suggest that weak gel deep profile control is a crucial method for improving the sweep efficiency of injected water and enhancing the ultimate recovery of high-water content reservoirs. While extensive research has been conducted on evaluating the effectiveness and mechanisms of weak gel flooding, the distribution patterns of the remaining oil after weak gel flooding and its development potential have not been sufficiently addressed.

At the same time, as oil fields on land or at sea re-enter a period of high-water content following chemical EOR processes, the direction for further EOR techniques remains uncertain. Hence, after chemical flooding such as weak gel or polymer, the research direction of further improving oil recovery has gradually attracted people's attention. What is more, the distribution of the remaining oil becomes more complex after chemical EOR, posing challenges to the implementation of additional recovery methods. Therefore, to explore directions for further EOR techniques, it is essential to first understand the distribution patterns and development potential of the remaining oil after weak gel flooding. Only by comprehending these distribution patterns can we accurately investigate how to mobilize the remaining oil.

The LD oilfield is located in the Liaodong Bay of the Bohai Sea, in the middle and southern part of the Liaoxi Low Bulge. It belongs to the fractured semi-dorsal tectonic structure developed on the submerged mountains. The provenance is from a northwesterly direction. The depositional environment is the front edge of the Braid River delta, with a predominantly submerged diversion channel developing. The reservoir type is a stratified tectonic reservoir influenced by lithology, with a large reservoir thickness and good physical properties. It is a highly porous, medium-high permeability sandstone reservoir. The reservoir porosity is mainly distributed between 25% and 31%. The permeability is mainly distributed between 400 mD and 2500 mD. As a typical offshore field, it has an exploitation history of twenty years. The development history of LD oilfield showed that when the water cut was 9%, weak gel flooding was carried out to slow down the water cut rise rate. However, it is now entering a period of high-water content. The unclear understanding of distribution patterns of the remaining oil post-weak gel flooding leads to uncertainties regarding further EOR methods post-weak gel flooding.

In addition, according to the usual experimental specifications, the experiment of water flooding generally ends after the water cut reaches 99.95% or displaces 30 PV. However, for offshore reservoirs of high porosity and high permeability, the displacement efficiency could be greatly enhanced through high-intensity and high-multiplicity (far beyond 30 PV) water flooding processes. Meanwhile, there is an increasing trend in the core recovery degree as displacement multiplies. In this case, accurately assessing the development potential of offshore reservoirs is difficult based on general experimental specifications. Li et al. [32] pointed out that for heavy oil reservoirs with bottom water in Bohai Oilfield, especially in the high-water content stage, water flooding with a high multiplicity of 2000 PV can significantly improve the oil displacement efficiency. Therefore, the development potential of the Bohai LD oilfield can be more comprehensively analyzed after weak gel flooding via limited water flooding (2000 PV high-multiple water flooding).

In response to the challenges faced by the LD oilfield during the high-water content period following weak gel flooding (such as the unclear distribution patterns of remaining oil), the current potential for further development, and the ambiguous direction for further

EOR techniques, this study employed experimental simulations to trace the developmental history: at a water cut of 9%, a 0.4 PV of weak gel was injected.

Firstly, this study employed nuclear magnetic resonance online displacement experiments and microscopic displacement experiments to simulate the weak gel flooding process in the offshore LD oilfield. It aimed to determine the effects and mechanisms of weak gel flooding under different permeabilities and understand the microscopic distribution patterns of the remaining oil post-weak gel flooding. Furthermore, the development potential was analyzed through 2000 PV high-multiple water flooding and numerical simulations in the offshore LD oilfield. By clarifying the distribution patterns of the remaining oil post-weak gel flooding and the development potential in the offshore LD oilfield, it lay on a significant foundation for further EOR techniques in offshore oil fields following weak gel flooding.

2. Results and Discussion

2.1. Evaluation of Weak Gel Flooding Effect

2.1.1. Weak Gel Displacement NMR T_2 Spectrum

This study selected representative natural core samples from the LD oilfield, with permeabilities of 500 and 1500 mD, extracted from a depth of 1581 m, to simulate the development process of weak gel profile control. When the water cut reached 9% during the initial water flooding, 0.4 PV of weak gel was injected, followed by subsequent water flooding. The results of the weak gel flooding NMR online displacement experiment are depicted in Figure 1.

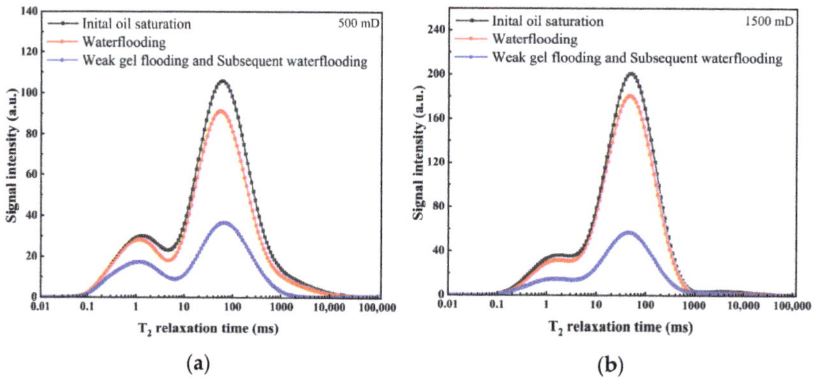

Figure 1. T_2 spectrum of weak gel flooding: (**a**) 500 mD core; (**b**) 1500 mD core.

From Figure 1a, it is evident that at a permeability of 500 mD, there is a significant reduction in the T_2 spectrum curve after water flooding of 0.4 PV, mainly around a relaxation time of 100 ms, indicating a relatively low degree of water drive. The oil displacement efficiency of water flooding is 13.8%. After injecting 0.4 PV of weak gel, the overall T_2 spectrum curve shows a marked decline, with the oil displacement efficiency during the weak gel flooding stage increasing to 60.62%, an improvement of 46.82%. This efficiency boost is attributed to the selective nature of the weak gel, which preferentially enters medium to large pores, increasing the viscosity of the displacing phase and exhibiting viscoelastic properties, effectively displacing the crude oil from medium and large pores. The significant decrease in signal quantity for relaxation times of 10~100 ms and 100~1000 ms supports this observation.

Furthermore, the reduction in signal quantity for relaxation times of 1~10 ms is mainly due to the weak gel increasing the flow resistance in high-permeability channels, forcing the injected fluid to flow towards low-permeability channels, thus enhancing oil displacement efficiency in smaller pores. During the subsequent water flooding stage, the signal quantity for relaxation times of 1~10 ms further decreases as the weak gel blocks high-permeability channels, forcing subsequent injected water towards lower-permeability channels, thereby further improving oil displacement efficiency in smaller pores.

For the core with a permeability of 1500 mD, the oil displacement efficiency during waterflooding was 10.6%. After injecting 0.4 PV of weak gel, the displacement efficiency increased to 68.8%, an improvement of 58.2%. The trend of the T_2 spectrum curve during different stages was essentially consistent with that observed in the 500 mD core. Due to the more severe heterogeneity in the high-permeability core, the effect of weak gel profile control was more pronounced, with the oil displacement efficiency during the weak gel flooding stage being 9% higher compared to the 500 mD core. Consequently, the weak gel has a significant profile control effect in both 500 mD and 1500 mD cores, effectively controlling the channeling pathways and improving oil displacement efficiency.

2.1.2. Utilization of Crude Oil in Pores of Different Sizes

In the nuclear magnetic resonance (NMR) T_2 spectrum, the horizontal axis representing relaxation time corresponds to the sizes of pores within the core, while the vertical axis signal amplitude indicates the oil content in pores of different sizes. The relaxation times of 0.1~1, 1~10, 10~100, and 100~1000 ms are defined as micropores, small pores, medium pores, and large pores, respectively. By analyzing the signal quantity in pores of various sizes, the effect of weak gel on mobilizing crude oil in different pore sizes can be assessed.

Figures 2 and 3 illustrate the mobilization of crude oil in different pore sizes during water flooding and weak gel flooding stages in cores with different permeabilities. From Figure 2, for a core with a permeability of 500 mD, the overall degree of mobilization during the water flooding stage is relatively low, with micropores, small pores, medium pores, and large pores showing mobilization of 2.57%, 14.09%, 13.93%, and 17.14%, respectively. During the weak gel flooding stage, there is noticeable mobilization of oil in micropores and small pores, increasing to 32.21% and 37.57%, respectively. The degree of oil mobilization in medium and large pores is higher, at 53.24% and 45.76%, respectively. This indicates that the weak gel preferentially enters high-permeability channels, fully mobilizing oil in medium and large pores. Subsequently, by increasing the flow resistance in high-permeability channels, the fluid is redirected towards lower-permeability channels, enhancing the mobilization of oil in medium and small pores. For a core with a permeability of 1500 mD, as shown in Figure 3, the degrees of mobilization during the water flooding stage for micropores, small pores, medium pores, and large pores are 17.72%, 12.25%, 10.11%, and 9.14%, respectively. During the weak gel flooding stage, the degrees of mobilization in micropores, small pores, medium pores, and large pores increase to 34.46%, 50.04%, 60.93%, and 62.03%, respectively.

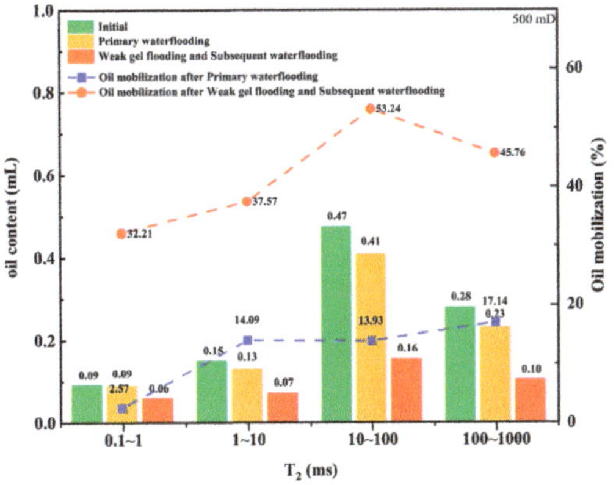

Figure 2. Oil content and mobilization degree of various pores in different stages (500 mD).

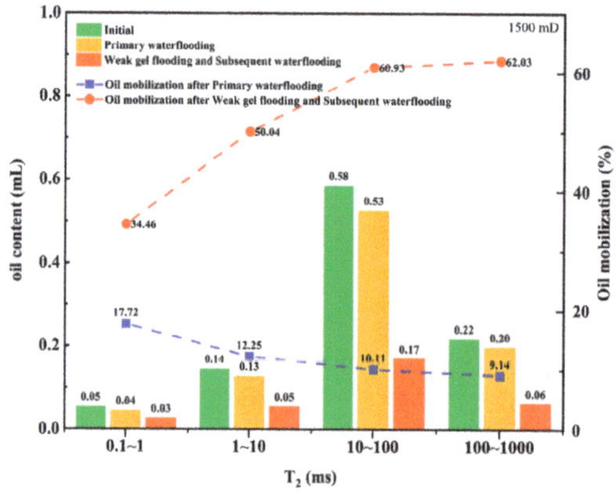

Figure 3. Oil content and mobilization degree of various pores in different stages (1500 mD).

Furthermore, the contribution of oil production from different pore sizes during the water flooding and weak gel flooding stages was analyzed. This refers to the proportion of oil produced from each pore relative to the total oil produced. As shown in Figure 4a, for a core with a permeability of 500 mD during the water flooding stage, the majority of the produced oil comes from medium and large pores, accounting for 47.94% and 34.86%, respectively, followed by small pores at 15.48%. The contribution from micropores is the lowest at only 1.73%. After the injection of weak gel, as depicted in Figure 4b, during the weak gel flooding stage for a 500 mD core, due to the higher initial oil content in medium and large pores, the majority of produced oil still comes from these pores, with contribution rates of 54.02% and 27.45%, respectively. The contribution from small pores increased to 12.17%, and the contribution from micropores increased from 1.71% to 6.37%. This indicates that the weak gel effectively blocked high-permeability channels, forcing the fluid to flow towards lower-permeability channels, thereby enhancing the oil displacement efficiency in smaller pores.

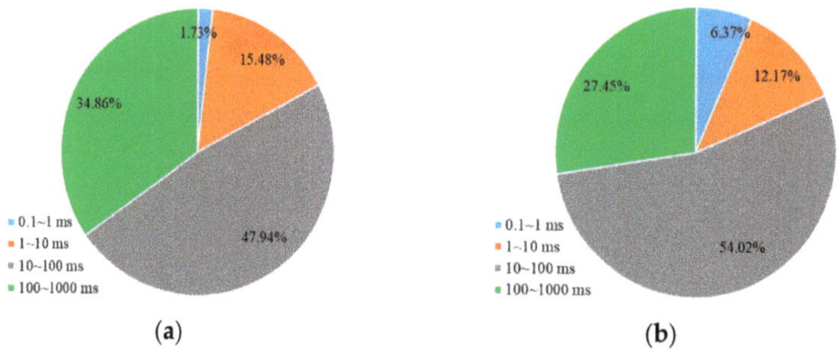

Figure 4. The percentage of oil production from different pore sizes relative to the total oil production (500 mD): (**a**) waterflooding; (**b**) weak gel and subsequent waterflooding.

For the core with a permeability of 1500 mD, as shown in Figure 5, during the water flooding stage, the majority of the produced oil comes from medium pores, accounting for 55.70% of the total. The contributions from small and large pores are relatively close at 16.64% and 18.71%, respectively, with micropores contributing 8.95%. During the weak gel flooding stage, as indicated in Figure 5b, the contribution from medium pores increases to 61.25%, while large pores contribute 23.16%. The contributions from small pores and micropores are 12.41% and 3.18%, respectively. This suggests that in a higher-permeability core, medium and large pores play a more significant role in oil production, and the use of weak gel enhances the mobilization and production of oil from these pores.

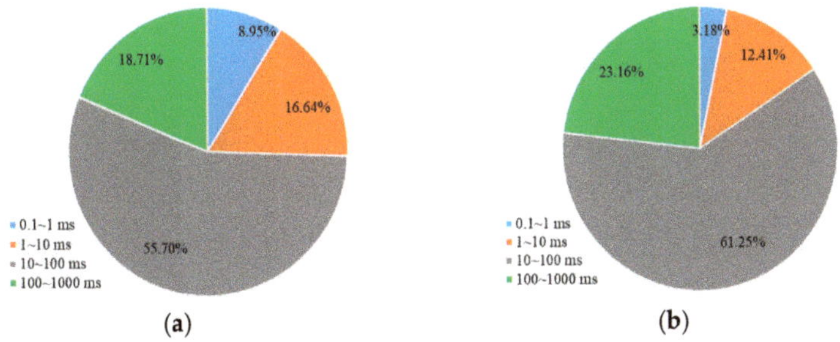

Figure 5. The percentage of oil production from different pore sizes relative to the total oil production (1500 mD): (**a**) waterflooding; (**b**) weak gel and subsequent waterflooding.

2.2. *Mechanism of Weak Gel Flooding*

2.2.1. Profile Control

Initially, due to the viscosity properties of the weak gel, the capillary number was increased during the weak gel injection phase, further mobilizing the remaining oil in high-permeability channels, with the mobilization degree of remaining oil in large pores increasing from 17.14% to 45.76% (Figure 2). Theoretically, the capillary number during weak gel flooding should only increase by a few times compared to water flooding and is effective only during the weak gel injection phase, which would not significantly enhance oil displacement efficiency. However, the experimental results (Figures 2 and 3) show a significant improvement in displacement efficiency across cores with varying permeabilities. Thus, a more crucial mechanism by which weak gel flooding significantly enhances recovery is through its deep profile modification capabilities. The weak gel preferentially enters high-permeability channels with lower flow resistance, further mobilizing crude oil in medium and large pores. Additionally, due to the higher viscosity of weak gel compared to water, the flow resistance in high-permeability channels is increased, potentially even blocking these channels, leading to an increased injection pressure gradient. This forces the subsequently injected water to overcome the capillary forces in small pores, thereby mobilizing the remaining oil in micropores and small pores of the core. In the 500 mD core, the mobilization degree of the remaining oil in micropores and small pores increased from 2.57% to 32.21% and from 14.09% to 37.57%, respectively (Figure 2). In the 1500 mD core, the mobilization degree of the remaining oil in micropores and small pores increased from 17.72% to 34.46% and from 12.25% to 50.04%, respectively (Figure 3).

2.2.2. Microscopic Mechanism

Based on the thin cast sections of natural cores from the LD oilfield, a microscopic visualization model was created to conduct weak gel microscopic oil displacement experimental research. As indicated in Figure 6, during the initial water flooding, the injected water first flows along high-permeability channels, forming dominant channels in the central part of the model. After the injection of weak gel, the remaining oil in the peripheral regions of the model begins to be mobilized. In the subsequent water flooding stage, the remaining oil in these peripheral areas is further effectively mobilized. A detailed analysis of the microscopic model, particularly after the injection of weak gel, reveals the following, as shown in Figure 7a,b: the weak gel initially prefers to enter larger pores, thereby increasing the flow resistance in dominant channels. This alteration forces the subsequent fluid to flow towards areas that have not yet been reached by the initial water flooding (as depicted in Figure 7c,d), initiating the mobilization of remaining oil in the peripheral regions of the model. This process illustrates the effectiveness of weak gel in altering the flow paths within the reservoir and enhancing oil recovery.

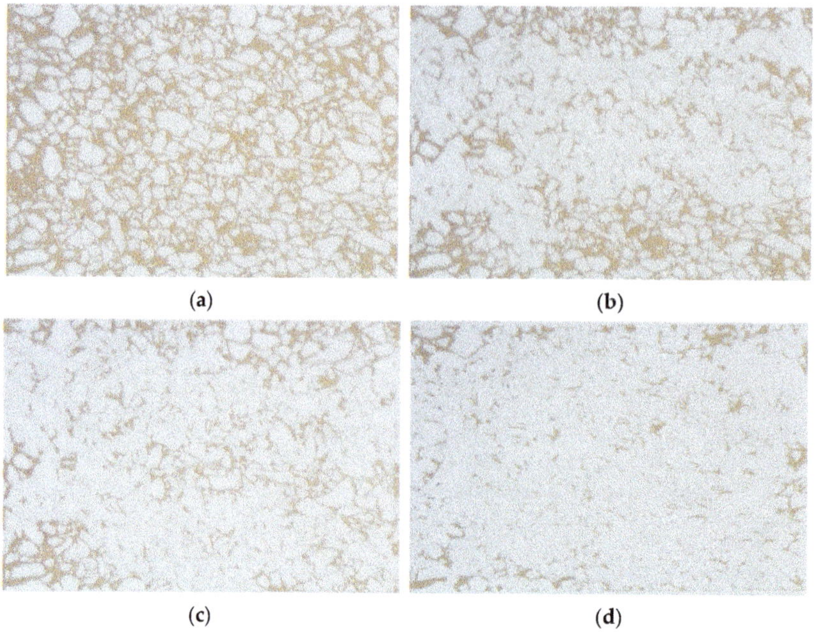

Figure 6. Weak gel micro-displacement process: (**a**) initial state; (**b**) water flooding forms a dominant channel; (**c**) weak gel flooding; (**d**) subsequent water flooding.

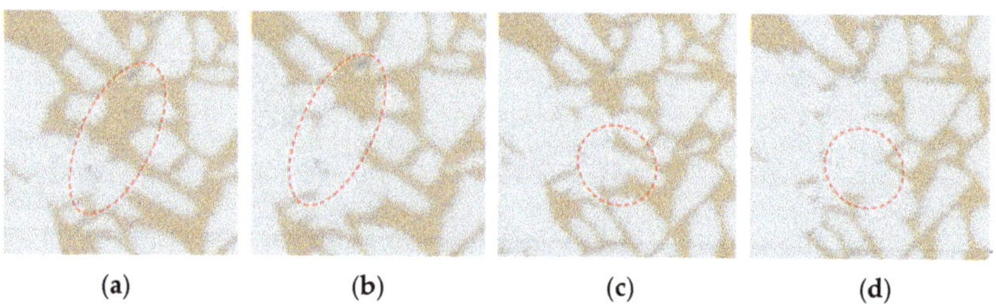

Figure 7. Microscopic flow process of weak gel: (**a**,**b**) weak gel selects into large pore; (**c**,**d**) the flow is diverted.

2.3. Distribution Patterns of Remaining Oil after Weak Gel Flooding

The online nuclear magnetic resonance (NMR) oil displacement and microscopic oil displacement experiments demonstrate that weak gel profile control is significantly effective. However, after weak gel profile control, some of the remaining oil remains undisturbed in the core. Understanding the distribution pattern of this remaining oil after weak gel profile control is crucial to guiding further research and application in oil field development. An analysis of the remaining oil post-weak gel flooding, as shown in Figure 8, reveals that the distribution of remaining oil is scattered, with types including clustered, columnar, membranous, and dead-end remaining oil. A statistical analysis, following the identification of different types of remaining oil, indicates that clustered remaining oil is the predominant form in the core's pore channels post-weak gel profile control, accounting for 16.49%. This is followed by membranous remaining oil at 9.55%, columnar remaining oil at 5.89%, and dead-end remaining oil, constituting only 0.72%. The distribution pattern of microscopic remaining oil suggests that the primary target for mobilization post-weak gel profile control should be clustered remaining oil.

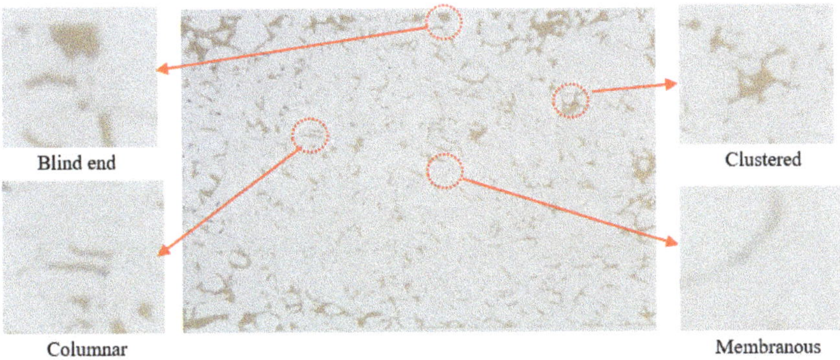

Figure 8. Types of microscopic remaining oil after weak gel flooding. (The red circles represent different types of remaining oil).

Further analysis was conducted on the remaining oil content in different pore sizes of cores with varying permeabilities after weak gel profile control. This analysis was similar to the method described in Section 2.1.2, as depicted in Figures 9 and 10. For a core with a permeability of 500 mD, during the water flooding stage, the produced crude oil accounted for 13.80% of the initial saturated oil. The remaining oil in micropores, small pores, medium pores, and large pores constituted 9.02%, 13.03%, 23.27%, and 40.88% of the initial saturated oil, respectively. During the weak gel profile control stage, the cumulative produced crude oil accounted for 60.62%, with the remaining oil in micropores, small pores, medium pores, and large pores constituting 6.03%, 7.33%, 15.60%, and 10.42%, respectively. For a core with a permeability of 1500 mD, during the water flooding stage, the produced crude oil accounted for 10.62% of the initial saturated oil. The remaining oil in micropores, small pores, medium pores, and large pores constituted 4.41%, 12.66%, 52.57%, and 19.74% of the initial saturated oil, respectively. During the weak gel profile control stage, the cumulative produced crude oil accounted for 68.80%, with the remaining oil in micropores, small pores, medium pores, and large pores constituting 2.56%, 5.44%, 16.94%, and 6.26%, respectively.

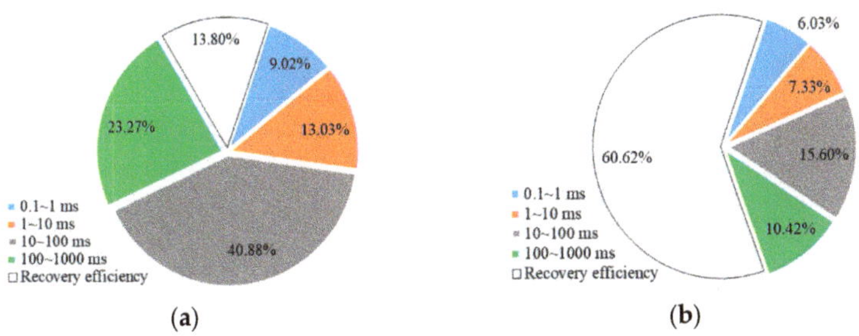

Figure 9. Distribution of remaining oil in pores at different stages and stage recovery efficiency (500 mD): (**a**) waterflooding; (**b**) weak gel and subsequent waterflooding.

After weak gel profile control, the remaining oil is still primarily distributed in medium and large pores. This is because the weak gel's displacement distance is limited, and its profile control action is mainly concentrated in the front half of the core, resulting in a limited effect in the latter half and leaving a substantial amount of remaining oil. Additionally, the proportion of remaining oil in micropores and small pores increases. For a core with a permeability of 500 mD, during the water flooding stage, the remaining oil in micropores and small pores accounted for 9.02% and 13.03% of the total remaining oil, respectively. At the end of the weak gel profile control stage, these percentages increased to 15.34% and 18.61%. For a core with a permeability of 1500 mD, the proportion of remaining oil in micropores increased from 4.93% to 8.21%, and the proportion in small pores increased from 14.16% to 17.44%. Therefore, after weak gel profile control, the primary targets for further enhanced oil recovery efforts are medium and large pores. However, attention should also be given to the remaining oil in micropores and small pores, indicating a need for more refined strategies to address these challenging areas for improved oil recovery.

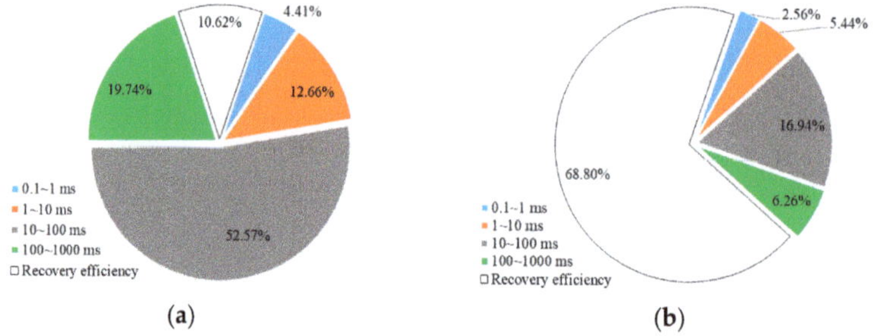

Figure 10. Distribution of remaining oil in pores at different stages and stage recovery efficiency (1500 mD): (**a**) waterflooding; (**b**) weak gel and subsequent waterflooding.

2.4. *Analysis of Development Potential after Weak Gel Flooding in Offshore LD Oilfield*

After implementing weak gel profile control in the Bohai LD oilfield, significant increases in injection pressure and enhanced oil recovery were observed. However, the field has now entered a mid-to-high water cut stage, with a water cut of 87.4% as of February 2023. It is imperative to analyze the development potential of the LD oilfield. By further analyzing the ultimate oil displacement efficiency and target recovery rates post-weak gel profile control, the long-term development effects can be predicted. Understanding the development potential post-weak gel profile control in the Bohai LD oilfield, combined with the distribution patterns of remaining oil after such control, is crucial for guiding further research on enhancing recovery rates.

2.4.1. Evaluation of the Ultimate Oil Displacement Efficiency

Simulating the weak gel flooding process of the LD oilfield, 0.4 PV of weak gel was injected when the water cut reached 9%, followed by a high multiple water flooding experiment of 2000 PV. The change in oil displacement efficiency with the injected PV number was recorded, with results shown in Figure 11. After the injection of weak gel, the oil displacement efficiency rapidly increased. After 2000 PV of high multiple water flooding, the ultimate oil displacement efficiencies for 500 mD and 1500 mD were 71.85% and 80.69%, respectively. The ultimate oil displacement efficiency for single water flooding ranged between 68.8% and 76.5% [23]. After weak gel flooding, the increase in ultimate oil displacement efficiency was not significant. However, the corresponding PV number for reaching a near-flat efficiency curve significantly reduced from 100 PV in water flooding to 14~20 PV post-weak gel flooding [23]. Given the limited operational life of offshore platforms, weak gel flooding can substantially reduce development time costs.

Furthermore, the current water cut of the Bohai LD Oilfield is 87.4%, with oil displacement efficiencies for 500 mD and 1500 mD being 60.86% and 69.12%, respectively. When the water cut reaches 98%, the displacement efficiencies could increase to 62% and 73.12%, respectively. In the ultimate water flooding scenario, these efficiencies could further rise to 71.85% and 80.69%. Considering the current development status of the LD Oilfield, the oil displacement efficiency could potentially be improved by approximately 11%, indicating that the Bohai LD Oilfield still has significant development potential post-weak gel profile control.

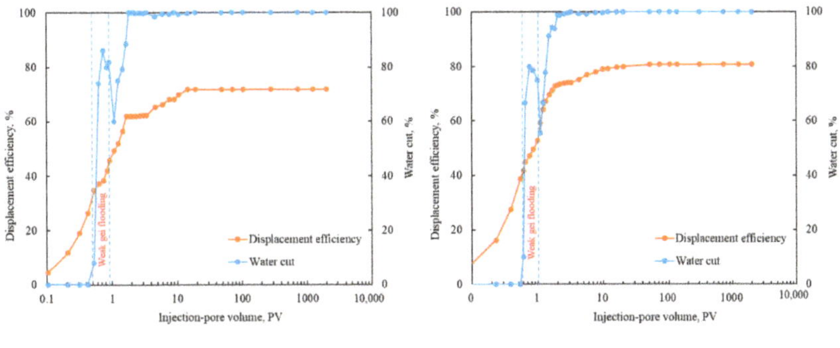

(a) **(b)**

Figure 11. A high-multiple water flooding experiment of 2000 PV: (**a**) 500 mD core; (**b**) 1500 mD core.

2.4.2. Evaluation of Target Recovery Rate

An in-depth analysis of the LD oilfield's development potential post-weak gel flooding was conducted using CMG numerical simulation methodology. This analysis focused on crucial indicators such as the target recovery rate and maximum sweep coefficient. Leveraging geological, lithological, and reservoir properties, as well as development methods data specific to the LD oilfield (refer to Table 1), a conceptual pattern of a row–column staggered well network model was established. The foundational geological parameters comprised a five-spot well pattern, well spacing of 200 m, an 80 m reservoir thickness, and vertical division into eight sub-layers (from top to bottom: 20 m, permeability of 1720 mD; 15 m, 1320 mD; 10 m, 970 mD; 5 m, 1040 mD; 6 m, 970 mD; 8 m, 1280 mD; 5 m, 850 mD; 11 m, 790 mD). The model's total volume was 6200 × 10^4 m^3, as illustrated in Figure 12. This model serves as a basis for simulating the weak gel profile control process.

Table 1. Conceptual model parameters.

Parameters	Value
Grid step size (m)	1550 × 500 × 80
Number of grids	91 × 50 × 8
Porosity (%)	26
Permeability variation coefficient	0.275
Mean permeability (mD)	1117.5
Oil saturation	0.65
Reservoir thickness (m)	80
Crude oil viscosity (mPa·s)	14
Formation water viscosity (mPa·s)	0.46
Weak gel injection (PV)	0.4

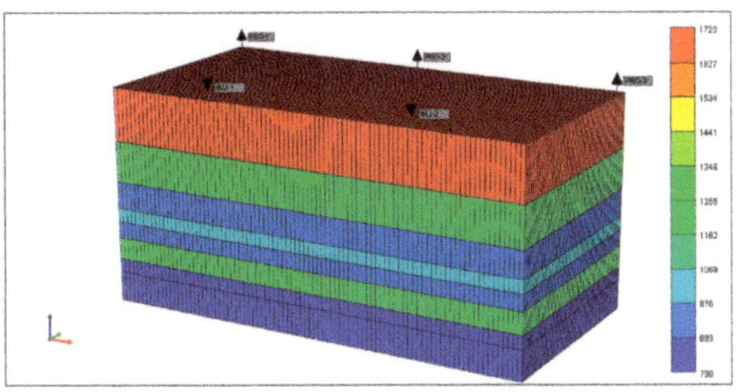

Figure 12. Establishment of the conceptual model.

Simulations of water flooding and weak gel flooding were conducted to determine the target recovery rates and maximum sweep efficiencies under different development methods, thereby assessing the development potential of the LD Oilfield. The results of the simulations for water flooding and weak gel flooding are depicted in Figure 13.

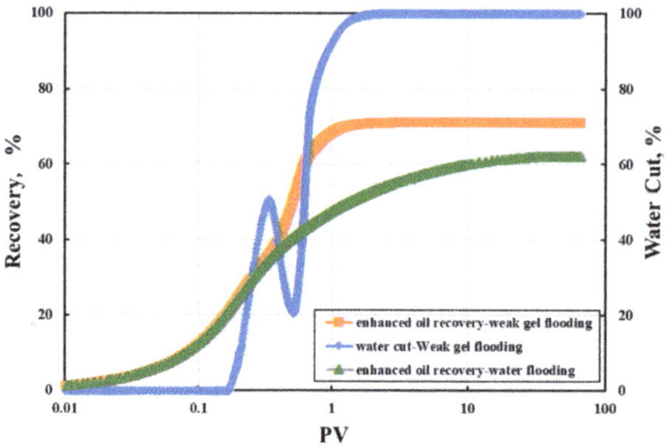

Figure 13. Recovery efficiency and water cut under water flooding and weak gel flooding.

During the initial phase of water flooding, the recovery rate increases slowly, followed by a rapid increase. The recovery rate reaches 11.05% when the cumulative injection volume is 0.09 PV. It continues to increase swiftly, reaching 38.9% at a cumulative injection of 0.45 PV. Subsequently, the rate of increase slows down, and the target recovery rate ultimately reaches 62.04% when the cumulative injection volume is 31.41 PV. The long-term water flooding simulation of the LD Oilfield indicates that the increase in recovery rate slows down in the later stages of development. Additionally, after the limit of water flooding is reached, a significant amount of remaining oil remains unexploited. Given the constraints of the operational lifespans of offshore oilfield platforms, long-term water flooding in the LD Oilfield has certain limitations.

Further assessment of weak gel flooding's effect on the LD Oilfield was conducted by simulating the reservoir history. Upon reaching a water cut of 9% during water flooding, 0.4 PV of weak gel was injected, followed by continued water flooding. The simulation results are illustrated in Figure 13. It is observable from Figure 13 that after the initial injection of weak gel, the rate of recovery increased gradually but then rapidly escalated, ultimately reaching a maximum recovery rate of 71.3%. This represents a 9.26% increase compared to water flooding. Notably, the water cut rose rapidly to 50.23% upon reaching an accumulated injection volume of 0.33 PV, followed by a decline to 20.93% due to the weak gel blocking the high-permeability channels. During this phase, the rapid increase in the recovery rate was attributed to the weak gel entering the porous media of the reservoir, primarily infiltrating the high-permeability layers. Chemical adsorption and mechanical trapping caused the weak gel to be retained in these high-permeability layers, generating additional flow resistance and thereby expanding the affected volume. As the single-phase water cut reached 95%, employing reservoir engineering techniques computed a strong affected volume of 23.65% and a moderate affected volume of 76.35% during water flooding, indicating limited strong affected areas due to the formation of water dominant channels. In contrast, during weak gel flooding at a 95% water cut, the calculated strong affected volume was 93.69%, with a weak affected volume of 6.31%. Compared to water flooding, weak gel flooding expanded the affected range by 70.04%.

Moreover, the minimum PV required to approach the target recovery rate for weak gel flooding was 1.28 PV, significantly lower than the 15 PV for water flooding, substantially reducing development time. Results from ultimate oil displacement experiments and numerical simulations indicate that compared to pure water flooding, weak gel flooding exhibits better development potential. Post-weak gel flooding, the oil recovery efficiency of the LD oilfield can be enhanced to 71.85% and 80.69%, consequently increasing the target recovery rate from 62.04% to 71.3%.

3. Conclusions

Understanding the remaining oil distribution pattern and development potential post-weak gel flooding in the offshore LD oilfield holds crucial significance in guiding further studies aimed at EOR. The effects of weak gel flooding in the LD oilfield are remarkable.

1. Experimental findings indicate that weak gel preferentially enters the intermediate to large pores (10~100 ms, 100~1000 ms), amplifying the flow resistance in high-permeability channels. This compels injected fluids to flow towards low-permeability channels, thereby enhancing the utilization of smaller pores (1~10 ms).
2. Post-weak gel flooding, the remaining oil is predominantly found in the intermediate to large pores while the distribution of microscopic remaining oil is scattered, predominantly consisting of cluster-shaped oil, accounting for 16.49%, followed by columnar, membranous, and dead-end oil.
3. To further increase the oil recovery, the focus should be on exploiting the remaining oil in the intermediate pores or predominantly targeting cluster-shaped and membranous residual oil. In comparison to the current development status of the LD Oilfield, the oil displacement efficiency can be enhanced by approximately 11%.
4. Moreover, the maximum target oil recovery can reach 71.3%, indicating that even post-weak gel flooding, the offshore LD oilfield still retains a certain level of development potential.

4. Materials and Methods

To investigate the distribution patterns and development potential of remaining oil during the high water cut period after weak gel flooding in the LD oilfield, this study explores the distribution of remaining oil on both the macroscopic (using real core models) and microscopic scales. Initially, displacement experiments were conducted using natural cores and crude oil to simulate the displacement history of the LD oilfield, ensuring experimental repeatability. The first phase involved water flooding until the water cut at the core outlet reached 9%, marking the transition to the second phase of injecting 0.4 PV of weak gel. The third phase resumed water flooding until the water cut reached 98%, at which point the experiment was terminated. Nuclear magnetic resonance scanning tests were performed at different displacement stages to analyze the T_2 spectrum curves, where the relaxation time on the x-axis represented pores of different sizes in the core and the signal intensity on the y-axis indicated oil content, facilitating the characterization of the distribution and changes in remaining oil within different pores.

Furthermore, based on the thin cast sections of the LD oilfield's natural cores, the pore throat structures were extracted to create microscopic models that represent the core's real microscopic structure, enhancing the experiment's repeatability. Subsequent microscopic visual displacement experiments involved water flooding, weak gel flooding, and subsequent water flooding to identify the types and patterns of remaining oil distribution at different stages.

Lastly, to further explore the development potential of the LD oilfield, considering the uniqueness of offshore fields where platforms have limited lifespans, often leading to aggressive injection and production strategies, using a water cut of 98% as the endpoint for displacement experiments may not be suitable for offshore fields. In other words, when the water cut reaches 98%, the water injection volume is generally a few PVs. Typically, the water injection multiples at the end of development in offshore fields far exceed those in the experiments. Thus, to more accurately assess the development potential of the LD oilfield under extreme conditions, this study extended water flooding beyond a 98% water cut until reaching an injection multiple of 2000 PV, thereby determining the ultimate recovery rates under extreme water flooding conditions in the LD oilfield. The detailed experimental steps are as follows.

4.1. Experimental Materials and Instruments

The experimental oil was simulated to be composed of crude oil and kerosene, and the simulated oil viscosity was 14 cP at 65 °C. The experimental water was simulated formation water with a salinity of 1924 mg/L. The experimental weak gel was composed of a polymer (1200 mg/L) and a chromium cross-linking agent, with a mass ratio of polymer to Cr^{3+} of 180:1. When the shear rate was 7.34 s^{-1}, the apparent viscosity of weak gels was 8 mPa·s. For a more detailed description of the rheological properties, refer to Reference [33]. The polymer utilized was partially hydrolyzed polyacrylamide (HPAM) with a molecular weight of 2000 $\times$ 10^4, an effective cost content of 90%, and a hydrolysis degree of 25%, procured from Beijing ZKCHEM Co., LTD, Beijing, China. The cross-linker employed was organic chromium, with an active Cr^{3+} concentration of 3.6%. Firstly, Cr^{3+} formed polynuclear olation complex ions through a complexation reaction, hydrolysis reaction, and olation reaction. Then, polynuclear olation complex ions coordinated with $-CONH_2$ and $-COO^-$ in HPAM to fabricate a weak gel possessing a reticular structure. For a more detailed description of the reaction process, refer to Reference [31]. A schematic diagram of the weak gel molecular structure is depicted in Figure 14.

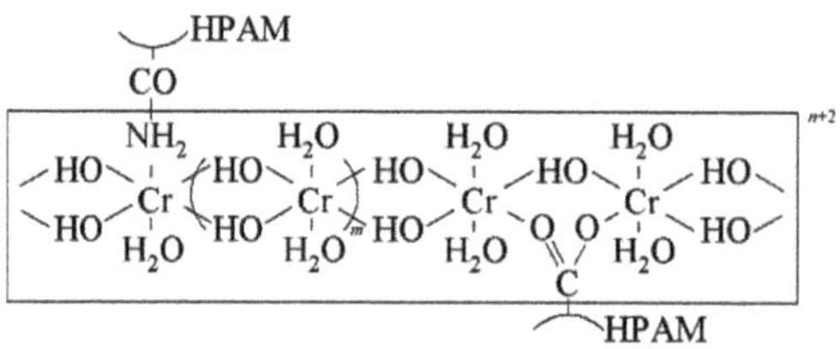

Figure 14. A schematic diagram of the weak gel molecular structure [31].

The experimental sandstone cores were from the lower part of the East Second Section of the LD oilfield. The termination depth of the core was 1581 m in the lower part of the East Second Section. The average permeabilities were 500 mD and 1500 mD, respectively. The LD oilfield belongs to the fractured semi-dorsal tectonics developed on the submerged mountains. The main oil-containing reservoir in the lower part of the East Second Section is deposited on the front edge of the braided river delta. It is dominated by the development of submerged, diverging river sand bodies.

The experimental equipment includes a microscopic displacement apparatus comprising a microscopic model holder, an intermediate container, a micro displacement pump, and a microscope. Additionally, a nuclear magnetic resonance imaging analyzer (MesoMR23-60H, Newmay Instruments Co., Ltd., Suzhou, China) was used for analysis. The displacement experimental setup involved a temperature-controlled chamber, core holder, ISCO high-precision displacement pump, pressure sensor, and intermediate container, among other components.

4.2. Experimental Methods

4.2.1. Online Nuclear Magnetic Resonance (NMR) Oil Displacement Experiment

1. Natural core samples were cleaned, dried, and had their basic parameters measured. The cores were subjected to vacuum treatment, followed by saturation with simulated formation water. After saturation, the T_2 spectrum of the nuclear magnetic resonance was scanned.
2. The cores were displaced with simulated formation water containing Mn^{2+} ions at a constant rate to thoroughly displace the original water and eliminate the water signal within the cores.

3. Simulated formation water was displaced with simulated oil at a constant rate until the oil content at the outlet of the core reached 100%. This established the initial oil saturation, and the nuclear magnetic resonance T_2 spectrum was measured to determine the original oil saturation.
4. Water flooding was conducted using simulated formation water containing Mn^{2+} ions at a constant flow rate of 0.2 mL/min, simulating the oil field process. After injecting the weak gel, water flooding was continued until the water saturation reached 98%. The NMR T_2 spectrum was measured during the displacement process, alongside recording the liquid production and oil recovery at different time intervals.

4.2.2. Microscopic Displacement Experiment

1. Based on the cast thin sections of natural cores from Field LD, pore throat distribution characteristics were extracted to simulate etched microscopic glass models.
2. The model was treated with dimethyldichlorosilane for 48 h at 60 °C, treating the pore structure of the microscopic model to alter its wettability.
3. The model was saturated with simulated oil at a constant rate of 0.02 mL/min. Subsequently, the simulation of the field displacement process began with a constant injection rate of 0.05 mL/min, involving water flooding, weak gel flooding, and subsequent water flooding.

4.2.3. Evaluation of the Ultimate Oil Recovery Efficiency Experiment

Following step 4 of Section 4.2.1, water flooding continued until 2000 PV were displaced while recording the liquid production at the outlet of the core.

Author Contributions: Conceptualization, L.G. and X.C.; methodology, G.W., G.Z. and J.L.; validation, W.Y. and L.X.; formal analysis, Y.W. and W.Y.; investigation, L.X., W.Y. and Y.L.; data curation, Y.L. and W.Y.; writing—original draft preparation, M.Q. and W.Y.; writing—review and editing, L.G. and M.B.; visualization, W.Y.; project administration, L.G.; funding acquisition, L.G. All authors have read and agreed to the published version of the manuscript.

Funding: This research was funded by the National Natural Science Foundation of China, grant number No. 52174046.

Institutional Review Board Statement: Informed consent was obtained from all subjects involved in the study.

Informed Consent Statement: Not applicable.

Data Availability Statement: Data are available from the corresponding author upon reasonable request.

Acknowledgments: The authors gratefully acknowledge the graduate students in the Research Institute of Unconventional Petroleum Science and Technology for their help in experimental data analysis and the experts of Tianjin Branch of CNOOC Ltd. for their support to the research of this project.

Conflicts of Interest: L.Z.G., X.M.C, G.W., G.H.Z., J.Y.L. was employed by Tianjin Branch of CNOOC Ltd. The remaining authors declare that the research was conducted in the absence of any commercial or financial relationships that could be construed as a potential conflict of interest.

References

1. Li, J.; Wang, Z.; Yang, H.; Wu, T.; Wu, H.; Wang, H.; Yu, F.; Jiang, H. Compatibility evaluation of in-depth profile control agents in dominant channels of low-permeability reservoirs. *J. Pet. Sci. Eng.* **2020**, *194*, 107529. [CrossRef]
2. Zhang, X.; Su, Y.; Li, L.; Hao, Y.; Wang, W.; Liu, J.; Gao, X.; Zhao, A.; Wang, K. Microscopic remaining oil initiation mechanism and formation damage of CO_2 injection after waterflooding in deep reservoirs. *Energy* **2022**, *248*, 123649. [CrossRef]
3. Hua, Z.; Lin, M.; Dong, Z.; Li, M.; Zhang, G.; Yang, J. Study of deep profile control and oil displacement technologies with nanoscale polymer microspheres. *J. Colloid Interface Sci.* **2014**, *424*, 67–74. [CrossRef] [PubMed]

4. Yang, Z.; Jia, S.; Zhang, L.; Wu, X.; Dou, H.; Guo, Z.; Zeng, L.; Li, H.; Guo, Z.; Jia, Z.; et al. Deep profile adjustment and oil displacement sweep control technique for abnormally high temperature and high salinity reservoirs. *Pet. Explor. Dev.* **2016**, *43*, 97–105. [CrossRef]
5. Cao, W.; Xie, K.; Lu, X.; Liu, Y.; Zhang, Y. Effect of profile-control oil-displacement agent on increasing oil recovery and its mechanism. *Fuel* **2019**, *237*, 1151–1160. [CrossRef]
6. Sun, Z.; Lu, X.; Sun, W. The profile control and displacement mechanism of continuous and discontinuous phase flooding agent. *J. Dispers. Sci. Technol.* **2017**, *38*, 1403–1409. [CrossRef]
7. Cheng, T.; Hou, J.; Yang, Y.; You, Z.; Liu, Y.; Zhao, F.; Li, J. Study on the plugging performance of bilayer-coating microspheres for in-depth conformance control: Experimental study and mathematical modeling. *Ind. Eng. Chem. Res.* **2019**, *58*, 6796–6810. [CrossRef]
8. Skauge, A.; Zamani, N.; Gausdal Jacobsen, J.; Shaker Shiran, B.; Al-Shakry, B.; Skauge, T. Polymer flow in porous media: Relevance to enhanced oil recovery. *Colloids Interfaces* **2018**, *2*, 27. [CrossRef]
9. Nazar, M.F.; Shah, S.S.; Khosa, M.A. Microemulsions in enhanced oil recovery: A review. *Pet. Sci. Technol.* **2011**, *29*, 1353–1365. [CrossRef]
10. Nazar, M.F.; Shah, S.S.; Eastoe, J.; Khan, A.M.; Shah, A. Separation and recycling of nanoparticles using cloud point extraction with non-ionic surfactant mixtures. *J. Colloid Interface Sci.* **2011**, *363*, 490–496. [CrossRef]
11. Nazar, M.F.; Myakonkaya, O.; Shah, S.S.; Eastoe, J. Separating nanoparticles from microemulsions. *J. Colloid Interface Sci.* **2011**, *354*, 624–629. [CrossRef] [PubMed]
12. Xiao, L.; Hou, J.; Wen, Y.; Qu, M.; Wang, W.; Wu, W.; Liang, T. Imbibition mechanisms of high temperature resistant microemulsion system in ultra-low permeability and tight reservoirs. *Pet. Explor. Dev.* **2022**, *49*, 1398–1410. [CrossRef]
13. Al-Shakry, B.; Skauge, T.; Shaker Shiran, B.; Skauge, A. Impact of mechanical degradation on polymer injectivity in porous media. *Polymers* **2018**, *10*, 742. [CrossRef]
14. Liu, S.; Ott, W.K. Sodium silicate applications in oil, gas & geothermal well operations. *J. Pet. Sci. Eng.* **2020**, *195*, 107693.
15. Bai, B.; Leng, J.; Wei, M. A comprehensive review of in-situ polymer gel simulation for conformance control. *Pet. Sci.* **2022**, *19*, 189–202. [CrossRef]
16. Zhu, D.Y.; Fang, X.Y.; Sun, R.X.; Xu, Z.H.; Liu, Y.; Liu, J.Y. Development of degradable pre-formed particle gel (DPPG) as temporary plugging agent for petroleum drilling and production. *Pet. Sci.* **2021**, *18*, 479–494. [CrossRef]
17. Gallo, G.; Erdmann, E.; Cavasotto, C.N. Evaluation of silicone fluids and resins as CO_2 thickeners for enhanced oil recovery using a computational and experimental approach. *ACS Omega* **2021**, *6*, 24803–24813. [CrossRef]
18. Smith, J.; Inc, T. Performance of 18 Polymers in Aluminum Citrate Colloidal Dispersion Gels. In Proceedings of the SPE International Symposium on Oilfield Chemistry, San Antonio, TX, USA, 14–17 February 1995. [CrossRef]
19. Zhu, D.; Bai, B.; Hou, J. Polymer gel systems for water management in high-temperature petroleum reservoirs: A chemical review. *Energy Fuels* **2017**, *31*, 13063–13087. [CrossRef]
20. Kang, W.; Kang, X.; Lashari, Z.A.; Li, Z.; Zhou, B.; Yang, H.; Sarsenbekuly, B.; Aidarova, S. Progress of polymer gels for conformance control in oilfield. *Adv. Colloid Interface Sci.* **2021**, *289*, 102363. [CrossRef]
21. Abdulbaki, M.; Huh, C.; Sepehrnoori, K.; Delshad, M.; Varavei, A. A critical review on use of polymer microgels for conformance control purposes. *J. Pet. Sci. Eng.* **2014**, *122*, 741–753. [CrossRef]
22. Formanek, M.; Rovigatti, L.; Zaccarelli, E.; Francesco, S.; Angel, J. Moreno Gel formation in reversibly cross-linking polymers. *Macromolecules* **2021**, *54*, 6613–6627. [CrossRef]
23. Zhao, G.; Dai, C.; You, Q. Characteristics and displacement mechanisms of the dispersed particle gel soft heterogeneous compound flooding system. *Pet. Explor. Dev.* **2018**, *45*, 481–490. [CrossRef]
24. Elsharafi, M.O.; Bai, B. Effect of weak preformed particle gel on unswept oil zones/areas during conformance control treatments. *Ind. Eng. Chem. Res.* **2012**, *51*, 11547–11554. [CrossRef]
25. Wu, W.X.; Song, X.; Fu, Y.; Yuan, P.F. Priority selection of weak gel flooding formula on heterogeneous reservoir. *Adv. Mater. Res.* **2015**, *1073*, 2248–2251. [CrossRef]
26. Wang, W.; Gu, Y.; Liu, Y. Applications of weak gel for in-depth profile modification and oil displacement. *J. Can. Pet. Technol.* **2003**, *42*, 54–61. [CrossRef]
27. Seright, R. Impact of dispersion on gel placement for profile control. *SPE Reserv. Eng.* **1991**, *6*, 343–352. [CrossRef]
28. Cheng, H.; Zheng, X.; Wu, Y.; Zhang, J.; Zhao, X.; Li, C. Experimental and numerical investigation on oil displacement mechanism of weak gel in waterflood reservoirs. *Gels* **2022**, *8*, 309. [CrossRef] [PubMed]
29. Cao, G.; Wu, J.; Bai, Y.; Zhang, N.; Xing, P.; Xu, Q.; Li, D.; Cong, X.; Liu, J. Research on the Formulation System of Weak Gel and the Influencing Factors of Gel Formation after Polymer Flooding in Y1 Block. *Processes* **2022**, *10*, 1405. [CrossRef]
30. Yin, H.; Yin, X.; Cao, R.; Zeng, P.; Wang, J.; Wu, D.; Luo, X.; Zhu, Y.; Zheng, Z.; Feng, Y. In situ crosslinked weak gels with ultralong and tunable gelation times for improving oil recovery. *Chem. Eng. J.* **2022**, *432*, 134350. [CrossRef]
31. Wang, C.; Zhong, L.; Liu, Y.; Han, Y.; Zhao, P.; Yuan, Y.; Han, X. Characteristics of weak-gel flooding and its application in LD10-1 oilfield. *ACS Omega* **2020**, *5*, 24935–24945. [CrossRef]

32. Li, J.; He, Y.; Liu, B.; Zhang, Y.; Hu, Z.; Hong, X. Ultimate Displacement Efficiency of Heavy Oil Reservoir with Bottom Water Formed by Fluvial Deposition in Bohai Oilfiled at Ultra-high Water Cut Stage. *Offshore Oil* **2022**, *42*, 24–30.
33. Zhao, P.; Han, Y.; Meng, X.; Yuan, Y.; Zhang, X.; Zhong, L.; Yu, Z. Experimental investigation on the injection profile variation in the weak gel injection process. *J. Phys. Conf. Ser.* **2020**, *1707*, 012020. [CrossRef]

gels

Article

Evaluation of Profile Control and Oil Displacement Effect of Starch Gel and Nano-MoS_2 Combination System in High-Temperature Heterogeneous Reservoir

Lianfeng Zhang [1,2], Yanhua Liu [1,2], Zhengxin Wang [1,2], Hao Li [1,2], Yuheng Zhao [3], Yinuo Pan [3], Yang Liu [3], Weifeng Yuan [3,*] and Jirui Hou [3,*]

1 Key Laboratory of Enhanced Oil Recovery of Henan Province, Nanyang 473000, China; zhlf1109@126.com (L.Z.); yjycsllyh.hnyt@sinopec.com (Y.L.); yjycslwzx.hnyt@sinopec.com (Z.W.); lih7327.hnyt@sinopec.com (H.L.)
2 Exploration and Development Research Institute of Henan Oilfield Branch Company, Sinopec, Nanyang 473000, China
3 Research Institute of Unconventional Petroleum Science and Technology, China University of Petroleum (Beijing), Beijing 102249, China; z13633831585@163.com (Y.Z.); 2022211726@student.cup.edu.cn (Y.P.); 2022216642@student.cup.edu.cn (Y.L.)
* Correspondence: yuanweifeng_0121@163.com (W.Y.); houjirui@cup.edu.cn (J.H.)

Abstract: The Henan Oilfield's medium-permeability blocks face challenges such as high temperatures and severe heterogeneity, making conventional flooding systems less effective. The starch gel system is an efficient approach for deep profile control in high-temperature reservoirs, while the nano-MoS_2 system is a promising enhanced oil recovery (EOR) technology for high-temperature low-permeability reservoirs. Combining these two may achieve the dual effects of profile control and oil displacement, significantly enhancing oil recovery in high-temperature heterogeneous reservoirs. The basic performance evaluation of the combination system was carried out under reservoir temperature. Displacement experiments were conducted in target blocks under different permeabilities and extreme disparity core flooding to evaluate the combination system's oil displacement effect. Additionally, the displacement effects and mechanisms of the starch gel and nano-MoS_2 combination system in heterogeneous reservoirs were evaluated by simulating interlayer and intralayer heterogeneity models. The results show that the single nano-MoS_2 system's efficiency decreases with increased core permeability, and its effectiveness is limited in triple and quintuple disparity parallel experiments. After injecting the starch gel–nano-MoS_2 combination system, the enhanced oil recovery effect was significant. The interlayer and intralayer heterogeneous models demonstrated that the primary water flooding mainly affected the high-permeability layers, while the starch gel effectively blocked the dominant channels, forcing the nano-MoS_2 oil displacement system towards unswept areas. This coordination significantly enhanced oil displacement, with the combination system improving recovery by 15.33 and 12.20 percentage points, respectively. This research indicates that the starch gel and nano-MoS_2 combination flooding technique holds promise for enhancing oil recovery in high-temperature heterogeneous reservoirs of Henan Oilfield, providing foundational support for field applications.

Keywords: starch gel; nano-MoS_2; profile control and oil displacement; enhanced oil recovery; high-temperature heterogeneous reservoirs

1. Introduction

Due to reservoir heterogeneity and long-term water flooding, injected water tends to flow along preferential channels, leading to issues like water flooding in production wells and ineffective circulation between injection and production wells [1–3]. This makes it challenging to mobilize the oil in the reservoir effectively. Reservoir heterogeneity is one of the key factors affecting the efficiency of water flooding and subsequent enhanced oil recovery efforts [4–6]. Interlayer heterogeneity refers to significant permeability differences between individual layers, varying by several times, tens of times, or even hundreds of times [7,8]. These permeability differences cause the injected water to rapidly advance through layers with good connectivity and high permeability, leading to early water breakthrough in production wells and rapid increases in water cut, even resulting in water flooding and shutdown. Meanwhile, significant residual oil remains in the low-permeability layers, highlighting the contrast between layers. Intralayer heterogeneity refers to differences in reservoir properties within a single sand layer [9,10]. For instance, in rhythmically deposited reservoirs, the bottom high-permeability segments are prone to intralayer advancement, while the top low-permeability segments accumulate residual oil, emphasizing the intralayer contrast.

In response to the contradictions of interlayer and intralayer water flooding in reservoirs, profile modification and plugging agents, like polymers, frozen gels, gels, and particulates, have shown effectiveness in improving reservoir heterogeneity, with good field application results [11–13]. However, polymers and conventional frozen gel water plugging agents tend to degrade and lose effectiveness in high-temperature reservoirs, leading to a decrease in strength and failing to meet plugging requirements [14,15]. Gel systems block formations by swelling upon water absorption, but the gelation time is influenced by many factors, and the plugging measures become ineffective once water breaks through the gel layer, reducing the profile modification effect [16,17]. Polymer/nano-microsphere systems need to be selected based on the actual pore sizes of the reservoir, and their plugging effectiveness in high-permeability channels is limited [18–20].

In contrast, starch gels can address these issues effectively. Previous research has demonstrated that starch gel systems exhibit strong plugging effects. Tang et al. [21] introduced starch graft copolymer plugging agents as a novel, efficient, and cost-effective solution for water plugging. Laboratory experiments demonstrated that these starch strong gel plugging agents possess selective plugging capabilities and achieve excellent plugging results. Furthermore, field tests of modified starch strong gel plugging agents were conducted in well N12-P6 in the JIDONG Oilfield. The field test results indicated that prior to implementation, well N12-P6 had a water cut as high as 98%, but after implementation, the daily water cut significantly decreased, reaching below 45%. The effective production increase period exceeded one year, indicating the long-term stability of the starch gel. In addressing the issue of CO_2 gas channeling, Zhao et al. [22] proposed the use of ethylenediamine and modified starch gel as plugging agents to plug the gas channels. Laboratory experiments demonstrated that the injection of modified starch gel and ethylenediamine rapidly increased injection pressure, effectively plugging high-capacity gas channels and fractures. Building on the research by Zhao et al., Hao et al. [23] further evaluated the static and plugging performance of gel systems. They identified a high-strength starch gel composed of 8% modified starch + acrylamide, 0.05% crosslinker, 0.15% initiator, and 0.15% stabilizer as an optimal plugging agent. Under acidic conditions, this gel effectively sealed fractures, leading to a rapid decrease in water cut, reaching a minimum of 29%. The displacement pressure drop sharply increased from 64.525 psi to over 145 psi, indicating complete fracture plugging and forcing the injected gas into lower-permeability rock layers.

The primary characteristics of modified starch gel plugging agents are as follows: Before gelation, starch gels are purely viscous fluids, with a pressure index that linearly increases with the injection volume and exhibits a small increase in amplitude and ease of injection. After gelation, they transform into viscoelastic fluids with strong plugging capabilities [22]. Luo et al. [24] developed an in situ starch grafted copolymer gel system (ISGCG) as an effective plugging material and systematically evaluated its gelation properties. The experimental results showed that the ISGCG system exhibits shear thinning behavior before gelation. The higher the shear rate, the more orderly the alignment of modified starch molecules along the shear flow direction, resulting in lower flow resistance and, consequently, lower viscosity. Notably, at a shear rate of 7.34 s^{-1}, the viscosity of the gelant is only 107 mPa·s, demonstrating its excellent injectability in formations. Furthermore, the gelation time can be controlled within 4–8 h, indicating very good pumping performance. Additionally, Luo et al. [24] incorporated sand grains of various sizes into modified starch gels. All of them can be completely crosslinked, and the sand particles can be cemented in the gel as a whole and have strong gel strength. Furthermore, following the completion of sand column plugging experiments, the gel system can ultimately cement the gravel together, making the gravels form a larger overall structure. This demonstrated the modified starch gel system's ability to form strong gels within subsurface pore structures, exhibiting commendable adhesion to rocks and sand grains. Finally, the ISGCG was successfully applied in the Xinjiang oilfield, where the daily water cut of the treated wells decreased from 92.7% to 75.8%, and daily oil production increased from 8.1 to 18.8 t/d, signifying remarkable effectiveness.

Furthermore, Leng et al. [25] systematically compared the deep profile control effects of modified starch gels with polymer gels in sandstone reservoirs. Their study revealed that the injectability and plugging properties of modified starch gels are superior to those of viscoelastic polymer gels. During the gel flooding stage, the modified starch gels maintained their rigid shape, unlike the deformation and migration behavior observed with polymer gels. In the subsequent water flooding stage, the modified starch gels remained immobile, effectively sealing high-permeability layers and enhancing the sweep efficiency in low-permeability layers by 60.00%. In contrast, polymer gels continued to flow through the core, increasing the sweep efficiency in low-permeability layers by 37.26%.

In a word, starch gels have moderate viscosity, ease of injection, high underground gelation strength, strong adhesion to rock, and long-term stability, effectively plugging high-strength leakage channels. However, when using starch gel alone for profile modification, the subsequent water flooding oil washing capability is generally limited, only moderately improving oil recovery in heterogeneous reservoirs.

Combining starch gel with oil displacement agents is expected to achieve the dual effects of profile control and oil displacement, significantly enhancing oil recovery in heterogeneous reservoirs. Researchers like Zhao et al. [22] and Hao et al. [23] used starch gel systems to plug escaping reservoirs and conducted subsequent displacement with CO_2. Their results showed that the gelation time of the starch gel system is controllable, with good injection performance and appropriate plugging strength, making it feasible for field implementation. Li et al. [26] combined dispersed particle gel with surfactant for combination profile modification and displacement. Their experiments showed that this combination has been developed as a cost-effective EOR method due to highly beneficial synergistic behavior of improving both sweep efficiency and displacement efficiency. Currently, there are no studies on the combination profile control and oil displacement of starch gel with nanomaterials for high-temperature heterogeneous reservoirs.

Nano-displacement technology is a highly promising enhanced oil recovery technique for high-temperature, low-permeability, or ultra-low-permeability reservoirs. Modified nanomaterials maintain good stability in high-temperature, high-salinity environments and have certain flow control capabilities [27–29]. Nanomaterials reported to enhance oil recovery include SiO_2 [30], TiO_2 [31], Al_2O_3 [32], CuO [33], ZnO [34], and graphene [35]. Nanofluids, created by dispersing these nanomaterials in specific solvents, enhance oil recovery by reducing oil–water interfacial tension [36], altering rock surface wettability [37], forming in situ Pickering emulsions [38], creating structural disjoining pressure (SDP) [39], and reducing oil viscosity [40]. Nano-MoS_2 is a flexible, plate-like material with nano-scale dimensions, approximately 60 nm $\times$ 80 nm in size and an average thickness of 1.2 nm [41]. Amphiphilic nanosheets, especially, exhibit higher interfacial activity compared to spherical and rod-shaped nanomaterials [42,43]. Consequently, these amphiphilic nanosheets demonstrate greater potential in reducing interfacial tension (IFT), stabilizing emulsions, and altering interfacial properties. Furthermore, compared with the SDP formation when using spherical nanoparticles (20 vol%), Qu et al. [44] proposed that far fewer nanosheets (0.005 wt%) are needed to generate SDP.

Infant Raj et al. [41] previously demonstrated the potential of two-dimensional molybdenum disulfide (MoS_2) nanosheets in enhancing crude oil recovery. The synthesized amphiphilic MoS_2 nanosheets, even at extremely low concentrations (0.005 wt%), increased recovery by 18.25% in 25 mD permeability cores saturated with crude oil. Liang et al. [45] investigated the mechanisms of enhanced recovery by modified MoS_2 nanosheets, focusing on interfacial tension reduction, wettability alteration, and emulsion stabilization. The results show that ultra-low-concentration MoS_2 nanofluid (50 mg/L) can decrease the IFT to 2.6 mN/m, change the contact angle (CTA) from 131.2° to 51.7°, and significantly enhance emulsion stability. Similarly, ultra-low-concentration MoS_2 nanofluids could increase oil displacement efficiency by 14% after water flooding. Qu et al. [44] synthesized amphiphilic MoS_2 nanosheets through a one-step simple hydrothermal method and conducted a comprehensive study of their physicochemical properties. The results showed that these nanosheets, with a distinctly ultra-thin lamellar structure, could stably disperse in water at ultra-low concentrations (50 mg/L), reducing oil–water interfacial tension, altering solid surface wettability, and stabilizing emulsions. Further, Ming Qu et al. [46] undertook both laboratory studies and field applications of amphiphilic MoS_2 nanosheets for enhanced oil recovery. Following the injection of ODA-MoS_2 nanofluid, oil production notably increased from 0.8 t/d to 1.4 t/d, while the water cut decreased significantly from 88% to 78.8%. The MoS_2 oil displacement system displays excellent dispersion stability and robust temperature resistance. Previous studies have produced amphiphilic nano-MoS_2 with good dispersion stability through chemical modification [44–46]. Liang et al. [47] reported that nano-MoS_2 solutions were applied in enhanced oil recovery tests at the Shengli Oilfield (Shandong, China) Xin 154 well group (reservoir temperature 114 °C) and the Tahe Oilfield (Xinjiang, China) TK7-459 well group (reservoir temperature 130 °C), demonstrating a significant oil production increase and affirming the exceptional thermal stability of MoS_2. Numerous laboratory studies and field applications have demonstrated that nano-MoS_2 oil displacement technology is an effective successor technology for enhancing oil recovery in high water-cut reservoirs or after chemical flooding, significantly increasing oil recovery [44–48].

In light of previous studies, MoS_2 nanosheets have been selected for recovery enhancement. Firstly, MoS_2 nanosheets, being lamellar nanomaterials with a 2 nm thickness and flexible properties, possess a thinner profile and greater deformability compared to spherical nanoparticles with larger diameters and rigidity. This makes them more effective in penetrating nanoscale pores and throats. Secondly, MoS_2 nanosheets, unlike other lamellar nanomaterials such as graphite, can be synthesized through a simple, direct hydrothermal method, making them suitable for industrial production. Finally, the concentration of MoS_2 used is 0.005 wt% to 0.01 wt%, significantly lower than other nanofluids.

However, due to characteristics like low viscosity, nano-MoS_2 has difficulty effectively increasing the sweep efficiency in heterogeneous formations. If nano-MoS_2 is used for oil displacement without profile modification, its flow control capabilities may not meet the requirements in reservoirs with strong heterogeneity and high permeability, leading to leakage and affecting the nano-MoS_2's oil recovery effect while increasing usage costs. Starch gel is an effective means of deep reservoir plugging. Combining starch gel with nano-MoS_2 is expected to achieve the dual effects of profile control and oil displacement, significantly improving sweep efficiency and oil displacement efficiency in heterogeneous reservoirs.

The Anpeng Block in Henan Oilfield exhibits significant heterogeneity, with oil sand body permeabilities ranging from 40×10^{-3} μm^2 to 453×10^{-3} μm^2, averaging 168×10^{-3} μm^2, and an average porosity of 15.3%. The reservoir temperature ranges from 91.1 °C to 95.0 °C.

In response to the high temperatures and severe heterogeneity in the medium-permeability blocks of the Henan Oilfield's Anpeng area, a study was conducted on the adaptability of the starch gel and nano-MoS_2 combination system in high-temperature medium-permeability reservoirs. Firstly, the basic properties of the starch gel and nano-MoS_2 combination system at high temperature were evaluated. Secondly, experiments were conducted under reservoir temperatures of 95 °C with different permeabilities and parallel core flooding to evaluate the displacement effect of the starch gel and nano-MoS_2 combination system. Finally, using the interlayer and intralayer heterogeneous models, the profile modification effect of the combination system in interlayer or intralayer heterogeneous reservoirs was assessed, clarifying the adaptability of the starch gel and nano-MoS_2 combination profile control and oil displacement system in high-temperature reservoirs.

2. Results and Discussion

2.1. Evaluation of Basic Properties of Starch Gel and Nano-MoS_2

A certain amount of modified starch, acrylamide monomer, crosslinker, and initiator was proportionately taken to prepare a starch gel solution. The starch gel solution was placed in a constant temperature oven set at 95 °C, and the gelation time and gel strength of the starch gel solution were recorded. Experimental results show that the gelation time of the starch gel is 10~12 h, and its gel strength reaches the H level (Sydansk's Gel Strength Code, GSC) [49,50], as shown in Figure 1. This indicates that the prepared starch gel has good temperature resistance and can stably form gel at 95 °C, with rigid gel strength.

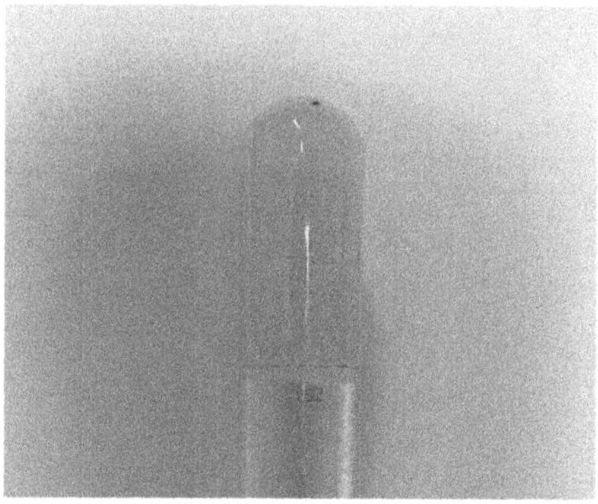

Figure 1. Appearance of starch gel after gelation at 95 °C.

Previous studies have shown that the oil displacement system of MoS_2 exhibits excellent dispersion stability [44–46] and good temperature resistance [47]. It also produces structuring disjoining pressure at ultra-low concentrations [44], altering wettability [44–46], reducing interfacial tension, and emulsifying crude oil, among other functions [47].

Based on previous research, preliminary assessments of its compatibility with Henan oilfield crude oil were conducted. The measurement results at 95 °C show that the oil–water interfacial tension in formation water is 2.955 mN/m, which could be the result of a combination of temperature, formation water, and crude oil properties [51–53]. And a 0.005% concentration of nano-MoS_2 can reduce the oil–water interfacial tension to 0.297 mN/m (as shown in Figure 2). Furthermore, the system exhibited a good emulsification capacity for crude oil, consistently forming O/W emulsions across varying oil–water ratios. Emulsions with oil–water ratios of 3:7, 4:6, and 5:5 demonstrate water separation rates of 100%, 100%, and 90%, respectively (see Figure 3). Notably, at an oil–water ratio of 7:3, a stable O/W-type emulsion is formed without visible water phase separation or significant stratification. This shows that the nano-MoS_2 solution has the ability of emulsification and rapid demulsification, which avoids the difficult problem of oil–water demulsification in the produced solution. This is a characteristic of nanofluids. Nanoparticles can improve emulsion stability by forming single- or multilayered nanoparticle interfacial films at the oil–water interface [54–56]. Therefore, the nano-MoS_2 solution has better emulsification ability than water [57]. However, the viscosity of the nanofluid is the same as that of saline, so when the shear action stops, the emulsion is broken more quickly than that of surfactant emulsions.

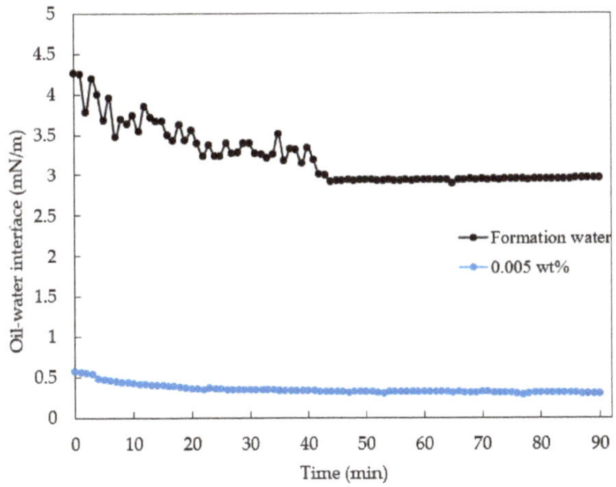

Figure 2. Tension–time curve of oil–water interface.

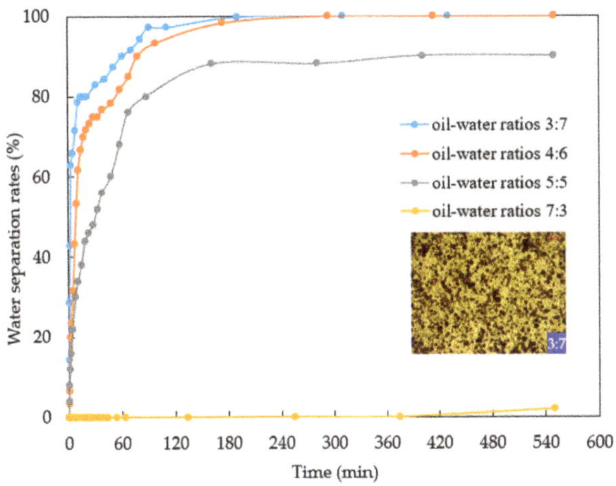

Figure 3. The water separation rate of emulsion.

Basic performance evaluation experiments demonstrate that nano-MoS_2 possesses commendable thermal resistance. Even at high temperatures of 95 °C, it retains its functionality in reducing interfacial tension and emulsifying crude oil. Preliminary results suggest that the combination system of starch gel and nano-MoS_2 is adaptable to reservoir temperatures and the properties of crude oil. This compatibility was further explored through displacement experiments assessing the oil displacement efficiency of the starch gel and nano-MoS_2 system.

2.2. *Evaluation of Oil Displacement Efficiency of Nano-MoS_2 under Different Core Permeabilities*

To verify the oil displacement effect of the nano-MoS_2 system, considering economic factors and past field application experiences, a system with a mass concentration of 0.005% nano-MoS_2 was selected. Experiments were conducted at a reservoir temperature of 95 °C to evaluate the oil displacement efficiency within the target permeability range of the reservoir. The results under different permeabilities (air-measured permeabilities of 100, 200, 300, and 500 mD) are presented in Table 1.

Table 1. The oil displacement effect of nano-MoS_2 in cores with different permeabilities.

Core ID	Gas Permeability (mD)	Initial Water Flooding Recovery Rate (%)	Final Recovery Rate (%)	Enhance Recovery Rate (%)
1	100	36.07	54.10	18.03
2	200	45.72	65.78	20.06
3	300	47.95	62.47	14.52
4	500	50.86	63.17	12.31

From Table 1, it is evident that the nano-MoS_2 system significantly enhances oil recovery in the cores of different permeabilities. Specifically, the improvements in oil recovery were 18.03%, 20.06%, 14.52%, and 12.31% for cores with permeabilities of 100 mD, 200 mD, 300 mD, and 500 mD, respectively. These results indicate that within a certain range, as core permeability increases, the nano-MoS_2 system penetrates more pore throats, activating oil in areas not reached by primary water flooding. This enhances sweep efficiency and washing efficiency. However, as permeability further increases, the nano-MoS_2 system tends to channel through preferential pathways, reducing sweep efficiency and leading to a decrease in oil recovery with increasing permeability.

2.3. *Evaluation of Combination System Oil Displacement Efficiency under Different Heterogeneity Extents Using a Parallel Model*

To validate the oil displacement effect of a combination system of starch gel and nano-MoS_2 in heterogeneous reservoirs and compare it with the use of nano-MoS_2 alone, cores with air-measured permeabilities of 100 mD, 300 mD, and 500 mD were selected. Experiments were conducted under reservoir conditions at 95 °C to evaluate the displacement efficiency of the combination system in parallel models with threefold (100 mD and 300 mD) and fivefold (100 mD and 500 mD) heterogeneity extents. The results are shown in Figures 4 and 5 and Table 2.

Table 2. The oil displacement effect of combination system in cores with different permeabilities.

Core ID	Gas Permeability (mD)	Initial Water Flooding Recovery Rate (%)	Nano-MoS_2 Flooding Enhance Recovery Rate (%)	Combination Displacement Enhance Recovery Rate (%)	Final Recovery Rate (%)
3	100	37.47	5.90	4.72	48.10
	300	44.77	7.44	6.34	58.55
Whole		41.18	6.69	5.54	53.41
4	100	30.15	2.48	6.80	39.43
	500	46.63	7.80	5.84	60.27
Whole		38.58	5.20	6.31	50.09

From Figure 4 and Table 2, during the water flooding phase, cores with higher permeability showed a rapid increase in oil recovery, with higher water flooding recovery rates than those in lower-permeability cores. In the nano-MoS_2 phase, the oil recovery in the threefold heterogeneity model increased by 7.44 and 5.90 percentage points for high- and relatively low-permeability layers (the term "relatively low-permeability" mentioned here refers to cores with permeability lower than that of 300 mD or 500 mD cores, and the following applies), respectively. In the fivefold heterogeneity model, the increases were 7.80 and 2.48 percentage points. This suggests that the nano-MoS_2 exhibits a certain degree of profile control and oil displacement. However, with increasing heterogeneity, the improvement in relatively low-permeability layers diminishes.

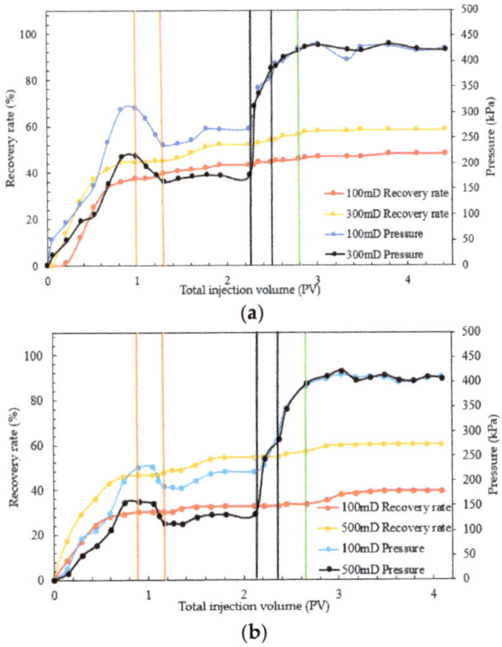

Figure 4. Experimental results of parallel model combination system: (**a**) threefold; (**b**) fivefold.

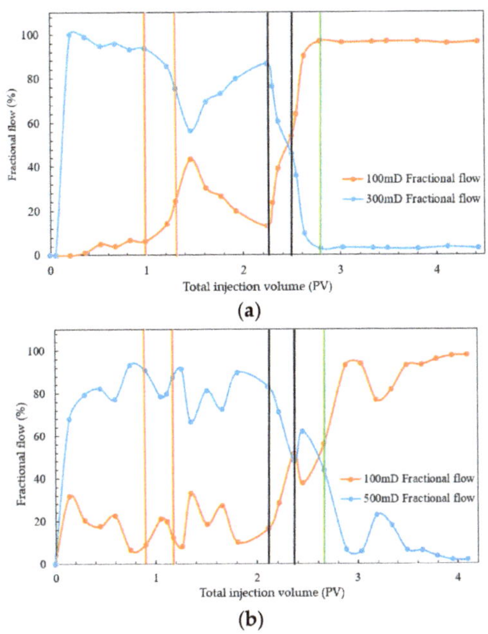

Figure 5. Fractional flow of oil displacement experiment of combination system with parallel mode: (**a**) threefold; (**b**) fivefold.

In the combination displacement phase with starch gel and nano-MoS_2, significant improvements were observed in both heterogeneity models. In the threefold model, oil recovery increased by 6.34 and 4.72 percentage points in high- and relatively low-permeability layers, respectively; in the fivefold model, the increases were 5.84 and 6.80 percentage points. This indicates that after injection of the starch gel system and a three-day static period, the gel formed within the porous medium effectively plugged preferential channels, altering the flow direction of the subsequent displacement fluids. This allowed the nano-MoS_2 system to fully interact with the oil in relatively low-permeability cores, thereby enhancing oil recovery.

Furthermore, during the entire displacement process, the injection pressure was initially low, increased slightly during the starch gel phase, and rapidly rose during the nano-MoS_2 phase, stabilizing after reaching a peak. As shown in Figure 4, during the stable pressure phase of water flooding, the pressures in the high-/relatively low-permeability layers of the threefold and fivefold heterogeneity models were 215.05/308.95 kPa and 165.66/246.80 kPa, respectively. Compared to relatively low-permeability layers, formation water in high-permeability layers during the oil displacement stage tends to flow more easily, leading to channeling phenomena. This results in the formation of preferential pathways in high-permeability layers, causing subsequent displacement fluids to flow through these lower-resistance channels. Consequently, water cut at the outlet increases, oil phase continuity is lost, and difficult-to-recover dispersed residual oil remains trapped in the porous medium. In the combination displacement stage, the injection pressures in the high-/low-permeability layers of the threefold model slowly increased from 176.77/267.49 kPa to 385.22/366.75 kPa, and in the fivefold model, from 132.05/218.59 kPa to 283.57/298.41 kPa, indicating effective plugging of preferential channels by the starch gel.

During fluid injection, due to the differing permeabilities of the two cores, diversion of the injected fluid occurred [58]. The relative flow [59] or fractional flow [60,61] will be higher in the core with higher permeability (the percentage of flow through high-permeability rock cores to the total flow).

Further analysis of the diversion in different heterogeneity parallel cores revealed that during the water flooding stage (Figure 5), diversion in high-permeability cores was predominant. In the nano-MoS_2 phase, a slight increase in fractional flow in relatively low-permeability cores in the threefold heterogeneity model indicated some autonomous profile control and displacement adjustment by the nano-MoS_2. The nano-MoS_2 peeled off the oil film and formed an "oil wall", changing the displacement phase from water to oil–water phases, thereby increasing the flow resistance of the displacement phase in the porous medium. This led to a reduction in the relative flow in high-permeability cores and an increase in relatively low-permeability cores, forcing the nano-MoS_2 oil displacement system to turn towards secondary channels for profile control and displacement adjustment. In the fivefold heterogeneity experiment, although a similar self-adjusting phenomenon was observed, its stronger heterogeneity meant that the decrease in the diversion rate in high-permeability cores was small and short-lived. This further indicated that the nano-MoS_2 exhibits a self-adjusting profile control and displacement effect, albeit limited. In the combination displacement phase, the starch gel entered preferential channels and gelled, causing the subsequent nano-MoS_2 solution to flow towards relatively low-permeability layers, rapidly increasing the diversion rate in these layers. The experiments showed that after the injection of starch gel, its profile control function weakened the heterogeneity of the parallel cores, improving the displacement efficiency in relatively low-permeability layers.

In conclusion, after primary water flooding, the injection of the nano-MoS_2 system in the threefold heterogeneity model resulted in significantly higher oil recovery in relatively low-permeability layers compared to the fivefold heterogeneity model, indicating a certain self-adjusting profile control and oil displacement ability of the nano-MoS_2 solution. The injection of a combination system of starch gel and nano-MoS_2, with the gel blocking high-permeability layers and forcing the nano-MoS_2 solution towards relatively low-permeability layers, effectively mobilized oil in unused reservoirs. This suggests that the combination displacement system can effectively enhance oil recovery in strongly heterogeneous reservoirs.

2.4. Evaluation of Oil Displacement Efficiency of Starch Gel and Nano-MoS_2 in Inter- and Intralayer Heterogeneous Model

To further verify the feasibility of the combination displacement technology of starch gel and nano-MoS_2 in inter- and intralayer heterogeneous reservoirs, and to compare it with the use of the nano-MoS_2 system alone, displacement experiments were conducted using inter- and intralayer heterogeneity models, simulating one injection and four production wells.

2.4.1. Evaluation of Oil Displacement Effectiveness in Intralayer Heterogeneous Reservoirs

For the intralayer heterogeneous model, the total production of water and oil in three layers was measured, and the overall oil recovery was calculated to analyze the displacement effects of each system. The results of the intralayer heterogeneous displacement experiments are presented in Figure 6 and Table 3.

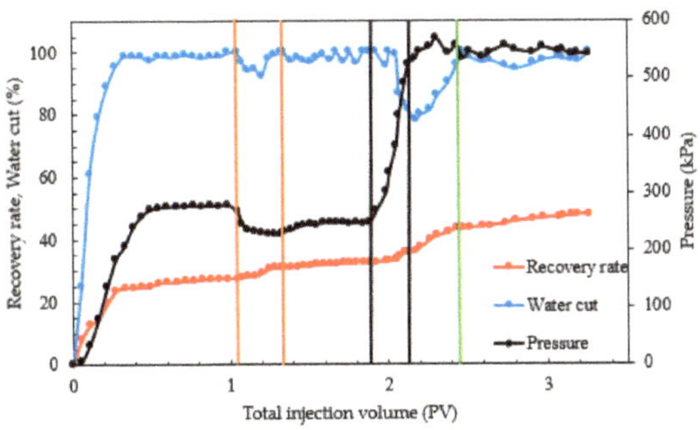

Figure 6. Results of intralayer heterogeneous displacement experiment.

Table 3. The oil displacement effectiveness in intralayer heterogeneous model.

Core ID	Gas Permeability (mD)	Initial Water Flooding Recovery Rate (%)	Nano-MoS_2 Flooding Enhance Recovery Rate (%)	Combination Displacement Enhance Recovery Rate (%)	Final Recovery Rate (%)
5	80 200 400	27.47	5.07	15.33	47.87

From Table 3 and Figure 6, it is observed that the oil recovery rate during primary water flooding in the intralayer heterogeneous model was 27.47%, which is relatively low and led to the formation of preferential channels. When the nano-MoS_2 system was used alone, the oil recovery increased by only 5.07 percentage points, primarily due to the nano-MoS_2 through preferential pathways. During the combination displacement phase, the water cut decreased to about 75%, and the oil recovery increased by 15.33 percentage points, significantly higher than with the nano-MoS_2 system alone. The starch gel successfully entered high-permeability layers or preferential channels in the combination system, plugging water breakthrough channels and improving sweep efficiency. This forced the nano-MoS_2 system towards uninvaded areas, enhancing oil displacement efficiency.

2.4.2. Evaluation of Oil Displacement Effectiveness in Interlayer Heterogeneous Reservoirs

The interlayer heterogeneous model enabled the separate measurement of water and oil production in relatively low-, medium-, and high-permeability layers, as well as the analysis of remaining oil activation and improvement in interlayer heterogeneity. The experimental results are shown in Figure 7 and Table 4.

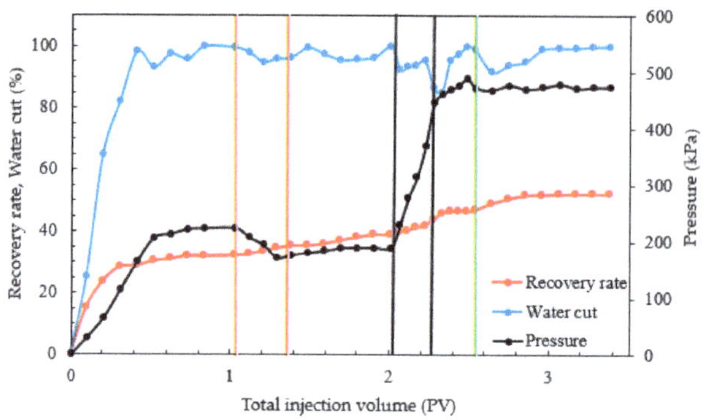

Figure 7. Results of interlayer heterogeneous displacement experiment.

Table 4. The oil displacement effectiveness in interlayer heterogeneous model.

Core ID	Gas Permeability (mD)	Initial Water Flooding Recovery Rate (%)	Nano-MoS_2 Flooding Enhance Recovery Rate (%)	Combination Displacement Enhance Recovery Rate (%)	Final Recovery Rate (%)
	80	15.16	3.59	2.05	21.82
6	200	12.70	1.84	5.64	20.59
	400	4.51	1.02	4.51	10.04
Whole		32.77	6.45	12.20	52.45

From Table 4 and Figure 7, during the water flooding phase in the rhythmically designed interlayer heterogeneous model, the water cut in the high-permeability layer increased rapidly. The total oil recovery rate of the water flooding phase was 32.77%, with the middle- and high-permeability layers being the main contributors, at 12.70% and 15.16% respectively, while the relatively low-permeability layer was almost uninvaded, with only a 4.51% recovery rate. After the injection of the 0.3 PV nano-MoS_2 system and a 12 h static period, the total oil recovery rate in the nano-MoS_2 phase was 6.45%, mainly through the high-permeability layer at 3.59%, followed by the middle- and relatively low-permeability layers at 1.84% and 1.02%, respectively. This indicated that due to strong heterogeneity, the nano-MoS_2 mainly flowed through high-permeability layer channels, exerting a wedging permeation effect, stripping oil films, and further activating remaining oil in the high-permeability layer water flooding, with a lower impact on middle- and relatively low-permeability layers. After the injection of 0.2 PV starch gel and a 72 h static period, followed by another 0.3 PV nano-MoS_2 injection, the pressure increased sharply. The total oil recovery rate in the combination displacement phase was 12.20%, mainly through the middle- and low-permeability layers at 5.64% and 4.51%, with a significant decrease in water cut, while the high-permeability layer contributed 2.05%. This showed that starch gel effectively plugged high-permeability layers and preferential channels, forcing the subsequent nano-MoS_2 solution to enter middle- and relatively low-permeability layers and secondary channels, effectively mobilizing oil in unswept reservoirs and significantly enhancing oil displacement efficiency.

In summary, the patterns of oil recovery rate changes during the displacement phases in both intra- and interlayer heterogeneity models were similar. During primary water flooding, the injection water had a limited impact on the uppermost low-permeability layer. When the nano-MoS_2 system was injected alone, it primarily flowed through high-permeability channels, mainly further activating remaining oil films after high-permeability layer water flooding. When the starch gel system was injected and blocked these channels, it effectively plugged high-permeability channels, forcing subsequent nano-MoS_2 towards middle- and relatively low-permeability layers, significantly enhancing oil recovery and showing a significant synergistic enhancement in oil displacement efficiency.

3. Conclusions

For the Henan Oilfield's high-temperature heterogeneous reservoirs, laboratory experiments demonstrated the significant application potential of the combination displacement system of starch gel and nano-MoS_2. The following results and conclusions were obtained.

1. The combination system comprising starch gel and nano-molybdenum disulfide exhibits notable thermal stability. At 95 °C, the starch gel consistently forms a rigid gel. Concurrently, nano-MoS_2 exhibits significant capabilities in reducing interfacial tension and effectively emulsifying crude oil.
2. Under high-temperature (95 °C) heterogeneous reservoir conditions, the injection of 0.2 PV starch gel, followed by 0.3 PV of a 0.005% mass concentration nano-MoS_2 solution, effectively plugged dominant channels. This forced the nano-MoS_2 towards uninvaded areas, simultaneously enhancing sweep efficiency and displacement efficiency and significantly improving oil recovery by over 10 percentage points.
3. This study indicates that this combination displacement technology holds promise for enhancing oil recovery in high-temperature heterogeneous reservoirs of the Henan Oilfield, providing foundational support for further field applications.

4. Materials and Methods

4.1. Experimental Materials and Instruments

The main study area, H3IV2 layer of Henan Oilfield Anpeng block (Henan, China), provided formation water and crude oil (viscosity of 3.04 mPa·s). The pH of the formation water from the H3IV2 layer is 8.3, indicating it is slightly alkaline. The total salinity is 7755.66 mg/L, with a chloride ion content of 2512.25 mg/L, and the water type is $NaHCO_3$ type. The ion composition of the H3IV2 layer formation water is provided by Henan Oilfield, as shown in Table 5.

Table 5. Ion composition of the formation water in H3IV2 formation, Henan Oilfield.

Ion(s)	HCO_3^-	Cl^-	SO_4^{2-}	Ca^{2+}	Mg^{2+}	K^+	Na^+	Salinity
Concentration, mg/L	1476.68	2512.25	777.87	23.42	4.03	86.23	2524	7755.66

Outcrop cores, 10 cm in length and 2.5 cm in diameter, with air-measured permeabilities of 100, 200, 300, and 500 mD, were used. The dimensions of the interlayer heterogeneous model were 30 × 30 × 6 cm, with permeability combinations of 80/200/400 mD and uniform layer thicknesses of 2 cm. Similarly, the intralayer heterogeneous model was of the same dimensions and permeability combinations. The nano-MoS_2 concentrate (1% mass concentration, self-developed) and the modified starch were obtained from Henan Hengrui Starch Technology Co., Ltd (Luohe, China). Acrylamide (AM, 98%), N, N′-methylene bisacrylamide (crosslinker), and potassium persulfate (initiating agents) were analytical grade and purchased from Shanghai Macklin Biochemical Co., Ltd (Shanghai, China).

The experimental apparatus primarily included the following: interfacial tensiometer (SVT20N, Dataphysics, Filderstadt, Germany), pressure monitoring equipment, a high-temperature constant temperature oven (Jiangsu Hai'an Petroleum Instrument Co., Ltd,

Nantong, China), an electronic balance, a vacuum pump, a double-cylinder piston pump (ISCO, Teledyne, Lincoln, NE, USA), a confining pressure hand pump and a core holder, a high-speed disperser (T 18 digital ULTRA-TURRAX, IKA, Staufen, Germany), among others.

4.2. Experimental Methods

4.2.1. Preparation and Evaluation of Basic Properties of Starch Gel and Nano-MoS_2 Solution

The MoS_2 nanosheets were synthesized as reported in previous work [41,44–46]. Formation water from the H3IV2 layer of Anpeng and the 1% mass concentration nano-MoS_2 concentrate were used to prepare a 0.005% mass concentration nano-MoS_2 solution. The solution was subjected to ultrasonic oscillation for 15 min post-preparation, ensuring uniform and stable dispersion of nano-MoS_2 in the formation water.

Interfacial tension between nano-MoS_2 solution and the crude oil from layer H3IV2 in the Anpeng main area of the Henan Oilfield was determined at 95 °C using a rotating drop interfacial tensiometer.

Different oil–water ratio emulsions were prepared by mixing crude oil from layer H3IV2 in the Anpeng area with a 0.005% mass concentration nano-MoS_2 solution (oil–water ratios of 3:7, 4:6, 5:5, 7:3) using a high-speed disperser (3000 r/min) for 3 min. The emulsions were then placed in a constant-temperature chamber at 95 °C, and the process of oil and water separation was observed and recorded; the separation rate was calculated. Microscopic images of the initial emulsion were captured using a microscope.

The starch gel was synthesized by modified starch, acrylamide (AM), a crosslinking agent, and initiator. Modified starch serves as a rigid framework within the starch gel structure. AM (acrylamide) is a typical monomer and acts as a flexible side chain within the starch gel structure. Crosslinkers can control the rate and extent of crosslinking reactions, thus influencing the strength of the gel system. Potassium persulfate can control the polymerization reaction rate, thereby regulating the gelation time.

The starch gel system is composed of 40,000 mg/L modified starch, 40,000 mg/L acrylamide, 1000 mg/L crosslinker, and 100 mg/L initiating agents.

Modified starch, acrylamide monomer, N, N′-methylenebisacrylamide, and potassium persulfate were weighed and slowly poured into deionized water with continuous stirring until complete dissolution. The stirring continued for 4 h to yield the experimental starch gel solution. The starch gel solution was placed in a constant-temperature oven set at 95 °C, using Sydansk's gel strength code (GSC) [49,50] for the gelation time and gel strength of the starch gel solution.

4.2.2. Evaluation of Oil Displacement Efficiency of Nano-MoS_2 Solution in Cores

Basic parameters of the cores were measured, followed by vacuum saturation with water to calculate pore volume and water-measured permeability. The cores were then saturated with oil and aged for 7 days, with initial oil saturation recorded. Primary water flooding was conducted at a constant rate until the water cut reached 98%, followed by the injection of 0.3 PV nano-MoS_2 oil displacement system. Subsequent water flooding continued until the water cut again reached 98%.

4.2.3. Evaluation of Oil Displacement Efficiency of Starch Gel and Nano-MoS_2 in Parallel Cores

After measuring basic core parameters, vacuum and saturating with water, the pore volume and water-measured permeability were calculated. The cores were oil-saturated and aged for 7 days, with initial oil saturation noted. Different permeability cores were paralleled, and primary water flooding was conducted at a constant rate until the water cut reached 98%. This was followed by nano-MoS_2 displacement, injecting 0.3 PV nano-MoS_2 oil displacement system and waiting for 12 h before continuing subsequent water flooding until the water reached 98%. Finally, a combination displacement with starch gel and nano-MoS_2 was carried out, injecting 0.2 PV starch gel system and allowing it to sit for

72 h. Then, 0.3 PV nano-MoS_2 oil displacement system was injected and left for 12 h before continuing with subsequent water flooding until the water cut reached 98%.

4.2.4. Evaluation of Oil Displacement Efficiency of Starch Gel and Nano-MoS_2 in Heterogeneity Models

The intralayer and interlayer heterogeneous models, geometrically and physically similar to the main reservoirs of H3IV2 layer in Henan Oilfield Anpeng, were utilized (as shown in Figure 8). The interlayer heterogeneity model represents variations in permeability between different layers. This model comprises three independent rock cores, each with a thickness of 2 cm and separated by a 1 cm gap, with designated permeabilities of 80, 200, and 400 mD, respectively. The intralayer heterogeneity refers to variations in reservoir properties within a single sand layer, featuring a gradient of permeability values (80, 200, and 400 mD) from bottom to top. This layer is formed by bonding three separate sub-layers with sand and cement, representing a single heterogeneous sand layer.

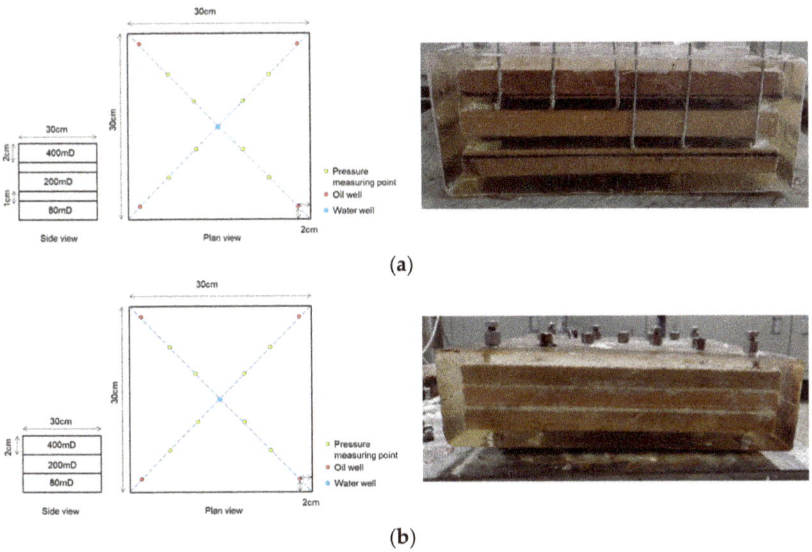

Figure 8. Schematic and physical models of interlayer (**a**); and intralayer (**b**) heterogeneity.

The models, measuring 30 × 30 × 6 cm with permeability combinations of 80/200/400 mD and layer thicknesses of 2 cm, were sealed with epoxy resin and designed with one injection well and four production wells.

According to the experimental method, at a reservoir temperature of 95 °C, oil displacement effectiveness experiments were conducted for both intralayer and interlayer heterogeneous models. The models were vacuum saturated with water and the pore volumes calculated. After oil saturation and a 7-day aging period, initial oil saturations were recorded. Primary water flooding was then carried out at a constant rate until the water cut reached 98%. This was followed by nano-MoS_2 displacement, injecting 0.3 PV nano-MoS_2 oil displacement system and waiting for 12 h before continuing with subsequent water flooding until the water cut reached 98%. Finally, a combination displacement with starch gel and nano-MoS_2 was performed, injecting 0.2 PV starch gel system and allowing it to sit for 72 h, then injecting 0.3 PV nano-MoS_2 oil displacement system and waiting for 12 h before continuing with subsequent water flooding until the water cut reached 98%.

In the interlayer heterogeneous model oil displacement experiments, the production volumes of oil and liquid from the four producing wells in each layer were recorded. In the five-spot intralayer heterogeneous model experiments, the total production volumes of oil and liquid from the four producing wells across the three layers were noted.

Author Contributions: Conceptualization, L.Z., Z.W. and J.H.; data curation, H.L. and Y.L. (Yang Liu); formal analysis, H.L.; investigation, Y.L. (Yanhua Liu), Y.Z., Y.P. and W.Y.; methodology, Y.L. (Yanhua Liu), Z.W. and W.Y.; project administration, L.Z. and J.H.; visualization, Y.Z., Y.P. and Y.L. (Yang Liu); writing—original draft, Y.Z. and W.Y.; writing—review and editing, J.H. All authors have read and agreed to the published version of the manuscript.

Funding: The authors gratefully appreciate the financial support of the National Natural Science Foundation of China (Grant No. 52174046).

Institutional Review Board Statement: Informed consent was obtained from all subjects involved in the study.

Informed Consent Statement: Not applicable.

Data Availability Statement: All data and materials are available on request from the corresponding author. The data are not publicly available due to ongoing researches using a part of the data.

Acknowledgments: The authors gratefully acknowledge the graduate students in the research institute of unconventional petroleum science and technology for their help in experimental data analysis and the experts of Henan Oilfield Exploration Institute for their support in the research of this project.

Conflicts of Interest: Authors Lianfeng Zhang, Yanhua Liu, Zhengxin Wang, Hao Li were employed by the company Exploration and Development Research Institute of Henan Oilfield Branch Company, Nanyang, China. The remaining authors declare that the research was conducted in the absence of any commercial or financial relationships that could be construed as a potential conflict of interest.

References

1. Wang, L.; Tian, Y.; Yu, X.; Wang, C.; Yao, B.; Wang, S.; Winterfeld, P.H.; Wang, X.; Yang, Z.; Wang, Y.; et al. Advances in improved/enhanced oil recovery technologies for tight and shale reservoirs. *Fuel* **2017**, *210*, 425–445. [CrossRef]
2. Yang, Y.; Yuan, W.; Hou, J.; You, Z. Review on physical and chemical factors affecting fines migration in porous media. *Water Res.* **2022**, *214*, 118172. [CrossRef] [PubMed]
3. Zheng, J.; Wang, Z.; Ju, Y.; Tian, Y.; Jin, Y.; Chang, W. Visualization of water channeling and displacement diversion by polymer gel treatment in 3D printed heterogeneous porous media. *J. Pet. Sci. Eng.* **2021**, *198*, 108238. [CrossRef]
4. Alajmi, A.F.; Gharbi, R.; Algharaib, M. The effect of heterogeneity and well configuration on the performance of hot water flood. *J. Pet. Sci. Eng.* **2014**, *122*, 524–533. [CrossRef]
5. Yuan, B.; Wood, D.A. A comprehensive review of formation damage during enhanced oil recovery. *J. Pet. Sci. Eng.* **2018**, *167*, 287–299. [CrossRef]
6. Kang, W.-L.; Zhou, B.-B.; Issakhov, M.; Gabdullin, M. Advances in enhanced oil recovery technologies for low permeability reservoirs. *Pet. Sci.* **2022**, *19*, 1622–1640. [CrossRef]
7. Yu, H.; Wang, L.; Zhou, D.; Wang, F.; Li, S.; Li, J.; Chen, X.; Cao, A.; Han, H. Experimental study on sweep characteristics of gas gravity drainage in the interlayer oil reservoir. *Front. Energy Res.* **2021**, *9*, 760315. [CrossRef]
8. Tang, H.; Wen, X.; Zhang, X.; Ren, X.; Liu, H. Water-oil displacing modeling experiment of interlayer heterogeneous conglomerate reservoir. *J. Southwest Pet. Univ. Sci. Technol. Ed.* **2014**, *36*, 129.
9. Zhang, W.; Zhang, J.; Xie, J. Research on reservoir bed heterogeneity, interlayers and seal layers and controlling factors of 2+ 3 sands of upper second member, Shahejie Formation, in the west of the Pucheng Oilfield. *Pet. Sci.* **2008**, *5*, 135–144. [CrossRef]
10. Jiang, L.; Zhao, W.; Zhang, B.; Zheng, C.; Hong, F.; Hao, J. Characterization method of tight sandstone reservoir heterogeneity and tight gas accumulation mechanism, Jurassic Formation, Sichuan Basin, China. *Geofluids* **2022**, *2022*, 9420835. [CrossRef]
11. Li, J.; Wang, Z.; Yang, H.; Yang, H.; Wu, T.; Wu, H.; Wang, H.; Yu, F.; Jiang, H. Compatibility evaluation of in-depth profile control agents in dominant channels of low-permeability reservoirs. *J. Pet. Sci. Eng.* **2020**, *194*, 107529. [CrossRef]
12. Chen, X.; Li, Y.; Liu, Z.; Zhang, J.; Chen, C.; Ma, M. Investigation on matching relationship and plugging mechanism of self-adaptive micro-gel (SMG) as a profile control and oil displacement agent. *Powder Technol.* **2020**, *364*, 774–784. [CrossRef]
13. Hua, Z.; Lin, M.; Guo, J.; Xu, F.; Li, Z.; Li, M. Study on plugging performance of cross-linked polymer microspheres with reservoir pores. *J. Pet. Sci. Eng.* **2013**, *105*, 70–75. [CrossRef]
14. Yu, B.; Zhao, S.; Long, Y.; Bai, B.; Schuman, T. Comprehensive evaluation of a high-temperature resistant re-crosslinkable preformed particle gel for water management. *Fuel* **2022**, *309*, 122086. [CrossRef]
15. Zhao, G.; Dai, C.; You, Q. Characteristics and displacement mechanisms of the dispersed particle gel soft heterogeneous compound flooding system. *Pet. Explor. Dev.* **2018**, *45*, 481–490. [CrossRef]

16. Yin, H.; Yin, X.; Cao, R.; Zeng, P.; Wang, J.; Wu, D.; Luo, X.; Zhu, Y.; Zheng, Z.; Feng, Y. In situ crosslinked weak gels with ultralong and tunable gelation times for improving oil recovery. *Chem. Eng. J.* **2021**, *432*, 134350. [CrossRef]
17. Elsharafi, M.O.; Bai, B. Effect of weak preformed particle gel on unswept oil zones/areas during conformance control treatments. *Ind. Eng. Chem. Res.* **2012**, *51*, 11547–11554. [CrossRef]
18. Hua, Z.; Lin, M.; Dong, Z.; Li, M.; Zhang, G.; Yang, J. Study of deep profile control and oil displacement technologies with nanoscale polymer microspheres. *J. Colloid Interface Sci.* **2014**, *424*, 67–74. [CrossRef]
19. Abdulbaki, M.; Huh, C.; Sepehrnoori, K.; Delshad, M.; Varavei, A. A critical review on use of polymer microgels for conformance control purposes. *J. Pet. Sci. Eng.* **2014**, *122*, 741–753. [CrossRef]
20. Zhang, F.; Hou, B.; Wang, S.; Chen, H.; Yang, J.; Fan, H.; Gui, L. Study on the Profile Control and Oil Displacement Mechanism of a Polymer Nano-microsphere for Oilfield. *Energy Fuels* **2023**, *37*, 2692–2701. [CrossRef]
21. Yongqiang, T.; Jirui, H.; Chenghui, L. Water shut off in a horizontal well: Lab experiments with starch graft copolymer agent. *J. Pet. Sci. Eng.* **2013**, *108*, 230–238. [CrossRef]
22. Zhao, F.; Hao, H.; Hou, J.; Hou, L.; Song, Z. CO_2 mobility control and sweep efficiency improvement using starch gel or ethylenediamine in ultra-low permeability oil layers with different types of heterogeneity. *J. Pet. Sci. Eng.* **2015**, *133*, 52–65. [CrossRef]
23. Hao, H.; Hou, J.; Zhao, F.; Song, Z.; Hou, L.; Wang, Z. Gas channeling control during CO_2 immiscible flooding in 3D radial flow model with complex fractures and heterogeneity. *J. Pet. Sci. Eng.* **2016**, *146*, 890–901. [CrossRef]
24. Luo, Q.; Tang, K.; Bai, L.; Li, K.; Sun, P.; Xu, C.; Zhao, Y.; Zhu, D. Development of in-situ starch grafted copolymerized gels for conglomerate reservoir conformance control and oil recovery improvement. *J. Pet. Sci. Eng.* **2021**, *210*, 110005. [CrossRef]
25. Leng, G.; Yan, W.; Li, F.; Wang, B.; Yuan, G.; Chu, C.; Li, Y.; Gao, C. Improved oil recovery in sandstone reservoirs with deep profile control technology: A comparative study between modified starch gel and polymer gel. *Geoenergy Sci. Eng.* **2023**, *226*, 211697. [CrossRef]
26. Li, W.; Dai, C.; Jan, O.; Hafiz, A.; Tao, J.; He, X.; Zhao, G. Adsorption and retention behaviors of heterogeneous combination flooding system composed of dispersed particle gel and surfactant. *Colloid Surface A.* **2018**, *538*, 250–261. [CrossRef]
27. Hadia, N.J.; Ng, Y.H.; Stubbs, L.P.; Torsæter, O. High salinity and high temperature stable colloidal silica nanoparticles with wettability alteration ability for eor applications. *Nanomaterials* **2021**, *11*, 707. [CrossRef]
28. Gizzatov, A.; Mashat, A.A.; Kosynkin, D.V.; Alhazza, N.; Kmetz, A.; Eichmann, S.L.; Abdel-Fattah, A.I. Nanofluid of petroleum sulfonate nanocapsules for enhanced oil recovery in high-temperature and high-salinity reservoirs. *Energy Fuels* **2019**, *33*, 11567–11573. [CrossRef]
29. Zhu, D.; Wei, L.; Wang, B.; Feng, Y. Aqueous hybrids of silica nanoparticles and hydrophobically associating hydrolyzed polyacrylamide used for EOR in high-temperature and high-salinity reservoirs. *Energies* **2014**, *7*, 3858–3871. [CrossRef]
30. Lu, T.; Li, Z.; Zhou, Y.; Zhang, C. Enhanced oil recovery of low-permeability cores by SiO_2 nanofluid. *Energy Fuels* **2017**, *31*, 5612–5621. [CrossRef]
31. Murshed, S.M.S.; Tan, S.H.; Nguyen, N.T. Temperature dependence of interfacial properties and viscosity of nanofluids for droplet-based microfluidics. *J. Phys. D Appl. Phys.* **2008**, *41*, 085502. [CrossRef]
32. Rezvani, H.; Panahpoori, D.; Riazi, M.; Parsaei, R.; Tabaei, M.; Cortés, F.B. A novel foam formulation by Al_2O_3/SiO_2 nanoparticles for EOR applications: A mechanistic study. *J. Mol. Liq.* **2020**, *304*, 112730. [CrossRef]
33. Bahraminejad, H.; Manshad, A.K.; Riazi, M.; Ali, J.A.; Sajadi, S.M.; Keshavarz, A. CuO/TiO_2/PAM as a novel introduced hybrid agent for water—Oil interfacial tension and wettability optimization in chemical enhanced oil recovery. *Energy Fuels* **2019**, *33*, 10547–10560. [CrossRef]
34. Alnarabiji, M.S.; Yahya, N.; Nadeem, S.; Adil, M.; Baig, M.K.; Ben Ghanem, O.; Azizi, K.; Ahmed, S.; Maulianda, B.; Klemeš, J.J.; et al. Nanofluid enhanced oil recovery using induced ZnO nanocrystals by electromagnetic energy: Viscosity increment. *Fuel* **2018**, *233*, 632–643. [CrossRef]
35. AfzaliTabar, M.; Alaei, M.; Bazmi, M.; Khojasteh, R.R.; Koolivand-Salooki, M.; Motiee, F.; Rashidi, A. Facile and economical preparation method of nanoporous graphene/silica nanohybrid and evaluation of its Pickering emulsion properties for Chemical Enhanced oil Recovery (C-EOR). *Fuel* **2017**, *206*, 453–466. [CrossRef]
36. Esmaeilzadeh, P.; Hosseinpour, N.; Bahramian, A.; Fakhroueian, Z.; Arya, S. Effect of ZrO_2 nanoparticles on the interfacial behavior of surfactant solutions at air–water and n-heptane–water interfaces. *Fluid Phase Equilibria* **2014**, *361*, 289–295. [CrossRef]
37. Hill, D.; Barron, A.R.; Alexander, S. Controlling the wettability of plastic by thermally embedding coated aluminium oxide nanoparticles into the surface. *J. Colloid Interface Sci.* **2020**, *567*, 45–53. [CrossRef] [PubMed]
38. Alnarabiji, M.S.; Husein, M.M. Application of bare nanoparticle-based nanofluids in enhanced oil recovery. *Fuel* **2020**, *267*, 117262. [CrossRef]
39. Zhang, H.; Ramakrishnan, T.S.; Nikolov, A.; Wasan, D. Enhanced oil recovery driven by nanofilm structural disjoining pressure: Flooding experiments and microvisualization. *Energy Fuels* **2016**, *30*, 2771–2779. [CrossRef]
40. Patel, H.; Shah, S.; Ahmed, R.; Ucan, S. Effects of nanoparticles and temperature on heavy oil viscosity. *J. Pet. Sci. Eng.* **2018**, *167*, 819–828. [CrossRef]
41. Raj, I.; Qu, M.; Xiao, L.; Hou, J.; Li, Y.; Liang, T.; Yang, T.; Zhao, M. Ultralow concentration of molybdenum disulfide nanosheets for enhanced oil recovery. *Fuel* **2019**, *251*, 514–522. [CrossRef]

42. Wei, P.; Luo, Q.; Edgehouse, K.J.; Hemmingsen, C.M.; Rodier, B.J.; Pentzer, E.B. 2D particles at fluid–fluid interfaces: Assembly and templating of hybrid structures for advanced applications. *ACS Appl. Mater. Interfaces* **2018**, *10*, 21765–21781. [CrossRef] [PubMed]
43. Nonomura, Y.; Komura, S.; Tsujii, K. Adsorption of disk-shaped Janus beads at liquid–liquid interfaces. *Langmuir* **2004**, *20*, 11821–11823. [CrossRef]
44. Qu, M.; Hou, J.; Liang, T.; Qi, P. Amphiphilic rhamnolipid molybdenum disulfide nanosheets for oil recovery. *ACS Appl. Nano Mater.* **2021**, *4*, 2963–2972. [CrossRef]
45. Liang, T.; Hou, J.; Xi, J. Mechanisms of nanofluid based modification MoS_2 nanosheet for enhanced oil recovery in terms of interfacial tension, wettability alteration and emulsion stability. *J. Dispers. Sci. Technol.* **2023**, *44*, 26–37. [CrossRef]
46. Qu, M.; Liang, T.; Hou, J.; Liu, Z.; Yang, E.; Liu, X. Laboratory study and field application of amphiphilic molybdenum disulfide nanosheets for enhanced oil recovery. *J. Pet. Sci. Eng.* **2022**, *208*, 109695. [CrossRef]
47. Liang, T.; Yang, C.; Zhang, Y.; Li, P.; Qu, M.; Hou, J. Research and Application Progress of Nanofluid for Enhanced Oil Recovery. *Xinjiang Oil Gas* **2023**, *19*, 29–41.
48. Xu, H.; Li, Y.; Yu, G.-L.; Yang, S.-S.; Li, B.-J.; Zhou, F.-J.; Yao, E.-D.; Bai, H.; Liu, Z.-Y. The enhancement of performance and imbibition effect of slickwater-based fracturing fluid by using MoS_2 nanosheets. *Pet. Sci.* **2023**, *20*, 2187–2201. [CrossRef]
49. Sydansk, R.D.; Argabright, P.A. Conformance Improvement in a Subterranean Hydrocarbon-Bearing Formation Using a Polymer Gel. U.S. Patent No. 4683949, 4 August 1987.
50. Sydansk, R.D. A newly developed chromium (lll) gel technology. *SPE Reserv. Eng.* **1990**, *5*, 346–352. [CrossRef]
51. Motraghi, F.; Manshad, A.K.; Akbari, M.; Ali, J.A.; Sajadi, S.M.; Iglauer, S.; Keshavarz, A. Interfacial tension reduction of hybrid crude-oil/mutual-solvent systems under the influence of water salinity, temperature and green SiO_2/KCl/Xanthan nanocomposites. *Fuel* **2023**, *340*, 127464. [CrossRef]
52. Hassan, M.E.; Nielsen, R.F.; Calhoun, J.C. Effect of pressure and temperature on oil-water interfacial tensions for a series of hydrocarbons. *J. Pet. Technol.* **1953**, *5*, 299–306. [CrossRef]
53. Gaonkar, A.G. Effects of salt, temperature, and surfactants on the interfacial tension behavior of a vegetable oil/water system. *J. Colloid Interf. Sci.* **1992**, *149*, 256–260. [CrossRef]
54. Arab, D.; Kantzas, A.; Bryant, S.L. Nanoparticle stabilized oil in water emulsions: A critical review. *J. Pet. Sci. Eng.* **2018**, *163*, 217–242. [CrossRef]
55. Stancik, E.J.; Kouhkan, M.; Fuller, G.G. Coalescence of particle-laden fluid interfaces. *Langmuir* **2004**, *20*, 90–94. [CrossRef] [PubMed]
56. Simovic, S.; Prestidge, C.A. Nanoparticles of varying hydrophobicity at the emulsion droplet—water interface: Adsorption and coalescence stability. *Langmuir* **2004**, *20*, 8357–8365. [CrossRef] [PubMed]
57. Zhu, D.-Y.; Zhao, Y.-H.; Zhang, H.-J.; Zhao, Q.; Shi, C.-Y.; Qin, J.-H.; Su, Z.-H.; Wang, G.-Q.; Liu, Y.; Hou, J.-R. Combined imbibition system with black nanosheet and low-salinity water for improving oil recovery in tight sandstone reservoirs. *Pet. Sci.* **2023**, *20*, 1562–1571. [CrossRef]
58. Guillen, V.R.; Carvalho, M.S.; Alvarado, V. Pore scale and macroscopic displacement mechanisms in emulsion flooding. *Transport Porous Med.* **2012**, *94*, 197–206. [CrossRef]
59. Ding, M.-C.; Li, Q.; Yuan, Y.-J.; Wang, Y.-F.; Zhao, N.; Han, Y.-G. Permeability and heterogeneity adaptability of surfactant-alternating-gas foam for recovering oil from low-permeability reservoirs. *Pet. Sci.* **2022**, *19*, 1185–1197. [CrossRef]
60. Gong, H.; Zhang, H.; Xu, L.; Li, K.; Yu, L.; Li, Y.; Dong, M. Further enhanced oil recovery by branched-preformed particle gel/HPAM/surfactant mixed solutions after polymer flooding in parallel-sandpack models. *RSC Adv.* **2017**, *7*, 39564–39575. [CrossRef]
61. Le, N.N.H.; Sugai, Y.; Vo-Thanh, H.; Nguele, R.; Ssebadduka, R.; Wei, N. Experimental investigation on plugging performance of CO_2 microbubbles in porous media. *J. Pet. Sci. Eng.* **2022**, *211*, 110187. [CrossRef]

gels

MDPI

Article

Analysis of the Influencing Factors on the Extraction of Residual Oil through the Gel Foam Flooding of Underground Reservoirs in the Tahe Oilfield

Chang-Ming Li [1,2], Ji-Rui Hou [1,2,*], Yu-Chen Wen [1,2,*] and Tuo Liang [3]

1 Research Institute of Unconventional Petroleum Science and Technology, China University of Petroleum, Beijing 102249, China; lichangming0213@163.com
2 Key Laboratory of Petroleum Engineering, China University of Petroleum, Beijing 102249, China
3 College of Petroleum Engineering, Xi'an Shiyou University, Xi'an 710065, China; ta.liang@foxmail.com
* Correspondence: houjirui@cup.edu.cn (J.-R.H.); yc.wen@foxmail.com (Y.-C.W.)

Abstract: Fractured-vuggy reservoirs are mainly composed of three types: underground rivers, vugs, and fractured-vuggy structures. Based on the similarity criterion, a 3D model can truly reflect the characteristics of the multi-scale space of a fractured-vuggy reservoir, and it can reflect fluid flow laws in the formation. Water flooding, gas flooding, and gel foam flooding were carried out in the model sequentially. Based on gas flooding, the enhanced recovery ratio of gel foam flooding in the underground river was approximately 12%. By changing the injection rate, the average recovery ratio of nitrogen flooding was 6.84% higher than that of other injection rates at 5 mL/min, and that of gel foam flooding was 1.88% higher than that of other injection rates at 5 mL/min. The experimental results showed that the gel foam induced four oil displacement mechanisms, which selectively plugged high-permeability channels, controlled the mobility ratio, reduced oil-water interfacial tension, and changed the wettability of rock surfaces. With different injection-production methods, gel foam flooding can spread across two underground river channels. Two cases of nitrogen flooding affected one underground river channel and two underground river channels. By adjusting the injection rate, it was found that after nitrogen flooding, there were mainly four types of residual oil, and gel foam flooding mainly yielded three types of remaining oil. This study verified the influencing factors of extracting residual oil from an underground river and provides theoretical support for the subsequent application of gel foam flooding in underground rivers.

Keywords: underground river; gel foam flooding; residual oil; enhanced oil recovery (EOR)

1. Introduction

Among the types of oil and gas reservoirs discovered worldwide, fractured-vuggy carbonate reservoirs play a crucial role [1]. Geological reserves account for 60% of the total reserves of conventional reservoirs, and their production occupies approximately 50% of the total production of conventional reservoirs [2,3]. Tahe oilfield, located in the Tarim Basin in western China, is a typical fractured-vuggy carbonate reservoir [4]. Based on the development characteristics of the fractured-vuggy system of the Ordovician carbonate reservoir in Tahe oilfield, the fractured-vuggy system can be categorized into three types: underground river system, vugs, and fractured-vuggy [5,6].

The subsequent energy in the reservoir is mainly supplied by water injection in different ways [7,8]. Many scholars have carried out extensive research on water flooding in fractured-vuggy reservoirs and put forward a variety of water flooding production models to improve the effect of water flooding [5,9,10]. For example, Khormali proposed an increase in displacement efficiency by injecting asphaltene inhibitors, which reduce the amount of asphaltene adsorbed on the rock surface [11]. Due to the strong heterogeneity of fractured-vuggy carbonate reservoirs, water flows tend to form dominant channels in fractures or vugs during water flooding, leading to water channeling and a poor water flooding effect [12,13]. Karimaie studied the oil recovery ratio of nitrogen gas and carbon dioxide in a fractured-vuggy carbonate reservoir and carried out physical simulation experiments to record the oil recovery ratio changes in the two gases with time. The research showed that gas flooding could displace residual oil after water flooding and reduce residual oil saturation [14]. Hui proved that nitrogen huff and puff replacement of remaining oil is a feasible EOR technology for fracture-vuggy carbonate reservoirs. [15].

The viscosity of injected gas is much lower than that of formation water and crude oil; the injected gas is prone to fingering during gas flooding. This fingering phenomenon leads to gas bypassing crude oil, reducing the sweep efficiency of gas flooding, and eventually causing gas channeling [16]. Foam can effectively increase gas phase viscosity and reduce gas fingering. Additionally, surfactants contained in the foam can effectively reduce the interfacial tension between oil and water; thus, foam has a good research prospect in the enhanced oil recovery (EOR) of fractured-vuggy carbonate reservoirs [17]. Li developed a visual-physical model based on a fractured-vuggy cross-section of the Tahe oilfield and carried out gas flooding and foam flooding experiments after water flooding. The experiments proved that foam can expand the sweep volume, emulsify, and carry residual oil to enhance the recovery ratio [18]. Yuan developed a 3D simulation model and carried out nitrogen gas flooding and foam flooding experiments. The research showed that foam flooding can effectively enhance the oil recovery of fractured-vuggy carbonate reservoirs, which is because the gas in the foam could start attic oil extraction. The surfactant could peel off the oil film and carry the oil droplets out through emulsification [9]. Yang performed experiments on nitrogen gas flooding and foam-assisted nitrogen gas flooding through a 2D visual model and found that the recovery factor of foam-assisted nitrogen gas flooding is 12% higher than that of nitrogen gas flooding. Foam can effectively control gas mobility and delay gas channeling [19]. Xu established a microscopic fractured-vuggy reservoir model and analyzed the flow characteristics and the enhanced oil recovery (EOR) mechanism of foam. The experiment showed that foam flooding can displace residual oil near the top of the vugs, high-speed foam flooding after low-speed foam flooding can better displace the oil in dead-end pores, and foam flooding can improve oil recovery from fractured vuggy reservoirs [20].

The migration behavior of foam in the experimental process is quite different from that in the actual reservoir. Especially for fractured-vuggy carbonate reservoirs, the mechanism of foam displacing the residual oil in vuggy is not clear, and the impact of various factors on enhanced oil recovery (EOR) needs to be further explored. Among the three fractured-vuggy structures in fractured-vuggy carbonate reservoirs, the underground river is a vuggy underground channel with the main characteristics of a river, representing the largest vuggy system. Research on it is of great significance for the subsequent development of fractured-vuggy carbonate reservoirs.

In this study, a 3D visual underground river fractured-vuggy reservoir model was designed by combining the actual data of oilfield production with visualized physical materials. We carried out experimental research on water flooding, nitrogen gas flooding, and gel foam flooding in the underground river model. By adjusting the injection speed and injection-production method of nitrogen gas flooding and gel foam flooding, the effects of enhanced oil recovery (EOR) under different factors were observed, the existing forms of residual oil in different stages were summarized, the mechanism of gel foam flooding in the start-up of the residual oil in the underground river model was analyzed, and the influencing factors of gel foam flooding on initiating residual oil were studied. This paper provides a valuable reference for gel foam flooding in the later stages of fractured-vuggy carbonate reservoirs.

2. Results and Discussion

2.1. Experimental Study of Gel Foam Flooding in the 3D Visual Underground River Physical Model

2.1.1. Analysis of the Experimental Results of the 3D Visual Underground River Physical Model

1. Water flooding stage

The 3D visual physical model was designed and developed based on the TK425CH unit and was a typical underground river reservoir physical model. The model mainly included two underground river channels, one located between TK425CH-TK467 and the other located in the TK410 well area. The bottoms of the TK410 and TK425 wells were located in the vuggy area, and the bottoms of the TK467 wells were located in the underground river channel area. In the water flooding stage, after water injection in the TK425CH well, the injected water first displaced the residual oil in the vuggy at the bottom of the well, then entered the underground river. The injected water migrated along the underground river channel in the direction of TK425CH-TK410 and then entered the vuggy at the bottom of well TK410. A water cone phenomenon occurred, resulting in the first water breakthrough in well TK410. The other underground river was not effectively swept by flooding, as shown in Figure 1, the yellow arrows in the figure indicate the flow direction of water flooding. The experimental results showed that the overall oil recovery of water flooding was low and there was still a large amount of residual oil at the bottom of the well; however, the water flooding exerted a strong displacement effect on the fractured-vuggy area around the injection well and the structure of the underground river.

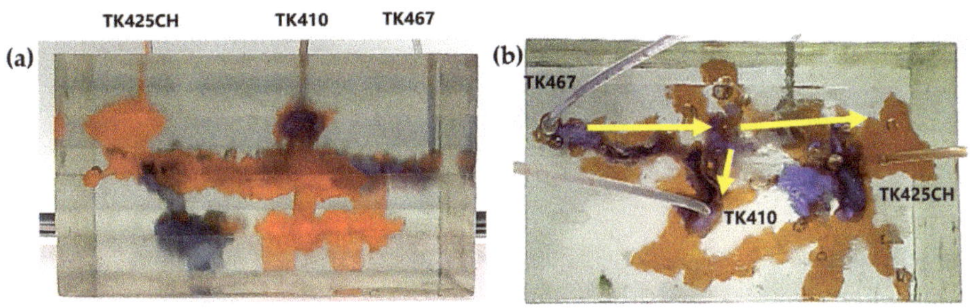

Figure 1. Water flooding stage of the 3D underground river model: (**a**) front view; (**b**) top view.

2. Gas flooding stage

Gas flooding experiments were conducted on the 3D visual underground river model of the TK425CH unit. The gas injection began in the TK425CH well; the injected gas entered the bottom of the TK425CH well, then replaced the residual oil in the surrounding vugs first. Subsequently, the gas entered the underground river and migrated along the underground river channel in the direction of TK425CH-TK467. As shown in Figure 2, yellow arrows in the figure indicate the flow direction of gas flooding, the effect of gas flooding was different from that of water flooding. The TK467 well on this channel was affected first, and then gas entered the vuggy at the bottom of the TK410 well along the underground river channel. The TK410 well was affected, and gas formed a gas cap in the vuggy to replace the attic oil at the top of the vuggy; however, the residual oil below the vuggy was not affected effectively. Finally, the dominant channel of gas channeling was formed in the direction of TK425CH-TK467, and the recovery ratio after gas flooding was 68.1%. As shown in Figures 1b and 2b, both water flooding and gas flooding effectively affected only one underground river channel, and the main flow channels were different. Neither water nor gas flooding fully utilized the residual oil in the underground channel. After the formation of fluid flow-dominant channels in the underground river, a large amount of residual oil remained in the structure of the underground river and could not be affected. After the formation of dominant channels for fluid flow in the underground river, a large amount of residual oil existed in the underground river structure and could not be affected. Through comparison, it was found that the experimental phenomenon was consistent with the actual situation in the mine, which indicates that the experimental results have good accuracy and reference value.

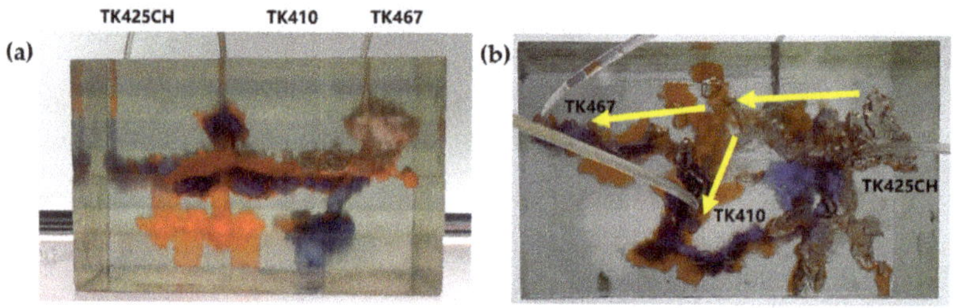

Figure 2. Gas flooding stage of the 3D underground river model: (**a**) front view; (**b**) top view.

3. Gel foam flooding stage

The gel foam flooding experiment was carried out on the 3D visual underground river model in the TK425CH unit. Firstly, the gel foam was injected into the TK425CH well. As shown in Figure 3, water flooding and gas flooding only formed a single flow dominant channel, and the subsequent gel foam flooding could effectively start two underground river channels at the same time, the yellow arrows in the figure indicate the flow direction of gel foam flooding. After gel foam injection in the TK425CH well, the injected gel foam exhibited a good displacement effect on the bottom vugs. Then, after the gel foam entered the underground channel, the channeling flow after water and gas flooding was plugged, and the subsequent fluid flow direction was adjusted. Two different underground channels were started, which had a good displacement effect on the upper and lower parts of the underground river. Finally, the recovery ratio of gel foam flooding was 79.6%. This proved that gel foam can control the direction and rate of fluid in the displacement of fractured-vuggy reservoirs and can prevent the fluid from migrating along the water, inhibiting gas channeling. Finally, gel foam initiated contact with the fractured-vuggy structure that could not be reached by water flooding and gas flooding and carried out uniform displacement in multiple directions with good displacement efficiency.

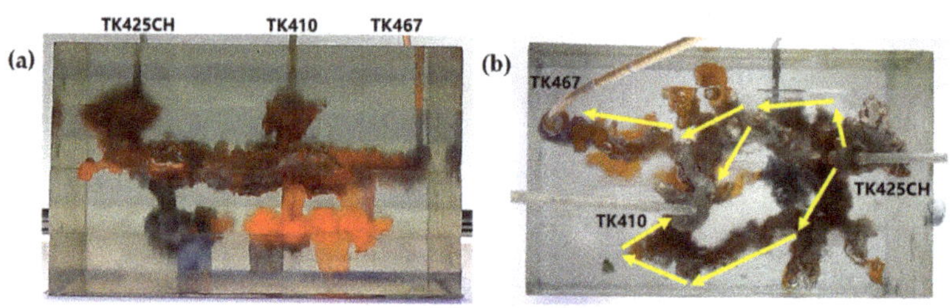

Figure 3. Gel foam flooding stage of the 3D underground river model: (**a**) front view; (**b**) top view.

After water flooding had formed a dominant channel in the TK467 well, the recovery ratio of the TK425CH unit 3D visual underground river model was 39.7%. Subsequently, gas injection was carried out in the TK425CH well, and after gas channeling, the final recovery ratio reached 68.2%, with an increase of 28.5%. In the actual production process, the average water flooding utilization degree in Tahe District 4 was 37.43%, and the gas flooding utilization degree was 30.3%, which is consistent with the experimental results.

As shown in Figure 4, 0.4 PV gel foam was injected into the underground river unit model; the final recovery ratio was 79.6%, and the recovery ratio was increased by 11.4%.

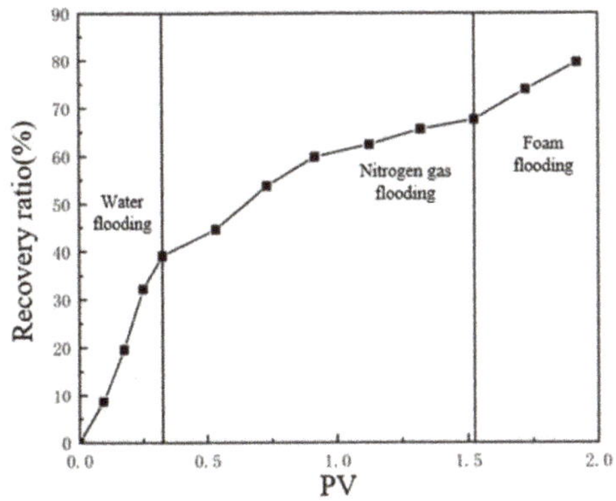

Figure 4. Relationship curve between the recovery ratio and PV in a displacement experiment of a 3D underground river model.

2.1.2. Impact of the Injection-Production Method on the Displacement Effect of the 3D Visual Underground River Physical Model

1. Nitrogen gas flooding at different well locations

Based on the TK425CH unit 3D visual underground river model, nitrogen gas flooding experiments were conducted at different well locations to study the nitrogen gas flooding displacement effect at different well locations in the fractured-vuggy reservoir. Experiments showed that when TK425CH was a gas injection well, the injected gas first gathered at the top of the vugs at the bottom of the TK425CH well. Most of the residual oil in the fractured-vuggy structure near the TK425CH well was replaced. Then, the injected gas entered the underground river channel. Nitrogen gas mainly migrated forward along the TK425CH-TK467 channel, as shown in Figure 5. When the TK410 well was a gas injection well, the injected gas first replaced most of the attic oil in the TK425CH well-bottom vugs. After entering the underground river channel, nitrogen gas mainly migrated in two directions: TK410-TK467 and TK410-TK425CH. After arriving at the TK425CH well, nitrogen gas flooding had a certain displacement effect on the residual oil in the vugs at the bottom of the well, as shown in Figure 6. Compared with TK425CH, during the gas injection process in the TK410 well, more underground river channels were started, the nitrogen gas swept volume was wider, and the recovery ratio was higher. After water flooding, the final recovery ratio for gas flooding was 71.3%. However, after water flooding in TK425CH, the final recovery ratio for gas flooding was 67.9%.

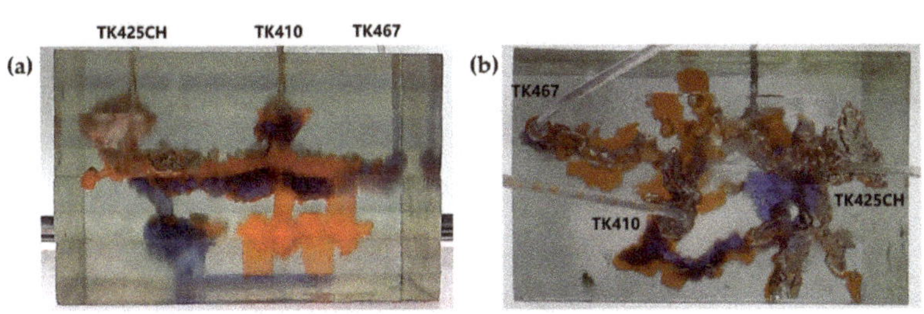

Figure 5. Gas injection in the TK425CH well of the 3D underground river model: (**a**) front view; (**b**) top view.

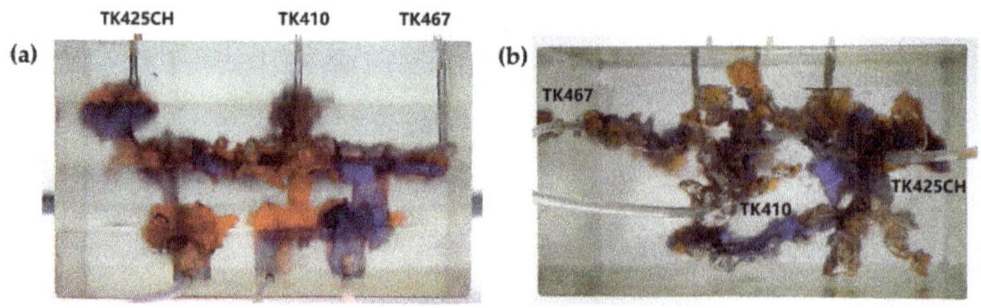

Figure 6. Gas injection in the TK410 well of the 3D visual physical model: (**a**) front view; (**b**) top view.

2. Gel foam flooding at different well locations

Based on the TK425CH well unit 3D visual underground river model, gel foam flooding experiments at different well locations were carried out to study the gel foam flooding displacement effect at different well locations in fractured-vuggy reservoirs. The experimental results are shown in Figures 7 and 8. After gel foam was injected into the TK425CH well, gel foam first entered the vugs at the bottom of the TK425CH well, effectively displacing residual oil in the vugs at the bottom of the TK425CH well. After entering the underground channel, the gel foam migrated along the two directions of TK425CH-TK467 and TK425CH-TK410. Gel foam exhibited a good displacement effect on the underground channel and the fractured-vuggy structure connected by the channel. However, the displacement effect of gel foam on the underground river channel in the TK410-TK467 direction was poor, and the final recovery ratio of gel foam flooding was 79.6%. After gel foam was injected into well TK410, gel foam effectively displaced the residual oil in the vuggy at the bottom of well TK410. After entering the underground channel, gel foam effectively started two underground river channels, TK410-TK467 and TK410-TK425CH.

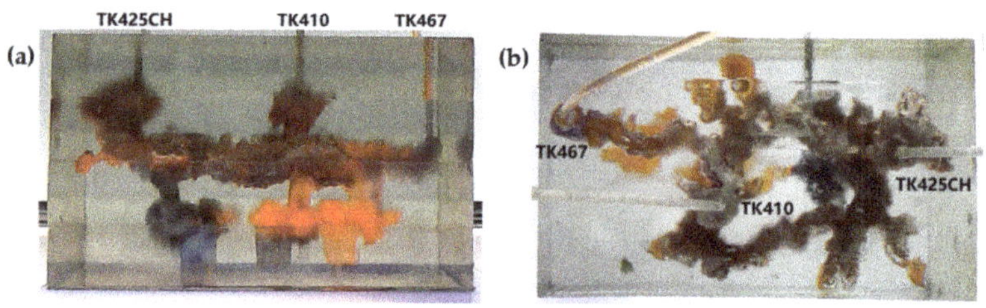

Figure 7. Gel foam injection in the TK425CH well of the 3D underground river model: (**a**) front view; (**b**) top view.

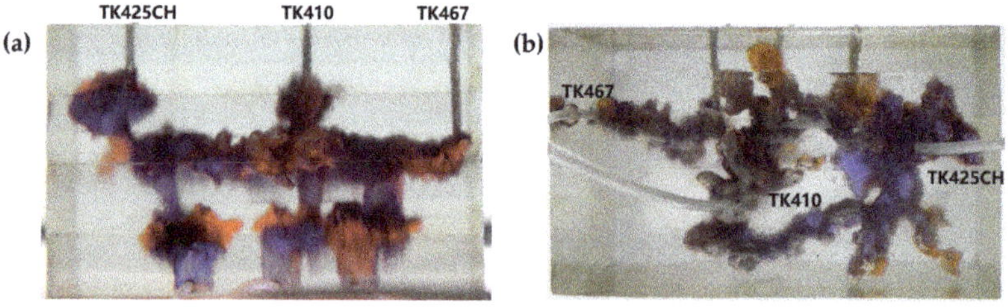

Figure 8. Gel foam injection in the TK410 well of the 3D underground river model: (**a**) front view; (**b**) top view.

The displacement effect of the TK410-TK467 channel was better than that of gel foam injection in the TK425CH well, which could effectively displace the corner residual oil of the underground river tributaries. The gel foam flooding final recovery ratio was 86.1%, 6.5% higher than that of gel foam injection in the TK425CH well, as shown in Figure 9.

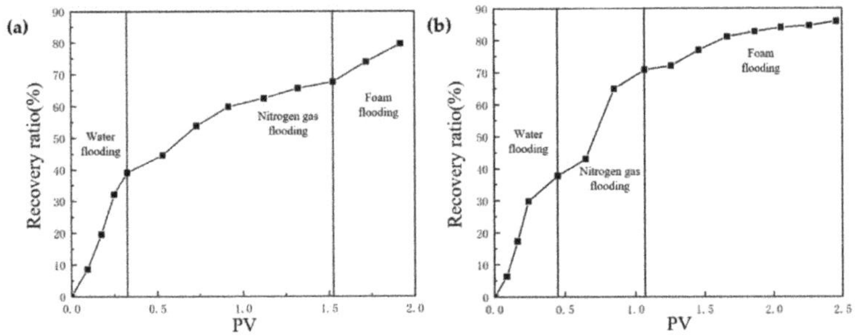

Figure 9. Relationship curve between the recovery ratio and PV in a displacement experiment of the 3D underground river model: (**a**) TK425CH well; (**b**) TK410 well.

2.1.3. Impact of the Injection Rate on the Displacement Effect of the 3D Visual Underground River Physical Model

1. Nitrogen flooding gas with different injection rates

Based on the TK425CH unit 3D underground river visual model, nitrogen gas flooding experiments with different injection rates were carried out to study the nitrogen gas flooding displacement effect at different rates in fractured-vuggy reservoirs. Through experiments, it was found that when TK410 was a gas injection well with a nitrogen gas injection rate of 5 mL/min, the injected gas first migrated to the bottom of the production well through the vugs located in the upper part connecting the fracture. The gas gathered at the top of the vugs at the bottom of TK410, and the flow component of the injected gas continued to increase. Then, the residual oil in the surrounding vugs was quickly replaced, and after entering the underground river, it migrated along the underground river channel in the direction of TK410-TK467. The TK467 well on this channel was affected first. After the attic oil around the TK410 well was completely displaced, gas entered the vugs at the bottom of the TK425CH well along the underground channel. The injected gas formed a gas cap in the vugs to replace the attic oil at the top of the vugs. The TK410 well was also affected; however, the residual oil in the lower part of the vugs was not effectively affected. Finally, a dominant gas channel was formed in the TK410-TK425CH direction. The experimental results are shown in Figure 10, the yellow arrows in the figure indicate the flow direction of gas flooding. When TK410 was a gas injection well and the nitrogen gas injection rate was 10 mL/min, the gas channeling component of the injected gas increased. The displacement component of the injected gas decreased along the model channel, and the injected gas migrated along the underground river channel in the direction of TK410-TK467, resulting in oil production from well TK467. The experimental results are shown in Figure 11, the yellow arrows in the figure indicate the flow direction of gas flooding. Compared with the two injection rates of nitrogen gas, when the injection rate of nitrogen gas was 5 mL/min, the dominant channel in the TK410-TK467 direction was formed, gas could continue to migrate in the direction of TK410-TK425CH, and residual oil around the TK425CH well was displaced. Compared with the nitrogen gas injection rate of 10 mL/min, more underground river channels were displaced, the sweep effect was better, and the recovery ratio was higher. The final recovery ratio after gas flooding was 66.4%, as shown in Figure 12.

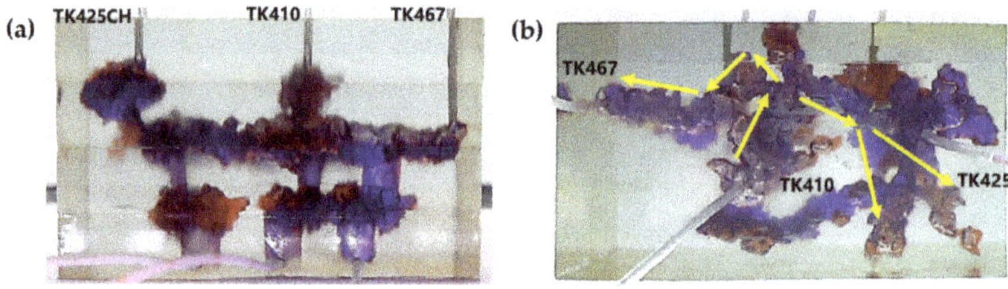

Figure 10. Gas injection in the TK410 well of the 3D underground river model at 5 mL/min: (**a**) front view; (**b**) top view.

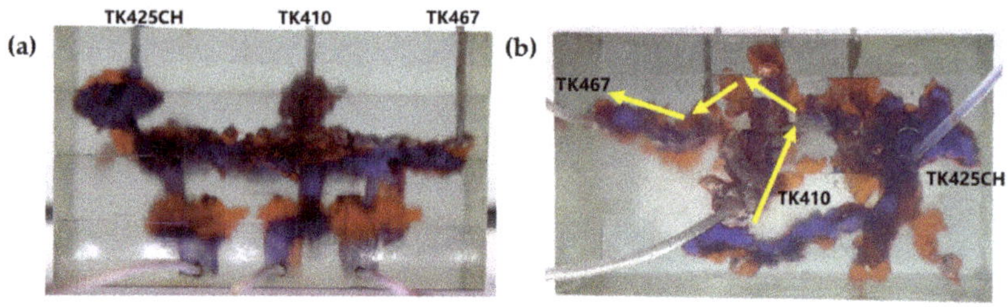

Figure 11. Gas injection in the TK410 well of the 3D underground river model at 10 mL/min: (**a**) front view; (**b**) top view.

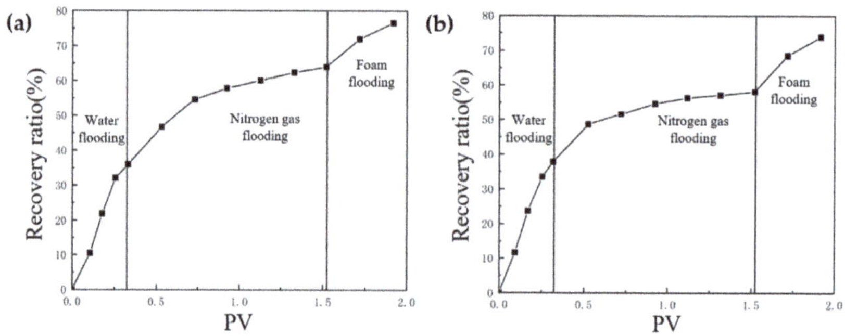

Figure 12. Relationship curve between the recovery ratio and PV in the displacement experiment of the 3D underground river model: (**a**) gas rate of 5 mL/min; (**b**) gas rate of 10 mL/min.

It can be concluded that in the process of gas flooding after water flooding, it is necessary to ensure that the injection rate of nitrogen gas is not too fast, and too high a gas injection rate cannot fully enable nitrogen gas flooding in the EOR. To prevent gas channeling and ensure that the displaced volume is maximized, a reasonable injection rate can achieve the optimal oil recovery effect. As shown in Figure 13, a consistent water flooding rate was ensured, the optimal nitrogen injection rate was 5 mL/min, and the average recovery rate was increased by 6.84%. Therefore, the optimal injection rate should be controlled at approximately 5 mL/min.

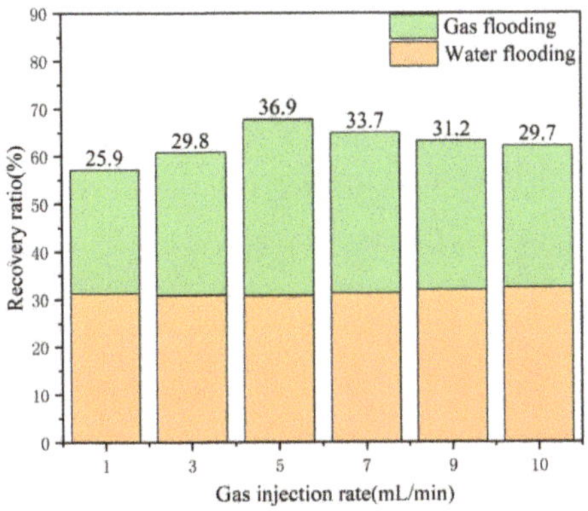

Figure 13. Diagram of the nitrogen flooding recovery ratio with different injection rates.

2. Gel foam flooding with different injection rates

Based on the TK425CH unit 3D underground river visual model, gel foam flooding experiments with different injection rates were carried out to study the gel foam flooding displacement effects at different rates in fractured-vuggy reservoirs. Through experiments, it was found that when TK410 was a gel foam injection well with a gel foam injection rate of 5 mL/min, the gel foam effectively displaced the residual oil in the vugs at the bottom of the TK410 well. After entering the underground channel, gel foam migrated along TK410-TK467 and TK410-TK425CH, demonstrating a good displacement effect on the underground channel and the connected fractured-vuggy structure, as shown in Figure 14, the yellow arrows in the figure indicate the flow direction of gel foam flooding. When TK410 was an injection well and the injection rate of gel foam was 10 mL/min, the gel foam effectively displaced the residual oil in the vugs at the bottom of the TK410 well and migrated along the two directions of TK410-TK467 and TK410-TK425CH; however, the sweet effect on the TK410-TK425CH underground river channel was poor, as shown in Figure 15, the yellow arrows in the figure indicate the flow direction of gel foam flooding. Compared with the two gel foam injection rates, when the gel foam injection rate was 5 mL/min, the sweep effect on the underground river channel was better, which could effectively displace some corner residual oil on the underground river tributaries. Gel foam also had a certain effect on the residual oil disturbed by bottom water below the vugs; the final recovery ratio after gel foam flooding was 79.6%, as shown in Figure 16.

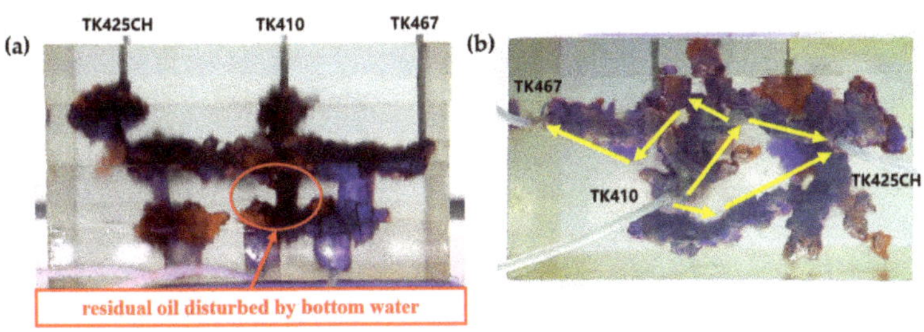

Figure 14. Gel foam injection in the TK410 well of the 3D underground river model at 5 mL/min: (**a**) front view; (**b**) top view.

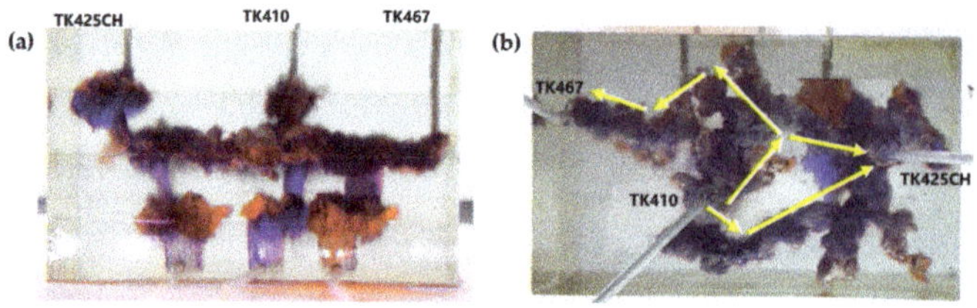

Figure 15. Gel foam injection in the TK410 well of the 3D underground river model at 10 mL/min: (a) front view; (b) top view.

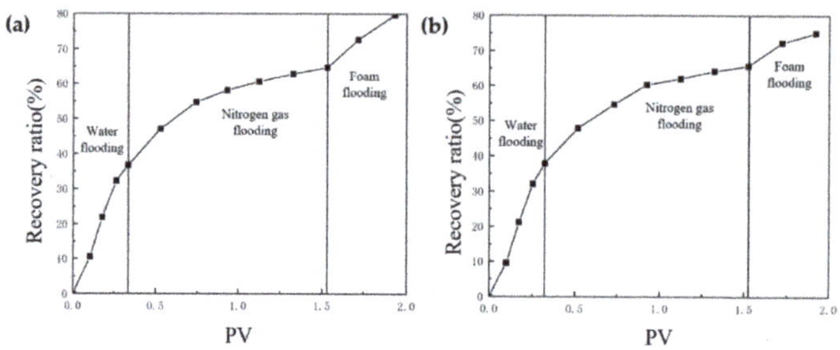

Figure 16. Relationship curve between the recovery ratio and PV in the displacement experiment of the 3D underground river model: (a) gel foam rate with 5 mL/min; (b) gel foam rate with 10 mL/min.

Due to the bursting and regeneration of gel foam in the process of formation migration, when the gel foam is injected into vugs, the gas will exist in the form of a dispersed phase. When the injection rate of gel foam was too high, the injection pressure rose rapidly. The gas in the gel foam changed from the dispersed phase to the continuous phase, forcing the gas phase to flow separately from the liquid phase. The flow capacity of the gas phase was higher than that of the liquid phase; therefore, the phenomenon of channeling was prone to occur, leading to a decrease in the oil recovery ratio. According to the experimental results, gel foam flooding had a strong displacement effect after gas flooding; however, the selection of gel foam injection rate was also crucial. As shown in Figure 17, when consistent water flooding and nitrogen gas rates were ensured, the optimal gel foam injection rate was 5 mL/min, and the average recovery rate increased by 6.84%. Therefore, the optimal injection rate should be controlled at approximately 5 mL/min.

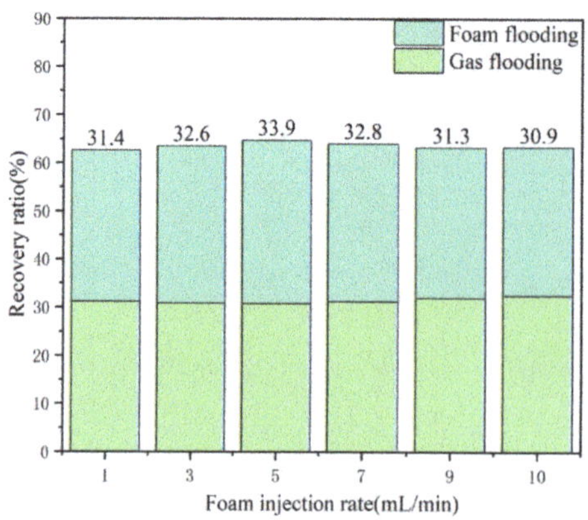

Figure 17. Graph of the gel foam flooding recovery ratio with different injection rates.

2.1.4. The Distribution of Residual Oil in Different Displacement Stages of the 3D Visual Underground River Physical Model

Residual oil in the visual underground river physical model displacement experiment can be divided into the following types: residual oil disturbed by bottom water, bypassed residual oil in the connected fractured-vuggy structure, residual oil in the isolated fractured-vuggy structure, residual oil shielded by channeling and residual oil at the blind end vug. Residual oil in isolated fractured-vuggy structures mainly existed in small reservoirs, one of the main reservoirs of residual oils after gas flooding. Taking the top view of Figures 18 and 19 as an example, there is an isolated fractured-vuggy structure above the middle of the image. Due to the effect of gas gravity differentiation, gas flooding could not displace residual oil in the fractured cavity, and the residual oil existed in the isolated fractured-vuggy structure Due to the presence of bottom water, a portion of the fractured-vuggy structure was submerged. The residual oil disturbed by bottom water was the residual oil submerged in the fractured-vuggy structure, which is hardly affected by conventional displacement methods. Taking the effect of gas flooding swept in the front view of Figure 19 as an example, gas flooding only affected the underground river channel above the model, and the residual oil surrounded by bottom water below the model was not affected. This residual oil is disturbed by bottom water. When the displacement medium was injected into the fractured-vuggy structure, the flow channel was formed. In the underground river channel, some residual oil was still adsorbed on the rock surface after the displacement medium was displaced; the residual oil that had not been swept was the bypassed residual oil in the connected fractured-vuggy structure. Taking the effect of gas flooding swept in the front view of Figure 19 as an example, an underground river channel was displaced after gas flooding. After gas flooding, there is still some residual oil on the wall around the underground river channel on the left side of the model; this residual oil was bypassed in the connected fractured-vuggy structure. When the displacement medium was injected into the fractured-vuggy structure, the channeling channel was formed, and the residual flow channels could not be effectively affected, forming residual oil shielded by channeling. When the displacement medium was injected into the fracture-cave structure, a channeling flow channel was formed, which could not effectively influence another flow channel, and residual oil in the other flow channel was shielded by channeling. In the top view of Figure 19, for example, an underground river channel was used after gas flooding. Another underground river channel on the left of the top view was not affected, and the residual oil in this underground river channel was shielded by channeling. The residual oil at the blind end of the vug was mainly caused by the complex fractured-vuggy structure, which generated some unconnected pores containing residual oil. Conventional displacement media could not replace the oil in the blind end and corner structures; residual oil existed at the blind end.

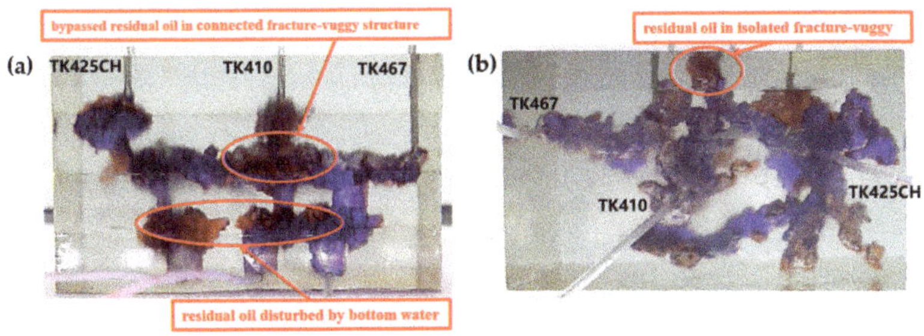

Figure 18. The distribution of residual oil after gas flooding in the 3D underground river model (gas rate of 5 mL/min): (**a**) front view; (**b**) top view.

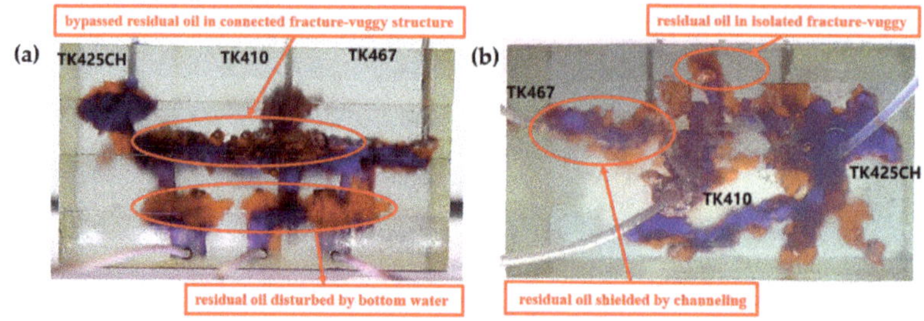

Figure 19. The distribution of residual oil after gas flooding in the 3D underground river model (gas rate of 10 mL/min): (**a**) front view; (**b**) top view.

1. The types of residual oil after gas flooding

Figure 18 shows the distribution diagram of residual oil after nitrogen gas displacement at 5 mL/min in a 3D visual underground river reservoir. After gas flooding in the 3D visual underground river fractured-vuggy reservoir, there were three main types of residual oil: a small portion of bypassed residual oil in the connected fractured-vuggy structure, residual oil in the isolated fractured-vuggy structure, and a small portion of residual oil disturbed by bottom water.

Figure 19 shows the distribution diagram of residual oil after nitrogen gas displacement at 10 mL/min in a 3D visual underground river reservoir. After gas flooding in the 3D visual underground river fractured-vuggy reservoir, there were four main types of residual oil: bypassed residual oil in the connected fractured-vuggy structure, residual oil in the isolated fractured-vuggy structure, residual oil shielded by channeling, and residual oil disturbed by bottom water.

2. The types of residual oil after gel foam flooding

Figure 20 shows the distribution diagram of residual oil after gel foam displacement at 5 mL/min in a 3D visual underground river reservoir. After gel foam flooding in the 3D visual underground river fractured-vuggy reservoir, there were three types of residual oil: residual oil shielded by channeling, residual oil at the blind end vug, and a small portion of residual oil disturbed by bottom water.

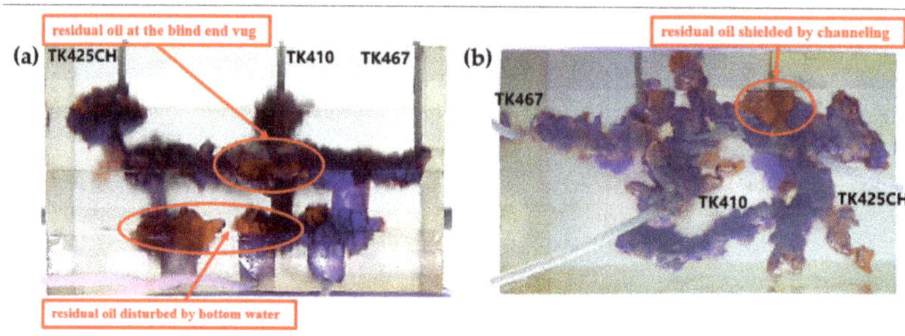

Figure 20. The distribution of residual oil after gel foam flooding in the 3D underground river model (gel foam rate of 5 mL/min): (**a**) front view; (**b**) top view.

Figure 21 shows the distribution diagram of residual oil after gel foam displacement at 10 mL/min in a 3D visual underground river reservoir. After gel foam flooding in the 3D visual underground river fractured-vuggy reservoir, there were three main types of residual oil: residual oil shielded by channeling, residual oil at the blind end vug, and residual oil disturbed by bottom water.

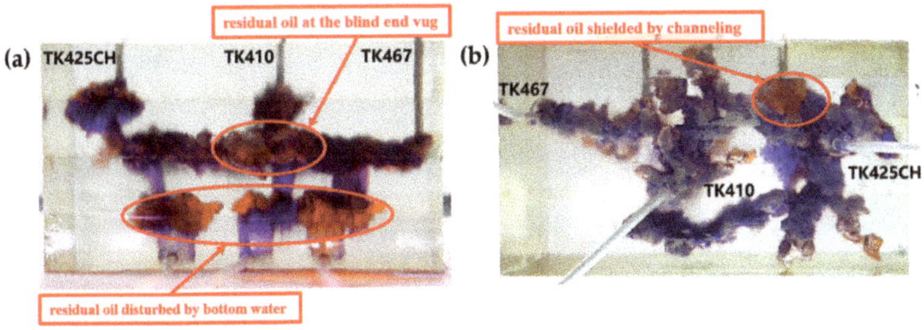

Figure 21. The distribution of residual oil after gel foam flooding in the 3D underground river model (gel foam rate of 10 mL/min): (**a**) front view; (**b**) top view.

In the experiment, the residual oil displacement scenarios were observed by simulating different injection rates during nitrogen flooding and foam flooding in fractured-vuggy reservoirs. The experiment showed that different nitrogen gas injection rates cause differences in the utilization of the remaining oil. Compared with the injection rate of 10 mL/min, when nitrogen gas was injected at 5 mL/min, it could effectively displace the residual oil shielded by channeling, and the utilization degree of residual oil disturbed by bottom water and bypassed residual oil in the connected fractured-vuggy structure was also relatively high. Compared with the injection speed of 10 mL/min for gel foam, when gel foam was injected at 5 mL/min, the utilization of residual oil disturbed by bottom water was also relatively high.

2.2. Mechanical Mechanism of Gel Foam Leaving Residual Oil in the Underground River System

2.2.1. Wettability Alternation

After the gel foam was injected into the model, as seen from the top view, the gel foam migrated forward in the form of pistons through irregular channels. The surfactant molecules in the gel foam were able to change the rock surface from lipophilic to hydrophilic, changing the wettability of the rock surface and increasing the wettability angle, θ, as shown in Figure 22. In Figure 22, the schematic diagram of the yellow circle is shown in the middle picture, while the oil displacement process of the gel foam at the yellow circle is shown in the right picture. The formula for calculating adhesion work was obtained from Equation (1). When the wetting angle increased, it can be seen from the formula that the adhesion work decreased. When the adhesion work is less than the cohesion value of crude oil, S > 0, the oil droplets contract on the rock surface. Oil droplets are easily displaced by the gel foam and eventually migrate out of the gel foam.

$$W_{adhesion} = \sigma_{S/O} + \sigma_{W/O} - \sigma_{W/S} = \sigma_{W/O}(COS\theta + 1) \quad (1)$$

$$S = W_{cohesion} - W_{adhesion} \quad (2)$$

where $W_{adhesion}$ is the adhesion work, $\sigma_{S/O}$ is the interfacial tension between the rock surface and oil, $\sigma_{W/O}$ is the interfacial tension between oil and water, $\sigma_{W/S}$ is the interfacial tension between the rock surface and water, θ is the contact angle, $W_{cohesion}$ is the cohesion work, and S is the spreading coefficient.

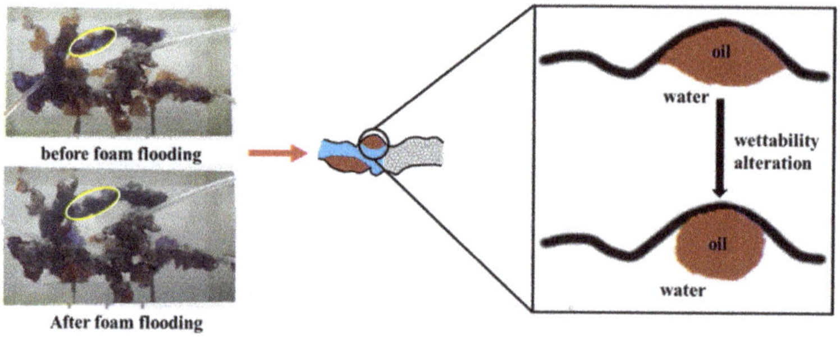

Figure 22. Schematic diagram of rock surface wettability alterations.

2.2.2. Selective Plugging Effect

As seen from the top view, after gel foam was injected into the model from well TK410, it first gathered in the vugs at the bottom of the well. The flow resistance of the underground river between well TK410 and well TK467 was low; therefore, gel foam first entered the large underground river, displacing most of the residual oil in the big channel from the produced well. Oil saturation in the large underground river was reduced, and the stability of the gel foam was enhanced. With the continuous injection of gel foam, gel foam gathered in the large underground river and increased its flow resistance. When gel foam gathered to a certain extent, the flow resistance of the large underground river gradually increased, inducing changes in the flow direction of the injection of gel foam. Subsequently, gel foam entered the small underground river between well TK410 and well TK425CH and finally displaced the residual oil shielded by channeling and residual oil at the blind end vug, as shown in Figure 23, the grey arrows in the figure indicate the flow direction of gas flooding, and the yellow arrows in the figure indicate the flow direction of gel foam flooding.

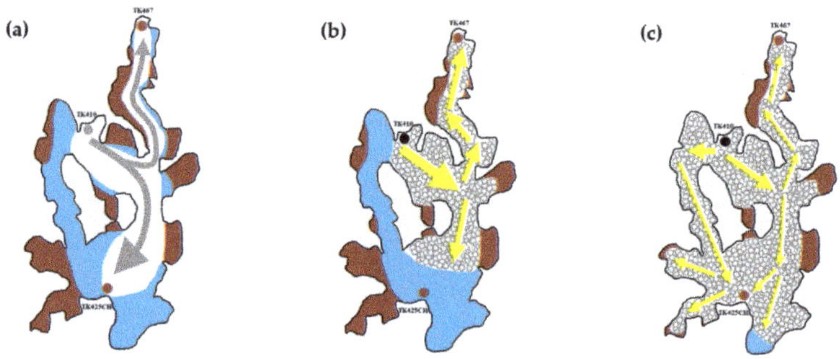

Figure 23. Schematic diagram of gel foam selective plugging: (**a**) late stage of gas flooding; (**b**) initial stage of gel foam flooding; and (**c**) late stage of gel foam flooding.

2.2.3. Mobility Control Ability

After the gel foam was injected from the injection well, it first gathered near the well, and then the gel foam slowly migrated along the large channel. The movement of the gel foam trapped the gas, and the rich liquid film in the gel foam restricted the flow of the gas. In the process of gel foam migration in the underground river, new bubbles were constantly generated, which reduced the relative permeability of gas, increased the apparent gas viscosity, and delayed gas channeling.

$$M = \frac{\lambda_D}{\lambda_d} = \frac{K_D \mu_d}{K_d \mu_D} \quad (3)$$

where M is the mobility ratio, λ_D is the mobility of the displacement phase, λ_d is the mobility of the displaced phase, K_D is the permeability of the displacement phase, K_d is the permeability of the displaced phase, μ_d is the viscosity of the displaced phase, and μ_{dD} is the viscosity of the displacement phase.

2.2.4. Gravity Differentiation

Gel foam exhibits the characteristics of defoaming when encountering oil and stabilizing when encountering water. When the gel foam came into contact with oil in the large channel, the gel foam became unstable and burst. Due to gravity differentiation, the gas in the gel foam gathered above the channel after the gel foam burst, forming a gas cap. As more and more gel foam burst, gas gathered and squeezed the residual oil below, thus forcing the residual oil out of the outlet, as shown in Figure 24, the red arrows in the figure indicate the flow direction of gas, and the yellow arrows in the figure indicate the flow direction of oil.

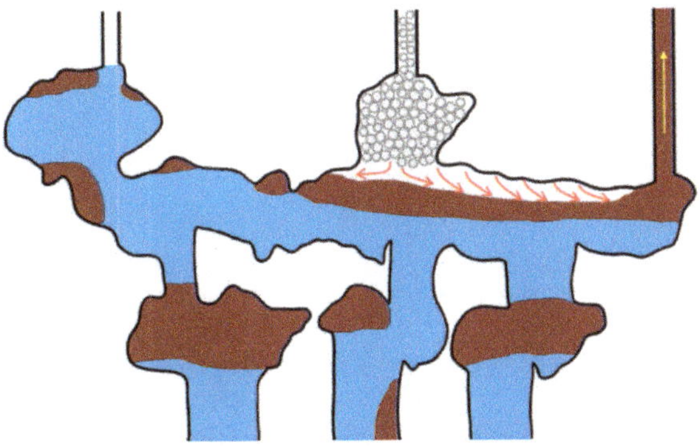

Figure 24. Schematic diagram of gel foam gravity differentiation.

3. Conclusions

In this study, a 3D visual underground river physical model of a fractured-vuggy reservoir was established based on similarity criteria, and experimental simulations of gel foam flooding were carried out. This study verified the feasibility of gel foam flooding in the EOR of fractured-vuggy carbonate reservoirs.

1. Regarding the impact of injection and production methods on oil recovery in underground river fractured-vuggy reservoirs, the results showed that in the process of oil displacement, actual reservoir conditions should be considered. For the well group with good connectivity in the fractured-vuggy reservoir, the effect of gas flooding on EOR is poor, and the structure of the fractured-vuggy reservoir contains a large amount of residual oil. Gel foam flooding can be carried out for wells connecting multiple vugs, and the recovery ratio of the well group can be improved by combining multiple displacement methods. Based on gas flooding, the enhanced recovery ratio of gel foam flooding in the underground river was approximately 12%.
2. The impact of injection speed on oil recovery was studied. Gel foam flooding can spread across two underground river channels; two cases of nitrogen flooding affected one underground river channel and two underground river channels with different injection rates. In underground rivers, increasing the injection rate sufficiently can improve oil recovery by increasing the injection pressure. When the gas injection rate reaches 10 mL/min, the difference in oil and gas mobility quickly causes gas channeling, shortening the gas injection time and yielding a lower final recovery ratio than that for a gas injection rate of 5 mL/min. The average recovery ratio of nitrogen flooding was 6.84% higher than that of other injection rates at 5 mL/min. In the process of gel foam injection, gel foam plugged the high-permeability channel

and delayed gas channeling to a certain extent. The average recovery ratio of gel foam flooding was 1.88% higher than that of other injection rates at 5 mL/min. When the injection rate of gel foam is too high, the gas will break through the gel foam to form a channeling channel, which will lead to a decline in the oil displacement effect. Therefore, whether gas flooding or gel foam flooding, a reasonable injection rate can achieve the comprehensive oil displacement effect of delaying gas channeling and increasing swept volume.

3. Through oil displacement experiments at different stages on 3D visual models, it was found that after nitrogen flooding, there were mainly four types of residual oil, and gel foam flooding mainly yielded three types of remaining oil. Research showed that, compared with 10 mL/min nitrogen gas displacement, 5 mL/min nitrogen gas displacement can displace residual oil shielded by channeling, most of the bypassed residual oil in connected fractured-vuggy structures, and most of the residual oil disturbed by bottom water. The effect of 5 mL/min nitrogen gas displacement was better. Compared with 10 mL/min gel foam flooding, 5 mL/min gel foam flooding can displace most of the residual oil disturbed by bottom water, and 5 mL/min gel foam flooding had a better displacement effect.
4. The experiments demonstrated that gel foam exerts an oil displacement mechanism that changes the wettability of rock surfaces, a selective plugging effect, mobility control ability, and gravity differentiation in fractured-vuggy carbonate reservoirs.

4. Materials and Methods

4.1. Similarity Design of the Physical Model

The flow laws of fluid in a fractured-vuggy carbonate reservoir are complicated. To simulate the geological condition and reservoir space of oilfields as accurately as possible, the physical model should reflect the characteristics of multi-scale fractured-vuggy reservoir space and reflect the coexistence of multiple flows. The theoretical basis of physical simulation is the similarity theory; the similarity criterion guides the fabrication of physical models. There are two common methods for deriving similarity criteria, one of which is dimension analysis, and the other is equation analysis. To ensure the suitability and applicability of the derived results, dimension analysis was chosen to deduce the similarity criterion.

According to the similarity theory of physical simulation, to achieve similarity between the model and the prototype, the physical model used in the experiment should meet the three conditions of geometric, kinematic, and dynamic similarity principles in design. Due to the complex heterogeneity of fractured-vuggy reservoirs, it is impossible to achieve complete similarity criteria. Therefore, in the process of experimental analysis, it is necessary to focus on the key points, select the main similarity criteria, and ignore the secondary similarity criteria; as such, the laboratory physical simulation experiment can reflect the actual reservoir to a certain extent. This experiment focuses on the selection of kinematic and dynamic similarity criteria to study the mechanism. An oilfield in Tahe was selected as the research object, the geological data were extracted, and the similarity design was defined by combining the parameters of fracture size, tortuosity, vuggy size, etc. A model similar to the actual reservoir was produced by scaling down. By selecting different parameters as references and combining them with the flow law of fluid in the actual reservoir, the similarity criterion formula was calculated using the dimension rule.

The correlation data of the actual reservoir parameters and physical models are included in the similarity criterion formula for calculation. The similarity coefficient is defined as the ratio of the physical model parameters to the actual reservoir parameters. The similarity index is defined as the ratio of the calculated results obtained by bringing two parameters into the formula. The similarity index represents the similarity between the physical model and the actual reservoir. The closer the similarity index is to 1, the higher the similarity between the physical model and the actual reservoir. Therefore, the similarity index was set to 1 to calculate the parameters of the physical model. The main similarity criteria of physical simulation are shown in Table 1, and the parameter comparison and similarity coefficient between the reservoir prototype and physical model are shown in Table 2.

Table 1. The main similarity criteria of physical simulation.

Similarity Conditions	Similarity Criteria	Physical Significance	Similarity Index
Dynamic similarity	$F_G = \Delta P/(\rho_o g d)$	Ratio of the injected pressure to gravity	1.01–1.04
	$Re = \rho u l/\mu$	Ratio of the moving inertial force to the viscous force	1
Kinematic similarity	$F_Q = Q/(r^2 u)$	Ratio of the recovery volume to the injection volume	1.01–1.04
Characteristic similarity	$\pi_2 = \xi$	Coordination number	1
	$\pi_3 = \eta$	Filling degree	1

Table 2. Parameter comparison and similarity coefficient between the reservoir prototype and the physical model.

Parameters	Reservoir Model	Physical Model	Similarity Coefficient
Cavity diameter (d)/cm	500–2500	3–12	166.667–200
Pressure difference (Δp)/kPa	2000–14,000	10–60	200–233.333
Viscosity of crude oil (μ)/(mPa·s)	19.7–28.5	65	0.3038–0.4377
Density of crude oil (ρo)/(g·cm^{-3})	0.92	0.8	1.15
Gravity acceleration (g)/(m·s^{-2})	9.8	9.8	1
Seepage velocity (u)/(m·s^{-1})	0.0147–0.147	0.007–0.049	2.103–2.993
Injection velocity (Q)/(m^3·d^{-1})	20–60	0.006–0.015	3334–4000
Well diameter (r)/mm	120	3	40
Fracture aperture(B)/mm	0.5–5	0.2–4.5	0.25–1.11
Filling degree	0–100%	0–100%	1

In the experiment, kerosene and liquid paraffin were used to prepare simulated oil with the same viscosity as crude oil under actual reservoir conditions, and the simulated oil was dyed with Sudan red. The simulated formation water was prepared according to the salinity and ion content of reservoir formation water, which ensured that the fluid medium used in the physical simulation experiment was the same as the actual fluid. The experimental fluid was closer to the actual fluid flow law, ensuring the reliability of the physical simulation results.

4.2. Physical Model Design and Fabrication

To study the flow law and oil displacement effect of gel foam in different well locations and injection rates, as well as the residual oil production in different stages in fractured-vuggy reservoirs, we designed a 3D visual fractured-vuggy reservoir model. By taking the TK425CH well group unit of the Tahe fractured-vuggy reservoir as the research object, based on the similarity criteria, we extracted the 3D geological carving data volume and obtained the overall geological characteristics. As shown in Figures 25 and 26, nine cross-sections of the fractured-vuggy body of the well group were equidistantly intercepted to obtain the cross-section of multiple simulated oil layers. The obtained fractured-vuggy structure data were imported into 3D simulation software. Each simulated fractured-vuggy unit was arranged in the same section according to the cross-section of each oil layer. After optimizing the design, each simulated oil layer was produced using a 3D printer. Then, the simulated oil layers were superimposed and bonded according to the overall geological characteristics, and it was ensured that the adjacent joints of fractured-vuggy units were interconnected, resulting in a 3D visual fractured-vuggy reservoir model. The fractures, or vugs, in the 3D visual fractured-vuggy reservoir model, were formed using transparent materials. The organic glass material used was polymethyl methacrylate, which has good temperature resistance and chemical stability. The oil's wettability was consistent with the wettability of the core. The above characteristics can reduce experimental errors. We added the location of injection-production well groups to the physical profile map, as shown in Figure 27. The 3D visual fractured-vuggy reservoir physical model of well group TK425CH was 15 cm × 8 cm × 8 cm.

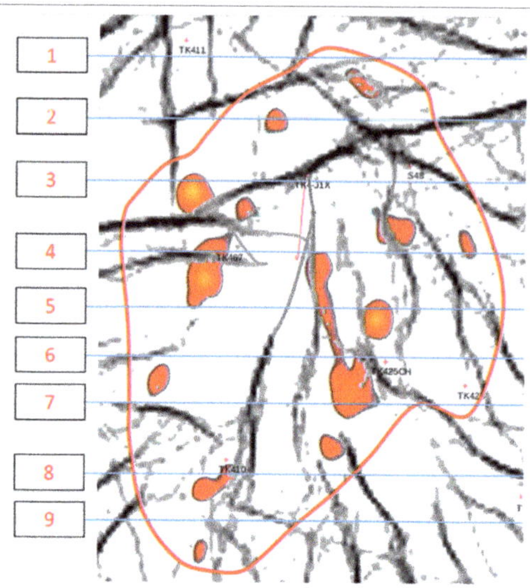

Figure 25. Schematic drawing of the cross-section of the TK425CH well group. The TK425CH well group was divided into nine equal parts. The numbers 1–9 represent geological drawing of longitudinal sections at different locations in the well group.

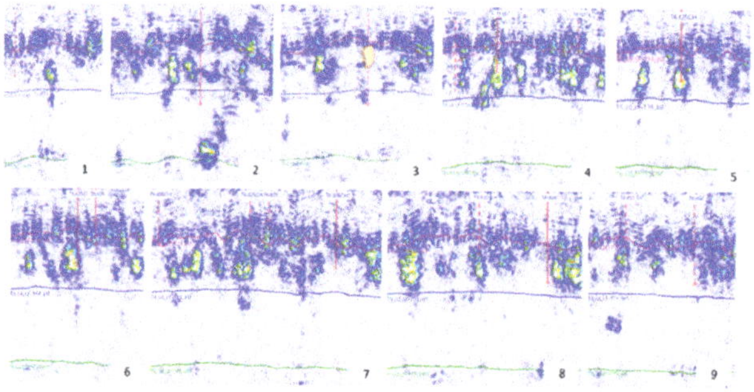

Figure 26. Geological drawing of longitudinal sections of the TK425CH well group. The numbers 1–9 represent different geological drawings of longitudinal sections. The location number of each longitudinal section in Figure 25 corresponds to the number of its corresponding geological drawing in Figure 26.

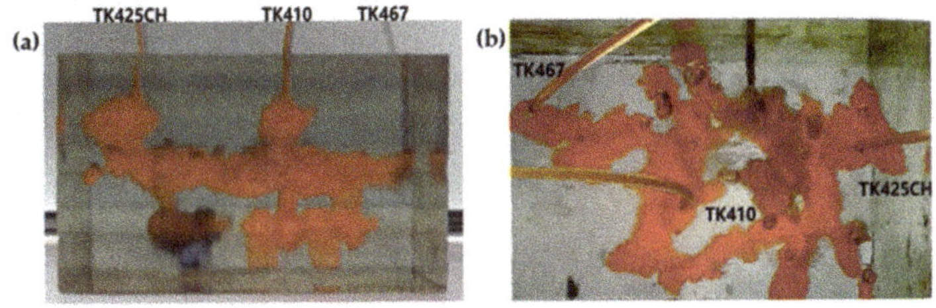

Figure 27. Schematic diagram of TK425CH well group locations in a 3D underground river model: (**a**) front view; (**b**) top view.

4.3. Experimental Materials and Equipment

The experimental simulated oil was configured using liquid paraffin and kerosene in proportion. The viscosity of the simulated oil was 23.9 mPa·s, and the density of the simulated oil was 0.821 g/mL at 25 °C. To facilitate observation under visual conditions, the simulated oil was dyed red with Sudan III. The experimental water was configured based on the actual formation water ionic composition of Tahe oilfield, with a viscosity of 1 mPa·s, a density of 0.98 g/mL, and a mineralization degree of about 250,000 mg/L. To facilitate observation under visual conditions, the experimental water was dyed blue with methylene blue dye. The composition and specific content of ions in water are shown in Table 3. The foam used in the experiment was gel foam, and α-starch nanogel particles were added to the foam, which yielded the foam with highly stable properties during the migration process in the fractured-vuggy reservoir. The chemical reagents used to prepare the highly stable gel foam included the foaming agent (SS-163, Qingdao Changxing High-tech Development Co., Ltd., Qingdao, China), sodium dodecyl sulfate, α-modified starch, acrylamide, N, N′-methylenebisacrylamide, and potassium persulfate. The preparation method was detailed in our previous study [21]. The gas used in the experiment was industrial nitrogen (99% purity).

Table 3. Formation of water ion composition.

Total Salinity/mg·L^{-1}	Ionic Concentration/mg·L^{-1}					
	Na^+	Ca^{2+}	Mg^{2+}	Cl^-	HCO_3^-	SO_4^{2-}
253,225	82,023	14,849	905	154,748	418	282

The gel foam flooding physical simulation experiment equipment mainly consisted of three parts: a visual physical model system, an experimental control system, and an image and data acquisition system, as shown in Figures 28 and 29. The experimental control system was composed of an injection pump (the working pressure was 0~30 MPa and the flow rate was 0.001~20 mL/min), a nitrogen gas bottle, a gas flow controller, a six-way valve, a gel foam generator, and intermediate containers containing gel foam liquid, simulated oil, and simulated formation water. The image and data acquisition system included a metering installation of liquid and a high-definition camera with a maximum resolution of 1080P. The model used in the experiment was the 3D visual underground river fractured-vuggy reservoir model.

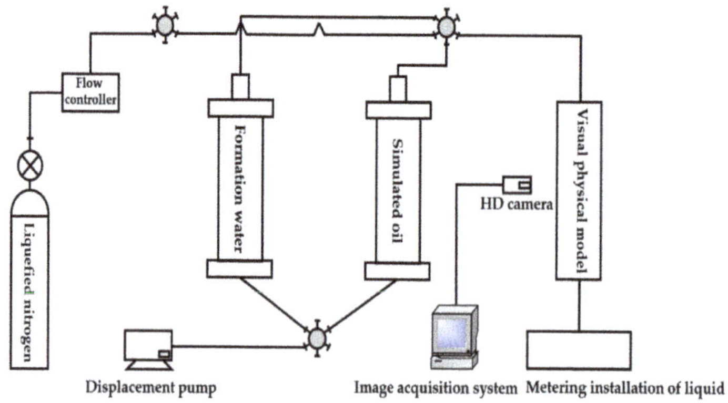

Figure 28. Flow chart of a physical simulation experiment of gas displacement.

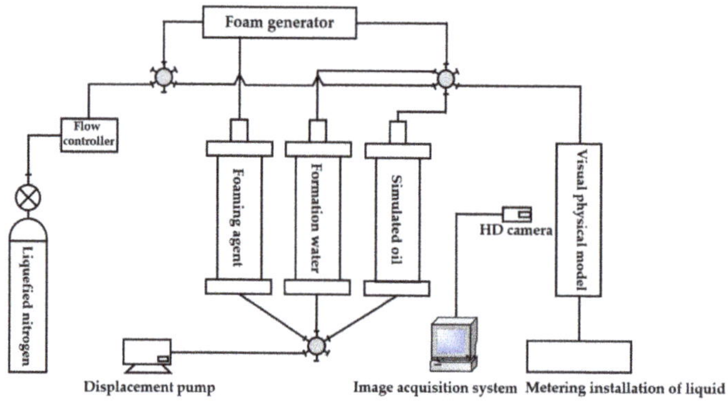

Figure 29. Flow chart of a physical simulation experiment of gel foam displacement.

4.4. Experimental Methods and Procedures

Under normal temperature and pressure (25 °C and 0.1 MPa, respectively), water flooding, gas flooding, and gel foam flooding physical simulation experiments were carried out successively in the 3D visual underground river physical mode. By observing the 3D visual physical model, the effect of EOR in the fractured-vuggy reservoir can be observed, and the remaining oil displacement law of fractured-vuggy structures can be analyzed. The specific experimental process is as follows:

a. The 3D visual underground river fractured-vuggy reservoir physical mode was saturated with oil.
b. Water flooding was carried out in an injection TK425CH well with a water injection rate of 2 mL/min (the production wells were TK410 and TK467). Water injection was stopped when the water cut at the production channel reached 98% or the water channeling was complete.
c. Gas flooding was carried out in the TK425CH well with a gas injection rate of 5 mL/min. The gas injection was stopped until gas channeling occurred at the production channel.
d. Gel foam flooding was carried out in a TK425CH well with an injection rate of 2 mL/min. Gel foam flooding was stopped until no oil was produced at the production channel.
e. The experimental process was recorded using an HD camera. The experimental schemes are shown in Table 4.

Table 4. Physical experiment schemes.

	Injection Well	Water Flooding	Nitrogen Gas Flooding	Foam Flooding
1	TK425CH TK410 TK467	2 mL/min	5 mL/min	2 mL/min
2	TK425CH TK410 TK467	2 mL/min	5 mL/min	2 mL/min
3	TK425CH TK410 TK467	5 mL/min	5 mL/min	1, 3, 5, 7, 9, 10 mL/min
4	TK425CH TK410 TK467	5 mL/min	1, 3, 5, 7, 9, 10 mL/min	10 mL/min

Author Contributions: Conceptualization, C.-M.L. and Y.-C.W.; methodology, C.-M.L.; validation, C.-M.L., Y.-C.W. and J.-R.H.; formal analysis, T.L.; investigation, C.-M.L.; data curation, C.-M.L. and Y.-C.W.; writing—original draft preparation, C.-M.L.; writing—review and editing, J.-R.H. and Y.-C.W.; visualization, C.-M.L.; project administration, J.-R.H.; funding acquisition, J.-R.H. All authors have read and agreed to the published version of the manuscript.

Funding: This research was funded by National Natural Science Foundation of China (Grant No. 52174046), the Major National Science and Technology Projects of China (Grant No. 2016ZX05014-004-004) and the Shaanxi Province Postdoctoral Research Project (2023BSHTBZZ45). The APC was funded by National Natural Science Foundation of China (Grant No. 52174046).

Institutional Review Board Statement: Not applicable.

Informed Consent Statement: Not applicable.

Data Availability Statement: Data are available from the corresponding author upon reasonable request.

Acknowledgments: The authors gratefully acknowledge the graduate students in research institute of unconventional petroleum science and technology for their help in experimental data analysis and video recording.

Conflicts of Interest: The authors declare no conflict of interest.

References

1. Guo, W.; Fu, S.; Li, A.; Xie, H.; Cui, S.; Nangendo, J. Experimental research on the mechanisms of improving water flooding in fractured-vuggy reservoirs. *J. Pet. Sci. Eng.* **2022**, *213*, 110383. [CrossRef]
2. Li, Y. The theory and method for the development of carbonate fractured-cavity reservoirs in Tahe oilfield. *Acta Petrol. Sin.* **2013**, *34*, 115–121.
3. Yu, Z.; Wang, Z.; Wei, H.; Xiao, F.; Li, Y. Characterization of fracture-cave karst reservoirs with different genesis in Tahe Oilfield. *Pet. Geol. Recovery Effic.* **2019**, *26*, 53–61.
4. Tian, F.; Di, Q.; Jin, Q.; Cheng, F.; Zhang, W.; Lin, L.; Wang, Y.; Yang, D.; Niu, C.; Li, Y. Multiscale geological-geophysical characterization of the epigenic origin and deeply buried paleokarst system in Tahe Oilfield, Tarim Basin. *Mar. Pet. Geol.* **2019**, *102*, 16–32. [CrossRef]
5. Li, Y.; Fan, Z. Developmental pattern and distribution rule of the fracture-cavity system of Ordovician carbonate reservoirs in the Tahe Oilfield. *Acta Petrol. Sin.* **2011**, *32*, 101–106.
6. Lu, G.; Zhang, L.; Liu, Q.; Xu, Q.; Zhao, Y.; Li, X.; Deng, G.; Wang, Y. Experiment analysis of remaining oil distribution and potential tapping for fractured-vuggy reservoir. *J. Pet. Sci. Eng.* **2022**, *208*, 109544. [CrossRef]
7. Zheng, S.; Yang, M.; Kang, Z.; Liu, Z.; Long, X.; Liu, K.; Li, X. Controlling factors of remaining oil distribution after water flooding and enhanced oil recovery methods for fracture-cavity carbonate reservoirs in Tahe Oilfield. *Pet. Explor. Dev.* **2019**, *46*, 746–754. [CrossRef]
8. Liu, Z. Enhanced oil recovery in Tahe karstic/fractured carbonate reservoir. *Pet. Geol. Recovery Effic.* **2012**, *19*, 66–68.
9. Rong, Y.; Tu, X.; Liu, X. Single well water injection production in Tahe fracture-vuggy reservoir, Tahe oilfield. *Pet. Geol. Recovery Effic.* **2010**, *17*, 97–100.

10. Ren, Y.; Li, J.; Huang, X. Study on developmental technologic strategies in Tahe carbonate oilfield. *Pet. Geol. Recovery Effic.* **2004**, *5*, 57–59.
11. Khormali, A. Effect of water cut on the performance of an asphaltene inhibitor package: Experimental and modeling analysis. *Pet Sci Technol.* **2022**, *40*, 2890–2906. [CrossRef]
12. Wang, J.; Zhao, W.; Liu, H.; Liu, F.; Zhang, T.; Dou, L.; Yang, X.; Li, B. Inter-well interferences and their influencing factors during water flooding in fractured-vuggy carbonate reservoirs. *Petrol. Explor. Dev.* **2020**, *47*, 1062–1073. [CrossRef]
13. Zeng, S.; Liu, Z.; Qiu, L. Evaluation of Injected Water Utilization for Fractured–Vuggy Reservoirs in Tahe Oilfield. *Xinjiang Pet. Geol.* **2017**, *38*, 452–454.
14. Karimaie, H.; Darvish, G.; Lindeberg, E.; Torsæter, O. Secondary and tertiary gas injection in fractured carbonate rock: Experimental study. *J. Pet. Sci. Eng.* **2008**, *62*, 45–51. [CrossRef]
15. Hui, J.; Liu, X.; Wang, Y.; Yang, Z.; Chen, Y. Mechanism and Practice of Nitrogen Injection for EOR in Fractured-Vuggy Carbonate Reservoir in Tahe Oilfield, Tarim Basin. *Mar. Pet. Geol.* **2015**, *36*, 75–77.
16. Wang, J.; Ji, Z.; Liu, H.; Huang, Y.; Wang, Y.; Pu, Y. Experiments on nitrogen assisted gravity drainage in fractured-vuggy reservoirs. *Pet. Explor. Dev.* **2019**, *46*, 355–366. [CrossRef]
17. Zhang, C.; Li, Z.; Li, S.; Lv, Q.; Wang, P.; Liu, J.; Liu, J. Enhancing Sodium Bis(2-ethylhexyl) Sulfosuccinate Injectivity for CO_2 Foam Formation in Low-Permeability Cores: Dissolving in CO_2 with Ethanol. *Energy Fuels.* **2018**, *32*, 5846–5856. [CrossRef]
18. Li, H.; Hou, J.; Li, W.; Yuan, D.; Zhang, L.; Jiang, Y.; Yuan, W. Laboratory research on nitrogen foam injection in fracture-vuggy reservoir for enhanced oil recovery. *Pet. Geol. Recovery Effic.* **2014**, *21*, 93–96.
19. Yang, J.; Hou, J.; Qu, M.; Liang, T.; Wen, Y. Experimental study the flow behaviors and mechanisms of nitrogen and foam assisted nitrogen gas flooding in 2-D visualized fractured-vuggy model. *J. Pet. Sci. Eng.* **2020**, *94*, 107501. [CrossRef]
20. Xu, Z.; Li, Z.; Cui, S.; Li, B.; Chen, D.; Zhang, Q.; Zheng, L.; Husein, M. Flow characteristics and EOR mechanism of foam flooding in fractured vuggy reservoirs. *J. Pet. Sci. Eng.* **2022**, *211*, 110170. [CrossRef]
21. Qu, M.; Hou, J.; Liang, T.; Raj, I.; Yang, Y.; Qi, P. Synthesis of α-starch based nanogel particles and its application for long-term stabilizing foam in high-salinity, high-temperature and crude oil environment. *J. Pet. Sci. Eng.* **2020**, *191*, 107185. [CrossRef]

gels

Article

Experimental Study on Multi-Dimensional Visualization Simulation of Gas and Gel Foam Flooding in Fractured-Vuggy Reservoirs

Yuchen Wen * and Jirui Hou *

Key Laboratory of Petroleum Engineering, Research Institute of Unconventional Petroleum Science and Technology, China University of Petroleum (Beijing), Beijing 102249, China
* Correspondence: yc.wen@foxmail.com (Y.W.); houjirui@126.com (J.H.)

Abstract: Gas flooding and foam flooding are potential technologies for tertiary oil recovery in fractured-vuggy reservoirs. The development and mechanism research of fractured-vuggy reservoirs is difficult due to the complex structures and the strong heterogeneity of fractured-vuggy reservoirs. Visualization simulation is one of the effective methods to study the flow behavior of fluid in fractured-vuggy reservoirs. In this study, an upscaling method of visualization simulation from one dimension (1D) to three dimensions (3D) was established, and the physical models of fractured-vuggy reservoirs were designed and fabricated. Water flooding, gas flooding, and gel foam flooding were carried out in the models. The experimental results showed that gas flooding has a single flow channel and water flooding has multiple flow channels in fractures and vugs. Gel foam with an excellent capability of mobility control and a high microscopic displacement efficiency swept in all directions at a uniform velocity. The EOR mechanisms of gel foam in fractured-vuggy reservoirs were mainly as follows: reducing interfacial tension, increasing mobility ratio, selectively plugging high permeability channels, and discontinuous flow. In the displacement process of fractured-vuggy reservoirs, water should be injected from the well at the bottom of the reservoir, and gas should be injected from the well located in the vug at the high part of the reservoir. Gel foam with strong stability and high viscosity should be selected and injected in most kinds of injection wells in fractured-vuggy reservoirs. This study provides a complete method of visualization simulation for the study of flow behavior in fractured-vuggy reservoirs and provides theoretical support for the application of gas flooding and gel foam flooding in fractured-vuggy reservoirs.

Keywords: fractured-vuggy reservoirs; foam flooding; gas flooding; visualization simulation; EOR

1. Introduction

Carbonate reservoirs play an important role in the world's oil and gas resources, and their oil and gas reserves account for 70% of the world's total oil and gas reserves [1]. Carbonate reservoirs have the characteristics of large-scale and high production. Among them, fractured-vuggy reservoirs account for more than 30% of carbonate reservoirs [2]. Carbonate sedimentary rocks in China are mainly distributed in the Tarim Basin, the Sichuan Basin, the Ordos Basin, and North China, among which the most typical fractured-vuggy reservoir is the Ordovician carbonate reservoir in the Tahe Oilfield [3]. The main structures of the reservoir are fractures and secondary pores, indicating that the reservoir space of oil is mainly fractures and vugs. The matrix permeability of the carbonate reservoir is extremely low (less than 1 mD). The flow ability of fluid in the matrix is extremely poor, and fractures are the main channels [4,5].

After multiple tectonic processes, many fractures and vugs are crisscross developed in fracture-vuggy carbonate reservoirs. The complexity of the connection between fractures and vugs results in the strong heterogeneity of fracture-vuggy carbonate reservoirs [6]. There is a big difference in the size of the fractured-vuggy structures and the regularity of the connection mode. The strong heterogeneity in the spatial structure distribution of fractured-vuggy carbonate reservoirs resulted in great difficulty in the development and low oil recovery of fracture-cave carbonate reservoirs [7]. The Tahe Oilfield was mainly developed by natural energy in the early stage of development, and the average oil recovery ratio was 12.4% [8]. Subsequent water flooding was carried out in different ways, but laboratory experiments and field applications have proved that the development effect of water flooding in fractured-vuggy reservoirs was extremely limited [9,10]. By 2009, the oil recovery ratio in the stage of water flooding increased by 2.45% [8]. By drawing lessons from the successful cases of gas flooding in the United States and other countries, field tests of gas flooding were carried out in the Tahe Oilfield in 2017 [11]. Gas replaced the residual oil in the high part of fractured-vuggy reservoirs by gravity differentiation. Due to the special structure of the fractured-vuggy reservoir, gas channeling easily occurs in reservoirs [12,13]. By 2019, the oil recovery ratio in the stage of gas flooding increased by 2.16%, which is not satisfactory [8]. Foam and nanoparticles are potential technologies for enhanced oil recovery in fractured-vuggy reservoirs [14–16]. Laboratory experiments showed that foam could not only expand the swept volume but also improve the displacement efficiency, thus significantly improving the oil recovery of fractured-vuggy reservoirs [17]. In recent years, some progress has been made in the application of foam flooding in fractured-vuggy reservoirs. Foam has been proven to be beneficial in improving the gas flooding effect in fractured-vuggy reservoirs [18].

With the continuous development and understanding of carbonate reservoirs, the theory of the "fractured-vuggy unit" was put forward to reasonably classify the reservoir space of the fractured-vuggy reservoir. The basic unit is a separate fractured-vuggy unit, and the fractured-vuggy reservoir system is composed of multiple basic units through different combinations [7,19]. This classification method not only provided a geological basis for the development of fractured-vuggy carbonate reservoirs but also provided a theoretical basis for experimental research and the establishment of physical models.

On this basis, a large number of studies were carried out by researchers on numerical simulations [20–23], physical simulations [24,25] and other methods. However, because of the unique geological characteristics and complex reservoir environment of the fractured-vuggy reservoir, it is very difficult to study the flow law of fluids and the development effect of reservoirs in the fractured-vuggy reservoir. Visualization simulation is still one of the most intuitive and effective methods to study the law of fluid flow in fractured-vuggy reservoirs.

Table 1 was a partial statistic on visualization simulation of fractured-vuggy reservoirs. Y. Wen. used a one-dimensional (1D) visual model to reveal the microscopic flow characteristics and displacement effect of foam in fractures [15]. Yang used a 1D visual model to reveal the influence of the shape of fracture-vuggy structure on fluid flow and residual oil distribution [26]. A two-dimensional (2D) model was established by Hou [27], and the EOR effect of water flooding and nitrogen gas flooding in fractured-vuggy structures was verified. On this basis, Qu designed a similar 2D visual model to study foam flooding and revealed that the main EOR mechanism of foam flooding was the reduction of oil–water interfacial tension and gravity differentiation [28]. Wang verified the displacement law and the EOR effect of gas flooding in fractured-vuggy reservoirs by using a 2D visual model [29]. Liang and Hou also established a new kind of 2D visual model and clarified the flow characteristics and phase behavior of nitrogen and foam-assisted nitrogen in the fractured-vuggy reservoir [17]. A 2D model was established by Xu to reveal the flow characteristics and the EOR mechanism of foam in fractured-vuggy reservoirs [30]. The design and manufacture of the three-dimensional (3D) visual model of the fractured-vuggy reservoirs were complicated, and related research is rare. A 3D visual model of the fractured-vuggy reservoir was established by Qu, and the flow characteristics of oil, gas, and water in the process of oil displacement were studied systematically and comprehensively [31]. Lu also established a 3D visual model and analyzed the flow characteristics and EOR mechanism of foam flooding in fractured-vuggy reservoirs [32].

Table 1. A brief review of visual models of fractured-vuggy reservoirs.

Dimension	Model	Materials	Flooding	Size	Ref.
1D	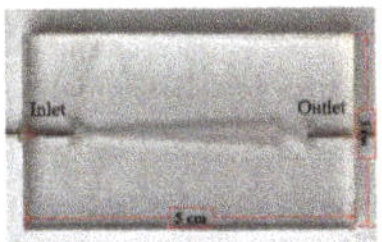	Plexiglass	Nitrogen gas flooding and foam flooding	3 cm × 5 cm	[15]
1D	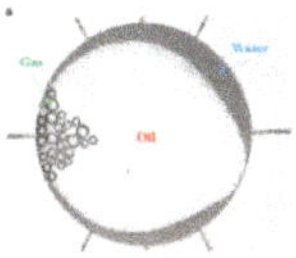	Photopolymer Vero Clear and Fullcure 706	Water flooding and nitrogen gas flooding	340 mm × 340 mm × 200 mm	[26]
2D		Quartz sand, calcium carbonate and epoxy resin	Water flooding and nitrogen gas flooding	150 mm × 120 mm × 20 mm	[27]
2D		Quartz sand, calcium carbonate, and epoxy resin	Foam flooding	15 cm × 7 cm × 0.68 cm	[28]
2D	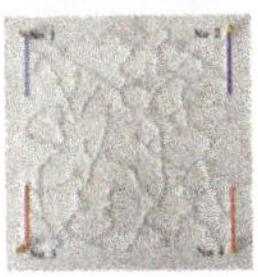	Glass	Nitrogen gas flooding	25 cm × 25 cm	[29]
2D	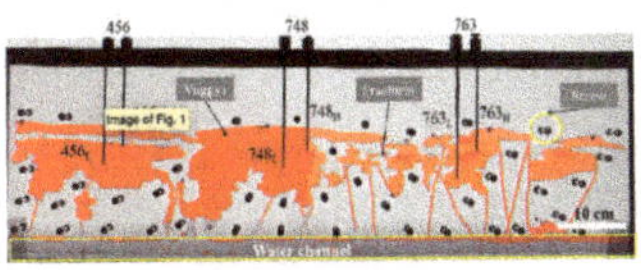	Plexiglass	Nitrogen gas flooding and foam flooding	480 mm × 480 mm × 10 mm	[17]
2D	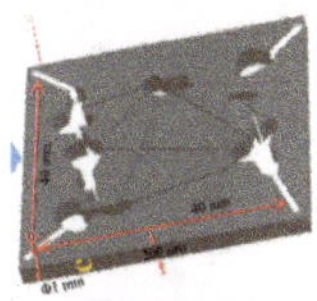	Plexiglass	Foam flooding	40 mm × 40 mm × 1 mm	[30]
3D	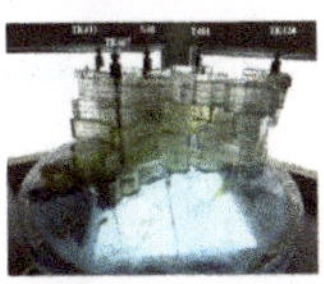	Polymethyl methacrylate	Water flooding and nitrogen gas flooding	1571.29 cm^3	[31]
3D	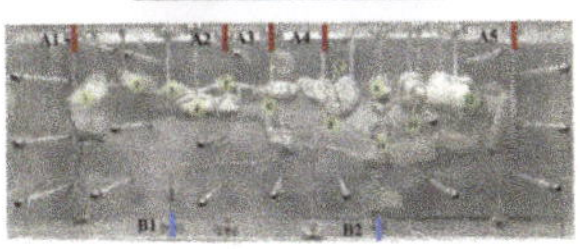	Plexiglass	Water flooding	80 cm × 30 cm × 20 cm	[32]

In this paper, the model parameters and experimental parameters of the physical simulation are designed according to the similarity criterion. Three kinds of visual models of fractured-vuggy reservoirs with different dimensions were fabricated. The experiments of water flooding, nitrogen gas flooding, and gel foam flooding were carried out in these models. The experimental results revealed the flow rules of water, gas, and foam in the fractured-vuggy structure. The main EOR mechanisms of gas and gel foam in fractured-vuggy reservoirs were analyzed. The EOR effects of different flooding methods were evaluated, and the best injection modes of different flooding methods in fractured-vuggy reservoirs were put forward. It provided a theoretical basis and technical support for the application of gas flooding and gel foam flooding in the Tahe Oilfield and other fractured-vuggy reservoirs.

2. Results and Discussion

2.1. Experimental Results and Analysis of Physical Simulation in the 1D Visual Models

The flow of viscous fluid conforms to the Boundary Layer theory. The greater the difference in viscosity between the displacement phase and the displaced phase, the more serious the viscous fingering was and the easier it was to form a dominant channeling. The fluid flow in the fracture conforms to the Poiseuille Flow law. During the water flooding process, the flow velocity of fluids at different locations on the cross-section of the fracture was different due to the irregular shape of the fracture and the viscous difference of fluids. The first dominant channeling of water formed at the cross-section where the velocity of fluid was fastest. After that, the flow resistance in the first dominant channeling became the flow resistance of water and remained unchanged. The other places on the cross-section of the fracture were swept by subsequent injection water and gradually formed multiple water channels. After the dominant channelings were formed, all the injected water flowed along the dominant channelings, as shown in Figure S1a and Figure 1a. The flow velocity in the undisplaced area on the cross-section of the fracture was 0.

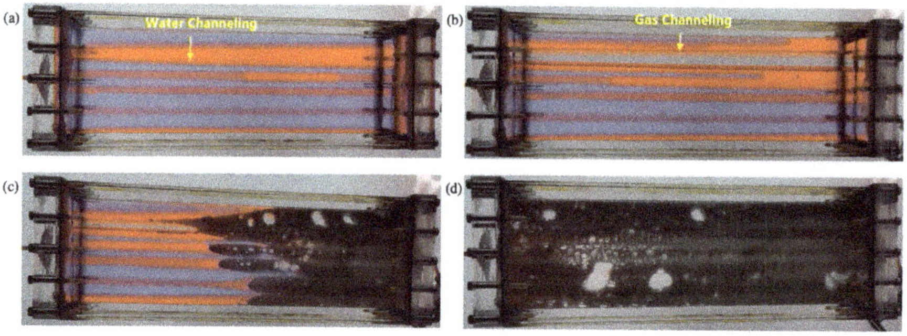

Figure 1. Experiments in the 1D visual model of a single fracture: (**a**) water flooding; (**b**) gas flooding after water flooding; (**c**) gel foam flooding after gas and water flooding; (**d**) experimental result after gel foam flooding.

The viscous fingering was more serious during gas flooding than water flooding because the viscosity ratio between gas and oil is much lower than that between oil and water. After a dominant channeling of gas was formed, the injected gas flowed along the dominant channeling preferentially. It was difficult for the gas to displace residual oil in the cross-section of the fracture at a constant flow rate. Finally, only one dominant channeling of gas was formed in the 1D visual model, as shown in Figure S2a. The gas flooding experiment after water flooding revealed that the injected gas did not necessarily flow along the dominant channeling of water but formed a new dominant channeling, as shown in Figure 1b. This phenomenon was caused by the difference in fluid viscosity and fracture shape.

Since the viscosity of gel foam is much greater than that of water and gas, gel foam displaced all the oil on the cross-section of the fracture, as shown in Figure 1d. All subsequent experiments revealed that gel foam can control fluid mobility in fractures, such as gel foam flooding after water flooding (see Figure S1b), gel foam flooding after gas flooding (see

Figure S2b), and gel foam flooding after water flooding and gas flooding (see Figure 1c). Gel foam can uniformly displace fractures with irregular cross-sections, expand the swept volume, and enhance the oil recovery of fractures.

2.2. Experimental Results and Analysis of Physical Simulation in the 2D Visual Models

2.2.1. Experimental Results of the 2D Visual Model of Complex Fractures

Water flooding, gas flooding, and gel foam flooding experiments were carried out in the 2D visual model of complex fractures, and the experimental results are shown in Figure 2. Compared to water flooding, the large viscosity difference between gas and oil, as well as the large oil-gas interfacial tension, lead to the extremely poor microscopic displacement efficiency of gas flooding in fractures. The injected gas rapidly broke through along the porous section in the large fracture (U–T). Then, it successively migrated along the fracture (V–W), the fracture (W–X), the fracture (X–Z), and the fracture (Z–Y). Finally, the injected gas formed a channel from the inlet to the outlet, and the result of the gas flooding experiment is shown in Figure 2b. Gas flooding with a recovery ratio of 39% did not sweep the small fracture (0.5 mm) in the 2D visual model of complex fractures.

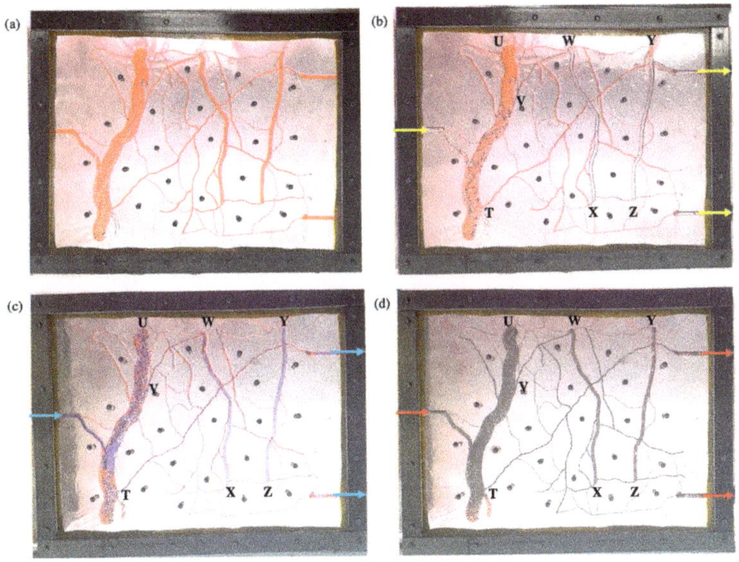

Figure 2. Experimental results of the 2D visual model of complex fractures: (**a**) saturated with oil; (**b**) gas flooding; (**c**) water flooding; (**d**) gel foam flooding.

The interfacial tension and viscosity difference between water and oil are smaller than those between gas and oil, resulting in the displacement effect of water flooding in the porous section of the large fracture (U–T) media, which is much better than gas drive. After the injected water swept the fracture (U–T), it migrated to the fracture (W–X) along small fractures (1 mm). Then, the subsequent water swept through fractures (0.5 mm and 1 mm) between fracture (W–X) and fracture (Z–Y). Finally, injected water migrated to the outlet along the fracture (Z–Y). Water flooding with a recovery ratio of 68% swept some of the small fractures between the large fractures, as shown in Figure 2c.

The larger viscosity difference and lowest interfacial tension between gel foam and oil compared to the others enable the gel foam to sweep most of the connected fractures in this model. The small fractures (0.5 mm and 1 mm) that cannot be swept by the water or gas can be effectively displaced by gel foam. Gel foam flooding with a recovery ratio of

89% is shown in Figure 2d. The residual oil after gel foam flooding mainly existed in the fractures with blind ends.

2.2.2. Experimental Results of the 2D Visual Model of Complex Vugs

Water flooding, gas flooding, and gel foam flooding experiments were carried out in the 2D visual model of complex fractures, and the experiment result of the unfilled model is shown in Figure 3. The results indicated that the gravitational differentiation was obvious during the water flooding, gas flooding, and gel foam flooding process in the fracture-vuggy structure. Gas flowed along the upper parts, water flowed along the lower parts, and gel foam flowed along the middle parts of the fracture-vuggy structure. Some of the gel foam released gas when it burst in vugs. Then, the gas and water stratified under the influence of gravitational differentiation. Gas migrated upward to replace the oil in the upper vugs and formed a gas cup (see Figure 3b). The final recovery ratio of water flooding was 34.8% (see Figure 3a), the recovery ratio of gel foam flooding after water flooding was 68.2% (see Figure 3b), the recovery ratio of gas flooding was 38.5% (see Figure 3c), and the recovery ratio of gel foam flooding after gas flooding was 54.8% (see Figure 3d).

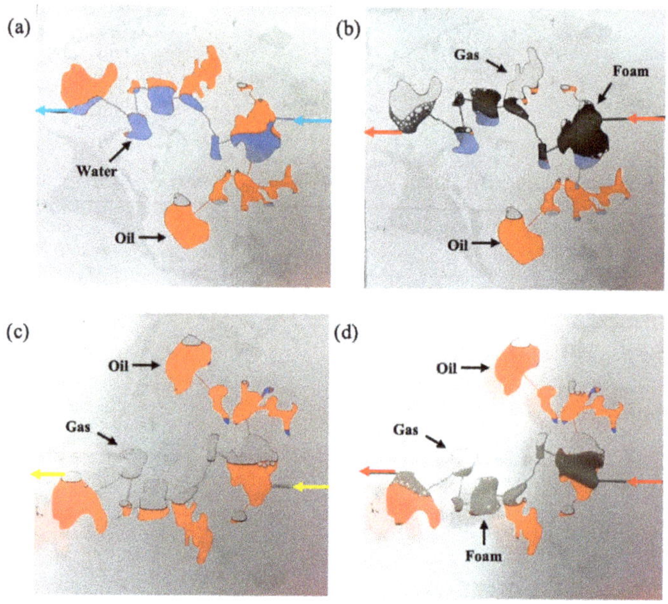

Figure 3. Experimental results of the 2D visual unfilled model of complex vugs: (**a**) water flooding; (**b**) water flooding after gas flooding; (**c**) gas flooding; (**d**) gel foam flooding after gas flooding.

The experiment result of the filled model is shown in Figure 4. The results indicated that the swept volume of water flooding and gel foam flooding in filled vugs was larger than that of unfilled vugs. The swept volume of gas flooding was just the opposite. Due to the difference in fluid density, gas was extremely easy to channel in filled vugs, but gel foam could control the mobility very well. The final recovery ratio of water flooding was 45.8% (see Figure 4a), and the recovery ratio of gel foam flooding after water flooding was 66.7% (see Figure 4b). The final recovery ratio of gas flooding was 31.2% (see Figure 4c), and the recovery ratio of gel foam flooding after gas flooding was 58.9% (see Figure 4d).

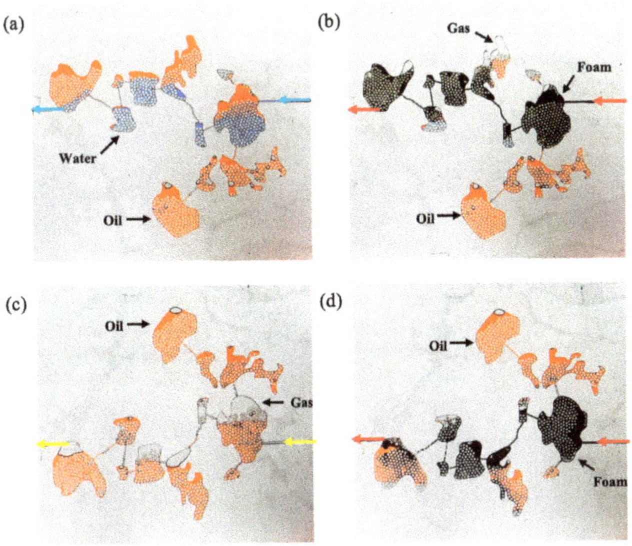

Figure 4. Experimental results of the 2D visual filled model of complex vugs: (**a**) water flooding; (**b**) water flooding after gas flooding; (**c**) gas flooding; (**d**) gel foam flooding after gas flooding.

2.3. Experimental Results and Analysis of Physical Simulation in the 3D Visual Model

The TK671 well-group unit is a typical epikarst reservoir with complex connection relationships. This unit has six wells, all of which are located in connected vugs. Since the TK661 well is located at the bottom of the vug and is the deepest of all wells, the TK661 well was selected as the water injection well.

2.3.1. Gas Flooding and Gel Foam Flooding in TK692X

Flooding experiments were carried out in the 3D visual model. The TK661 well was chosen to be a water injection well at first. The TK692X well was chosen to be a gas injection well or gel foam injection well. The TK661 well, TK696X well, and TK671 well were chosen to be production wells. After the water was injected into the TK661 well, it displaced the residual oil in the bottom vug and then migrated to other vugs through the fractures connected with the bottom vug. The injected water flowed forward along the main channel (TK661-TK696X-TK604-TK671). Under the influence of gravity, water displaced the oil in the lower part of the model, which led to an increase in the oil–water interface. The water effectively swept the large fracture channel between the TK696X well and the TK692X well but did not sweep other flow channels (see Figure 5b). The experiment of water flooding was stopped when water was produced from the TK696X well at the lower part of the model. The complex structure of the fracture-vugs reservoir leads to an inconsistent height of the oil–water interface. After water channeling occurred in the model, the oil–water interface was high in the swept vugs and low in a large number of unswept vugs (see Figure 5a). Water flooding (with a recovery ratio of 29%) swept the residual oil in large fractures and the lower part of the vugs, but it cannot effectively displace the residual oil in other channels, blind-end vugs, and poorly connected vugs. Multiple flow channels were formed during water flooding.

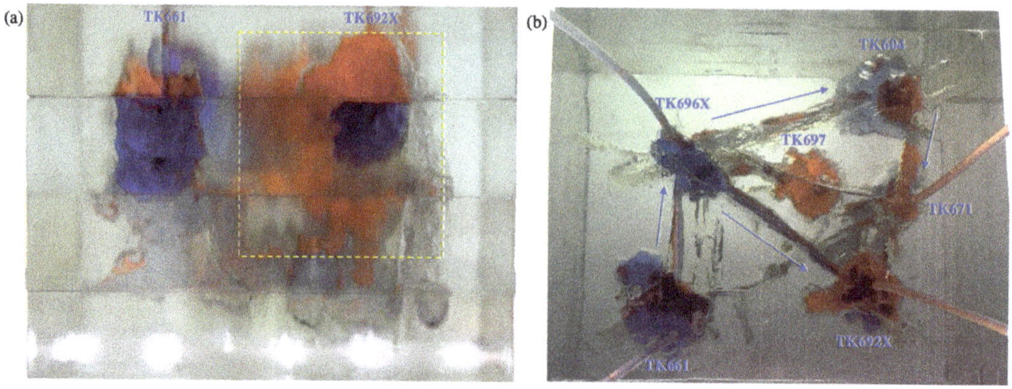

Figure 5. Experimental results of water flooding in the 3D visual model: (**a**) the front view and (**b**) the overhead view.

Gas was injected into the TK692X well after water flooding. It displaced the residual oil of the vug under the TK962X well and then broke through along the large fracture between the TK692X well and the TK696X well (see Figure 6a). After that, gas swept the vug under the TK966X well and then migrated along the fracture to the TK611 well. Gas was produced from the TK611 well and formed a gas channeling in the model (see Figure 6b). The well shut-in measure was taken for the TK661 well, which led to the subsequent gas beginning to migrate along another fracture channel (TK696X-TK604-TK671). The residual oil at the top of the TK604 well and the TK671 well was gradually displaced by gas. Finally, a gas channeling occurred in the TK671 well (see Figure 7). Under the influence of gravity, gas flooding swept the large fractures and vugs at the upper part of the model but could not effectively displace the residual oil in the fractures and vugs at the lower part of the model. Gas flooding, with a recovery ratio of 56%, was 27% higher than water flooding. Only one flow channeling was formed during gas flooding.

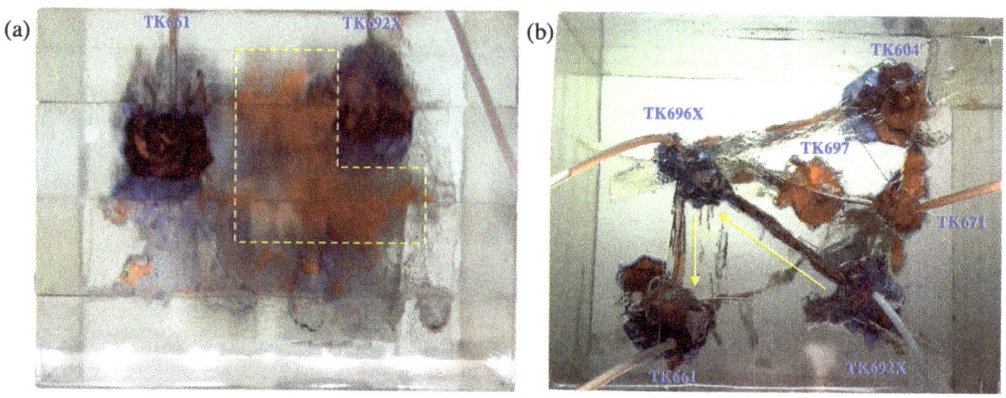

Figure 6. Experimental results of gas flooding in the 3D visual model (gas channeling occurred in TK661 well): (**a**) the front view and (**b**) the overhead view.

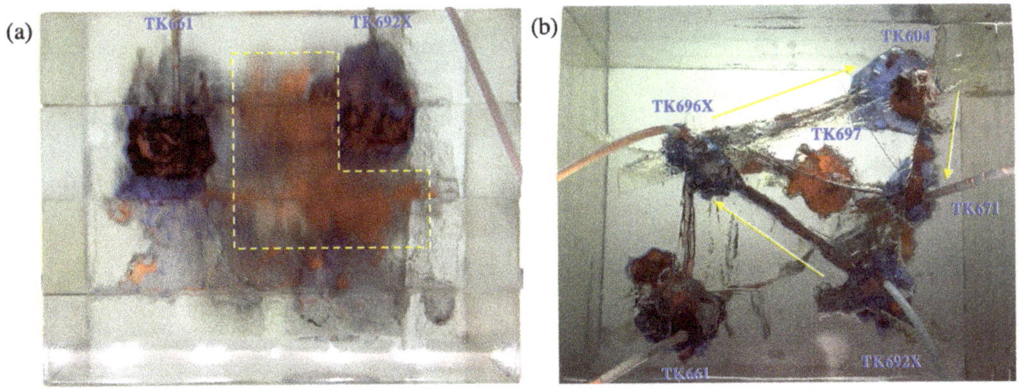

Figure 7. Experimental results of gas flooding in the 3D visual model (gas channeling occurred in TK671 well): (**a**) the front view and (**b**) the overhead view.

Gel foam flooding was carried out in the model after gas flooding. After the gel foam was injected into the TK692X well, the vug at the bottom of the TK692X well was filled by gel foam, and all remaining oil in it was displaced (see Figure 8a). Gel foam migrated along the large fracture between the TK692X well and the TK696X well and effectively displaced the residual oil in the vug at the bottom of the TK696X well. After that, the gel foam moved at a constant speed in three channels (TK696X-TK604, TK696X-TK697, and TK696X-TK661) at the same time, as shown in Figure 8b. The vugs at the bottom of the TK604 well and the TK661 well were effectively displaced by gel foam. The residual oil in the vug at the bottom of the TK697 well has not been completely displaced by gel foam because of its complex structure. The experiment of gel foam flooding was stopped after gel foam swept the vug at the bottom of the TK671 well. Gel foam has an excellent capability of mobility control and high microscopic displacement efficiency, which can uniformly sweep most of the connected channels and effectively displace the residual oil. Gel foam flooding, with a recovery ratio of 71.5%, was 15.5% higher than gas flooding (see Figure 9).

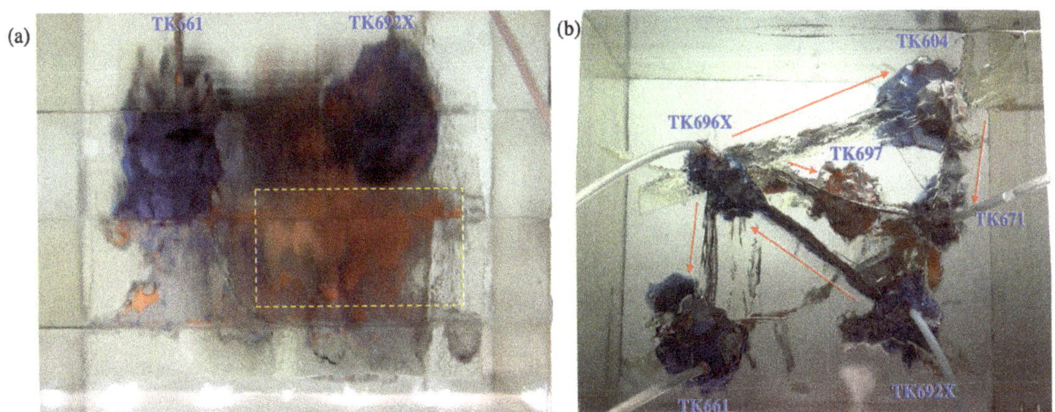

Figure 8. Experimental results of gel foam flooding in the 3D visual model: (**a**) the front view and (**b**) the overhead view.

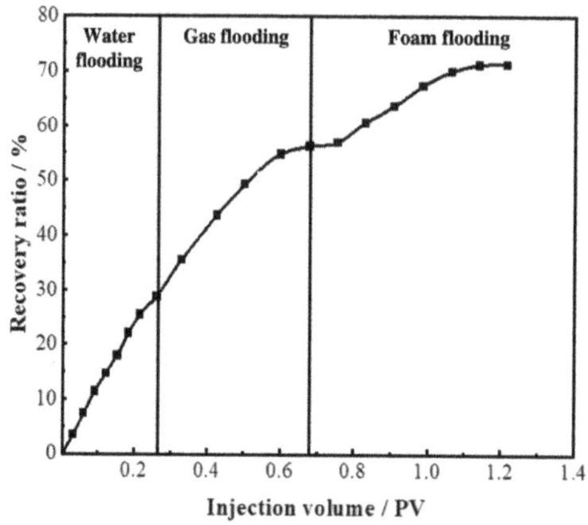

Figure 9. The recovery ratio of the 3D visual model (injection well: TK692X well).

2.3.2. Gas Flooding and Gel Foam Flooding in TK671

Flooding experiments were carried out in the 3D visual model. The TK661 well was chosen to be a water injection well at first. The TK671 well was chosen to be a gas injection well or gel foam injection well. The TK661 well, TK696X well, and TK692X well were chosen to be production wells.

During the gas flooding process, gas displaced the residual oil at the top of the vug under the TK671 well and then broke through along the main channel (TK671-TK604-TK696X-TK661). The residual oil at the top of the fractured-vuggy structure on this channel was swept by gas, and the gas channeling finally occurred in the TK661 well. Gas flooding did not sweep the fractured-vuggy structure on other channels (TK692X-TKTK696X and TK671-TK697-TKTK696X). The experimental result of gas flooding is shown in Figure S3, and the recovery ratio of gas flooding was 52.1%.

During the gel foam flooding process, gel foam not only displaced the main channel (TK671-TK604-TK696X-TK661) effectively but also swept a new flow channel (TKTK696X-TK692X). However, due to the poor connectivity of the vug under the TK697 well, the gel foam was unable to sweep the residual oil in this vug (see Figure S4). Gel foam flooding with a recovery ratio of 69.7% is shown in Figure 10.

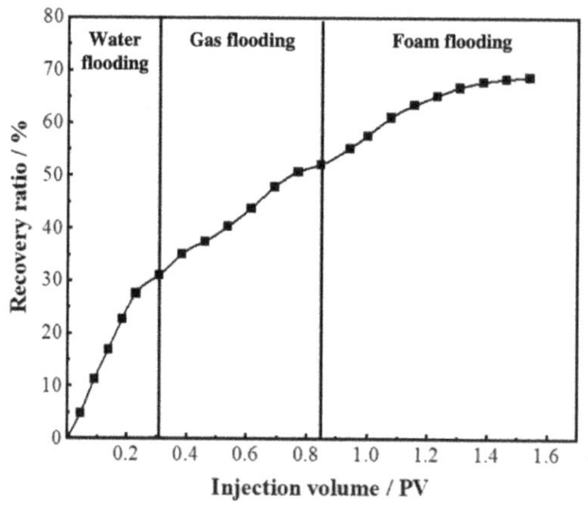

Figure 10. The recovery ratio of the 3D visual model (injection well: TK671 well).

2.3.3. Gas Flooding and Gel Foam Flooding in TK697

Flooding experiments were carried out in the 3D visual model. The TK661 well was chosen to be a water injection well at first. The TK697 well was chosen to be a gas injection well or gel foam injection well. The TK661 well, TK671 well, and TK692X well were chosen to be production wells.

During the gas flooding process, gas displaced the residual oil at the top of the vug under the TK697 well and then broke through along a flow channel (TK697-TK696X-TK661). Gas channeling finally occurred in the TK661 well. Gas flooding did not sweep the fractured-vuggy structure on other channels (TK692X-TKTK696X, TK692X-TK604-TK671, and TK671-TK697). The experimental result of gas flooding is shown in Figure S5, and the recovery ratio of gas flooding was 53%.

During the gel foam flooding process, the gel foam uniformly swept the fractures and vugs around the TK697 well and depressed the oil-gas interface near the bottom of the TK697 well. The residual oil submerged by the bottom water and that could not be swept by water or gas was effectively displaced by gel foam. Compared with gas flooding, gel foam flooding finally swept two new channels (TK696X-TK692X and TK696X-TK604-TK671), as shown in Figure S6. Gel foam flooding with a recovery ratio of 73.3% is shown in Figure 11.

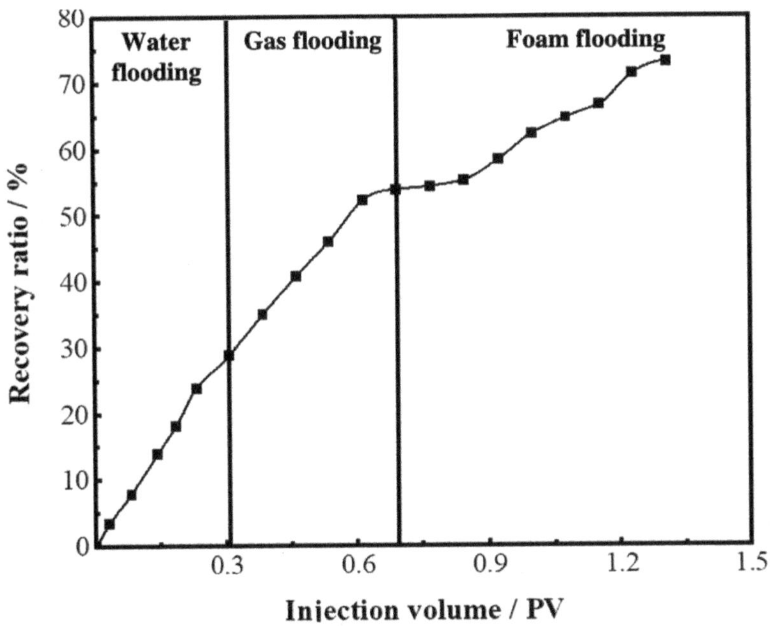

Figure 11. The recovery ratio of the 3D visual model (injection well: TK697 well).

The experimental results of the 3D visual model indicated that injected water flowed along the lower part of the fractured-vuggy reservoir and could form multiple flow channels. Gas flows along the upper part of the fractured-vuggy reservoir and usually forms only one flow channel. The complex fractured-vuggy structure led to the disunity difference of the oil–water interface in different vugs. Once the channeling was formed, the residual oil in a large number of fractured-vuggy structures was difficult to replace.

During the development process of fractured-vuggy reservoirs, the well at the bottom of the reservoir should be selected for water injection to prevent some poorly connected vugs from being submerged by water and unable to be swept by subsequent displacement media. The well located in a large vug at the upper part of the reservoir should be selected for gas injection to better displace the attic oil at the top of the vugs. Gel foam flooding flowed uniformly in different directions. Therefore, gel foam flooding has no special requirements for the location of the injection well but has higher requirements for the stability and viscosity of gel foam. Gel foam with strong stability and high viscosity might have a better displacement effect in the fractured-vuggy reservoir.

2.4. EOR Mechanism of Gel Foam in Fractured-Vuggy Carbonate Reservoirs

The capillary number is a dimensionless number that represents the ratio of viscous force to capillary force in the oil phase. It can reflect the displacement effect in porous media. It can be seen from the formula that the displacement effect of displacement medium on fractures is affected by interfacial tension, viscosity, and flow velocity. The decrease in the interfacial tension between the displacement phase and the displaced phase, the increase in the viscosity of the displacement phase, and the increase of the displacing velocity all have a positive effect on the sweep efficiency. In fractured-vuggy reservoirs, the interfacial tension and fluid viscosity of gas (except CO_2) and water are constant, and the recovery ratio can be improved by increasing velocity. Gel foam can improve the mobility ratio by increasing its viscosity, and the surfactant contained in the gel foam can reduce the interfacial tension between oil and water. Thus, gel foam has a better potential to improve oil recovery.

$$N_c = \frac{v\mu}{\sigma} \tag{1}$$

where the N_c is the capillary number; the v is the flow velocity of displacement media, m/s; μ is the viscosity of displacement media, mPa·s; and the σ is the interfacial tension between displacement media and displaced media, mN/m.

2.4.1. Reducing Interfacial Tension

The surfactant in gel foam can reduce the interfacial tension between oil and water and improve the displacement efficiency of gel foam in the fractured-vuggy structure. It can be seen from Figure 12 and Table S1 that P_{c1} was larger than P_{c2}. The resistance that needed to be overcome when gas entered the fracture (P_{c1}) was larger than that when water entered the fracture (P_{c2}). Thus, water could sweep some small fractures that could not be swept by gas. The displacement front of gel foam flooding was formed by a large number of small-size gel foam in fractures and vugs. A plateau boundary formed between gel foam and rock, but there was no obvious two-phase interface at the displacement front. There was no two-phase boundary, and the resistance that needed to be overcome when gel foam entered the fracture was far less than that of water and gas. Therefore, gel foam could sweep most of the fractures that could not be swept by gas or water. At the same time, the surfactant in the gel foam changed the wettability of the rock and effectively peeled off the oil film in the fractured-vuggy structure. The displacement effect of different displacement media is shown in Figures S7 and S8.

$$P_c = \frac{2\sigma\cos\theta}{r} \tag{2}$$

where the P_c is the capillary force, mPa; the σ is interfacial tension, mN/m; the θ is the contact angle, °; and the r is the capillary radius, m.

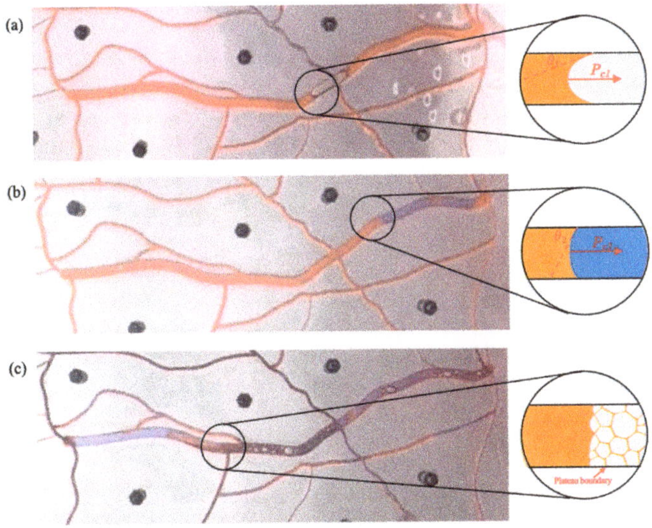

Figure 12. The sweep process of gas flooding, water flooding, and gel foam flooding in fractures. (**a**) the sweep process of gas flooding, (**b**) the sweep process of water flooding and (**c**) the sweep process of foam flooding. (θ_1 = 18°, θ_2 = 58°, r_1 = 1.5 mm, r_2 = 2.6 mm, σ_1 = 27.49 mN/m, σ_2 = 21.75 mN/m, Then $P_{c1} \approx$ 34.86 Pa, $P_{c2} \approx$ 8.87 Pa).

2.4.2. Increasing Mobility Ratio

Formula 3 shows that viscosity and permeability determine the mobility of the fluid in the fractured-vuggy structure. The smaller the mobility ratio is, the more stable the displacement front of the fluid in the fracture-vuggy structure is, and the larger the swept volume is. During the experiment, the viscosity of gel foam was much higher than that of water and gas, and the mobility of gel foam flooding was much lower than that of gas flooding and water flooding. Gel foam effectively restrained viscous fingering and stabilized the displacement front. In the filling medium of the fractured-vuggy structure, the gel foam was greatly affected by the Jamin effect, resulting in the permeability of the gel foam being much less than that of oil and water. As shown in Figures S7 and S8, the gel foam had a good sweep effect in the fractured-vuggy reservoir with varying apertures, different sizes, and different filling degrees.

$$M = \frac{\lambda_D}{\lambda_d} = \frac{K_D \mu_d}{K_d \mu_D} \tag{3}$$

where the λ_D is the mobility of the displacement phase, the K_D is the permeability of the displacement phase, mD; the μ_D is the viscosity of the displacement phase, mPa·s; the λ_d is the mobility of the displaced phase, the K_d is the permeability of the displaced phase, mD; the μ_d is the viscosity of the displaced phase, mPa·s.

2.4.3. Selectively Plugging High Permeability Channels

It was found that gel foam has the ability to selectively plug high permeability channels and start small fracture channels so as to achieve the effect of expanding the swept volume. A gas channeling was formed in the large fracture (WX), and the residual oil in small fractures could not be swept by the subsequent fluid, as shown in Figure S9a. After gel foam entered the large fracture (WX), the flow resistance in the large fracture (WX) was increased, and gel foam began to enter the small fracture (RS), as shown in Figure S9b. After the large fracture (WX) was filled with gel foam, the small fracture (RS) was effectively swept by the subsequent gel foam, as shown in Figure S9c. Gel foam played an effective role in plugging high permeability fracture channels in fractured-vuggy reservoirs.

2.4.4. Discontinuous Flow

The microscopic velocity of the continuous phase is the same, but the microscopic velocity of the discontinuous phase is different between different phases during the flow process. This difference in velocity can lead to a fluctuation of the flow field and has a positive effect on expanding the swept volume. As shown in Figure S10, gas and water were continuous phases. Gas flowed in a continuous state in both the filled fractured-vuggy structure and the unfilled fractured-vuggy structure because of its low viscosity. The viscosity of water is usually higher than that of gas but less than that of oil. When the viscosity difference between oil and water was small, it flowed in a discontinuous state in the filling fractured-vuggy structure but in a continuous state in the unfilled fractured-vuggy structure. Gel foam is a kind of discontinuous fluid that flows in a discontinuous state when flowing in both the filled fractured-vuggy structure and the unfilled fractured-vuggy structure. In addition, the change in viscosity and size was beneficial to the discontinuous flow of gel foam.

The viscous resistance of gel foam is usually characterized by apparent viscosity. There is a positive correlation between gel foam quality and apparent viscosity [33]. In the process of flow, the adsorption loss of surfactant and the increase of gel foam liquid phase will lead to the decrease of gel foam quality. The change in gel foam quality will lead to a change in the apparent viscosity of gel foam in different positions and then affect the flow velocity of gel foam at different positions, resulting in fluctuations in the flow field.

In porous media, the flow velocity of gel foam is negatively related to the size of the gel foam. The smaller the size of the gel foam is, the faster the flow velocity is [33]. The coarsening of gel foam is inevitable, resulting in a gradual increase in the size of the gel foam [34]. In addition, gel foam is easy to burst in the presence of oil. Under the influence of multiple factors, the size of the gel foam gradually varies in different positions in the reservoir. Due to the superposition of the Jamin Effect, the velocity of the gel foam in different locations will be different, which will also result in fluctuations in the flow field.

3. Conclusions

In this study, the visual and physical models of a fractured-vuggy reservoir were designed and fabricated innovatively based on the real reservoir geological data. The experimental study on improving oil recovery by water, gas, and gel foam flooding was carried out in these models. Combined with the displacement experiments from one-dimensional to three-dimensional, the following results and conclusions are obtained.

1. Water has multiple flow channels when flowing in the fractured-vuggy reservoir, but the recovery ratio of water flooding is limited. Gas channeling with a single flow channel is easy to form in fractures and vugs, and the EOR of gas flooding is unsatisfactory. Gel foam flooding with an excellent capability of mobility control and a high microscopic displacement efficiency has great potential to enhance oil recovery in fracture-cave reservoirs;
2. The EOR mechanisms of gel foam in fractured-vuggy reservoirs are mainly as follows: reducing interfacial tension, increasing mobility ratio, selectively plugging high permeability channels, and discontinuous flow;
3. By injecting water from the well at the bottom of the reservoir and gas from the well located in the vug at the high part of the reservoir, multiple technologies can be used together to enhance oil recovery;
4. Gel foam has no special restriction on the location of the injection well, but gel foam with strong stability and high viscosity should be selected in fractured-vuggy reservoirs.

4. Physical Model Design and Fabrication

In this study, a visual, physical simulation method is used to study the flooding behavior and sweep mechanism of gas flooding and gel foam flooding in fractured-vuggy carbonate reservoirs (as shown in Table 2). The physical models, including the 1D visual model, the 2D visual model, and the 3D visual model, were designed based on similarity criteria, as well as the experimental parameters. The matrix permeability of fractured-vuggy carbonate reservoirs is extremely low, which leads to poor flow ability of fluid in the matrix. The main flow channel of fluid is the fractured-vuggy structure, and the fluid motion in the fractured-vuggy structure conforms to Newton's law and the laws of thermodynamics. The fluid flow was similar as long as the various influencing factors in the fluid motion equation were similar [35]. To make the fluid flow pattern in the physical model of the fractures and vugs similar to that in the actual reservoir, the characteristic parameters of fractures and cavities are designed by geometric similarity, and the physical properties of fluids and injection parameters are designed by kinematic similarity and dynamic similarity in this study (as shown in Table 3). Combined with field production data in the Tahe Oilfield, the design of the kinematic similarity and dynamic similarity was carried out for the physical properties of the fluid, production pressure differences, and injection velocities.

Table 2. Visual models of fractured-vuggy reservoirs in this study.

Models	Materials	Size	Features	Flooding
One-dimensional visual model	Plexiglass	30 cm × 10 cm × 10 cm	Fracture model with irregular cross-section.	Water flooding, gas flooding, and foam flooding
Two-dimensional visual model	Plexiglass	26 cm × 20 cm × 2 cm; 10 cm × 10 cm × 2 cm	Filled fracture model with multi-scale fracture network.	Water flooding, gas flooding, and foam flooding
Three-dimensional visual model	Curable resin	18 cm × 14 cm × 14 cm	The fracture-vuggy reservoir model with complex fracture-vuggy structure and various connectivity.	Water flooding, gas flooding, and foam flooding

Table 3. The fracture parameters and similarity coefficients of the laboratorial physical model and actual fractures in reservoirs.

Parameter Source	Pressure Difference (kPa)	Fracture Aperture (mm)	Oil Density ($g \cdot cm^{-3}$)	Gravitational Acceleration ($m \cdot s^{-2}$)	Flow Rate ($m \cdot s^{-1}$)	Injection Rate ($m^3 \cdot d^{-1}$)	Wellbore Diameter (mm)
Reservoirs	2000–14,000	40–2500	0.92	9.8	0.0147–0.147	10–50	120
Laboratory	9.2–34.01	0.2–10	0.821	9.8	0.007–0.049	0.0015–0.002	2
Similarity coefficients	217–411	200–250	1.1	1	2.1–3	6666.7–25,000	60

According to the similarity coefficient, the similarity number $F_Q = \frac{Q}{r^2 u} = 0.99$ and $F_G = \frac{\Delta p}{\rho_0 g B} = 1.06$ were obtained. It was proved that the actual fractures in reservoirs and the laboratorial physical fracture model satisfy the above similarity criteria.

Note: F_Q is kinematic similarity; Q is injection rate, $m^3 \cdot d^{-1}$; r is wellbore diameter, mm; u is flow rate, $m \cdot s^{-1}$; F_G is dynamic similarity; Δp is the pressure difference, kPa; ρ_0 is the oil density, $g \cdot cm^{-3}$; g is gravitational acceleration, $m \cdot s^{-2}$; B is fracture aperture, cm.

4.1. One-Dimensional Visual Model of a Single Fracture

The 1D visual model of a single fracture with varying apertures was designed to study the flow behavior in wide fractures. The same shape of the fracture was etched on two pieces of plexiglass (with an oleophilic surface) through the laser etching technique based on the design diagram (see Figure 13a). Then, the plexiglass was covered by the other plexiglass and fixed by a steel frame, with a 10 mm distance between the two plates. Finally, the tightness of the 1D visual model is ensured by high-temperature sealing bonding. The 1D visual model of a single fracture with an apparent size of 30 cm × 10 cm × 10 cm and a 10 mm height fracture is shown in Figure 13b.

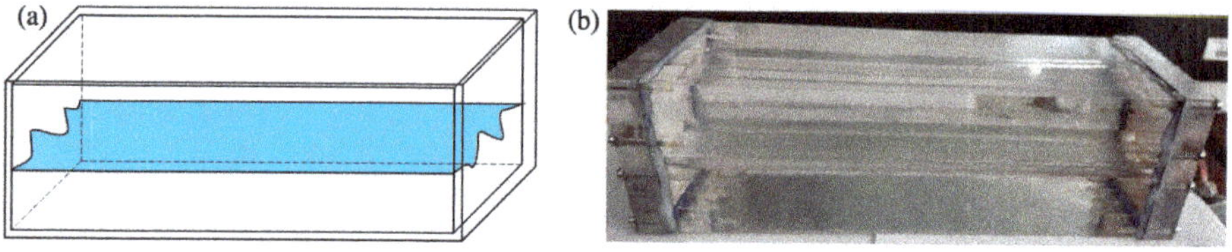

Figure 13. 1D visual model of a single fracture: (**a**) design diagram and (**b**) physical diagram.

4.2. Two-Dimensional Visual Model

Two kinds of 2D visual models were designed in this study, including a 2D visual model of complex fractures and a 2D visual model of complex vugs.

4.2.1. Two-Dimensional Visual Model of Complex Fractures

The 2D visual model of complex fractures was designed according to the photograph of the actual fractured-vuggy reservoir (Figure 14). Based on the model design diagram, the laser etching technique was used to etch the fracture on the plexiglass (oleophilic surface). After etching, another plexiglass was bonded onto it with a high-temperature sealing treatment and then fixed with a steel frame. The 2D visual model of complex fractures with an external dimension of 26 cm × 20 cm × 2 cm is shown in Figure 14c.

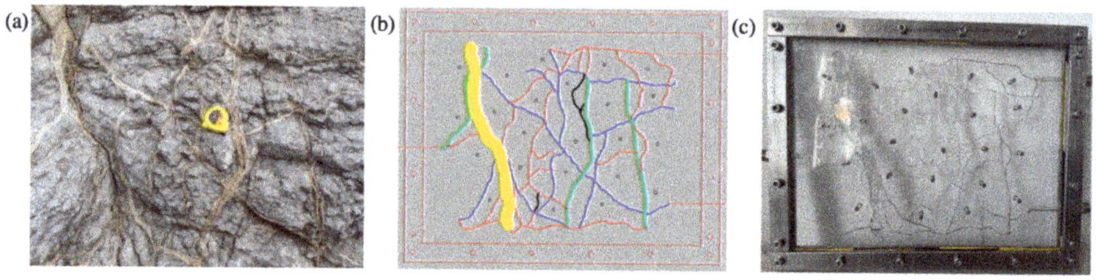

Figure 14. The 2D visual model of complex fractures: (**a**) the photo of the actual fractured-vuggy reservoir, (**b**) the design diagram, (**c**) the physical diagram. The size of the yellow fracture is 10 mm wide and 5 mm deep; the size of the green fracture is 3 mm wide and 3 mm deep; the size of the blue fracture is 1 mm wide and 3 mm deep; the size of the red fracture is 0.5 mm wide and 3 mm deep. The filler used in the model is oleophilic transparent acrylic beads (2 mm).

4.2.2. Two-Dimensional Visual Model of Complex Vugs

In the same way as the 2D visual model of complex fractures, the 2D visual model of complex vugs with an external dimension of 10 cm × 10 cm × 2 cm was produced through the laser etching technique. In addition, an unfilled model (Figure 15b) and a filled model (Figure 15c) were designed and fabricated to characterize the different filling degrees in vugs.

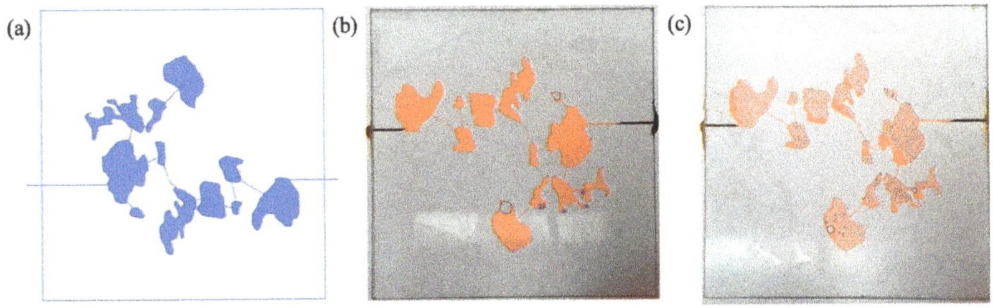

Figure 15. The 2D visual model of complex vugs: (**a**) the design diagram, (**b**) the physical diagram of the unfilled model, (**c**) the physical diagram of the filled model. The width of the fracture is 1 mm, and the depth of fractures and vugs is 3 mm in this model. The filler used in the model is oleophilic transparent acrylic beads (2 mm).

4.3. Three-Dimensional Visual Model of Fractured-Vuggy Carbonate Reservoir

The 3D visual model of the fractured-vuggy carbonate reservoir was made with 3D printing technology. The 3D geological data of fractures and vugs of the TK671 well-group unit in the Tahe Oilfield were selected to design this model. In this 3D geological data, 22 sections (line.1~line.22) were intercepted at equal distances, and 3 over-well sections (line.24~line.26) were intercepted according to the flow condition between wells (see Figures S11 and S12). After the obtained structure data of sections were imported into 3D simulation software and optimized, the 3D visual model of the fractured-vuggy carbonate reservoir was produced using a 3D printer. The 3D visual model of the TK671 well-group unit with an external dimension of 18 cm × 14 cm × 14 cm is shown in Figure 16.

Figure 16. The 3D visual model of the fractured-vuggy carbonate reservoir: (**a**) the 3D design diagram, (**b**) the front view of the physical diagram, (**c**) the overhead view of the physical diagram.

5. Experimental Materials and Process

5.1. Experimental Materials

The fractured-vuggy carbonate reservoir in the Tahe Oilfield is characterized by high temperature, high pressure, and high mineralization. It does not have a similar capability to foaming by shearing as porous media. For this reason, highly stable gel foam is injected into the Tahe Oilfield after foaming on the ground. This kind of gel foam can be transported stably in the fractured-vuggy reservoir. The chemical reagents used to prepare for highly stable gel foam include the foaming agent (SS-163, Qingdao Changxing High-tech Development Co., Ltd., Qingdao, China), Sodium dodecyl sulfate, α-modified starch, Acrylamide, N,N′-Methylenebisacrylamide, and Potassium persulfate. The preparation method can be referred to in our previous study [12]. The gel foam quality used in this experiment was 70%.

The simulated oil is a mixture of liquid paraffin and kerosene with a viscosity of 60 mPa·s and a density of 0.83 g/mL. The saltwater used in the experiment with a salinity of 250 g/L is prepared with deionized water and inorganic salts. The composition of inorganic salts is Na_2SO_4 (417 mg/L), $NaHCO_3$ (576 mg/L), NaCl (207,759 mg/L), $CaCl_2$ (41,106 mg/L), $MgCl_2$ (3549 mg/L). To better observe the experimental results, the simulated oil was dyed red with Sudan red, and the salt water was dyed blue with blue ink. The gas used in the experiment was industrial nitrogen (99.2% purity).

5.2. Experimental Instruments

The instruments used to prepare the gel foam liquid included a hydrothermal reactor with high temperature and high pressure, a B75 agitator, and beakers. The instruments used to simulate the core-flooding experiment under reaction temperature and pressure included a high-temperature dryer, a displacement pump, a gas flow controller, a thermostat, a foam generator, a nitrogen cylinder, an intermediate vessel, six-way valves, and connecting pipelines. The working pressure of the displacement pump is 0~20 MPa, and its working flow is 0~20 mL/min. The flow range of the gas flow controller is 0.2~20 mL/min, its working pressure difference is 0.1~0.4 MPa, and its working pressure is 0~20 MPa. The volume of the intermediate vessel is 2000 mL, and its working pressure is 0~40 MPa.

5.3. Experimental Process

The 1D visual model of a single fracture, the 2D visual model, and the 3D visual model of the fractured-vuggy carbonate reservoir were connected according to the flow chart (Figure 17). The experimental steps are as follows: a. Oil was saturated into the visual models, and the saturation volume was recorded. b. Water (gas/foam) was injected into the model at a rate of 4 mL/min. The oil production and the experimental process were recorded. c. The experiment was stopped when the oil production was 0 mL/min.

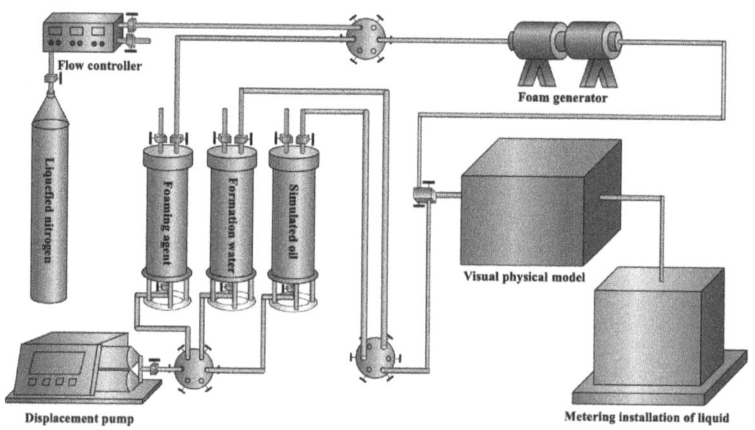

Figure 17. The flow chart of visual models.

During the experiments, the 1D visual model of a single fracture, the 2D visual model of complex fractures, and the 3D visual model of the fractured-vuggy carbonate reservoir were placed horizontally, and the 2D visual model of complex vugs was placed vertically.

Supplementary Materials: The following supporting information can be downloaded at: https://www.mdpi.com/article/10.3390/gels9090722/s1, Figure S1: Experiments results of the 1D visual model of a single fracture: (a) water flooding; (b) water flooding after gel foam flooding; Figure S2: Experiments results of the 1D visual model of a single fracture: (a) water flooding; (b) water flooding after gel foam flooding; Figure S3: Experimental results of gas flooding in the 3D visual model: (a) the front view and (b) the overhead view; Figure S4: Experimental results of gel foam flooding in the 3D visual model: (a) the front view and (b) the overhead view; Figure S5: Experimental results of gel foam flooding in the 3D visual model: (a) the front view and (b) the overhead view;

Figure S6: Experimental results of gel foam flooding in the 3D visual model: (a) the front view and (b) the overhead view; Figure S7: The displacement results of gas flooding, water flooding, and gel foam flooding in fractures; Figure S8: The displacement results of gas flooding, water flooding, and gel foam flooding in filled fracture; Figure S9: High permeability fracture channel plugged by gel foam: (a) before gel foam entered the fracture (WX), (b) gel foam entered the fracture (WX), (c) the fracture (WX) plugged by gel foam; Figure S10: The flow behavior of different fluid in fractured-vuggy structure: (a) gas flowed in unfilled model, (b) water flowed in unfilled model, (c) gel foam flowed in unfilled model, (d) gas flowed in filled model, (e) water flowed in filled model, (f) gel foamed flow in filled model. Figure S11: Schematic diagram of the sectional position of the TK671 well-group unit; Figure S12: The over-well sections of the TK671 well-group unit: (a) line.23, (b) line.24, (c) line.25; Table S1: The experimental values of water flooding and gas flooding in fractures.

Author Contributions: Conceptualization, Y.W. and J.H.; methodology, Y.W. and J.H.; validation, Y.W. and J.H.; formal analysis, Y.W.; investigation, Y.W. and J.H.; data curation, Y.W.; writing—original draft preparation, Y.W.; writing—review and editing, Y.W. and J.H.; visualization, Y.W.; project administration, J.H.; funding acquisition, J.H. All authors have read and agreed to the published version of the manuscript.

Funding: The authors gratefully appreciate the financial support of the National Natural Science Foundation of China (Grant No. 52174046) and the Major National Science and Technology Projects of China (Grant No. 2016ZX05014-004-004).

Institutional Review Board Statement: Informed consent was obtained from all subjects involved in the study.

Informed Consent Statement: Not applicable.

Data Availability Statement: Data are available from the corresponding author upon reasonable request.

Acknowledgments: The authors gratefully acknowledge the graduate students in research institute of unconventional petroleum science and technology for their help in experimental data analysis and video recording.

Conflicts of Interest: The authors declare no conflict of interest.

References

1. Li, Y.; Kang, Z.; Xue, Z.; Zheng, S. Theories and practices of carbonate reservoirs development in China. *Pet. Explor. Dev.* **2018**, *45*, 712–722. [CrossRef]
2. Xie, J.; Huang, C.; Wang, X. Distribution features of proved reserves of carbonate oil and gas pools in China. *Mar. Orig. Pet. Geol.* **2009**, *14*, 24–30.
3. Zou, C.; Du, J.; Xu, C.; Wang, Z.; Zhang, B.; Wei, G.; Wang, T.; Yao, G.; Deng, S.; Liu, J.; et al. Formation, distribution, resource potential, and discovery of Sinian–Cambrian giant gas field, Sichuan Basin, SW China. *Pet. Explor. Dev.* **2014**, *41*, 306–325. [CrossRef]
4. Li, Y.; Hou, J.; Li, Y. Features and classified hierarchical modeling of carbonate fracture-cavity reservoirs. *Pet. Explor. Dev.* **2016**, *43*, 655–662. [CrossRef]
5. Mao, C.; Zhong, J.; Li, Y.; Wang, Y.; Niu, Y.; Ni, L.; Shao, Z. Ordovician carbonate rock matrix fractured-porous reservoirs in Tahe Oilfield, Tarim Basin, NW China. *Pet. Explor. Dev.* **2014**, *41*, 745–753. [CrossRef]
6. Zhang, X.; Yang, J. Reservoir description andreserves estimation technique for fracture-cave type carbonatereservoir in Tahe Oilfield. *Acta Pet. Sin.* **2004**, *25*, 13–18. [CrossRef]
7. Chen, Z.; Dai, Y.; Lang, Z. Storage-percolation modes and production performance of the karst reservoirs in Tahe Oilfield. *Pet. Explor. Dev.* **2005**, *3*, 101–105. [CrossRef]
8. Jiao, F. Practice and knowledge of volumetric development of deep fractured-vuggy carbonate reservoirs in Tarim Basin, NW China. *Pet. Explor. Dev.* **2019**, *46*, 576–582. [CrossRef]
9. Hou, J. Three-dimensional physical simulation and optimization of water injection of a multi-well fractured-vuggy unit. *Pet. Sci.* **2016**, *2*, 259–271. [CrossRef]
10. Rong, Y. Discussion about pattern of water flooding development in multi-well fracture-cavity units of carbonate fracture-cavity reservoir in Tahe oilfeld. *Pet. Geol. Recovery Effic.* **2013**, *20*, 58–61.
11. Manrique, E.J.; Muci, V.E.; Gurfinkel, M.E. EOR Field Experiences in Carbonate Reservoirs in the United States. *SPE Reserv. Eval. Eng.* **2007**, *10*, 667–686. [CrossRef]

12. Qu, M.; Hou, J.; Wen, Y.; Liang, T. Nitrogen gas channeling characteristics in fracture-vuggy carbonate reservoirs. *J. Pet. Sci. Eng.* **2020**, *186*, 106723. [CrossRef]
13. Wang, J.; Jiao, B.; Zeng, W.; Li, D. Development Mode of Tahe Fracture- Cave Reservoirs in Late Water Flooding Stage. *Spec. Oil Gas Reserv.* **2015**, *22*, 125–128.
14. Singh, R.; Mohanty, K.K. Foam flow in a layered, heterogeneous porous medium: A visualization study. *Fuel* **2017**, *197*, 58–69. [CrossRef]
15. Wen, Y.; Qu, M.; Hou, J.; Liang, T.; Raj, I.; Ma, S.; Yuan, N. Experimental study on nitrogen drive foam assisted nitrogen drive in varying-aperture fractures of carbonate reservoir. *J. Pet. Sci. Eng.* **2019**, *180*, 994–1005. [CrossRef]
16. Surajudeen, S.; Rostami, A.; Soleimani, H.; Yahya, N.; Yusuf, A.; Aliu, O.; Yusuf, J.Y.; Oladosu, T.L. Graphene: Outlook in the enhance oil recovery (EOR). *J. Mol. Liq.* **2020**, *321*, 114519. [CrossRef]
17. Liang, T.; Hou, J. Fluids flow behaviors of nitrogen foam-assisted nitrogen floods in 2D visual fault-karst carbonate reservoir physical models. *J. Pet. Sci. Eng.* **2021**, *200*, 108286. [CrossRef]
18. Wen, Y.; Hou, J.; Qu, M.; Wu, W.; Liang, T.; Zhang, W.; Wu, W. Field Pilot Test of Micro-Dispersed Gel Foam in Fractured-Vuggy Carbonate Reservoirs. In Proceedings of the SPE Annual Technical Conference and Exhibition, Dubai, United Arab Emirates, 21–23 September 2021; OnePetro: Richardson, TX, USA, 2021. [CrossRef]
19. Chen, Z.; Ma, X.; Hang, G. Research on the fracture-vug unit division of fractured-vuggy carbonate rock oil pools—An example from the Ordovician oil pools in the principle producing area of Tahe oilfield. *Oil Gas Geol.* **2007**, *28*, 847–855. [CrossRef]
20. Gulbransen, A.F.; Hauge, V.L.; Lie, K.-A. A Multiscale Mixed Finite-Element Method for Vuggy and Naturally Fractured Reservoirs. *SPE J.* **2009**, *15*, 395–403. [CrossRef]
21. Kao, J.-W.; Wei, S.-M.; Wang, W.-Z.; Jin, Y. Numerical analysis of the hydraulic fracture communication modes in fracture-cavity reservoirs. *Pet. Sci.* **2022**, *19*, 2227–2239. [CrossRef]
22. Popov, P.; Qin, G.; Bi, L.; Efendiev, Y.; Kang, Z.; Li, J. Multiphysics and Multiscale Methods for Modeling Fluid Flow Through Naturally Fractured Carbonate Karst Reservoirs. *SPE Reserv. Eval. Eng.* **2009**, *12*, 218–231. [CrossRef]
23. Wei, S.; Jin, Y.; Xia, Y. Predict the mud loss in natural fractured vuggy reservoir using discrete fracture discrete vug network model. *J. Pet. Sci. Eng.* **2020**, *195*, 107626. [CrossRef]
24. Dai, C.; Fang, J.; He, L. Development of the research on EOR for carbonate fractured-vuggy reservoirs in China. *J. China Univ. Pet. Ed. Natrual Sci.* **2018**, *42*, 67–78. [CrossRef]
25. Yang, B.; He, J.; Lyu, D.; Tang, H.; Zhang, J.; Li, X.; Zhao, J. Production optimization for water flooding in fractured-vuggy carbonate reservoir—From laboratory physical model to reservoir operation. *J. Pet. Sci. Eng.* **2020**, *184*, 106520. [CrossRef]
26. Yang, W.; Zhang, D.; Lei, G. Experimental study on multiphase flow in fracture-vug medium using 3D printing technology visualization techniques. *J. Pet. Sci. Eng.* **2020**, *193*, 107394. [CrossRef]
27. Hou, J.; Luo, M.; Zhu, D. Foam-EOR method in fractured-vuggy carbonate reservoirs: Mechanism analysis injection parameter study. *J. Pet. Sci. Eng.* **2018**, *164*, 546–558. [CrossRef]
28. Qu, M. Mechanism and parameters of nitrogen foam flooding in cave reservoir bodies of fractured-cavity reservoirs. *Pet. Sci. Bull.* **2018**, *3*, 57–66.
29. Wang, J.; Ji, Z.; Liu, H.; Huang, Y.; Wang, Y.; Pu, Y. Experiments on nitrogen assisted gravity drainage in fractured-vuggy reservoirs. *Pet. Explor. Dev.* **2019**, *46*, 355–366. [CrossRef]
30. Xu, Z.; Li, Z.; Cui, S.; Li, B.; Chen, D.; Zhang, Q.; Zheng, L.; Husein, M.M. Flow characteristics EOR mechanism of foam flooding in fractured vuggy reservoirs. *J. Pet. Sci. Eng.* **2022**, *211*, 110170. [CrossRef]
31. Qu, M.; Hou, J.; Qi, P.; Zhao, F.; Ma, S.; Churchwell, L.; Wang, Q.; Li, H.; Yang, T. Experimental study of fluid behaviors from water nitrogen floods on a 3-D visual fractured-vuggy model. *J. Pet. Sci. Eng.* **2018**, *166*, 871–879. [CrossRef]
32. Lu, G.; Zhang, L.; Liu, Q.; Xu, Q.; Zhao, Y.; Li, X.; Deng, G.; Wang, Y. Experiment analysis of remaining oil distribution potential tapping for fractured-vuggy reservoir. *J. Pet. Sci. Eng.* **2022**, *208*, 109544. [CrossRef]
33. Omirbekov, S.; Davarzani, H.; Colombano, S.; Ahmadi-Senichault, A. Experimental and numerical upscaling of foam flow in highly permeable porous media. *Adv. Water Resour.* **2020**, *146*, 103761. [CrossRef]
34. Jones, S.A.; Getrouw, N.; Vincent-Bonnieu, S. Foam flow in a model porous medium: I. The effect of foam coarsening. *Soft Matter* **2018**, *14*, 3490–3496. [CrossRef] [PubMed]
35. Qu, M.; Hou, J.; Liang, T.; Raj, I.; Yang, Y.; Qi, P. Synthesis of α-starch based nanogel particles its application for long-term stabilizing foam in high-salinity high-temperature crude oil environment. *J. Pet. Sci. Eng.* **2020**, *191*, 107185. [CrossRef]

MDPI AG
Grosspeteranlage 5
4052 Basel
Switzerland
Tel.: +41 61 683 77 34

Gels Editorial Office
E-mail: gels@mdpi.com
www.mdpi.com/journal/gels

www.ingramcontent.com/pod-product-compliance
Lightning Source LLC
Chambersburg PA
CBHW061805210926
51850CB00006B/622
* 9 7 8 3 7 2 5 8 6 7 9 6 7 *